BEYOND THE MECHANICAL UNIVERSE

BEYOND THE MECHANICAL UNIVERSE

FROM ELECTRICITY TO MODERN PHYSICS

RICHARD P. OLENICK
ASSISTANT PROFESSOR OF PHYSICS, UNIVERSITY OF DALLAS
VISITING ASSOCIATE, CALIFORNIA INSTITUTE OF TECHNOLOGY

TOM M. APOSTOL
PROFESSOR OF MATHEMATICS, CALIFORNIA INSTITUTE OF TECHNOLOGY

DAVID L. GOODSTEIN
PROFESSOR OF PHYSICS AND APPLIED PHYSICS, CALIFORNIA INSTITUTE OF TECHNOLOGY
AND PROJECT DIRECTOR, *THE MECHANICAL UNIVERSE*

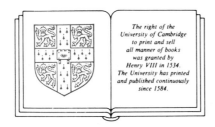

CAMBRIDGE UNIVERSITY PRESS
Cambridge
London   New York   New Rochelle
Melbourne   Sydney

Published by the Press Syndicate of the University of Cambridge
The Pitt Building, Trumpington Street, Cambridge CB2 1RP
32 East 57th Street, New York, NY 10022, USA
10 Stamford Road, Oakleigh, Melbourne 3166, Australia

© Cambridge University Press 1986

First published 1986

Printed in the United States of America

Library of Congress Cataloging-in-Publication Data
Olenick, Richard P.
Beyond the mechanical universe.

   1. Physics.   I. Apostol, Tom M.   II. Goodstein,
David L., 1939-   III. Title.
QC23.O485   1986   530   86-18862
ISBN 0-521-30430-X

CONTENTS

Preface   xi

Chapter 31 BEYOND THE MECHANICAL UNIVERSE (Program 27)   *1*

31.1 Science After Newton   *1*
31.2 Orders of Magnitude   *4*
31.3 A Final Word   *11*

Chapter 32 STATIC ELECTRICITY (Program 28)   *13*

32.1 The Beginnings of Electrical Science   *13*
32.2 The Electroscope   *16*
32.3 Charge Separation and Induction   *17*
32.4 Coulomb's Law   *20*
32.5 Electrostatic Machines   *26*
32.6 A Final Word   *29*

Chapter 33 THE ELECTRIC FIELD (Program 29)   *31*

33.1 Electric Fields of Isolated Point Charges   *31*
33.2 Electric Fields of Continuous Charge Distributions   *37*
33.3 Electric Flux   *44*
33.4 Gauss's Law   *47*
33.5 Applications of Gauss's Law   *51*
33.6 Electric Fields and Conductors   *57*
33.7 A Final Word   *61*

Chapter 34 POTENTIAL AND CAPACITANCE (Program 30)   *63*

34.1 A Great American Scientist   *63*
34.2 Electric Potential   *64*
34.3 Electric Energy of Systems of Charges   *74*
34.4 Capacitors   *77*
34.5 Combinations of Capacitors   *82*
34.6 Energy Storage in a Capacitor   *86*
34.7 A Final Word   *89*

Chapter 35 VOLTAGE, ENERGY AND FORCE (Program 31)   *91*

35.1 Electric Fields and Potentials   *92*
35.2 Equipotential Surfaces   *97*
35.3 Voltages in the World   *101*
35.4 Charge Distribution on Conductors   *107*
35.5 A Final Word   *110*

Chapter 36 THE ELECTRIC BATTERY (Program 32)   *113*

36.1 Frog Legs and Electricity   *113*
36.2 The Workings of Metals   *115*
36.3 Battery Basics   *123*
36.4 Real Batteries   *126*
36.5 A Final Word   *127*

Chapter 37 ELECTRIC CIRCUITS (Program 33)   *129*

37.1 The Invention of the Telegraph   *129*
37.2 Electrical Conduction in Wires   *131*
37.3 Ohm's Law   *134*
37.4 Resistors Connected in Series and in Parallel   *139*
37.5 Kirchhoff's Laws   *144*
37.6 *RC* Circuits with Variable Currents   *151*
37.7 A Final Word   *157*

Chapter 38 MAGNETISM (Program 34)   *159*

38.1 Lodestones and Magnetic Needles   *159*
38.2 Forces and Magnetic Fields   *162*
38.3 Magnets and Torques   *169*
38.4 Gauss's Law for Magnetism   *171*
38.5 Magnetic Force on Moving Charges   *174*
38.6 Magnetic Force on Currents   *183*
38.7 A Final Word   *188*

Chapter 39 THE MAGNETIC FIELD (Program 35)   *191*

39.1 The Connection between Electricity and Magnetism   *191*
39.2 The Law of Biot and Savart   *193*
39.3 Force between Current-Carrying Wires   *204*
39.4 Ampère's Law   *207*
39.5 A Final Word   *215*

Chapter 40 VECTOR FIELDS AND HYDRODYNAMICS (Program 36)   *217*

40.1 Action-at-a-Distance Revisited   *217*
40.2 Properties of Vector Fields   *218*
40.3 Flux of a Vector Field   *219*
40.4 Circulation of a Vector Field   *222*
40.5 Hydrodynamic Analogies for Energy and Forces   *232*
40.6 A Final Word   *233*

Chapter 41 ELECTROMAGNETIC INDUCTION (Program 37)   *235*

41.1 The Incomparable Experimentalist   *235*
41.2 Observations of Electromagnetic Induction   *237*
41.3 Faraday's Law   *245*
41.4 Lenz's Law   *251*
41.5 Self-Inductance   *256*
41.6 Mutual Inductance   *259*
41.7 *LR* Circuits   *263*

41.8 Energy and Magnetic Fields   *267*
41.9 A Final Word   *269*

Chapter 42 ALTERNATING CURRENTS (Program 38)   *273*

42.1 Two Great Inventors   *273*
42.2 Alternating Currents in Simple Circuits   *276*
42.3 *LC* Circuits   *280*
42.4 *LCR* Circuits   *282*
42.5 Power in ac Circuits   *288*
42.6 Transformers   *291*
42.7 A Final Word   *294*

Chapter 43 MAXWELL'S EQUATIONS (Program 39)   *295*

43.1 A Victorian Genius   *295*
43.2 The Link between Electricity and Magnetism   *296*
43.3 Maxwell's Equations in Free Space   *301*
43.4 Plane Waves Moving with Constant Speed   *302*
43.5 The Wave Equation   *303*
43.6 Electromagnetic Waves   *307*
43.7 Disturbances Caused by Accelerated Charges   *312*
43.8 A Final Word   *314*

Chapter 44 OPTICS (Program 40)   *317*

44.1 The Electromagnetic Spectrum   *317*
44.2 The Nature of Light   *323*
44.3 Reflection and Refraction   *325*
44.4 Interference of Light Waves   *332*
44.5 A Final Word   *341*

Chapter 45 THE MICHELSON–MORLEY EXPERIMENT (Program 41)   *343*

45.1 The Roots of Relativity   *343*
45.2 The Galilean Transformation   *345*
45.3 Space–Time Diagrams for Galilean Transformations   *351*
45.4 Relativity and the Nature of Light   *354*
45.5 The Michelson–Morley Experiment   *356*
45.6 A Final Word   *363*

Chapter 46 THE LORENTZ TRANSFORMATION (Program 42)   *367*

46.1 Interpreting the Michelson–Morley Experiment   *367*
46.2 The Postulates of the Special Theory of Relativity   *369*
46.3 The Lorentz Transformation   *372*

CONTENTS

46.4 Length Contraction   *378*
46.5 Space–Time Diagrams   *382*
46.6 A Final Word   *386*

Chapter 47 VELOCITY AND TIME (Program 43)   *389*

47.1 Proper Length and Proper Time   *389*
47.2 Combinations of Velocities in Special Relativity   *392*
47.3 The Fizeau Experiment   *397*
47.4 The Muon Experiment   *399*
47.5 The Twin Paradox   *402*
47.6 A Final Word   *405*

Chapter 48 MASS, MOMENTUM, ENERGY (Program 44)   *407*

48.1 Inertia and Relativity   *407*
48.2 Momentum and Mass   *408*
48.3 Relativistic Kinetic Energy   *417*
48.4 Applications of Conservation of Relativistic Energy and Momentum   *424*
48.5 A Final Word   *427*

Chapter 49 ATOMS (Program 49)   *431*

49.1 Early History of Atomic Theory   *431*
49.2 Experimental Evidence Supporting Atomic Theory   *432*
49.3 The Atomic Structure of Matter   *435*
49.4 Rutherford's Model of the Atom   *439*
49.5 Spectra of Electromagnetic Radiation   *444*
49.6 The Bohr Model of the Atom   *448*
49.7 A Final Word   *453*

Chapter 50 PARTICLES AND WAVES (Program 50)   *455*

50.1 Black Body Radiation   *455*
50.2 The Photoelectric Effect   *459*
50.3 The Dual Nature of Light   *463*
50.4 The De Broglie Model of the Hydrogen Atom   *466*
50.5 The Birth of Quantum Mechanics   *470*
50.6 The Quantum Mechanical Model of the Atom   *478*
50.7 The Heisenberg Uncertainty Principle   *483*
50.8 A Final Word   *488*

Chapter 51 ATOMS TO QUARKS (Program 51)   *491*

51.1 The Nature of Matter   *491*
51.2 Quantum States of the Hydrogen Atom   *492*

51.3 Wave Functions for the Hydrogen Atom   *495*
51.4 Atoms with Many Electrons   *500*
51.5 Nuclei and Radioactivity   *508*
51.6 Particles and More Particles   *516*
51.7 A Final Word   *525*

Chapter 52 THE QUANTUM MECHANICAL UNIVERSE (Program 52)   *527*

52.1 Introduction   *527*
52.2 Diffraction of Matter Waves   *528*
52.3 Electron Diffraction and the Dual Nature of Waves and Particles   *533*
52.4 Is Newtonian Mechanics Obsolete?   *540*
52.5 Quantum Mechanical Estimates   *543*
52.6 A Final Final Word   *546*

Appendix A THE INTERNATIONAL SYSTEM OF UNITS   *549*

Appendix B CONVERSION FACTORS   *551*

Appendix C THE PERIODIC TABLE OF THE ELEMENTS   *555*

Appendix D ASTRONOMICAL DATA   *559*

Appendix E PHYSICAL CONSTANTS   *561*

SELECTED BIBLIOGRAPHY   *563*

Index   *567*

PREFACE

I GENERAL INTRODUCTION (repeated from the first volume, *Introduction to Mechanics and Heat*)

The Mechanical Universe is a project that encompasses fifty-two half-hour television programs, two textbooks in four volumes (including this one), teachers' manuals, specially edited videotapes for high school use, and much more. It seems safe to say that nothing quite like it has been attempted in physics (or any other subject) before. A few words about how all this came to be seem to be in order.

Caltech's dedication to the teaching of physics began fifty years ago with a popular introductory textbook written by Robert Millikan, Earnest Watson and Duane Roller. Of the three, Millikan, whose exploits are celebrated in Chapter 12 of this book, was Caltech's founder, president, first Nobel prizewinner, and all-around patron saint. Earnest Watson was dean of the faculty, and both he and Duane Roller were distinguished teachers.

Twenty years ago, the introductory physics courses at Caltech were taught by Richard Feynman, who is not only a scientist of historic proportions, but also a dramatic and highly entertaining lecturer. Feynman's words were lovingly recorded, transcribed, and published in a series of three volumes that have become genuine and indispensable classics of the science literature.

The teaching of physics at Caltech, like the teaching of science courses everywhere, is constantly undergoing transition. Caltech's latest effort to infuse new life in freshman physics was instituted by Professor David Goodstein and eventually led to the creation of *The Mechanical Universe*. Word reached the cloistered Pasadena campus that a fundamental tool of scientific research, the cathode-ray tube, had been adapted to new purposes, and in fact could be found in many private homes. Could it be that a large public might be introduced to the joys of physics by the flickering tube that sells us spray deodorants and light beer?

As the idea of using television to teach physics started to reach serious proportions, a gift was announced by Walter Annenberg, publisher and former U.S. Ambassador to Great Britain, to support the use of broadcast means for teaching at the college level. Ultimately, nearly $6 million of Mr. Annenberg's funds, administered by the Corporation for Public Broadcasting, would be spent in support of *The Mechanical Universe*. That, in brief, is the story of how *The Mechanical Universe* came to be.

II PREFACE FOR STUDENTS

Just as in the first volume, each chapter of this book corresponds to one program of *The Mechanical Universe* television series. The book can also be used in the more traditional way as a physics textbook, without the television series. As before, we anticipate that you will read each chapter, view each program one or more times, and take advantage of further guidance, instruction, practice, and other help provided by institutions that offer this course for academic credit.

In the opening sequence of each television program, the viewer zooms into space, past asteroids, moons and planets, and beyond distant Pluto, pausing at the words *The Mechanical Universe*. For the second half of the series there also appear the words ... *and Beyond*, to indicate that we are now passing beyond mechanics, into other realms of physics.

And indeed we are. This second volume studies electricity and magnetism, their relation to each other and to light, and shows how the problem of light led to the special theory of relativity. Finally, we enter the world of modern physics, where particles may behave like waves and vice versa, and where some of the great verities of Newtonian physics appear less certain than they had.

In the course of all this, a few familiar mathematical tools from calculus are called into action and some new ones are introduced. For example, integrals along paths and integrals over surfaces are particularly useful for describing electric and magnetic fields. However, while it is important to understand the ideas expressed by these operations, they are seldom used for computation, and then only in the simplest cases.

That is not to say that our journey through this volume will be effortless. Our job is to go from the conclusion of one revolution in science – the discovery of Newtonian mechanics – to the beginning of another – the discovery of quantum mechanics. The end of the journey takes us close to the limits of human thought. If the journey is sometimes arduous, there is nevertheless quite a lot of remarkable scenery along the way, and considerable intellectual reward for reaching the goal.

Most of the important ideas in this course are presented in the television series, but many of them cannot be learned by simply watching television any more than they can be learned by simply listening to a classroom lecture. Mastering physics requires the active mental and physical effort of asking and answering questions, and especially of working out problems. The examples and questions interspersed through every chapter are intended to play an essential role in the process of learning.

III PREFACE FOR INSTRUCTORS AND ADMINISTRATORS

We expect that the ways in which *The Mechanical Universe* television series and textbooks are used will vary widely according to the circumstances and preferences of the institutions that offer it as a college course. The television programs can be viewed at home via broadcast or cable, presented in class, offered for viewing at the student's convenience at campus facilities, or even dispensed with altogether. However, we hope that no institution will imagine that the course can be presented without the services of live, flesh-and-blood college physics teachers. For most students, physics cannot be learned from a book alone, and it cannot be learned from a television screen either.

No laboratory component is offered as a part of *The Mechanical Universe* project. The reason is not because we judge a physics laboratory course to be unimportant or uninteresting, but rather that we judge its presentation by us to be impractical. We expect each institution offering the course to decide how it wishes to handle the laboratory component of learning physics.

This book is intended for use by students who have served their apprenticeship with the first volume of *The Mechanical Universe, Introduction to Mechanics and Heat*. We assume a level of mathematical sophistication attained by readers of that volume – basic skills with derivatives, integrals, and vectors, and some familiarity with differential equations. We do not assume that the student has read the unit on Heat (Chapters 15–18), which many schools prefer to offer after relativity (Chapters 45–48). The present volume allows this flexibility in the order of topics.

Throughout *The Mechanical Universe, and Beyond,* history is used as a means to humanize physics. It should go without saying that we don't expect students to memorize names and dates any more than we expect them to memorize detailed formulas and constants. *The Mechanical Universe* may or may not contribute to the vocational training of any given student. We hope it will contribute to the education of all of them.

IV ACKNOWLEDGMENTS

The Mechanical Universe textbooks, like the television series itself, would not have been possible without the cheerful and dedicated work of a long list of people who aided in its realization.

First of all, Professor Steven Frautschi, of Caltech, lead author of the companion volume for science and engineering majors, made contributions to this volume as well.

The authors benefited from comments made by *The Mechanical Universe* Local Advisory Committee: Keith Miller, Professor of Physics, Pasadena City College; Ronald F. Brown, Professor of Physics, California Polytechnic State University, San Luis Obispo; Eldred F. Tubbs, Member of the Technical Staff, Jet Propulsion Laboratory, Caltech; Elizabeth Hodes, Professor of Mathematics, Santa Barbara City College; and Eric J. Woodbury, Chief Scientist (retired), Hughes Aircraft Company.

In addition, Mario Iona (University of Denver) carefully reviewed the entire text and Judith Goodstein (Caltech) checked the manuscript for historical accuracy. The authors also received input from Dave Campbell (Saddleback Community College) and Jim Blinn (Jet Propulsion Laboratory), members of *The Mechanical Universe* team, and from Robert J. Sirko (Manager of Technology Planning for the Space Station Program at McDonnell Douglas Astronautics Company).

Special thanks go to Sharon Cox (University of California, Irvine) whose careful reading uncovered a number of technical errors in the original draft. She also made valuable suggestions for improving the exposition, and skillfully worked out the solutions to most of the problems.

Project Secretaries Renate Bigalke, Gwen Anastasi, Sarb Nam Khalsa, and Debbie Bradbury provided expert assistance in all phases of manuscript preparation, from word processing to obtaining permissions for reproducing copyrighted material. Carol Harrison sniffed out many photos and their sources for us, and Greg Borse located many historical references. Science Typographers, Inc., did a splendid job of copy editing, typesetting, and preparation of illustrations.

All of the work was watched over by Hyman Field of the Annenberg/CPB Project (sponsors of *The Mechanical Universe*) with the help of Peter Combes, and was gently prodded along by David Tranah and Peter-John Leone of Cambridge University Press. We are especially pleased that Cambridge, which published Newton's *Principia*, has decided to follow it up with *The Mechanical Universe*. Sally Beaty, Executive Producer of *The Mechanical Universe* television series, was present and instrumental at every important juncture in the creation of these books. Geraldine Grant and Richard Harsh supervised an extensive formal evaluation of various components of *The Mechanical Universe* project; the results of that effort have had their due effect on the final work.

Finally, special thanks are due Don Delson, Project Manager of *The Mechanical Universe*, who, through some miracle of organizational skill, cunning and compulsive worrying, managed to keep the whole show going.

CHAPTER 31

BEYOND THE MECHANICAL UNIVERSE

I do not know what I may appear to the world, but to myself I seem to have been only like a boy playing on the seashore, and diverting myself in now and then finding a smoother pebble or a prettier shell than ordinary, whilst the great ocean of truth lay all undiscovered before me.

Isaac Newton, from Brewster, *Memoirs of Newton* (1855)

31.1 SCIENCE AFTER NEWTON

The first volume of this series was largely devoted to the story of the first scientific revolution, and its consequence, the rise of the mechanical universe. It began with a Polish monk who wondered whether the universe might not look simpler from a different point of view, and it culminated with Isaac Newton, who gave us a wealth of scientific discoveries and a coherent view of the universe and how it ought to be described.

Not that Newton had discovered everything worth knowing – far from it. When he described himself as a child picking an occasional pretty pebble from the beach while a vast ocean of knowledge lay undiscovered before him, he meant precisely that science was in its infancy. But Newton had given us a way of organizing the unknown into questions that had answers. The world consists of matter and of forces, which together produce motion. The motion could be analyzed using his mechanics and his calculus. The questions that remained were, what are the forces of nature, and what is matter made of? One could add more detailed questions, such as, what is light, and what is life?

And so the world set out to provide Newtonian answers to all scientific questions. To be sure, not all the world undertook that task. Then, as always, most people were preoccupied with the more mundane aspects of simple survival, while the more powerful among them concerned themselves with the larger issues, such as making money, making war, spreading religion, and running slaves. But, throughout the 200 years that followed Newton's time, there were always a few who found the opportunity to pursue the goal of reducing natural phenomena to orderly, rational, Newtonian explanations.

They made a great deal of progress. Among the first issues taken up were the phenomena of electricity and magnetism. Both had been known to exist since antiquity, and had been observed and even made use of by various scholars. Their effects were examined with increasing precision and the results were quantified. Finally, the theory was described mathematically according to Newtonian precepts. As we have already seen in Chapter 11, this process reached its climax around the time of the American Civil War, when James Clerk Maxwell realized that the speed of light could be found by combining the static forces between electric charges with that between magnetic poles. Starting from this synthesis, he worked out an elegant theory that combined electricity, magnetism, and light into a single phenomenon. Maxwell's theory is the ultimate objective of the next 12 chapters of this book.

In the meantime, other Newtonian scientists had turned their attention to the properties of matter. Among these was a group of chemists who revived the ancient idea of the atomic theory of matter. Throughout the nineteenth century, those intrepid investigators broadened their knowledge of chemical elements and how they combined, and sought to learn how their properties could be explained from the nature of their constituent atoms. Another group of scientists perceived profound implications in the rules governing the flow of heat, and brought forth an astonishingly successful new axiomatic science called thermodynamics.

And then, at the dawn of the twentieth century, a strange thing happened. Suddenly, the whole Newtonian system didn't seem to work so well anymore. Of course, it was just as good as ever at explaining the orbit of the moon or the trajectory of a cannonball, but new phenomena appeared that were not so much beyond Newtonian science as inconsistent with it, or even contradictory to it.

For example, the law of inertia, which is at the very basis of Newtonian mechanics, makes perfect sense so long as any one speed is as good as any other, there being no state of absolute rest from which to measure absolute speed. But Maxwell had shown that there was an absolute speed – the speed of light. Was the law of inertia consistent with a universe in which light could propagate through the void? At first the question was too abstract, too philosophical, to be paid much attention to by the average working scientist, but in the end it was resolved by the theory of relativity, which changed forever our understanding of time and space. This theory is introduced in Chapters 45–48.

Another problem arose in the study of matter and its relation to light and electricity. As scientists probed deeper into the nature and behavior of the atom they found evidence that atomic structure could not be accounted for using the laws of Newtonian physics. The ultimate result was to formulate a new and more profound view of how things work. This gave birth to quantum mechanics, which is introduced in Chapters 49–51.

These events, which took place in the first quarter of the twentieth century, constitute a revolution in physics as far reaching as the first one. Our job in this book will be to bridge the gap, starting from Newton at the end of the first revolution, and ending with the result of the second: the quantum mechanical universe (Chapter 52). In this brave new world, Newton's laws are no less valid than they were before, but they turn out to be consequences of surprising and more profound laws that work not only in familiar realms, but also inside atoms, and at speeds close to the speed of light.

The foregoing paragraphs outlined the history of physics as a series of discoveries of fundamental laws, each to be expressed in precise mathematical form: Newton's laws, the laws of thermodynamics, Maxwell's equations, the theory of relativity, and quantum mechanics. But any brief historical survey is subject to bias, and this one is no exception. Our account is not so much a history of physics as a history of the content of physics textbooks. Of course, that bias is no accident, since it is designed to outline a physics textbook (this one).

Physics is a vigorous human endeavor, filled with confusion, misunderstanding, prejudice, and frustration, all justified by occasional illuminating insights that produce in its practitioners a kind of euphoria as addicting as any chemical substance. Finally, when the excitement is over, and the undeniable power of a given idea has created a solid consensus among scientists, its elevation to the status of "truth" is celebrated by interring its bones in a textbook. One consequence of that sequence of events is that the student or reader is likely to come to see science as a kind of graveyard, or at best a museum, of ideas, rather than as the lively battlefield it really is.

There is no simple antidote to that problem. Textbooks are written precisely for the purpose of propagating the received wisdom. Science is progressive. That is to say, this century's science isn't merely different from the last century's; it is better, because it subsumes the science of the past and goes on from there. In that sense, the second revolution in physics was radically different from the first. The first overthrew the Aristotelian world view, but the second really only expanded the Newtonian view. That's why, for better or for worse, a textbook such as this one must do the job of guiding the reader through the museum of old physics. The later discoveries could not have been made without the earlier ones, and they cannot be understood without them either. In this volume we resume that tour. But first let us pause a moment, and try to look at the whole picture from a slightly different perspective.

31.2 ORDERS OF MAGNITUDE

The history of physics can be seen as a gradually arising comprehension of the magnitudes of things. The formal equations of physics are essential for working out details, but an excellent first step toward getting the big picture is to master – and remember – certain essential orders of magnitude, and to learn how to use them to estimate others.

To take a specific and significant example, one of the most important magnitudes to be discovered was the size of an atom or molecule. The atomic theory of matter could be no more than a vague notion until one could say how big an atom is, or how many of them there are in a piece of matter. Even if we grant that, say, a liter of water can't be divided forever into smaller drops of water, how many water molecules *can* it be divided into? A million? 10^{10}? 10^{100}?

The answer to that question was a long time in coming (we'll discuss it further in Chapter 49). However, for those disposed to look for it there were solid hints. For example, Benjamin Franklin had the answer in his grasp when he estimated how large an oil slick would be formed on the surface of water by a given amount of oil. But he did not think of interpreting his result in terms of the size of an oil molecule. Example 1 shows that common experience holds clues to the answer.

Example 1

Estimate the average amount of rubber an automobile tire loses per revolution.

A good steel-belted radial tire may last for 60,000 km (40,000 miles). Its diameter is about 1 m and its tread is about 1-cm ($\frac{1}{2}$-in.) thick. Its circumference is π m = $10^{-3}\pi$ km, so it makes a total of

$$\frac{60{,}000 \text{ km}}{10^{-3}\pi \text{ km/revolution}} \approx 2 \times 10^7 \text{ revolutions}$$

during its lifetime. The tread is worn away at the rate of

$$\frac{1 \text{ cm}}{2 \times 10^7 \text{ revolutions}} \approx 5 \times 10^{-8} \text{ cm/revolution}.$$

This is about the thickness of one molecule of rubber. This calculation suggests that automobile tires get their traction by laying down a monomolecular layer of rubber everywhere they go.

Questions

1. Assume that an oil molecule is roughly the same size as a rubber molecule and that oil spreads on still water until it forms a monomolecular layer. Estimate the maximum area of the oil slick made by 10 cm³ of oil.

2. If water molecules are also roughly the same size as oil molecules (they're really a little smaller), estimate how many molecules there are in a liter of water.

31.2 ORDERS OF MAGNITUDE

Let's begin the task of mastering magnitudes by considering a few quantities that were known to Isaac Newton. Greek mathematicians had measured the radius of the earth (about 6000 km) and the distance to the moon (60 times the radius of the earth). That knowledge was essential for Newton's triumphant calculation (see Chapter 8) that the moon falls $\frac{1}{20}$ in. each second.

Newton did not know the mass of the earth, which is 6×10^{24} kg. Had he known it, he could have calculated his universal gravitational constant,

$$G = \frac{gR_E^2}{M_E} = 7 \times 10^{-11} \text{ N m}^2/\text{kg}^2,$$

where R_E is the radius of the earth, M_E is its mass, and g is the acceleration of a falling body (about 10 m/s²). But he did know the density of water, 1 g/cm³ (this is easier to remember than 10^3 kg/m³), and the density of heavy metals like iron (about 10 g/cm³). He then estimated the density of the earth, ρ_E, to be somewhere between these values,

$$\rho_E = \frac{\text{mass}}{\text{volume}} = \frac{M_E}{(4/3)\pi R_E^3} \approx 5 \text{ or } 6 \text{ g/cm}^3.$$

Newton obtained an excellent estimate of G a full century before Henry Cavendish was able to measure the quantity (see Chapter 10).

Neither Newton nor Copernicus before him knew with any certainty the distance to the sun (1 AU $\approx 150 \times 10^6$ km), much less the distance to the stars (at least 4 light years, or about 4×10^{13} km). But they both realized that because there is no apparent parallax (relative motion) of the stars as the earth goes around in its enormous orbit, the universe is very large indeed. In a sense, the ultimate lesson of the Copernican revolution was to realize just how small we are (about 2 m) compared to the size of things in the universe. The distance to the center of the galaxy is about 10^{17} km (10^4 light years), and the size of the universe is about 10^{23} km, or about 10^{10} light years. By no accident at all, the age of the universe is of the same order of magnitude as the time it takes light to travel that distance, about 10^{10} years.

Example 2

Estimate the mass and radius of the moon, given that the acceleration of gravity on the moon is $\frac{1}{6}$ that on Earth (see Chapter 8).

In Chapter 29 we found that the distance from the center of the earth to the center of mass of the earth–moon system is about $\frac{3}{4}R_E$, where R_E is the radius of the earth. Let r_m denote the mean distance from the surface of the earth to the center of the moon. Equating moments about the center of mass of the earth–moon system we obtain

$$M_m\left(r_m + \tfrac{1}{4}R_E\right) = M_E\left(\tfrac{3}{4}R_E\right).$$

But $\frac{1}{4}R_E$ is negligible compared to r_m, so this gives us the approximate formula

$$M_m = \frac{3}{4}\frac{R_E}{r_m}M_E.$$

Using $R_E = 6.4 \times 10^3$ km, $r_m = 3.84 \times 10^5$ km, and $M_E = 6 \times 10^{24}$ kg, we find $M_m \approx 7.5 \times 10^{22}$ kg.

Using Eq. (8.2) for the moon we find that the radius of the moon is given by the formula

$$R_m^2 = \frac{M_m}{g_m}G,$$

where $g_m = g/6 = (10/6)$ m/s^2, so

$$R_m^2 = \frac{(7.5 \times 10^{22} \text{ kg})(7 \times 10^{-11} \text{ N m}^2/\text{kg}^2)}{(10/6) \text{ m/s}^2} = 3.15 \times 10^{12} \text{ m}^2,$$

and hence

$$R_m \approx 1.7 \times 10^6 \text{ m}.$$

Example 3

Estimate the radius of the sun, R_s, using the fact that the moon is just large enough to eclipse the sun. (The distance from the earth to the sun is $r_s = 1.50 \times 10^8$ km, and the distance from the surface of the earth to the moon is $r_m = 3.8 \times 10^5$ km.)

When the moon is in a position to eclipse the sun we can use the accompanying

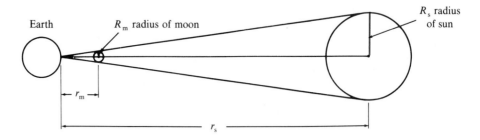

diagram and compare similar triangles to obtain

$$\frac{R_s}{r_s} = \frac{R_m}{r_m}.$$

Hence $R_s = R_m r_s/r_m$, and when we substitute the given values we find

$$R_s = 6.7 \times 10^5 \text{ km}.$$

31.2 ORDERS OF MAGNITUDE

Learning the age of things is also an essential step toward understanding the world. In the last century, it was believed on the basis of the best evidence available (the Bible) that the earth was a few thousand years old. That was not nearly long enough for either biological or geological evolution to have occurred. Later in the century, around 1862, a purely scientific estimate by William Thomson (who became Lord Kelvin), based on a calculation of the rate of cooling of an initially molten planet, also concluded that the earth was much younger than it really is, because the estimate was made before the discovery of the radioactivity that helps keep the earth warm. Kelvin estimated the age of the earth at between 20 and 400 million years. We know today that the earth is more than 4 billion years old. Natural radioactivity, the same phenomenon that helps warm the earth, also helps reveal its age.

As we learned in Chapter 11, a crucial discovery was that the speed of light is contained in the electric and magnetic force constants,

$$\frac{K_e}{K_m} = c^2,$$

where

$$c = 3 \times 10^8 \text{ m/s}.$$

This speed is enormous compared to anything in ordinary experience. For example, it's 10^6 times the speed of sound in air and 10^4 times the orbital speed of the earth around the sun. The theory of relativity that Albert Einstein worked out in 1905, revolutionary though it was, differs significantly from Newtonian mechanics only at speeds approaching the speed of light.

Questions

3. Using only magnitudes given in this chapter, estimate each of the following in units of light years, light minutes, or light seconds, whichever seems most appropriate.

   (a) The size of a molecule.
   (b) The size of the earth.
   (c) The distance from the earth to the moon.
   (d) The distance from the earth to the sun.

4. Assume that a computer transmits information from one working element to another at the speed of light. Also assume that after an element has performed a calculation, all other elements must be informed before the next calculation takes place. Determine the maximum volume of a computer capable of performing 10^9 calculations per second. (*Hint:* Use a spherical model.)

5. Telephone calls between continents are now routinely transmitted via geosynchronous artificial communication satellites. The signals are sent up and back at the speed of light. Estimate the time delay this causes at each end of the conversation. (The radius of a geosynchronous orbit was requested in Question 15 of Chapter 9.)

The quantities that we've discussed up to now fall into three general categories: fundamental constants, such as c and G, which are the same everywhere in the universe; quantities that aren't strictly constant, but are taken to be constant for most purposes, such as the density of water or the size of the universe; and quantities that are important to us only because we inhabit planet Earth (the radius of the earth, the distance to the sun, and so on). As we've already seen, among all these quantities, the fundamental constants tend to be associated with the great discoveries of physics.

Question

6. Identify which of the following quantities are fundamental quantities, which are nearly constant, and which are specific only to planet Earth.

   (a) The electric constant K_e.
   (b) The magnetic constant K_m.
   (c) The boiling point of water (100°C or 212°F).
   (d) The freezing point of water (0°C or 32°F).
   (e) The charge of an electron (1.6×10^{-19} C, see Chapter 12).
   (f) The area of a circular disk of radius 1.
   (g) The acceleration of gravity, g.

The discovery of the inner structure of the atom brought with it a whole set of new fundamental constants characteristic of the microscopic world. For example, as we saw in Chapter 12, Robert A. Millikan measured the charge e of the electron and found that

$$e = 1.6 \times 10^{-19} \text{ C}.$$

Earlier, J. J. Thomson had measured the ratio of the electron's charge to its mass, $e/m = 1.8 \times 10^{11}$ C/kg. Combining these measurements, it turns out that m_e is about 2000 times smaller than the mass of a proton or neutron, $m_e = 9 \times 10^{-31}$ kg. If M_n denotes the mass of a nucleon (a nucleon is either a proton or a neutron), we have

$$M_n = 1.7 \times 10^{-27} \text{ kg} \approx 2000 m_e.$$

Thus, nearly all of the mass of an atom is to be found in its nucleus.

Early in the nineteenth century, an Italian chemist named Amedeo Avogadro proposed that, at ordinary temperature and pressure, a given volume of any gas would have the same number of atoms or molecules (see Chapter 16 for the reason why). Knowing that number is equivalent to knowing the mass of a nucleon.

Example 4
Find the value of Avogadro's number.

A *mole* of any substance is defined to be M grams of the substance, where M is numerically equal to its atomic or molecular mass (which is the same as the number of nucleons in each atom or molecule). Avogadro's number is defined to be the

31.2 ORDERS OF MAGNITUDE

number of atoms or molecules in 1 mole of a substance. Because the mass (in grams) of one nucleon is

$$M_n = 1.7 \times 10^{-24} \text{ g},$$

Avogadro's number is

$$N_0 = \frac{1}{M_n} = 6 \times 10^{23} \text{ molecules/mole (or atoms/mole)}.$$

At standard temperature and pressure (0°C temperature, 1 atm pressure) 1 mole of any gas occupies 22.6 L of volume.

Questions

7. Estimate the number of molecules in 1 cm³ of air in the atmosphere.

8. Estimate the number of molecules in 1 cm³ of liquid water. The molecular mass of water is 18.

9. Estimate the number of atoms in the earth. (Your answer will depend on the assumptions you make concerning the composition of the earth, but it is not M_E/M_n.)

The discovery of quantum mechanics introduced one more fundamental constant of transcendent importance in physics. It is denoted by the symbol h and is called Planck's constant, after Max Planck whom we shall encounter in Chapter 50. It is most often divided by 2π and written in the form

$$\hbar = \frac{h}{2\pi} \approx 1 \times 10^{-34} \text{ J s}.$$

(The symbol $\hbar$ is read "h bar.") Planck's constant has the units of joule seconds. One of the first triumphs of quantum theory was the discovery by Niels Bohr that the radius of a hydrogen atom, a_0, is given by a combination of fundamental constants that includes $\hbar$:

$$a_0 = \frac{\hbar^2}{e^2 K_e m_e} = 0.5 \text{ Å},$$

where 1 Å = 10^{-10} m, a unit called the *angstrom*. This discovery was comparable to Maxwell's finding the speed of light imbedded in the electric and magnetic constants. The diameter of the hydrogen atom is about 1 Å.

The Copernican revolution not only tore us away from the center of the universe, it also began the process of teaching us relative sizes of objects that make up that universe. We now know that an average person is larger than an atom by a factor of about 2×10^{10}. On the other hand, the distance from the earth to the sun is about 10^{11} times greater than a person. So, perhaps in a logarithmic sense, we are once again close to the center of the universe. But modern science pushes relentlessly onward, outward to the galaxies and inward to the atomic nucleus and its

Table 31.1 Approximate Values of Important Quantities.

Quantity	Symbol	Value
1. Fundamental constants		
The speed of light	c	3×10^8 m/s
Planck's constant	$\hbar$	1×10^{-34} J s
Electron charge	e	1.6×10^{-19} C
Electron mass	M_e	$M_n/2000$
Nucleon mass	M_n	1.7×10^{-24} g
Avogadro's number	N_0	6×10^{23}
Gravitational constant	G	7×10^{-11} N m^2/kg^2
Diameter of a hydrogen atom	$2a_0$	1×10^{-8} cm = 1 Å
2. Quantities nearly constant		
Density of water		1 g/cm^3
Density of iron		10 g/cm^3
Freezing point of water		0°C
Distance to the center of the galaxy		10^4 light years
Size of the universe		10^{10} light years
3. Numbers of interest to Earthlings		
Boiling point of water at sea level		100°C
Radius of the earth	R_E	6×10^3 km
Distance from Earth to the moon		$60 R_E$
Distance from Earth to the sun		150×10^6 km
Distance to the next star after the sun		4 light years
Acceleration of gravity	g	10 m/s^2
Mass of the earth	$M = \dfrac{gR_E^2}{G}$	6×10^{24} kg

inner constituents, without any regard at all for our need to seem to be at the center of things.

The quantities discussed in this section are listed in Table 31.1. They are worth remembering. And, as you learn new subjects, you should add your own favorites to the list.

31.3 A FINAL WORD

The series of extraordinary events that led us beyond the mechanical universe, to the portals of the quantum mechanical universe, is yet another story full of human drama, and it too had its heroes, comparable to Copernicus and Kepler, Galileo and Newton.

One of them was Benjamin Franklin. A self-taught sophisticated Newtonian scientist, Franklin made a success of his every pursuit, starting with printing and ending with diplomacy and politics. He also stunned the European world of science with a series of penetrating discoveries and inventions that came from someone who delighted in playing the role of the sage from the wilderness.

Another hero was Michael Faraday, an Englishman with little education who remained mathematically illiterate throughout his life. Yet, again and again, in one field after another, it was Faraday who made the crucial discovery, Faraday who did the critical experiment, Faraday who had the deepest insight into the nature of things.

Then there was James Clerk Maxwell. Born into a wealthy family in Edinburgh, Maxwell had the best education Britain could offer. His work on the kinetic theory of gases helped establish the existence of atoms and made him a prominent scientist in his own time. But his grand synthesis of electricity, magnetism, and light into a set of electromagnetic equations was his crowning achievement. Yet he never saw the scientific world come to accept it because he died of cancer at the age of 48, 10 years before his theory was verified by experiment.

And finally, of course, we find the towering figure of Albert Einstein. In one year, 1905, while working as a patent clerk in Switzerland, Einstein wrote a brilliant doctoral thesis that helped establish the existence of atoms, published two papers on relativity that changed forever our understanding of space and time, and also found time for a short article on the photoelectric effect, for which he was later awarded the Nobel prize. This must have been the most glorious year for any one scientist since Isaac Newton's sojourn in the apple orchard during the great plague year of 1665. Einstein later became a world-famous folk hero, everybody's favorite uncle, the popular symbol of science in the twentieth century.

These, among others, are the heroes of the story that begins to unfold in the next chapter as we enter the world of electrical science.

CHAPTER 32

STATIC ELECTRICITY

Chance has thrown my way another principle, more universal and remarkable... which casts a new Light on the subject of electricity. This principle is that there are two distinct Electricities, very different from each other; one of these I call *vitreous Electricity*; the other *resinous Electricity*. The first is that of Glass, Rock-Crystal, Precious Stones, Hair of Animals, Wool, and many other bodies. The second is that of Amber, Copal, Gum-Lac, Silk, Thread, Paper, and a vast number of other substances. The characteristic of these two Electricities is that a body of the *Vitreous Electricity,* for example, repels all such as are of the same Electricity; and on the contrary, attracts all those of the *resinous electricity* .... This Principle very naturally explains why the ends of Thread, of Silk or Wool recede from one another in the form of a Pencil or Broom when they have acquired an electrick Quality.

Charles François de Cisternay Dufay, *Philosophical Transactions,* **38**, 258 (1734)

32.1 THE BEGINNINGS OF ELECTRICAL SCIENCE

Lightning storms were undoubtedly the first electrical phenomenon to be observed by man. But the first recorded observation of the effect of electric charge concerned a phenomenon much less spectacular than lightning. Around 600 B.C. a Greek mathematician and astronomer, Thales of Miletus, noted that amber, on being rubbed, attracts bits of straw, twigs, thread, and other light objects.

DE MAGNETE, LIB. II.

am, more indicis magnetici, cuius alteri fini appone succinum, vel

lapillum leniter fricatum, nitidum & politum, nam illico versorium conuertit se. Plura igitur attrahere videntur, tàm quæ à naturâ tantùm efformata, quàm quæ arte parata, aut conflata, & commixta sunt; nec ita vnius vel alterius singularis est proprietas (vti vulgò existimatur) sed plurimorum natura manifesta, tam simplicium suis tantùm formis consistentium, quàm compositorum; vt ceræ duræ sigillaris, & aliarum etiam quarundam ex pinguibus mixturarum. Sed vndè ista inclinatio fieret, & quænam sint vires illæ, (de quibus pauci paucissima, vulgus philosophantium nihil protulerunt) amplius inquirendum. A Galeno tria in vniuersum trahendi genera con-

Figure 32.1 Gilbert's versorium. (From Gilbert's book *De Magnete*. Courtesy of the Archives, California Institute of Technology.)

Twenty-two centuries later, in 1600, an Englishman named William Gilbert published a celebrated treatise on magnetism that contains a chapter devoted to the amber effect. To demonstrate a sharp distinction between magnetism and the amber effect he invented the first electrical instrument, the *versorium* that, in his own words is "...a rotating needle of any sort of metal, three or four fingers long, pretty light, and poised on a sharp point after the manner of a magnetic pointer. Bring near to one end of it a piece of amber or a gem, lightly rubbed, polished and shining: at once the instrument revolves." Gilbert's sketch of the versorium is shown in Fig. 32.1.

Before Gilbert's time, only a few substances were known to exhibit the amber effect. But with his versorium Gilbert demonstrated a host of substances capable of attracting objects when rubbed, and he called them "electrics" from the Greek word *elektron* for amber. Today, Gilbert's *electrics* are called *insulators*, and his *nonelectrics*, which are difficult to electrify by rubbing, are called *conductors*.

In the early eighteenth century, a rubbed electric was said to possess "the electric virtue." Another English scientist, Stephen Gray, discovered that electric virtue could be transferred from one object to another when the objects were joined by a piece of metal. In modern terminology, he recognized that conductors, such as copper or silver, readily transmit electricity, whereas insulators, such as rubber or silk, do not.

In 1734 the Frenchman Charles François de Cisternay Dufay discovered that a glass rod rubbed with silk had a different type of charge from a resinous rod rubbed with fur. After a series of experiments he concluded that charge created by friction was of two types: "vitreous," like that produced on glass, and "resinous," like that produced on resinous material such as rubber. Dufay imagined two distinct "electricities" – weightless fluids that would flow from one body to another to electrify it, and he thought that all bodies contained mixtures of the two fluids. Nowadays Dufay's "electricities" are known as electric charge.

Dufay also established the general law that like charges repel and opposite charges attract, and that any force between charged objects depends on the type and amount of charge.

Dufay's labels for vitreous and resinous electricities were soon replaced with the terms *positive* charge and *negative* charge introduced by Benjamin Franklin, whose work on electricity established him as an important American scientist. Franklin added to the confusion about the types of electricity by proposing a one-fluid theory in which negative electricity is the absence of the all pervasive positive electricity. This theory, however, contained the fruitful idea now called the principle of *conservation of charge*: In a closed or isolated system the algebraic sum of the positive and negative charges is constant.

Today we know that the type and amount of electricity is not dependent on any fluid but rather on the presence of charged particles that are usually electrons and protons, although other charged particles, such as mesons, are known to exist. Franklin's arbitrary sign convention is still used today. For example, an electron is said to have negative charge, whereas a proton has positive charge. Common materials have equal numbers of protons and electrons and are said to be electrically neutral because they have no net charge. When two different nonconducting substances such as glass and silk are rubbed together, electrons are transferred from the surface of the glass to the surface of the silk, and the glass rod is left with a positive charge. When a resinous rod is rubbed with fur, electrons are transferred from the fur to the surface of the rod, and the resinous rod acquires a negative charge.

In a conductor the electrons are free to move around in response to forces from other charges – electric forces. Whether a conductor is positively or negatively charged depends on whether it has a deficit or excess of electrons. If a conductor is charged, the excess charges repel each other and distribute themselves in such a way that all charges reside on the surface of the conductor, no matter what its shape. But as we'll see in Chapter 34, the shape of the conductor may affect the way the charges are distributed.

On the other hand, in an insulator the electrons are tightly bound to their atoms and cannot move about freely; that is why insulators do not readily conduct electricity. Once electrons are either added to or removed from an insulator through friction or some other means, the insulator remains charged until a transfer of charge alters the amount.

As various ideas concerning electricity evolved in the mid-eighteenth century, many lecturers on the subject devised spectacular lecture demonstrations that became quite popular. In one such demonstration, illustrated in Fig. 32.2, a young boy was suspended from the ceiling on silk cords and connected to a generator consisting of a rapidly spinning glass globe fitted with an attachment for rubbing it. Electric charge from the generator was conducted through the boy to a young lady

Figure 32.2 An eighteenth century electric demonstration. (Courtesy Burndy Library.)

who, it turn, attracted feathers or bits of paper. Sometimes a willing member of the audience would be invited to grasp the girl's hand or kiss her – with shocking results.

Professional scientists often earned extra income by presenting such demonstrations to the general public. Electricity became a popular rage and at some universities students complained that the public subscribers were crowding them out of their seats. Popular interest in electricity helped finance some of the research and sometimes also led to important developments.

32.2 THE ELECTROSCOPE

To estimate the type and amount of charge on an object, eighteenth century experimenters used a simple device known as an *electroscope*, an instrument consisting of a thin metal leaf, usually of gold, attached to a conducting post inside a metal box with two glass sides to permit viewing. The metal post extends through the top of the box but is insulated from it, as indicated in Fig. 32.3. We'll now describe how the electroscope detects alterations of charge on the leaf and the rod.

When a charged object is brought near the metal bulb at the top of the electroscope, the gold leaf moves outward. When the object is removed, the leaf collapses to its original position. To explain this action, imagine a positively charged glass rod placed near the electroscope without touching it. Negative charges (electrons) in the conducting post are attracted to the rod and quickly move to the top. That leaves a positive charge on the gold leaf and on the lower part of the post, so the gold leaf is repelled from the post. If the glass rod is removed, the negative

32.3 CHARGE SEPARATION AND INDUCTION

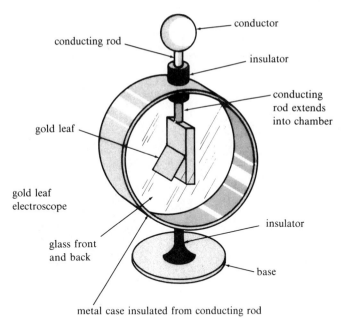

Figure 32.3 Schematic diagram of a gold leaf electroscope.

charges flow back to the lower part of the post, and the gold leaf collapses as the charge on it and on the post is neutralized.

Now suppose the positively charged glass rod is allowed to touch the bulb before it is removed. The gold leaf moves and stays repelled indicating that the electroscope is positively charged. This is an example of *charging by contact*. The glass rod loses some of its charge to the electroscope, which remains charged even after the rod is removed. What will happen if a negatively charged rubber rod is then brought near the bulb? The leaf collapses slightly, as illustrated in Fig. 32.4a, because some negative charges in the metal post are repelled by charges on the rubber rod and move down onto the leaf, neutralizing some of the positive charges there. If the rubber rod is removed, the leaf returns to its original deflected position. On the other hand, if a positively charged glass rod is brought near the initially charged electroscope, the gold leaf bends ever farther away from the central post, as Fig. 32.4b indicates. The sign and relative amount of charge can be measured on a scale etched on the glass window.

32.3 CHARGE SEPARATION AND INDUCTION

We have seen that an object can be electrified by friction and by contact. Another method discovered in the eighteenth century is that of charge separation, illustrated in Fig. 32.5. A negatively charged rod R is brought near to but not touching a neutral conductor that rests on an insulating stand. Tests with an electroscope will

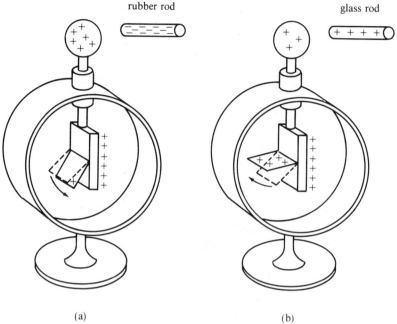

(a)   (b)

Figure 32.4 An initially charged electroscope indicates the sign of charge on an object by the changes in deflection of its gold leaf.

show that the insulated conductor exhibits *charge separation*. This means that the initially equal quantities of positive and negative charges have redistributed themselves so there is an excess of positive charge at end B near the rod R, and an equally large excess of negative charge at the opposite end A. The center C remains neutral. If rod R is then completely removed, the insulated conductor returns to its original neutral state. The separation has been only temporary. Sometimes charge separation is called *polarization*.

On the other hand, if rod R is not removed but instead the conductor is touched with a hand or is otherwise connected to the earth (all being good conductors) in a

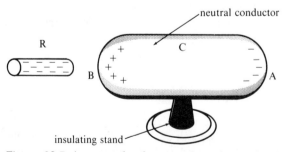

Figure 32.5 An example of a neutral conductor that has charges separated.

32.3 CHARGE SEPARATION AND INDUCTION

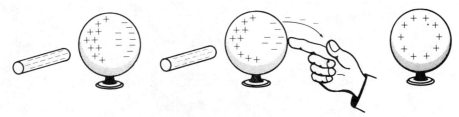

Figure 32.6 Charging a conductor by electrostatic induction.

process known as *grounding*, the charges redistribute themselves once more over the enlarged system (object + Earth) with an excess of positive charge at end B. If now the hand or grounding connection is removed, the insulated conductor will have a net positive charge and it keeps this charge even after rod R is removed. This method of charging a conductor is called *charging by electrostatic induction*.

Thus we have described three different methods of electrifying objects – charging by rubbing or friction, charging by separation, and charging by electrostatic induction. By any of these methods, charge is neither created nor destroyed, but is simply transferred from one body to another. Conservation of charge joins the list of fundamental laws that stretch our understanding beyond the mechanical universe.

Example 1
How does polarization help explain why a rubber comb passed through dry hair attracts bits of paper?

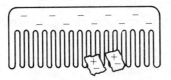

The comb becomes negatively charged through friction when it is passed through hair. The bits of paper are neutral. But when the comb is brought near a bit of paper, the positive charges in the paper are attracted to the comb and the negative charges are repelled. Although paper is an insulator and not a conductor, the negative charges move very slightly (a distance about the size of a molecule) and so the charge on the paper becomes separated. The force between the negative charges on the comb and those in the paper must be less than the force between the positive charges in the paper and the negative on the comb; that's why the paper is attracted to the comb. We do know that the distance between the positive charges in the paper and the comb is less than the distance between the negative charges and the comb. As we'll discuss later in more detail, the electrostatic force depends on the distance between charges.

Questions

1. Why does a balloon that's been rubbed on a wool sweater adhere to a wall?

2. Why do you suppose that on dry days you (a conductor) often get a shock when touching a metal doorknob after having walked across a carpet?

3. A phenomenon that baffled eighteenth century electricians was known as attraction–contact–repulsion. If a charged rod is brought near a small, electrically neutral pith ball that hangs at the end of a thread, the ball is attracted. However, if the ball is allowed to touch the rod, it is quickly repelled. How would you explain this?

4. Two neutral metal spheres are initially connected by a wire. A negatively charged rod is brought near one of the spheres without touching it and then the wire is removed. Then the charged rod is removed. Are the spheres still neutral?

32.4 COULOMB'S LAW

Following the unparalleled success of Newton's mechanical description of the universe, scientists in the eighteenth century naturally associated the electric interactions of charged objects with mechanical forces. Early experimental evidence concerning the behavior of the electric force was obtained by Benjamin Franklin in 1775. A curious and tireless observer of nature, Franklin noticed that a small cork hanging near the outside of an electrically charged metal can was strongly attracted to the can. However, if the cork were lowered inside the can by a thread, the cork experienced no force regardless of its location inside the can.

While in England, Franklin described this effect to Joseph Priestley, an English clergyman and schoolteacher, and asked him to repeat the experiment. Priestley is best known for his research in chemistry. He discovered that oxygen is involved in combustion and respiration and, amid other claims to fame, invented carbonated drinks. Priestley also made pioneer contributions to electrostatics. Upon repeating Franklin's experiment, Priestley was struck by the analogy to gravitational behavior. Having read Newton's *Principia*, he remembered that a mass inside a spherical shell experiences no gravitational force from the shell. Since the inverse-square nature of the gravitational force is responsible for this behavior, Priestley proposed that the electric force between two charges might also obey an inverse-square law regarding the distance between the charges.

The analogy was not complete because the gravitational force on an object in a hollow body is zero only if the body has uniform density and spherical shape, whereas the results of Franklin's experiment were the same for a hollow body of any shape provided the walls were made of conducting material.

Priestley's conjecture had to be tested by experiment. That task awaited the engineering talents of Charles Augustin Coulomb. In 1788 Coulomb published the conclusions of numerous experiments in which he used a torsion balance like the one in Fig. 32.7 to measure the force between two small, equally charged spheres. Coulomb assumed that electric force, like gravitational force, depends on the product of the two charges. Then by measuring the angle through which the torsion balance would turn he determined the force between the metallic balls at various

32.4 COULOMB'S LAW

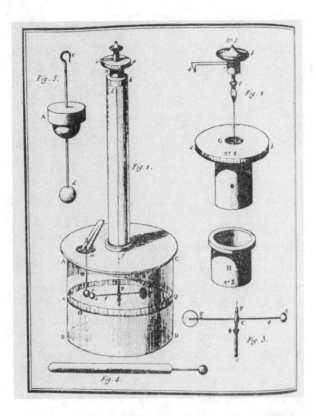

Figure 32.7 Coulomb's own drawing of the torsional balance used to determine the force law between two small metal spheres.

distances. Apparently Coulomb was unaware that 20 years earlier John Michell had used a similar instrument in England to measure the gravitational constant G. Coulomb's experiments verified Priestley's conjecture: The electric force follows an inverse-square law.

If a charge q_1 is located at position $\mathbf{r}_1$, and a charge q_2 at $\mathbf{r}_2$, then the electric force of q_1 on q_2 can be described mathematically as

$$\mathbf{F}_{12} = K_e \frac{q_1 q_2}{r_{12}^2} \hat{\mathbf{r}}_{12}, \tag{32.1}$$

where $\mathbf{r}_{12} = \mathbf{r}_2 - \mathbf{r}_1$, as indicated in Fig. 32.8. The factor K_e is the electric constant. This equation is known as *Coulomb's law*, and it holds only for point charges, ideal charges that are in a vacuum and have no physical extent in space. It expresses a fundamental principle in electrostatics.

Nearly 20 years before the experiments of Priestley and Coulomb, another talented Englishman, Henry Cavendish, was also convinced that an inverse-square law governed electrical force. He conducted an experiment similar to Priestley's and

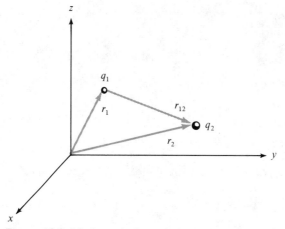

Figure 32.8 Mathematical quantities in Coulomb's law.

also made a masterful error analysis showing that if the force decreases like r^{-n}, then n cannot differ from 2 by more than $1/57$. However, the contents of Cavendish's manuscripts were not revealed until nearly a century later when most of his results had been rediscovered by others. Modern experiments have verified that the exponent n differs from 2 by less than 2 parts out of a billion. Chapter 33 will show that this $1/r^2$ dependence has important geometric consequences.

Coulomb's law is a simple vector equation that gives the force on one charge due to another. With other charges in the vicinity there will be additional forces present. Each of these forces is given by Coulomb's law and they combine by vector addition. That is, the Coulomb force obeys the *principle of superposition*. For example, the total force on charge q_1 in Fig. 32.9 due to four other charges q_2, q_3, q_4, and q_5, is given by

$$\mathbf{F} = K_e \frac{q_1 q_2}{r_{21}^2} \hat{\mathbf{r}}_{21} + K_e \frac{q_1 q_3}{r_{31}^2} \hat{\mathbf{r}}_{31} + K_e \frac{q_1 q_4}{r_{41}^2} \hat{\mathbf{r}}_{41} + K_e \frac{q_1 q_5}{r_{51}^2} \hat{\mathbf{r}}_{51},$$

where the distances and unit vectors are as indicated in the figure. Each pair of charges acts as if the others were not present.

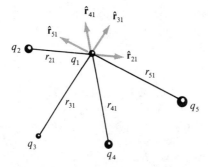

Figure 32.9 The total electric force on a point charge is equal to the vector sum of all the individual Coulomb forces acting on it.

32.4 COULOMB'S LAW

The direction of the force is also important. Because there are two kinds of charge possible, the electric force, unlike gravity, can be attractive or repulsive. If two charges q_1 and q_2 have the same sign, the product $q_1 q_2$ in Eq. (32.1) is positive, indicating a repulsive force that tends to push the charges apart. On the other hand, if the charges have opposite signs, their product is negative, indicating an attractive force.

Next, there is the matter of units to consider. In SI units the fundamental unit of charge is called the coulomb, abbreviated by C. Actually, the coulomb is defined in terms of the ampere, a unit of electric current that is easy to measure directly. This definition, to be discussed in Chapter 39, makes the coulomb equal to the magnitude of the charge carried by 6.2414×10^{18} electrons or protons. In a vacuum, a charge of 1 C repels a like charge, at a distance of 1 m, with a force of approximately 9×10^9 N, a huge force of about 1 million tons. Charges encountered in practice are much smaller than the coulomb. In SI units, F_{12}, q_1, q_2, and r_{12} in Coulomb's law are all defined independently and thus determine the value of K_e. In this system, the value of K_e turns out to be

$$K_e = 8.98742 \times 10^9 \text{ N m}^2/\text{C}^2,$$

which for most purposes can be rounded off to 9×10^9 N m²/C². A related constant, denoted by ε_0, is called the permittivity of free space,

$$\varepsilon_0 = \frac{1}{4\pi K_e},$$

so that

$$K_e = \frac{1}{4\pi \varepsilon_0}.$$

The value of ε_0 is 8.85415×10^{-12} C²/N m². As we shall see in a later chapter, the use of ε_0 helps simplify certain equations.

Example 2
Three point charges, $q_1 = 5.0$ μC, $q_2 = 3.0$ μC, and $q_3 = -8.0$ μC, are held fixed on the x axis at $x = 0$, 0.2, and 0.5 m, as shown. Calculate the net electric force on the charge q_2.

We use Coulomb's law to calculate the magnitude of each of the forces on q_2:

$$F_{12} = 9 \times 10^9 \text{ N m}^2/\text{C}^2 \frac{(5.0 \times 10^{-6} \text{ C})(3.0 \times 10^{-6} \text{ C})}{(0.2 \text{ m})^2} = 3.4 \text{ N},$$

$$F_{32} = 9 \times 10^9 \text{ N m}^2/\text{C}^2 \frac{(8.0 \times 10^{-6} \text{ C})(3.0 \times 10^{-6} \text{ C})}{(0.3 \text{ m})^2} = 2.4 \text{ N}.$$

Since both q_1 and q_2 are positive, q_1 repels q_2 to the right, giving the direction of $\mathbf{F}_{12}$. Because q_3 attracts q_2, $\mathbf{F}_{32}$ is also directed to the right. Therefore, the net electric force on q_2 is $F = 5.8$ N, directed along the positive x axis. Calculation of the force on q_3 is requested in Question 9.

Example 3

Three small particles having charges -1.0, 2.0, and 4.0 µC are fixed at the corners of an equilateral triangle having side $b = 0.1$ m, as shown in the first diagram. What is the magnitude of the electric force on q_1?

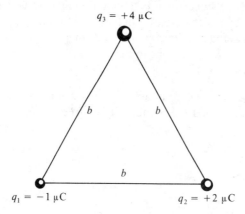

To calculate the total electric force acting on q_1, we first use Coulomb's law to find the force of q_2 on q_1 and the force of q_3 on q_1; then use vector addition to find the total force. The forces on q_1 have magnitudes

$$F_{21} = K_e \frac{q_2 q_1}{b^2} = 1.8 \text{ N}, \qquad F_{31} = K_e \frac{q_3 q_1}{b^2} = 3.6 \text{ N}.$$

The directions of these forces, determined by noting that q_1 is attracted to each of q_2 and q_3, are in the second diagram:

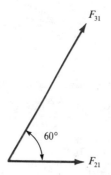

32.4 COULOMB'S LAW

The x component of the vector sum is

$$F_{1x} = F_{21} + F_{31} \cos 60° = 3.6 \text{ N},$$

and the y component is

$$F_{1y} = F_{31} \sin 60° = 3.1 \text{ N}.$$

Therefore, the magnitude of the total electric force on q_1 is given by

$$F_1 = \sqrt{F_{1x}^2 + F_{1y}^2} = 4.8 \text{ N}.$$

Questions

5. Two protons in a helium nucleus are about 10^{-15} m apart. Calculate the electrostatic force between them.

6. How much equal positive charge would have to be added to the earth and moon to nullify the gravitational force between them?

7. Two small metal spheres, each having a charge of 0.5 µC, are suspended by threads, as shown. The spheres are separated by 0.2 m and the threads make an angle of 60° with each other. Calculate (a) the mass of each sphere and (b) the tensions in the threads.

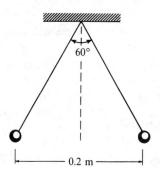

8. Two particles with charges $q_1 = 3.0$ µC and $q_2 = -5.0$ µC are held fixed at a separation distance $d = 0.2$ m. Where can a positive third charge q_3 be placed so that it has no net electrical force acting on it? Give a qualitative description of where to place q_3 if it is negative.

9. Refer to Example 2 and calculate the electric force on q_3.

10. A charge q is placed at each corner of a square of side b. Determine what charge q', placed at the center of the square, will make the total force on each corner charge equal to zero.

11. A particle of mass m and charge Q lies midway between two fixed identical charges q that are separated by a distance $2a$, and is constrained to move only along the x axis. If the particle is displaced a distance x from its original position, as shown, find the electric force acting on it.

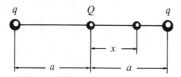

12. Refer to Question 11. Assuming that $x \ll a$, find the frequency of oscillations if the charge Q is released.

32.5 ELECTROSTATIC MACHINES

Coulomb's law is a fundamental principle of electrostatics, but it took a long time to discover because it involves point charges in a vacuum, entities that don't correspond to common everyday experience. Ordinarily we deal with matter rather than electricity. This made it difficult for the early experimenters to find the abstractions that simplify the study of electrical phenomena.

The important role played by electricity in the structure of matter was not realized in the eighteenth century. We now know that on a very large scale – the scale of the cosmos – the force that governs nature is gravity. On a very small scale – the scale of the nucleus of an atom – the nuclear force governs the behavior of nuclei. Everything in between is governed by the electric force. All the world accessible to our senses obeys the law of electricity. So electricity is much more than the mere curiosity it was thought to be in the eighteenth century. We'll find that the behavior of matter is intimately connected with the behavior of electricity.

Our understanding of electricity was greatly advanced by the invention of some important electrostatic devices, to which we now turn our attention. A typical electrostatic generating machine of the eighteenth century consisted of a large glass globe with a crank for rotating it, and a silk cloth that could be rubbed on it to charge it. Although that type of machine works, it's not very efficient. A better machine uses a conveyor belt scratching a screen-like device at the bottom. The friction produces separation of electric charge and the conveyor belt picks up charges as it moves, transporting them to a large metal sphere at the top, as shown in Fig. 32.10. As the belt moves past a metal comb inside the sphere, the charges are transferred to the metal dome and move to the outer surface. Then the belt runs down the other side uncharged, ready to pick up new charges. By this process the dome can accumulate sizable charge and huge impressive sparks can be made to leap from it. Such a machine is called a Van de Graaff generator.

Another electrostatic device that stores rather than produces electric charge, is called a capacitor. It was first discovered accidentally in 1745 by a German cathedral dean, E. J. von Kleist, while attempting to construct a portable sparking

32.5 ELECTROSTATIC MACHINES

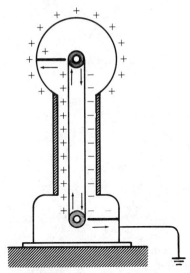

Figure 32.10 Schematic diagram of a Van de Graaff generator.

machine. Kleist never realized the importance of his discovery. A few months later, Professor Pieter van Musschenbroek of the University of Leyden independently pursued experiments that culminated in a similar discovery. Dufay had shown that water could play the part of a nonelectric body in drawing sparks from an electrified object. Responding to an idea suggested by G. M. Bose of the University of Leipzig, Musschenbroek attempted (unsuccessfully) to reverse the procedure by drawing "electrical fire" from water electrified in a glass vessel.

By chance, a friend of Musschenbroek's named Andreas Cunaeus, a lawyer who was intrigued by this experiment, tried to reproduce it at home. Not being aware of the so-called Rule of Dufay that required electrified bodies be properly insulated, Cunaeus electrified a water-filled jar in a manner most natural to him – holding it in his hand. He drew the spark himself and received a surprisingly powerful blow. He reported this discovery to Musschenbroek who substituted a globe for Cunaeus's jar and was himself painfully affected by the result. Musschenbroek reported his distressing experience to the Paris Academy.

These experiments resulted in a device to store electric charge, which became known as the *Leyden jar*. In its basic form, a Leyden jar consists of an outer conductor separated by a glass jar from an inner conductor, as illustrated in Fig. 32.11. (Tin foil is often used to coat the inner and outer surfaces of the jar.) The inner conductor is connected to a charging device, whereas the outer conductor is grounded. If the inner conductor is positively charged, for example, it attracts negative charges from ground to the outer conductor. The electric forces between the charges hold them there until some easy path is made available for them to recombine. In that way the Leyden jar stores electric charge. Whenever we speak of the charge stored in a Leyden jar we mean that charges equal in magnitude but opposite in sign are stored on the inner and outer conductor, respectively.

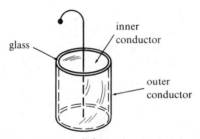

Figure 32.11 Schematic of a Leyden jar.

After the Leyden jar had been invented, a new kind of electrostatic generating machine became possible that could not only separate charge but could also accumulate charge in relatively large amounts. This device is known as a Wimshurst machine.

The principle behind the Wimshurst machine is that of straightforward electric induction concealed in some moderately complicated machinery. Two large counter-rotating circular glass plates with peripheral metallic tabs face each other on a

Figure 32.12 A simple Wimshurst machine. (Caltech photo by Robert Paz.)

common axle. As they rotate, a combination of rods and metallic brushes connect diametrically opposed tabs on opposite plates. Initially, a small difference in charge, created by chance events, exists between the two plates. This difference induces additional charge separation via the connecting rods. As the plates continue to rotate, the connection between tabs and brushes is broken and the induced charge is carried to additional brushes where it is removed and stored in Leyden jars. A detailed analysis shows that the process is regenerative. The larger the charge on the Leyden jars the greater the inductive effect. Eventually the growth is limited by electric breakdown. With several Leyden jars in place the machine is not only potent but dangerous. As we'll learn in Chapter 35, the Van de Graaff generator produces more spectacular sparks, but the Wimshurst machine is more dangerous.

Questions

13. If you wanted to make a Van de Graaff generator accumulate more charge, would you make the conveyor belt wider or the dome smaller?

14. The inner conductor of a charged Leyden jar is carefully removed with a rubber hook. The inner and outer conductors are connected together with a wire. The wire is then removed and the jar is reassembled using the rubber hook to insert the inner conductor. The experimenter observes that the jar will still produce a spark. Explain how this can happen.

32.6 A FINAL WORD

When physicists invent a machine that does something spectacular, they always want to construct a larger machine that does the same thing on an even more spectacular scale. The world's largest electrostatic generator was commissioned by a Dutch physicist, Martinus von Marum, director of the Teyler Foundation of Haarlem, and was built in 1785 by John Cuthbertson, an English instrument maker who lived in Holland. The Teylerian machine employed a glass plate 65 in. in diameter. With a bank of 100 Leyden jars to store charge, this machine could produce a spark 2-ft long and as thick as a quill pen. It was a lightning machine.

On the practical side, this colossal machine was not as useful as one of today's miniature penlight batteries. Nevertheless, the importance of research concerning sparks should not be underestimated because protection against lightning was the first practical application of the study of electricity. With his famous kite experiment, Benjamin Franklin showed that lightning was a giant electric spark, and that led him to invent the lightning rod.

Before the invention of the lightning rod, a thunderstorm was a terrifying phenomenon. When a thunderstorm approached a small town, people typically would try to break up the menacing clouds by vigorously ringing church bells. Unfortunately lightning has a tendency to strike a tall projecting object such as a church tower and the bell ringer was sometimes electrocuted. When Franklin showed how to make lightning rods for protection, it was the first demonstration that theoretical understanding of nature could help control the giant forces of nature. The psychological impact of that discovery cannot be overestimated.

CHAPTER 33

THE ELECTRIC FIELD

Many powers act manifestly at a distance; their physical nature is incomprehensible to us; still we may learn much that is real and positive about them, and amongst other things something of the condition of the space between the body acting and that acted upon, or between two mutually acting bodies. Such powers are presented to us by the phenomena of gravity, light, electricity, magnetism, etc. These when examined will be found to present remarkable difference in relation to their respective lines of force; and at the same time that they establish the existence of real physical lines in some cases, will facilitate the consideration of the question...

...All these points indicate the existence of physical lines of electric force—the absolutely essential relation of positive and negative surfaces to each other, and their dependence on each other contrasted with the known mobility of the forces, admit no other conclusion. The action also in curved lines must depend upon a physical line of force.

Michael Faraday, "On the Physical Lines of Magnetic Force," *Proceedings of the Royal Society,* 11 June 1851

33.1 ELECTRIC FIELDS OF ISOLATED POINT CHARGES

Coulomb's inverse-square law suggested that the electrical force, like gravity, was an action-at-a-distance force. The idea of force transmitted without bodies being in contact with one another was difficult for most scientists of the eighteenth and nineteenth centuries to accept, despite the fact that the same idea was inherent in Newton's visualization of gravitational forces acting through empty space. In the mid-nineteenth century, Michael Faraday helped resolve this dilemma by introducing the concept of an electric field.

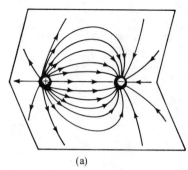 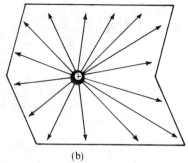

Figure 33.1 Faraday's lines of force emanate from positive charges, and end on negative charges.

Faraday had noticed that iron filings arrange themselves in definite curves around a bar magnet. Faraday called these curves *lines of force* and imagined that the filings followed these lines that spread out from one magnetic pole and converged on the other. In the same way, he visualized electric field lines reaching out from positive charges and ending on negative charges, as suggested by Fig. 33.1a. In his mind's eye, every positive electric charge radiates lines of force outward from itself. The lines emanating from an isolated charge such as that in Figure 33.1b can be thought of as terminating at an infinite distance from the positive charge.

If another positively charged body encounters one of Faraday's force lines, it experiences a force in the direction of the tangent to the line. The force would be stronger at one place than at another only if the density of lines were greater. This happens closer to the charge from which the lines radiate to or from. That's how Faraday explained that the force between electric charges becomes stronger as the charges are brought closer together. Faraday's picture was readily accepted because it could be used to argue that the force gets stronger in just the right quantitative way. In Faraday's visualization, attention was shifted from individual charges to the electric field as a whole.

Faraday's powerful insight provides a qualitative description of the behavior of electric forces without the use of mathematics. But mathematics is essential for obtaining quantitative information concerning forces arising from complex electrical structures, so we turn now to a mathematical formulation of Faraday's ideas. Faraday's qualitative description is obtained by geometrically interpreting the mathematical concepts.

To formally define the electric field in mathematical terms, imagine a single point charge q in space. Since that is the only charge present, there is, of course, no force acting on it. Yet we know that if another charge were placed near it, that charge would experience a force. Let's imagine a positive charge q_0, called a test charge, placed at a distance r from q. The Coulomb force on the test charge is

$$\mathbf{F} = q_0 \left(K_e \frac{q}{r^2} \hat{\mathbf{r}} \right). \tag{32.1}$$

The quantity in parentheses does not depend on the magnitude of the test charge,

33.1 ELECTRIC FIELDS OF ISOLATED POINT CHARGES

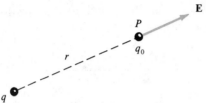

Figure 33.2 The direction of the electric field at a point P is that of the force acting on a positive test charge at P.

but only on the distance r from q to the point P where q_0 is placed. The test charge detects the force, but the quantity in parentheses exists whether or not q_0 is there to detect it. That quantity is denoted by **E** and is called the *electric field* generated by an isolated point charge q. Thus, by definition,

$$\mathbf{E} = K_e \frac{q}{r^2} \hat{\mathbf{r}}. \tag{33.1}$$

This equation defines a vector-valued function **E** that depends on the given point charge q and on the position P at which a test charge could be placed. The magnitude of **E** is called the electric field *strength* or *intensity*.

The dependence on P is expressed by the denominator r^2 and by the unit vector $\hat{\mathbf{r}}$, which has the direction from the given point charge q to the point P, as indicated by the example in Fig. 33.2. The SI unit of the electric field strength is N/C (newtons per coulomb).

Now suppose we place a small positive test charge at each of a large number of points P, measure the force **F**, and draw the corresponding field vector **E** at each of these points. The collection of vectors **E** so obtained is called a vector field and it provides a geometric map of the electric field. It is analogous to the configuration obtained by sprinkling iron filings around a bar magnet, except that the vector field is three-dimensional and is not confined to a plane. A knowledge of the electric field changes the focus of attention from electric forces between charges to a property at each point in space.

If we know the electric field at some point in space, the force acting on any charge q_0 placed there is simply equal to

$$\mathbf{F} = q_0 \mathbf{E}. \tag{33.2}$$

Example 1
A 0.02-kg sphere having a charge of -1.5×10^{-8} C is placed in a uniform (constant) electric field given by $\mathbf{E} = 7.2 \times 10^5 \hat{\mathbf{i}}$ N/C. Determine the acceleration of the sphere.

We use Eq. (33.2) for the force on the sphere together with Newton's second law, $\mathbf{F} = m\mathbf{a}$, to relate acceleration to force. Therefore, the acceleration is given by

$$\mathbf{a} = \frac{\mathbf{F}}{m} = \frac{q_0 \mathbf{E}}{m} = \frac{(-1.5 \times 10^{-8}\ \text{C})(7.2 \times 10^5\ \text{N/C})}{(0.02\ \text{kg})}\hat{\mathbf{i}} = -0.54\hat{\mathbf{i}}\ \text{m/s}^2.$$

The field concept can be generalized. If, instead of a single charge q, there are several point charges present, say $q_1, q_2, q_3, \ldots$, then the force on a positive test charge q_0 is the vector sum of all the Coulomb forces due to these charges,

$$\mathbf{F} = q_0 \left(K_e \frac{q_1}{r_1^2}\hat{\mathbf{r}}_1 + K_e \frac{q_2}{r_2^2}\hat{\mathbf{r}}_2 + K_e \frac{q_3}{r_3^2}\hat{\mathbf{r}}_3 + \cdots \right).$$

Again, the factor multiplying the test charge q_0 doesn't depend on the magnitude of q_0 but rather on the distance of the other charges from the point P at which q_0 is located. The factor multiplying q_0 is a vector quantity, a measure of the force that would act on a unit positive charge if one were placed at that particular point P. This factor is called the *electric field* at the point P and is denoted by $\mathbf{E}$. Thus, by definition,

$$\mathbf{E} = K_e \sum_i \frac{q_i}{r_i^2}\hat{\mathbf{r}}_i, \tag{33.3}$$

where $\mathbf{r}_i$ is the vector from the point charge q_i to the point at which $\mathbf{E}$ is being evaluated. The formula can be expressed more simply as

$$\mathbf{E} = \sum_i \mathbf{E}_i,$$

where $\mathbf{E}_i$ represents the electric field generated by q_i.

Equation (33.3) defines a vector field $\mathbf{E}$ that depends on the distribution of the given charges $q_1, q_2, \ldots$ and on the point P at which the test charge was placed. The magnitude of $\mathbf{E}$ is called the electric field *intensity* or *strength* at P. The direction of the electric field $\mathbf{E}$ at any point is the same as the direction of the force on a small positive test charge. We specify a *small* test charge so that it does not disturb the arrangement of other charges creating a force on it.

Example 2

Determine the electric field at an arbitrary point P on the xy plane arising from two positive point charges q at points $(0, 0, \pm a)$ on the z axis, as shown.

33.1 ELECTRIC FIELDS OF ISOLATED POINT CHARGES

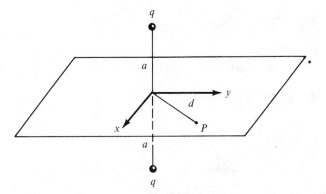

Let r denote the distance from either charge, and let d be the distance from P to the z axis. The Pythagorean theorem implies

$$r^2 = a^2 + d^2.$$

Thus the magnitude of the field from either charge is

$$E = K_e \frac{q}{a^2 + d^2}.$$

The direction of the electric force from each charge is along the straight line from the charge to point P, as indicated in the following diagram.

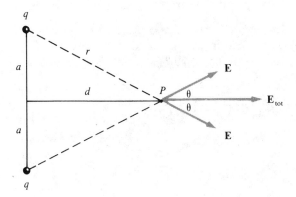

From symmetry we see that the z components of the electric forces cancel because they are equal and opposite. Therefore, the total electric field is directed radially toward P and has magnitude

$$E_{tot} = 2E \cos \theta.$$

The geometry of the problem implies that

$$\cos \theta = d/r,$$

so the magnitude of the field can be written as

$$E_{tot} = K_e \frac{2qd}{r^3} = K_e \frac{2qd}{(a^2 + d^2)^{3/2}}.$$

This can also be expressed in terms of the x and y coordinates of P. Because $d^2 = x^2 + y^2$, we have

$$E_{tot} = K_e \frac{2q\sqrt{x^2 + y^2}}{(a^2 + x^2 + y^2)^{3/2}}.$$

The field has the direction of the vector $x\hat{\mathbf{i}} + y\hat{\mathbf{j}}$, which points radially outward from the z axis.

Questions

1. An electron starts from rest in a uniform electric field of 30 N/C. (a) How far has it travelled in 0.6 µs? (b) What is its speed at that instant?

2. A water droplet of mass 3.2×10^{-8} kg floats inside the chamber where the electric field magnitude is 3000 N/C. How many excess electron charges must the droplet have to float?

3. Four point charges, each of 3.0 µC, are held fixed at the corners of a square of sides 0.15 m. Determine the electric field at the center of the square.

4. Two positive charges, $q_1 = 2.4$ µC and $q_2 = 4.8$ µC, are positioned at (3 cm, 0) and (5 cm, 0) on a plane rectangular grid. What are the components of the electric field at the following points? (a) (0, 0); (b) (0, 5 cm); and (c) (5 cm, 5 cm)?

5. An electric dipole consists of two equal and opposite charges Q and $-Q$ separated by a distance $2a$, as shown. Calculate the electric field along the perpendicular bisector of the axis through the charges at a distance $x \neq a$ from the center of the dipole.

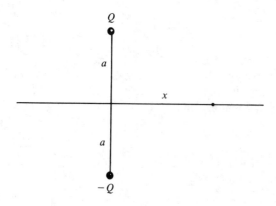

6. Referring to the dipole in Question 5, find the electric field at each point of the plane perpendicular to and bisecting the line segment joining the two charges.

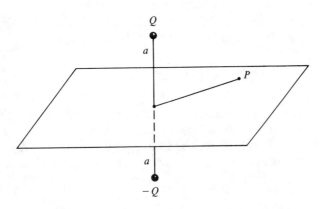

33.2 ELECTRIC FIELDS OF CONTINUOUS CHARGE DISTRIBUTIONS

So far we've only considered electric fields arising from isolated point charges that are idealized versions of real charged objects. To pass from a finite number of isolated point charges to continuously distributed charges spread out over some region we replace the sum in Eq. (33.3) by an appropriate integral. The type of integral depends on the type of region being considered. Instead of attempting to give a general description of the integral covering all possibilities, we show how to set up the integral in a number of different examples.

First we consider charge distributed along an infinitely long straight wire, which we take to be the x axis. Assume that the total amount of charge on the portion of the wire from 0 to x is $q(x)$. Then the amount of charge on a small portion of the wire, say from x to $x + \Delta x$, is

$$\Delta q = q(x + \Delta x) - q(x).$$

The quotient $\Delta q/\Delta x$ is called the average charge density on this portion, and the limiting value of this quotient as Δx shrinks to 0 is called the *linear charge density* at x and is denoted by $\lambda(x)$. Thus, by definition,

$$\lambda(x) = \lim_{\Delta x \to 0} \frac{\Delta q}{\Delta x} = \frac{dq}{dx}. \tag{33.4}$$

The charge is said to be distributed *uniformly* along the axis if $\lambda(x)$ is constant, $\lambda(x) = \lambda$ for all x.

A knowledge of the linear charge density is useful because we can use it to determine the total amount of charge from 0 to x by integration. If $q(0) = 0$, we have

$$q(x) = \int_0^x \lambda(x') \, dx'$$

for each x. Moreover, the amount of charge Δq in a short interval of length Δx is approximately

$$\Delta q = \lambda(x) \, \Delta x. \tag{33.5}$$

To set up the integral defining the field arising from charge distribution with linear density $\lambda(x)$ along the x axis we argue as follows. Consider a small portion of the wire at x with length Δx. The charge on this portion is $\Delta q = \lambda(x) \, \Delta x$. We treat Δq as if it were an isolated point charge that creates an electric field $\Delta \mathbf{E}$ given by the equation

$$\Delta \mathbf{E} = K_e \frac{\Delta q}{r^2} \hat{\mathbf{r}} = K_e \frac{\lambda(x) \, \Delta x}{r^2} \hat{\mathbf{r}}, \tag{33.6}$$

where r is the distance from the "point charge" Δq to the point P at which the electric field $\Delta \mathbf{E}$ is being evaluated. In this equation both r and $\hat{\mathbf{r}}$ depend on x and on P. Instead of adding all these fields to obtain the total electric field $\mathbf{E}$ we integrate over the entire axis and write

$$\mathbf{E} = K_e \int_{-\infty}^{\infty} \frac{\lambda(x)}{r^2} \hat{\mathbf{r}} \, dx. \tag{33.7}$$

This equation is taken as the definition of the electric field $\mathbf{E}$ arising from charge distribution with linear charge density $\lambda(x)$ along the x axis. The distance r and the unit vector $\hat{\mathbf{r}}$ must be expressed in terms of x before the integration can be carried out. The next example illustrates how the formula can be used in practice.

Example 3
Determine the electric field arising from uniform linear charge with constant density λ distributed along the x axis.

First we note that the resulting electric field at any point $P = (x, y, z)$ in space does not depend on the x coordinate of P. The reason for this is that a translation of the wire along the x axis has no effect on the charge distribution because the density is constant. Or, if we place a small charge $\Delta q = \lambda \, \Delta x$ at an arbitrary point on the x axis this charge depends only on the length Δx and not on its location on the axis. Moreover, the Coulomb force depends on the distance between charges, so the electric field at P will depend only on the radial distance R of P from the x

33.2 ELECTRIC FIELDS OF CONTINUOUS CHARGE DISTRIBUTIONS

axis. Therefore, it suffices to determine the field at an arbitrary point on the positive y axis, say at $(0, R, 0)$, where $R > 0$, as indicated on the following diagram.

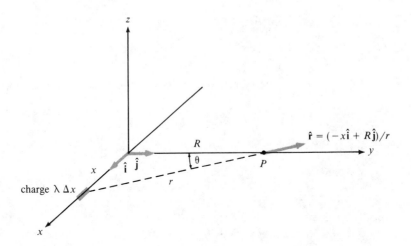

Applying Eq. (33.6), we find that the corresponding field at the point $P = (0, R, 0)$ arising from the small charge Δq is

$$\Delta \mathbf{E} = K_e \frac{\lambda \Delta x}{r^2} \hat{\mathbf{r}},$$

where, in this example, $r^2 = x^2 + R^2$ and $\hat{\mathbf{r}} = (-x\hat{\mathbf{i}} + R\hat{\mathbf{j}})/r$. Hence

$$\Delta \mathbf{E} = \lambda K_e \frac{-x\hat{\mathbf{i}} + R\hat{\mathbf{j}}}{r^3} \Delta x,$$

and the integral for $\mathbf{E}$ becomes

$$\mathbf{E} = \lambda K_e \int_{-\infty}^{\infty} \frac{-x\hat{\mathbf{i}} + R\hat{\mathbf{j}}}{r^3} dx.$$

The part of the integral multiplying $\hat{\mathbf{i}}$ is zero because the integrand x/r^3 is an odd function of x. (A function $f(x)$ is called *odd* if $f(-x) = -f(x)$. The integral of an odd function is zero over any interval symmetric about the origin.) This makes sense physically as well because the x components of $\mathbf{E}$ from the symmetrically located charges at x and $-x$ cancel. So, we are left with the integral

$$\mathbf{E} = \lambda K_e \int_{-\infty}^{\infty} \frac{R}{r^3} dx \, \hat{\mathbf{j}}.$$

To evaluate the integral we express x and r in terms of the constant distance R and the angle θ shown in the diagram, then integrate with respect to θ from $-\pi/2$ to

$\pi/2$. We have

$$x = R \tan \theta,$$

$$\frac{dx}{d\theta} = R \sec^2 \theta = \frac{R}{\cos^2 \theta}.$$

Using these relations together with $r = R/\cos \theta$, we find that the integral becomes

$$\mathbf{E} = \frac{\lambda K_e}{R} \int_{-\pi/2}^{\pi/2} \cos \theta \, d\theta \, \hat{\mathbf{j}} = \frac{2\lambda K_e}{R} \int_0^{\pi/2} \cos \theta \, d\theta \, \hat{\mathbf{j}}.$$

But $\int_0^{\pi/2} \cos \theta \, d\theta = 1$, so we obtain

$$\mathbf{E} = \frac{2\lambda K_e}{R} \hat{\mathbf{j}}. \tag{33.8}$$

This is a remarkably simple result. The electric field $\mathbf{E}$ is directed perpendicular to the line of charge and its magnitude varies inversely as the distance R from the wire (not the square of the distance). The field $\mathbf{E}$ is directed away from the wire if λ is positive and toward the wire if λ is negative.

Example 4

A portion of a wire with constant linear charge density λ is bent to form a circular loop of radius a. Determine the electric field at an arbitrary point along the axis of the loop.

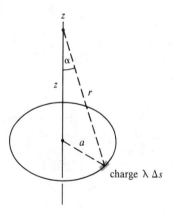

Place the loop in the xy plane with its axis along the z axis, as shown in the figure. A small portion of the wire of length Δs located at the point $(x, y, 0)$ carries a small charge $\Delta q = \lambda \, \Delta s$. By applying Eq. (33.6) we find that the field $\Delta \mathbf{E}$ at a

33.2 ELECTRIC FIELDS OF CONTINUOUS CHARGE DISTRIBUTIONS

point $(0, 0, z)$ on the z axis arising from the small charge Δq is given by

$$\Delta \mathbf{E} = K_e \frac{\lambda \Delta s}{r^2} \hat{\mathbf{r}}, \tag{33.9}$$

where $r^2 = a^2 + z^2$ and $\hat{\mathbf{r}} = (-x\hat{\mathbf{i}} - y\hat{\mathbf{j}} + z\hat{\mathbf{k}})/r$. Hence

$$\Delta \mathbf{E} = \lambda K_e \frac{-x\hat{\mathbf{i}} - y\hat{\mathbf{j}} + z\hat{\mathbf{k}}}{r^3} \Delta s.$$

Because the charge is distributed along a circle, it is natural to express the variable quantities x, y, and Δs in terms of polar coordinates. On the circle we have

$$x = a \cos \theta, \qquad y = a \sin \theta, \quad \text{and} \quad \Delta s = a \Delta \theta.$$

We insert these into Eq. (33.9) and then integrate with respect to θ from $\theta = 0$ to $\theta = 2\pi$ to go once around the loop. This gives us

$$\mathbf{E} = \frac{\lambda a K_e}{(a^2 + z^2)^{3/2}} \int_0^{2\pi} (-a \cos \theta \hat{\mathbf{i}} - a \sin \theta \hat{\mathbf{j}} + z\hat{\mathbf{k}}) \, d\theta.$$

But

$$\int_0^{2\pi} \cos \theta \, d\theta = \int_0^{2\pi} \sin \theta \, d\theta = 0,$$

so the terms multiplying $\hat{\mathbf{i}}$ and $\hat{\mathbf{j}}$ are equal to zero. This is to be expected because, by symmetry, the horizontal components of the field from the charge around the loop should cancel. Only the $\hat{\mathbf{k}}$ component survives, and, since $\int_0^{2\pi} z \, d\theta = 2\pi z$, the field is given by

$$\mathbf{E} = K_e \frac{2\pi \lambda a z}{(a^2 + z^2)^{3/2}} \hat{\mathbf{k}}.$$

But $2\pi a \lambda$ is the total charge on the loop. Calling this Q, we can write

$$\mathbf{E} = K_e \frac{Qz}{(a^2 + z^2)^{3/2}} \hat{\mathbf{k}}.$$

This result can be expressed more simply in terms of the angle α the unit vector $\hat{\mathbf{r}}$ makes with the positive z axis. Writing $r^2 = a^2 + z^2$ as above, we see from the diagram that $z/r = \cos \alpha$, so $\mathbf{E}$ becomes

$$\mathbf{E} = K_e \frac{Q \cos \alpha}{r^2} \hat{\mathbf{k}}. \tag{33.10}$$

Because $z = r \cos \alpha$, $\cos \alpha$ has the same algebraic sign as z. Thus, if Q is positive the field has the same direction as $\hat{\mathbf{k}}$ if z is positive and the opposite direction if z is negative. Note also that if z is positive and very large compared to a, then r is nearly z and $\cos \alpha$ is nearly 1, which gives a field

$$\mathbf{E} \approx K_e \frac{Q}{z^2} \hat{\mathbf{k}} \quad \text{for } z \gg a.$$

In other words, at points very far above the loop the field resembles that of an isolated point charge.

The foregoing examples indicate how Eqs. (33.6) and (33.7) can be used to determine the electric field arising from a continuous distribution of charge along a wire. First we set up the electric field $\Delta \mathbf{E}$ arising from a small charge Δq and then integrate over the charge distribution, taking advantage of any symmetry in the problem to simplify the calculation.

The next example determines the electric field arising from charge distributed over a plane region. First we need the concept of surface charge density. If a small amount of charge Δq is distributed over a small plane region of area ΔA at some point, we define surface charge density σ at that point by the limit relation

$$\sigma = \lim_{\Delta A \to 0} \frac{\Delta q}{\Delta A}.$$

Example 5

Determine the electric field generated by constant surface charge density σ distributed over the xy plane.

Because the density is constant, translation of the sheet in its own plane has no effect on the charge distribution. This means that the resulting field at any point (x, y, z) in space depends only on z and not on x and y. In other words, the field at (x, y, z) will have the same value as that at $(0, 0, z)$. Moreover, the Coulomb force depends only on the distance between charges, so the field will be symmetric about the xy plane. Therefore, it suffices to determine the field at an arbitrary point on the positive z axis, say at $(0, 0, z)$, where $z > 0$. To do this we imagine the plane as being decomposed into a collection of thin concentric rings centered at the origin, as indicated in the following diagram.

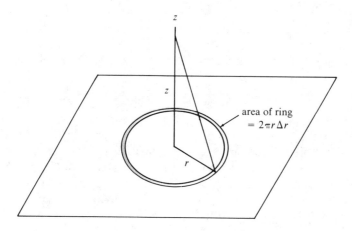

A ring of radius r and width Δr has area $2\pi r \Delta r$ and therefore carries a total charge $\Delta q = 2\pi \sigma r \Delta r$. Applying Eq. (33.10) to this ring with $Q = 2\pi \sigma r \Delta r$, we find that it

33.2 ELECTRIC FIELDS OF CONTINUOUS CHARGE DISTRIBUTIONS

generates a field

$$\Delta \mathbf{E} = K_e \frac{2\pi\sigma r z}{(r^2 + z^2)^{3/2}} \Delta r \, \hat{\mathbf{k}}.$$

The total field $\mathbf{E}$ is now obtained by integrating with respect to r from $r = 0$ to $r = \infty$,

$$\mathbf{E} = K_e(2\pi\sigma z) \int_0^\infty \frac{r}{(r^2 + z^2)^{3/2}} \, dr \, \hat{\mathbf{k}}.$$

The integrand is the derivative of the function $f(r) = -(r^2 + z^2)^{-1/2}$, so the value of the integral is simply $f(\infty) - f(0) = 0 - (-1/z) = 1/z$. The factor z cancels and we are left with

$$\mathbf{E} = K_e(2\pi\sigma)\hat{\mathbf{k}}, \tag{33.11}$$

which shows that the electric field is constant at each point on the positive z axis. If the constant K_e is expressed in terms of the permittivity $\varepsilon_0 = 1/(4\pi K_e)$ introduced in Section 32.4, the result in Eq. (33.11) becomes

$$\mathbf{E} = \frac{\sigma}{2\varepsilon_0}\hat{\mathbf{k}}. \tag{33.12}$$

If $z < 0$, symmetry requires the field to have the opposite direction, so there is simply a sign change in the right member of Eq. (33.12).

In summary, the field is directed perpendicular to and away from the xy plane with constant intensity everywhere. Of course, this is a highly idealized situation. There are no infinite sheets of charge in nature, but the foregoing analysis is a useful approximation to the field near a large sheet of charge.

Questions

7. A charge Q is uniformly distributed along a straight wire of finite length L.

   (a) Find the electric field a distance R from the wire along its perpendicular bisector.

   (b) Show that for $R \ll L$, the electric field in part (a) reduces to that of Example 3 for the infinitely along straight wire.

   (c) If $Q = 5.0 \times 10^{-4}$ C and $L = 2.0$ m, calculate the electric field at $R = 0.02$ m using both the result of part (a) and the result for the infinite wire in Eq. (33.8).

8. A long, thin nonconducting rod has length L and uniform linear charge density λ. Find the electric field at a point on the axis a distance x from the end of the line charge as shown.

9. A thin nonconducting rod with total positive charge Q uniformly distributed along its length is bent into a semicircular arc of radius a. Determine the electric field at an arbitrary point on an axis through the center of the circle and perpendicular to the plane of the circle.

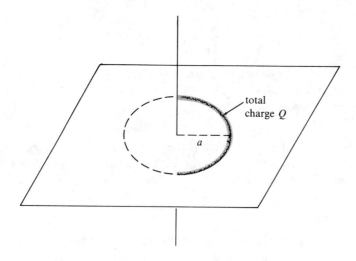

33.3 ELECTRIC FLUX

Coulomb's force law tells us how to determine the electric field from a prescribed configuration of charges. Now we consider the converse question of determining how much charge there is in a given region if the field is known. The answer to this question depends on a quantity called electric flux.

Flux is Latin for *flow*. Although nothing is actually "flowing" in an electric field, it is convenient to visualize the lines of force as being analogous to the lines of flow of a fluid, such as water flowing in a stream. We digress briefly to describe the concept of flux for flowing water, where the ideas seem quite natural, and then transfer the same ideas to electric fields.

At each particle of water in the stream we can attach a vector **v** that represents the velocity of that particular particle. This is the velocity field of the flow. Figure 33.3 shows an example in which **v** is constant. Now place a rectangular frame in the stream and ask for the volume of water that flows through the frame in unit time. This will depend on the size of the frame and on the orientation of the frame relative to the flow. Both the size and orientation can be described by an area vector **A** whose magnitude A is the area of the frame and whose direction is perpendicular to the frame, as indicated in Fig. 33.3. If the frame is perpendicular to the flow, that is, if **A** and **v** are parallel, the volume of water that flows through in unit time is vA. But if the frame is tilted with respect to the water flow, the effective area that is broadside to the flow is $A \cos \theta$, as shown in Fig. 33.3. So with **v** constant, the

33.3 ELECTRIC FLUX

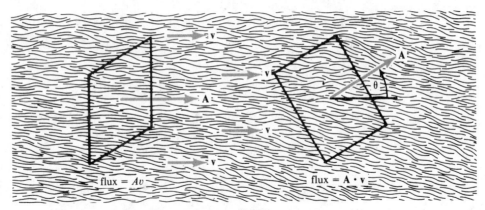

Figure 33.3 Flux is the volume of fluid flowing through the frame in unit time.

volume of water that flows through the frame in unit time is the dot product **v** · **A**. This scalar is called the flux through the frame.

It is not essential that the frame be rectangular – any reasonably shaped patch would serve equally well. And it's not essential that **v** be constant. If the patch is not too large then **v** will be nearly constant over the frame and the flux will be nearly equal to the volume of water flowing through the patch. The dot product **v** · **A** is still called the flux through the frame.

Now we do the same thing in an electric field. Imagine the field vector **E** as being analogous to the velocity vector of a moving fluid. Take a small patch described by an area vector **A**. The dot product **E** · **A** is called the electric flux through this patch.

Next consider a surface S immersed in the electric field, for example a portion of the surface of a sphere, an ellipsoid, or some balloon. Divide S into a large number of small patches with area vectors $\mathbf{A}_i$ pointing toward the same side of S (which we arbitrarily call the positive side) as suggested by Figure 33.4. (There are unusual surfaces such as Möbius bands and Klein bottles that are one-sided. We do not consider such surfaces here.) We assume the patches are small enough so the electric field vector **E** is nearly constant on each patch, say $\mathbf{E} = \mathbf{E}_i$ on patch i.

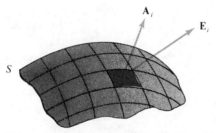

Figure 33.4 Flux toward the positive side of a surface.

Then the dot product $\mathbf{E}_i \cdot \mathbf{A}_i$ is the flux through patch i and the sum

$$\sum_i \mathbf{E}_i \cdot \mathbf{A}_i \qquad (33.13)$$

represents the approximate flux through the entire surface S toward the positive side.

The limiting value of the sum in (33.13), as the number of patches increases and the size of each patch shrinks to zero, is called the electric flux ϕ through the surface S and is denoted by the integration symbol

$$\phi = \int\!\!\int_S \mathbf{E} \cdot d\mathbf{A}. \qquad (33.14)$$

An integral of this type is called a *surface integral* and it can be defined not only for electric fields but for any vector field $\mathbf{F}$ by the process just described. There is a sophisticated mathematical theory related to surface integrals, but none of it is needed for our purposes. You will not be asked to do any elaborate calculations involving surface integrals. In fact, we will use surface integrals primarily for convenience of notation because they enable us to express certain physical principles in simple form. The next section introduces the first of these principles, called *Gauss's law*, a powerful tool that helps to simplify many problems.

Questions

10. Refer to Fig. 33.3. What orientation of the frame makes the flux through it (a) a maximum? (b) zero?

11. A uniform electric field $\mathbf{E} = 5.0 \times 10^4 \hat{\mathbf{j}}$ N/C exists in a region of space. Calculate the electric flux through the square S of side 0.03 m if (a) S lies in the xz plane and (b) the area vector of S makes an angle of 37° with the y axis, as shown.

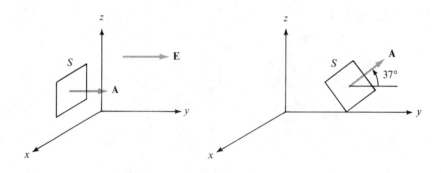

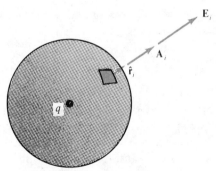

Figure 33.5 Calculating the outward flux through a sphere of radius r.

33.4 GAUSS'S LAW

By way of introduction to Gauss's law, we calculate the electric flux in a very simple case. Start with an electric field generated by an isolated positive point charge q and calculate the outward electric flux through the surface S of a sphere of radius r with center at q. For a *closed* surface, the surface normal is taken to be outward. Consequently, an outward electric field has a positive flux through the surface. The field is given by

$$\mathbf{E} = K_e \frac{q}{r^2} \hat{\mathbf{r}},$$

where $\hat{\mathbf{r}}$ is the outward unit normal at each point of S, as shown in Fig. 33.5.

To calculate the flux according to Eq. (33.13) we divide the surface S into patches as outlined in the foregoing section. The magnitude of $\mathbf{E}$ on each patch is constant, and the unit vector $\hat{\mathbf{r}}_i$ on patch i has the same direction as the area vector $\mathbf{A}_i$ (see Fig. 33.5) so the flux through patch i is given by

$$\mathbf{E}_i \cdot \mathbf{A}_i = EA_i = K_e \frac{q}{r^2} A_i.$$

Summing over all the patches we find

$$\sum_i \mathbf{E}_i \cdot \mathbf{A}_i = K_e \frac{q}{r^2} \sum_i A_i.$$

But $\sum A_i = 4\pi r^2$, the area of a sphere of radius r, hence the total flux through S is

$$\phi = 4\pi K_e q, \tag{33.15}$$

a quantity independent of the radius. Because of the inverse-square nature of the Coulomb force law, the factor r^2 in the denominator canceled the factor r^2 coming from the area of the sphere, giving a result independent of the radius of the sphere.

The result can also be expressed in terms of the permittivity constant $\varepsilon_0 = 1/(4\pi K_e)$, in which case it takes the simpler form

$$\phi = \frac{q}{\varepsilon_0}. \tag{33.16}$$

Now we come to an amazing fact discovered by the mathematician Carl Friedrich Gauss in 1839. Gauss showed that the equation for the total flux still holds if q is anywhere inside the sphere. Moreover, the same equation holds if the sphere is replaced by any closed surface S containing q in its interior. This general statement is known as *Gauss's law*. In integral form it states that

$$\boxed{\phi = \oiint_S \mathbf{E} \cdot d\mathbf{A} = \frac{q}{\varepsilon_0}} \tag{33.17}$$

for any closed surface S containing q in its interior. (It is implicitly assumed that the surface S has an inside and an outside, so objects such as Klein bottles are excluded.) The notation $\oiint$ indicates a surface integral taken over a closed surface.

To explain why Gauss's law works, an intuitive argument can be given by means of another often-used analogy. Imagine a point source of light. As light spreads out in all directions, its intensity decreases like $1/r^2$, where r is the distance from the source. This is because the total flux of light passing through any spherical surface concentric with the source must be equal to the flux emanating from the source; no light is gained or lost along the way. Thus the light intensity (flux per unit area) at a distance r from the source is

$$\text{intensity} = \frac{\text{flux}}{4\pi r^2}.$$

In other words, intensity of light has the same inverse-square dependence on distance as the electric field. Now, it's reasonable to conclude that the flux of light must be the same through *any* transparent surface enclosing the source, no matter what its shape, which is exactly what Gauss's law says.

For those who need to be convinced by a purely mathematical argument, a proof of Gauss's law for a special class of surfaces is outlined in the next example.

Example 6

Prove Gauss's law for a surface S having the property that each field line emanating from the point charge q intersects S exactly once, as shown in Fig. 33.6.

Draw a sphere of radius r lying inside the surface S with the charge q at its center. Divide the surface of this inner sphere into small patches as was done above in deriving Eq. (33.15), and let $\mathbf{a}_i$ denote the area vector of patch i. Lines radiating from the center through this patch trace out another patch on the outer surface S. Denote the area vector of this outer patch by $\mathbf{A}_i$ and let R be its distance from the

33.4 GAUSS'S LAW

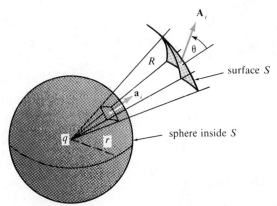

Figure 33.6 Proof of Gauss's law for a more general surface.

center. If the outer surface S were a concentric sphere, the area vector $\mathbf{A}_i$ would be parallel to $\mathbf{a}_i$ and its magnitude would simply be $(R/r)^2$ times that of $\mathbf{a}_i$. But in general the area vector $\mathbf{A}_i$ will be inclined at some angle θ with $\mathbf{a}_i$ so its component parallel to $\mathbf{a}_i$ is $A_i \cos \theta$. Therefore, the magnitudes of the two area vectors are related as follows:

$$A_i \cos \theta = (R/r)^2 a_i. \tag{33.18}$$

On the other hand, by the inverse-square law we also have

$$\mathbf{E}_i(R) = (r/R)^2 \mathbf{E}_i(r). \tag{33.19}$$

Now the outward flux through the inner patch is $\mathbf{E}_i(r) \cdot \mathbf{a}_i$, while that through the outer patch is $\mathbf{E}_i(R) \cdot \mathbf{A}_i$. Using the definition of dot product and then Eqs. (33.18) and (33.19), we find

$$\mathbf{E}_i(R) \cdot \mathbf{A}_i = E_i(R) A_i \cos \theta = (r/R)^2 E_i(r) A_i \cos \theta = (r/R)^2 E_i(r)(R/r)^2 a_i$$
$$= E_i(r) a_i = \mathbf{E}_i(r) \cdot \mathbf{a}_i.$$

In other words, the outward flux through the two patches is the same, so the sum over all the patches is also the same. Thus we have

$$\phi = \oiint_S \mathbf{E} \cdot d\mathbf{A} = \frac{q}{\varepsilon_0},$$

which means that the total electric flux through any surface enclosing a point charge q is independent of the position of the charge and of the shape of the surface S.

This argument also reveals why Gauss's law depends on the inverse-square nature of the Coulomb force. The inverse-square law produces the factor $(r/R)^2$ in Eq. (33.19) and this factor exactly cancels the factor $(R/r)^2$ relating the areas in Eq. (33.18).

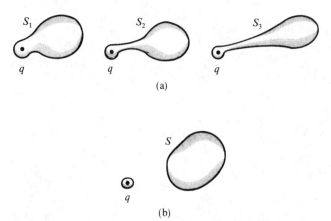

Figure 33.7 Gauss's law implies that the flux is zero if q is outside S.

As a corollary of Gauss's law it is easy to show that the total flux is zero if the point charge q lies *outside* the surface, as indicated in Fig. 33.7b. Using the light-source analogy, if S does not enclose a light source, the flux out of it will be equal to the flux into it, giving zero net flux. Another argument is to consider the surface S of Fig. 33.7b as being the limiting case of a collection of surfaces $S_1, S_2, S_3, \ldots$, each enclosing a single point charge q, as suggested by Fig. 33.7a. Because the net flux is constant and equal to q/ε_0 in each S_i, the same must be true in the limiting case, the constant value q/ε_0 being transferred to the portion containing q in its interior, leaving 0 for the net flux through S.

Gauss's law is easily extended to electric fields arising from more than one point charge. Equation (33.3) shows that the electric field $\mathbf{E}$ arising from a number of point charges $q_1, q_2, q_3, \ldots$, is the vector sum of the fields $\mathbf{E}_1, \mathbf{E}_2, \mathbf{E}_3, \ldots$, generated by the individual charges. Therefore, the total flux through some surface S is given by

$$\phi = \oint_S \mathbf{E} \cdot d\mathbf{A} = \oint_S (\mathbf{E}_1 + \mathbf{E}_2 + \mathbf{E}_3 + \cdots) \cdot d\mathbf{A}.$$

But we have just shown that each integral $\oint_S \mathbf{E}_i \cdot d\mathbf{A} = q_i/\varepsilon_0$ if q_i lies inside S and equals zero if q_i lies outside S. Therefore, each interior charge q_i contributes q_i/ε_0 to the integral and each exterior charge contributes zero. In other words, the total outward flux through S is equal to the total charge inside S divided by ε_0,

$$\phi = \oint_S \mathbf{E} \cdot d\mathbf{A} = q_{\text{encl}}/\varepsilon_0. \tag{33.20}$$

Figure 33.8 Carl Friedrich Gauss, German mathematician and physicist. (From *C. F. Gauss* by G. W. Dunnington (Exposition Press of Florida, 1955). Reproduced with permission of the publisher.)

This is the general statement of Gauss's law. It tells us how to determine the charge inside a surface if the electric field is known. So it provides a complement to Coulomb's law, which determines the field from a knowledge of the charge.

It should be noted that in deriving Gauss's law we used only the inverse-square nature of the Coulomb force law and the fact that forces add vectorially. Therefore, the same derivation would apply to any inverse-square force field, for example, the gravitational field.

33.5 APPLICATIONS OF GAUSS'S LAW

In many cases Gauss's law can be used to calculate the electric field due to a charge distribution more easily than by the method of direct integration discussed in Section 33.2. To illustrate, we shall use Gauss's law to rederive the results of Examples 3 and 5.

Example 7

Use Gauss's law to determine the electric field arising from a uniform linear charge with constant density λ distributed along the x axis.

As mentioned in our solution of Example 3, symmetry considerations suggest that the field will depend only on the radial distance R of the point from the x axis. In applying Gauss's law we exploit this symmetry by choosing for our closed surface S a cylindrical surface of radius R and axis of length L along the x axis, as shown.

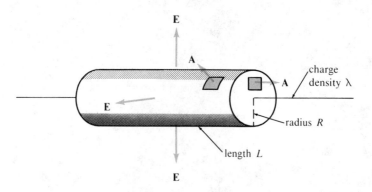

On the cylindrical part of S, the field $\mathbf{E}$ has constant magnitude and is directed outward away from the axis, in the same direction as the outer area vector. At each of the two ends, $\mathbf{E}$ is perpendicular to the area vector so there is no flux through the ends. Therefore, all the flux takes place through the cylindrical part and is equal to the magnitude of $\mathbf{E}$ times $2\pi RL$, the area of the cylindrical part,

outward flux = $2\pi RLE$.

But by Gauss's law the outward flux is also equal to $q_{\text{encl}}/\varepsilon_0$. The total charge enclosed inside the cylinder is λL, so

$$\text{outward flux} = \frac{\lambda L}{\varepsilon_0}.$$

Equating the two expressions for the outward flux and solving the E we find

$$E = \frac{\lambda}{2\pi\varepsilon_0 R} = \frac{2\lambda K_e}{R},$$

in agreement with Eq. (33.8), which was obtained by integration.

Example 8

Use Gauss's law to determine the electric field generated by constant surface charge density σ distributed over the xy plane.

33.5 APPLICATIONS OF GAUSS'S LAW

The symmetry of the problem tells us that the field is directly perpendicular to and away from the xy plane with magnitude depending only on the perpendicular distance of the point from the xy plane. To apply Gauss's law we exploit this symmetry and choose for S the surface of a cube with faces parallel and perpendicular to the xy plane and bisected by the xy plane as shown.

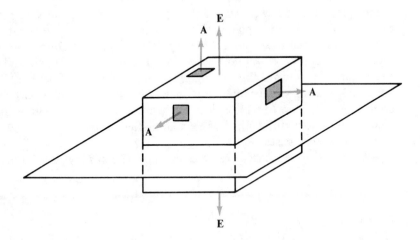

On four of the faces **E** is perpendicular to the outward area vector **A** so the flux through these faces is zero. Only the top and bottom faces make a nonzero contribution to the outward flux. Each contributes EA, where A is the area of the face, for a total contribution of $2EA$,

outward flux = $2EA$.

But by Gauss's law the total outward flux is equal to $q_{\text{encl}}/\varepsilon_0$, and the total charge inside S is σA, the surface charge density times the area, so

outward flux = $\sigma A/\varepsilon_0$.

Equating the two expressions for the outward flux and solving for E we find

$$E = \frac{\sigma}{2\varepsilon_0},$$

in agreement with Eq. (33.12).

The foregoing examples show how Gauss's law exploits symmetry as an aid in calculating electric fields without the use of integration. The surface integration inherent in Gauss's law was avoided in these examples because the area of the surfaces involved could be determined without integration. Example 8 used a surface with square faces and Example 7 used a cylindrical surface that could be unrolled to form a rectangle of the same area.

The solution of Example 7 suggests that a similar argument might be used to rederive the result of Example 4, the electric field on the axis of a uniformly charged circular loop of radius a and charge density λ. To exploit the symmetry in this problem the first thought might be to apply Gauss's law to a torus having its axis along the circular loop, the analog of the cylinder used in Example 7. By Gauss's law, the total outward flux through the torus is

outward flux = $q_{\text{encl}}/\varepsilon_0 = 2\pi a \lambda/\varepsilon_0$.

If the magnitude of **E** were constant on the surface of the torus we could calculate the outward flux by multiplying E by the area of the torus and equating this to the result just obtained to solve for E. However, **E** is *not* constant on the surface of the torus, so this approach will not succeed. In fact, there is no simple and obvious way to calculate the outward flux through the torus in terms of **E** without knowing **E** in advance and using surface integration over the torus. So for this particular example, applying Gauss's law to a torus does not help to determine **E**. Although Gauss's law seems to be a powerful method of calculating field intensities, it is actually useful in only a few special cases.

We turn now to further examples in which Gauss's law *is* helpful.

Example 9

Determine the electric field arising from charge distributed over the surface of a sphere of radius R with constant surface density σ.

The charged surface in this example is usually referred to as a charged spherical shell. The total charge on the shell is $Q = 4\pi R^2 \sigma$, the area of the shell times the surface density. The symmetry of the charge distribution suggests that we apply Gauss's law to the surface S of a sphere concentric with the charged shell.

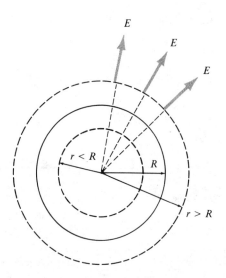

Choose a sphere S of radius $r \neq R$. If $r > R$ the total charge Q lies inside S, but if $r < R$ there is no charge inside S. In either case, by symmetry, the field **E** has constant magnitude everywhere on S and is radially outward, so the total outward flux through S is E times the area of the sphere,

outward flux = $4\pi r^2 E$.

By Gauss's law this is also equal to q_{encl}/ε_0, so solving for E we get

$$E = \frac{q_{encl}}{4\pi\varepsilon_0 r^2}. \tag{33.21}$$

Therefore,

$$E = \frac{Q}{4\pi\varepsilon_0 r^2} \quad \text{if } r > R \tag{33.22}$$

and

$$E = 0 \quad \text{if} \quad r < R. \tag{33.23}$$

In other words, the field outside of a charged spherical shell is identical to that of a point charge located at its center, while that inside the shell is zero.

When the argument of Example 9 is applied to the gravitational field it leads to a simple solution of a problem that caused Newton great difficulty. If, instead of point charges, we arrange a large number of mass points into a spherically symmetric mass, then this mass attracts outside bodies as though all the mass were concentrated at its center. It is thought that Newton delayed publication of his theory of gravitation partly because of the difficulty he had in convincing himself of the validity of this statement.

Example 10

A charge Q is distributed uniformly throughout a solid sphere of radius R. Determine the electric field both outside and inside the sphere. (This kind of charge distribution is difficult to achieve in practice, except perhaps in nuclei.)

First we need to define what is meant by uniformly distributed charge throughout a solid. If a small amount of charge Δq is distributed over a small portion of the solid of volume ΔV, the limit of the ratio $\Delta q/\Delta V$ as ΔV shrinks to zero is called the charge density. The charge is said to be uniformly distributed in a solid if the charge density has the same value at each point of the solid.

We proceed as in Example 9, constructing a concentric spherical surface S of radius $r \neq R$. By symmetry, the field lines are radially outward and the magnitude of the field is constant everywhere on S, so as in Example 9 we find

$$E = \frac{q_{encl}}{4\pi\varepsilon_0 r^2}. \tag{33.24}$$

Now if $r > R$, the total charge enclosed inside S is Q and hence

$$E = \frac{Q}{4\pi\varepsilon_0 r^2} \quad \text{if} \quad r > R, \tag{33.25}$$

the same result as for a charge spherical shell. But if $r < R$, the charge enclosed is only a fraction of Q, this fraction having the same value as the ratio of the two volumes, $(r/R)^3$, because the charge is uniformly distributed. Therefore, $q_{encl} = Q(r/R)^3$ so

$$E = \frac{Qr}{4\pi\varepsilon_0 R^3} \quad \text{if} \quad r < R. \tag{33.26}$$

The graph E as a function of r is shown.

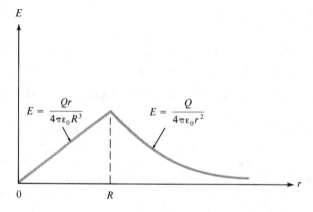

Questions

12. A uniform electric field of $4.3 \times 10^3 \hat{k}$ N/C flows through a hemispherical cup of radius 0.3 m as shown. Determine the flux through the cup.

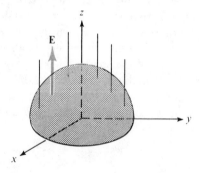

33.6 ELECTRIC FIELDS AND CONDUCTORS

13. For each of the following statements, give a proof if you think the statement is true, or exhibit a counterexample if you think it is false.

    (a) If the electric field is zero everywhere on a closed surface, then the net flux through the surface is necessarily zero.
    (b) If $\oint \mathbf{E} \cdot d\mathbf{A}$ is zero over a closed surface S, then $\mathbf{E}$ is zero everywhere on S.
    (c) If the net charge enclosed by a closed surface is zero, then the electric field is zero everywhere on the surface.

14. In the following figures, q denotes a positive charge. Among the closed surfaces shown, choose the one or ones that best satisfy the stated condition.

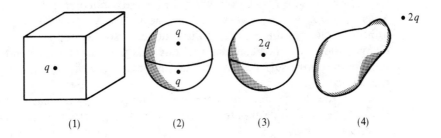

    (1)   (2)   (3)   (4)

    (a) The outward electric flux is zero.
    (b) The outward electric flux is twice that for surface 1.
    (c) Gauss's law is helpful for calculating E everywhere on the surface.

15. A very long solid cylinder of radius b carries a charge with uniform charge density ρ. Determine the electric field at a distance r from the axis of the cylinder for (a) $r > b$ and (b) $r < b$.

16. A thin-walled cylindrical shell of radius b and infinite length has a surface charge density σ. Use Gauss's law to show that (a) inside the shell $\mathbf{E} = \mathbf{0}$, and (b) outside the shell $E = \sigma b/(\varepsilon_0 r)$, where r is the perpendicular distance from the axis of the shell.

33.6 ELECTRIC FIELDS AND CONDUCTORS

As mentioned in Chapter 32, conductors are characterized by the presence of free electrons that can respond to electric forces. In electrostatic conditions all the free electrons are at rest; that's the origin of the term *static* in electrostatics. If there is an electric field in the conductor, the charges will rapidly move around until equilibrium is once again established. As the electrons move around, they tend to cancel the external electric field. Electrostatic equilibrium is impossible for a conductor unless the electric field inside the conducting material is zero. Hence we have an important property:

In electrostatics *the electric field is zero inside a conductor*.

We can discover additional properties of conductors with the aid of Gauss's law. Figure 33.9 shows a surface S that lies just a hair's breadth inside a charged

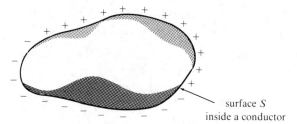

Figure 33.9 All charge on a conductor resides on its surface.

conductor. Because $E = 0$ everywhere inside the conductor, it is zero everywhere on S, so $\oint_S \mathbf{E} \cdot d\mathbf{A} = 0$. By Gauss's law this means that $q_{\text{encl}} = 0$ inside S. But S can be arbitrarily close to the surface of the conductor, so this implies that there is no net charge inside a charged conductor. If any charge is placed on a conductor, all of it must reside on the surface. Even if the conductor has cavities, the charge must lie on the surfaces.

If all the charge on a conductor resides on the surface, what then is the electric field at the surface? Suppose the conductor has a surface charge density σ, not necessarily constant. On the surface of the conductor the electric field must everywhere be perpendicular to the surface. Otherwise, if a component of the electric field existed tangent to the surface, electrons would experience a force given by $\mathbf{F} = q\mathbf{E}$ and consequently would accelerate. However, the conductor is in electrostatic equilibrium so no tangential component can exist.

Now imagine a small cylindrical surface S like that shown in Fig. 33.10, which partially lies inside the conductor. If the surface is small enough then σ is nearly constant on S and $\mathbf{E}$ is nearly parallel to the area vectors of the flat ends of the cylinder. Let's calculate the total flux through S.

Because $E = 0$ inside the conductor there is no contribution to the flux through the portion of S inside the conductor. There is also no flux through the cylindrical part because $\mathbf{E}$ is perpendicular to the area vectors on the cylindrical part. Thus, all

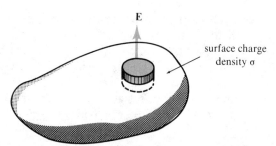

Figure 33.10 The electric field near the surface of a charged conductor depends on the charge density in the immediate neighborhood of each point.

33.6 ELECTRIC FIELDS AND CONDUCTORS

the flux comes through the protruding flat end of S. Since E is constant over that end, we have $\iint \mathbf{E} \cdot d\mathbf{A} = EA$, where A is the area of the flat end. But the total charge enclosed by S is just σA, so by Gauss's law we have

$$EA = \frac{\sigma A}{\varepsilon_0},$$

which implies

$$E = \frac{\sigma}{\varepsilon_0}. \qquad (33.27)$$

In other words, the electric field at the surface of a conductor depends only on the surface charge density in the immediate neighborhood of each point.

Example 11

A conducting spherical shell of innner radius $a = 0.20$ m and outer radius $b = 0.25$ m has a total charge of 5.0 µC on its surfaces. A point charge of -6.0 µC is placed at the center of the shell. If r denotes the distance from the center of the shell, determine the electric field in the regions where (a) $r < a$; (b) $a < r < b$; (c) $r > b$. (d) Determine the charges q_a and q_b on the inner and outer surfaces of the shell, and (e) make a sketch of the electric field intensity as a function of r.

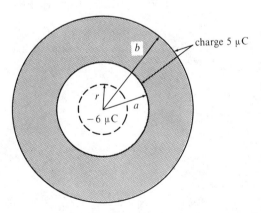

(a) We construct a spherical surface S of radius $r < a$ as shown. The charge enclosed by this surface is just the point charge, -6.0 µC. Because the charge is negative, the electric field is directed inward so the integral for the flux through S is $\oint \mathbf{E} \cdot d\mathbf{A} = -E(4\pi r^2)$. By Gauss's law this is equal to -6.0 µC$/\varepsilon_0$, so

$$E = (6.0 \text{ µC})/(4\pi\varepsilon_0 r^2) \quad \text{if} \quad r < a.$$

(b) If $a < r < b$, we are in the interior of the conductor where the electric field is zero: $E = 0$.

(c) To find the electric field outside the shell we construct a concentric spherical surface S of radius $r > b$. The total charge enclosed by S is that of the point charge plus that of the shell:

$$q_{encl} = -6.0\ \mu C + 5.0\ \mu C = -1.0\ \mu C.$$

Gauss's law in this case implies $-E(4\pi r^2) = -1.0\ \mu C/\varepsilon_0$, so

$$E = 1.0\ \mu C/(4\pi\varepsilon_0 r^2) \quad \text{if} \quad r > b,$$

with the direction of the electric field radially inward. In other words, outside the shell the field is the same as that arising from a point charge of $-1.0\ \mu C$ located at the center of the shell.

(d) To find the charges q_a and q_b on the two surfaces of the conducting shell we take a spherical surface S centered on the point charge and inside the shell. Since S lies within the conducting medium, $E = 0$ everywhere on S so, by Gauss's law, the total charge inside S must also be zero. But the charge enclosed by S is the point charge of $-6.0\ \mu C$ at the center plus the charge q_a on the inner surface, so

$$q_{encl} = -6.0\ \mu C + q_a = 0.$$

Solving for q_a we find $q_a = +6.0\ \mu C$. This is an *induced* charge on the inner surface of the conductor, and it is equal and opposite to the charge attracting it.

To find the charge q_b on the outer surface we use conservation of charge. There is a total charge of 5.0 μC on the two surfaces so $q_a + q_b = 5.0\ \mu C$. But $q_a = 6.0\ \mu C$, hence $q_b = -1.0\ \mu C$.

(e) Using our results for the electric field in the three regions, we construct the following graph of the field intensity E. The field $\mathbf{E}$ itself is directly radially inward if $r < a$ and also radially inward if $r > b$, with $\mathbf{E} = 0$ in the interval $a < r < b$.

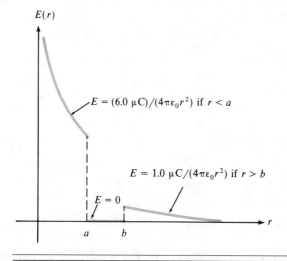

Questions

17. A solid conducting sphere of radius 0.15 m carries a net charge of 3.0 μC. Determine (a) the surface charge density σ and (b) the electric field just outside the surface.

18. A solid conducting sphere of radius a has a net charge of $+Q$ and is surrounded by a concentric conducting spherical shell with inner radius b and outer radius c. If the net charge on the shell is $-2Q$, determine the electric field at distance r from the center in each of the following cases: (a) $r < a$; (b) $a < r < b$; (c) $b < r < c$; (d) $r > c$. (e) Make a sketch of the electric field intensity as a function of r.

19. A point charge $+3q$ is located at the center of a spherical conducting shell with inner radius a and outer radius b. If the shell has a net charge of $-2q$, determine (a) the charge q_a on the inner surface of the shell, (b) the surface charge density on the outer surface of the spherical shell, and (c) the electric field just outside the outer surface of the shell.

20. A thin infinitely long conducting cylindrical shell of radius b has a uniform surface charge denisty σ. Along the axis of the cylinder there is an infinitely long line of uniform linear charge density $-\lambda$. Determine the electric field at a distance r from the axis of the cylinder for (a) $r < b$; (b) $r > b$.

33.7 A FINAL WORD

Michael Faraday did not understand mathematics. James Clerk Maxwell transformed Faraday's naive, unsophisticated picture of lines of force into an immensely sophisticated mathematical description of the field of force. Faraday's picture plays no role in this mathematical theory. Faraday believed that physical lines of force actually existed everywhere in space, but today we no longer believe that this is true. And so, why should we even remind ourselves that Faraday ever had that picture? Is there anything in modern physics today that provides a possible analogy to Faraday's picture of lines of force?

There is a theory concerning the ultimate constituents of matter – protons and neutrons. According to this theory, these particles are composed of even smaller particles called *quarks*. Unfortunately we can't smash a proton apart to study the quarks individually. As far as we know that is an impossible feat. According to theoreticians the quarks are forever hidden inside the protons and neutrons and can never be seen, in the same sense that Faraday's lines of force can never be seen. Nevertheless, the quark model is extremely successful because it guides experiments and correctly predicts many effects that can be observed. So the question arises, do quarks really exist?

The existence of quarks, like the existence of Faraday's lines of force, may turn out to be irrelevant. In the early development of electric science, Faraday's lines of force provided a kind of mental scaffolding – a structure that was constructed before the real edifice was built. After a real edifice is built the scaffolding is torn down and discarded, but that doesn't make it unimportant. It may very well be that quark theory plays exactly the same role as Faraday's model – a mental scaffolding needed to build the final edifice of the structure of matter.

CHAPTER 34

POTENTIAL AND CAPACITANCE

> So wonderfully are these two states of Electricity the plus and the minus combined and balanced in this miraculous bottle [Leyden jar]! Situated and related to each other in a manner that I can by no means comprehend!... Here we have a bottle, containing at the same time a *plenum* of electrical fire, and a *vacuum* of the same fire, and yet the equilibrium cannot be restored between them but by a communication *without*! though the plenum presses violently to expand, and the hungry vacuum seems to attract as violently in order to be filled.
>
> Benjamin Franklin, *New Experiments and Observations on Electricity* (1751)

34.1 A GREAT AMERICAN SCIENTIST

The study of natural philosophy in colonial America was promoted by Benjamin Franklin in the early 1740s, and eventually resulted in the establishment of the American Philosophical Society, which still exists today. Franklin became interested in electrical phenomena when he witnessed a demonstration by an itinerant lecturer in Boston in 1743. Some time thereafter he received from Peter Collinson, a Quaker friend in London, a copy of the *Gentleman's Magazine*, devoted to current happenings in European fashions, politics, and science. The April 1745 issue contained a description of remarkable electrical experiments in Germany. To encourage Franklin to repeat some of these experiments, Collinson sent a glass tube along with the journal. This gift launched Franklin on a new career, and by 1746 he was a full-time electrician, that is, a scientist investigating electricity. In a letter to Collinson in 1747 he wrote "I never before was engaged in any study that so totally engrossed my attention and my time as this has lately done."

Franklin performed many experiments but was not content merely to record his observations. Being a person of great insight and intelligence, he also sought to explain the phenomena he observed. For example, he proposed a one-fluid theory of electricity in which negative electricity was the absence of the all pervasive positive electricity. According to Franklin, rubbing fur along a rubber rod doesn't create electric charge. Instead, charge is simply transferred from one body to the other. When one body gains charge the other loses an equal amount, with the algebraic sum of the positive and negative charges being constant. This fundamental principle is known today as conservation of electric charge: If a positive charge is created, an equal amount of negative charge is also created in the same vicinity.

One of the earliest triumphs of Franklin's theory was its ability to explain the Leyden jar, which, as mentioned Sec. 32.5, was discovered accidentally in 1745. Franklin observed that if the inner conductor of a Leyden jar is positively charged, the outer conductor is negatively charged. The balance between charges can only be restored by connecting the inner conductor to the outer one. In another study, Franklin investigated whether the shape of the Leyden jar affected its ability to store charge. He showed that charge could be stored by placing a piece of glass between two parallel conducting sheets of lead. This device was the forerunner of today's *parallel-plate capacitor*, which has numerous technological applications. We turn now to an energy description of electricity that is intimately related to capacitors.

34.2 ELECTRIC POTENTIAL

The importance of work and energy in mechanics suggests that these concepts might also provide insights in the theory of electricity. We know that work provides a method of following the flow of energy in mechanics, so let's calculate the work required to move a test charge q_0 in an electric field $\mathbf{E}$.

If a test charge q_0 is placed in a field $\mathbf{E}$ it will experience a force

$$\mathbf{F} = q_0 \mathbf{E}. \tag{33.2}$$

To move the charge (without accelerating it) from point A to point B along some curve we apply an equal and opposite force $-\mathbf{F}$, as indicated in Fig. 34.1 The work done by $-\mathbf{F}$ is equal to the line integral

$$W_{AB} = \int_A^B -\mathbf{F} \cdot d\mathbf{r}, \tag{14.1}$$

where $\mathbf{r}$ denotes the vector function describing the curve from A to B. This is the work done *by* the force $-\mathbf{F}$, or *against* the force $\mathbf{F}$. (See Section 14.1.) Using (33.2)

34.2 ELECTRIC POTENTIAL

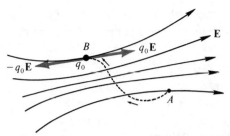

Figure 34.1 Work is done against the Coulomb force in moving a charge from one point to another in an electric field **E**.

this can also be written as

$$W_{AB} = -q_0 \int_A^B \mathbf{E} \cdot d\mathbf{r}. \tag{34.1}$$

In mechanics, if work W is done on a falling body against the gravitational force $\mathbf{F} = m\mathbf{g}$, the potential energy of the body increases by W. Guided by this example, we use the same principle to define the change in potential energy of any charge q_0. The *change in potential energy*, denoted by $U_B - U_A$ or ΔU, is equal to the work done against the Coulomb force $q_0\mathbf{E}$,

$$\Delta U = U_B - U_A = -q_0 \int_A^B \mathbf{E} \cdot d\mathbf{r}. \tag{34.2}$$

In SI units the potential energy difference is measured in joules.

By introducing the electric field in Chapter 33, we no longer had to worry about all the charges that created the field, but instead could focus on properties of the field itself. That idea can be extended to our energy description by simply dividing both members of (34.2) by q_0. In other words, we can consider the *change in potential energy per unit charge*, $(U_B - U_A)/q_0$. This is called the *electric potential difference* and is denoted by $V_B - V_A$ or by ΔV. Thus, by definition,

$$\Delta V = V_B - V_A = -\int_A^B \mathbf{E} \cdot d\mathbf{r}, \tag{34.3}$$

a quantity determined by the field.

Work, potential energy, and charge are all scalar quantities, so potential difference is also a scalar. In SI units the potential difference is measured in joules per coulomb, J/C, which is called the *volt* (V). If a potential difference of 1 V exists between two points, then an external agent does 1 J of work against the Coulomb force in moving a charge of 1 C from one point to the other. Since it is measured in volts, potential difference is often referred to as *voltage*.

Because the electric potential difference ΔV is defined as the work done per unit of positive charge, the corresponding change in potential energy ΔU of any charge q_0 is simply

$$\Delta U = q_0 \Delta V. \tag{34.4}$$

A convenient unit of potential energy for charges in electric fields is the *electronvolt*, abbreviated eV. It is defined as the work done in moving a charge of one electron through a potential difference of 1 V. The magnitude of the charge of the electron is 1.6×10^{-19} C, so the conversion between electronvolts and joules is

$$1 \text{ eV} = 1.6 \times 10^{-19} \text{ C V} = 1.6 \times 10^{-19} \text{ J}.$$

Just as in mechanics, Eq. (34.2) defines only *differences* in potential energy, not the potential energy itself. To speak of potential energy itself it is necessary to choose a reference point that is assigned zero potential energy. In many electrical problems the point assigned zero potential energy (and hence zero potential) is called the *ground* point and is denoted by the special symbol ⏚.

Now let's determine the potential difference between points A and B when the electric field is generated by a single point charge q. The field at distance r from q is

$$\mathbf{E} = K_e \frac{q}{r^2} \hat{\mathbf{r}}. \tag{33.1}$$

Because the Coulomb force has the same $1/r^2$ dependence as the gravitational force, the calculation is really the same as that done in Chapter 14 for the change in gravitational potential energy. The result is

$$V_B - V_A = -\int_A^B \mathbf{E} \cdot d\mathbf{r} = K_e q \left(\frac{1}{r_B} - \frac{1}{r_A} \right), \tag{34.5}$$

where r_A and r_B are the distances from q to points A and B, respectively.

In Chapter 14 we found that the work done by the gravitational force is independent of the path – it just depends on the initial and final points. So too is the work done against an electrostatic field. This means that the electrostatic field is *conservative* because the work done around a closed path is zero. We summarize this property mathematically by writing

$$\oint \mathbf{E} \cdot d\mathbf{r} = 0,$$

where the circle on the integral sign reminds us that the integral is taken around a *closed* path.

The result in Eq. (34.5) suggests that we assign zero potential to points that are infinitely far from the charge q. Thus, if we take $r_A = \infty$, then the *electric potential* at a distance r from a point charge q is simply

$$V(r) = K_e \frac{q}{r}. \tag{34.6}$$

This result is similar to that for the gravitational potential energy,

$$U(r) = -\frac{GMm}{r}, \tag{23.8}$$

34.2 ELECTRIC POTENTIAL

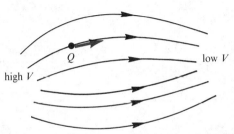

Figure 34.2 A positive charge Q released from rest moves toward a region of lower potential energy.

but unlike gravitational potential, which is always negative, electric potential can be positive or negative depending on the sign of q.

Let's bring the idea of electric potential down to earth by considering once more the analogy with the gravitational force. Suppose a rock is lifted toward the top of a hill. The higher it's lifted, the more gravitational potential energy it gains. When released, the rock moves downhill losing potential energy, which gets converted to kinetic energy and heat energy. Similarly, suppose a positive charge is brought near a positively charged Van de Graaff dome. The closer the charge is to the dome, the greater its electric potential becomes (because $1/r$ is greater). Consequently, when the charge is released it moves away from the dome toward lower potential. Even though, in this case, the force repels rather than attracts, charges are analogous to masses in the sense that they always move downhill, that is toward regions of lower potential.

This is consistent with Faraday's view that electric field lines always begin on positive charges and end on negative charges. If a positive charge, say a proton, is placed in an electric field and released, according to Faraday's picture the force on the proton at each point will be in the direction of the lines of force. So the proton will move away from positive charges and toward negative charges. The kinetic energy of the proton increases and, by conservation of energy, its potential energy must decrease. In other words, electric field lines point toward direction of decreasing potential, as suggested by Fig. 34.2.

Example 1
Inside an electron gun of a television set, electrons are accelerated from rest through a potential difference of 500 V. What is the speed of an electron as a result of this acceleration?

In accelerating through the potential difference, an electron gains kinetic energy and loses potential energy. The work done by the electric field is entirely converted into kinetic energy, so

$$q_0 V = \tfrac{1}{2} m v^2,$$

where q_0 is the magnitude of the electron charge. Solving for v, we find

$$v = \sqrt{\frac{2q_0 V}{m}} = \sqrt{\frac{2(1.6 \times 10^{-19} \text{ C})(500 \text{ V})}{9.1 \times 10^{-31} \text{ kg}}} = 1.3 \times 10^7 \text{ m/s}.$$

Example 2

In a hydrogen atom, an electron moves in a circular orbit at a distance of 0.53×10^{-10} m from the proton.

(a) What is the electric potential at the orbit?
(b) What is the potential energy of the electron at this separation distance?

(a) According to Eq. (34.6), the electric potential due to a proton of charge $q = 1.6 \times 10^{-19}$ C is simply

$$V = K_e \frac{q}{r} = \frac{(9 \times 10^9 \text{ N m}^2/\text{C}^2)(1.6 \times 10^{-19} \text{ C})}{(0.53 \times 10^{-10} \text{ m})} = 27.2 \text{ V}.$$

(b) The potential energy of the electron is $U = q_0 V$, and has the value of -27.2 eV.

We already know that if several point charges are present, the electric field at a particular point is the vector sum of the fields due to the individual charges. Therefore, in Eq. (34.2) the potential difference will be the integral of the sum of the electric fields, and since the integral of a sum is the sum of the integrals, we find that for a collection of point charges the electric potential at a point is the sum of the potentials due to each charge:

$$V = \sum_i K_e \frac{q_i}{r_i}. \tag{34.7}$$

Here r_i is the distance from q_i to the point at which the potential is to be calculated. As in (34.6), the potential is taken to be zero at infinity.

Now that we have defined the potential for a collection of point charges, let's apply it to a configuration that is useful for storing charge: two charged conducting sheets. Suppose two parallel flat conducting plates have equal and opposite surface charge densities $\pm \sigma$. To calculate the electric potential at an arbitrary point between the plates, we first need to know the electric field. If the plates are horizontal and infinite in extent, then by Eq. (33.12) the field due to the positive plate is as indicated in Fig. 34.3a, with the field directed away from the plate. Below the positive plate the field is directed downward and is given by

$$\mathbf{E}_+ = \frac{-\sigma}{2\varepsilon_0} \hat{\mathbf{k}}.$$

34.2 ELECTRIC POTENTIAL

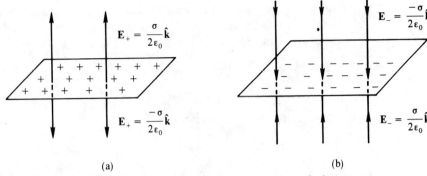

(a)   (b)

Figure 34.3 The electric field due to (a) a positively charged plate and (b) a negatively charged plate.

The field due to the negative plate is directed toward the plate, as indicated in Fig. 34.3b. Above the negative plate the field is directed downward with $\mathbf{E}_- = \mathbf{E}_+$.

A unit positive charge would be attracted to the negative plate and repelled from the positive plate so the resultant electric field between the plates is

$$\mathbf{E} = \mathbf{E}_+ + \mathbf{E}_- = -\frac{\sigma}{\varepsilon_0}\hat{\mathbf{k}}. \tag{34.8}$$

The resultant field is **0** above the positive plate and below the negative plate, as indicated in Fig. 34.4.

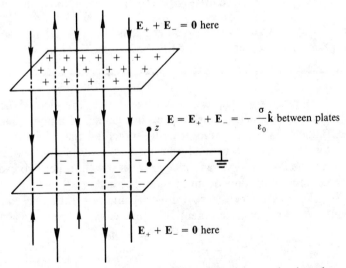

Figure 34.4 The electric potential between two conducting plates increases linearly with the distance from the grounded negative plate.

Calculating the potential between the plates is now an easy matter. Let z denote the distance above the negatively charged plate, as indicated in Fig. 34.4. Because the field is constant, Eq. (34.3) becomes

$$V(z) - V(0) = -E(z - 0) = \frac{\sigma}{\varepsilon_0} z.$$

If the negative plate is grounded, as indicated by the symbol ⏚ in Fig. 34.4, the potential there is zero, and we obtain

$$V(z) = \frac{\sigma}{\varepsilon_0} z. \tag{34.9}$$

Thus, the potential increases linearly with the distance from the negative plate.

Example 3

Find the electric potential at a distance r from a uniform line charge of density λ, assuming that $V(b) = 0$, where b is a fixed distance from the line charge.

From Eq. (33.8) we know that the electric field intensity at a distance R from a line of charge is given by

$$E = \frac{2\lambda K_e}{R} \tag{33.8}$$

and is radially directed. To calculate the potential we apply Eq. (34.3) and integrate in the radial direction because only in that direction is there a nonzero contribution to the integral. Hence, we have

$$V(r) - V(b) = -\int_b^r E\, dR = -2\lambda K_e \int_b^r \frac{dR}{R} = -2\lambda K_e \ln\left(\frac{r}{b}\right).$$

But $V(b) = 0$, so we can write the potential as $V(r) = -2\lambda K_e \ln(r/b)$ or, in terms of the permittivity constant, as $V(r) = -(\lambda/2\pi\varepsilon_0) \ln(r/b)$.

In Example 9 of Chapter 33 we found that the electric field outside of a uniformly charged spherical shell is the same as that of a point charge located at its center. If the field is the same, then the potential must be the same for a point charge located at the center of the shell. To verify this, consider a spherical conducting shell of radius R and charge Q, as shown in Fig. 34.5, and calculate the electric potential both outside and inside the shell.

Using Eq. (33.22) for the field intensity and integrating we find that outside of the shell (where $r > R$), the potential difference is

$$V(\infty) - V(r) = -\int_r^\infty E\, dr' = -\int_r^\infty \frac{Q\, dr'}{4\pi\varepsilon_0 r'^2} = \frac{-Q}{4\pi\varepsilon_0 r} = -K_e \frac{Q}{r}.$$

Taking $V(\infty) = 0$, we find the potential due to a spherical shell to be

$$V(r) = K_e \frac{Q}{r} \quad \text{if } r > R. \tag{34.10}$$

34.2 ELECTRIC POTENTIAL

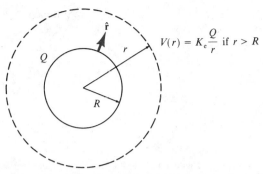

Figure 34.5 The potential outside a charged conducting spherical shell is the same as that of a point charge located at the center of the shell.

In other words, the potential *outside* of a spherical shell is precisely the same as that of a point charge located at its center.

We know from Eq. (33.23) that inside the shell the electric field is zero. Therefore, no work is done in moving a test charge inside the shell, and so the potential difference between a point inside the shell and one on the surface must be zero; Eq. (34.3) indicates that $V(r) - V(R) = 0$, or $V(r) = V(R)$ for $r < R$. The potential inside the shell is constant, equal to the potential on the surface of the shell. But what is the potential on the surface? From Eq. (34.10), we see that $V(r) \to K_e Q/R$ as $r \to R$ from the outside, and so it seems natural to call this limiting value the potential on the surface. This gives us the constant potential

$$V(r) = K_e \frac{Q}{R} \quad \text{for } r \le R. \tag{34.11}$$

The value of this constant is determined by the radius R and the amount of charge on the shell. Figure 34.6 shows graphs of both the electric field intensity and the potential for a uniformly charged spherical shell.

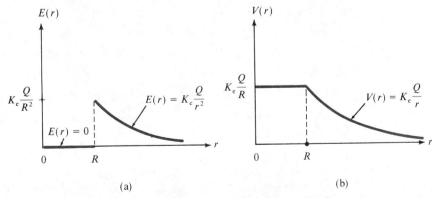

Figure 34.6 Graphs of (a) the electric field intensity and (b) the potential of a uniformly charged spherical shell of radius R.

A common misconception is to think that the potential must be zero inside the shell because the field is zero. As we've seen, zero electric field implies that the potential *difference* between any two points is zero. That, in turn, implies that the potential itself is constant everywhere inside the shell.

Questions

1. If the electric potential is zero in a region of space, must the electric field also be zero?

2. Suppose that an electron is placed in a vacuum chamber in which there is a uniform electric field. Ignoring gravity, which of the following properties does the subsequent motion of the electron possess?

    (a) Constant velocity in a direction opposite to the field.
    (b) Constant velocity in the direction of the field.
    (c) Constant acceleration in a direction opposite to the field.
    (d) Constant acceleration in the direction of the field.

3. A Van de Graaff generator is charged and an insulated conducting sphere is placed near it. Some representative electric field lines are shown in the figure. Which of the following statements is false?

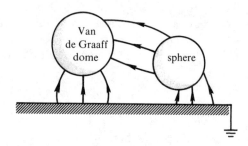

    (a) The Van de Graaff dome has a negative charge.
    (b) The sphere has a negative potential with respect to ground.
    (c) The sphere has a positive potential with respect to the dome.
    (d) The Van de Graaff dome has positive potential with respect to ground.

4. What is the speed of an electron with a kinetic energy of 2000 eV?

5. Along the x axis an electric field has intensity $E(x) = 3x^2 - 5x$. Calculate the work done in moving a charge from $x = 0$ to $x = b$.

6. An infinite plane sheet has a constant surface charge density $\sigma = 0.8 \ \mu C/m^2$. Calculate:

    (a) The magnitude of the electric field 0.02 m away from the sheet.
    (b) The electric potential difference between points A and B on the same side of the sheet if A is 0.02 m and B is 0.05 m away from the sheet.
    (c) The work required to move an electron from point A to B in part (b).

34.2 ELECTRIC POTENTIAL

7. Two point charges are fixed on a rectangular coordinate system: A 5.0-μC charge is at $(0, 0, 0.4\text{ m})$ and a -2.0-μC charge at $(0, 0.3\text{ m}, 0)$. Calculate the electric potential at the origin.

8. Four positive point charges, each of 0.4 μC, are held at the corners of a square of sides 0.1 m.
   (a) What is the electric field at the center of the square?
   (b) What is the electric potential at the center of the square?

9. Two positive charges q are fixed on the x axis at $x = -a$ and $x = a$.
   (a) Find the electric potential $V(x)$ on the x axis as a function of x.
   (b) Sketch the graph of $V(x)$ for $-a < x < a$, and explain the significance of the minimum of the curve.

10. In a particle accelerator used to treat cancer, protons are released from rest at a potential of 8.0 MeV (million electron volts) and travel in a vacuum to a region of zero potential. Find the speed of the protons when they are in the region of zero potential.

11. Refer to the accompanying charge configuration. Calculate each of the following:
    (a) The potential at the midpoint of the line joining the two charges.
    (b) The potential a distance $2a$ along the perpendicular bisector of the line joining the charges.
    (c) The work required to move a charge q_0 from the point in (b) to the point in (a).

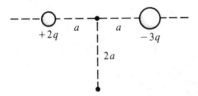

12. Two parallel conducting plates have equal and opposite surface charge densities and are separated by 0.15 m. One plate is grounded and the other has a potential of 500 V. Find the surface charge density on either conducting plate.

13. A long cylindrical shell of radius b is grounded. A line of uniform charge density $\lambda > 0$ is placed along its axis.
    (a) Calculate the electric potential at each point inside the shell.
    (b) Make a graph of the potential inside the shell as a function of the distance r from the axis.

14. Refer to Question 13. If a positive charge q is released at a distance $r = b/2$ inside the shell, find (a) the direction of its motion, (b) its acceleration, and (c) its final speed.

15. A positive charge Q is spread uniformly throughout a solid sphere of radius R.
   (a) What is the electric potential outside the sphere? (You should be able to answer this without calculation.)
   (b) Calculate the electric potential inside the sphere.
   (c) Plot the potential as a function of distance r from the center.

16. A spherical conducting shell of inner radius a and outer radius b has a net charge of q. A point charge of $-2q$ is placed at the center. Find the electric potential at a distance r from the center. Consider (a) $r > b$; (b) $a < r < b$; and (c) $r < a$.

17. Two concentric spherical conducting shells carry equal and opposite charges. The inner shell has radius a and positive charge q; the outer shell has radius b and charge $-q$. Find the potential difference between the shells.

34.3 ELECTRIC ENERGY OF SYSTEMS OF CHARGES

The Leyden jar is remarkable because it stores electric energy. How does it do this, and where does the energy come from?

In the last section we found how electric potential energy is related to the work done in moving a charge in an electric field. That's essentially all we need to know to determine the energy stored in a system of charges. Suppose there were only one charge, say q_1, in our galaxy. Now suppose a second charge, q_2, is brought from an infinite distance away and placed at a final distance r_{12} from the first charge, as illustrated in Fig. 34.7. How much work W_{12} is done in moving that charge?

The work W_{12} is equal to the product of q_2 and the change in potential V_1 of the field generated by q_1. Because the potential an infinite distance away is assigned the value zero, by Eq. (34.6) we have

$$V_1 = K_e \frac{q_1}{r_{12}},$$

and hence

$$W_{12} = K_e \frac{q_1 q_2}{r_{12}}. \tag{34.12}$$

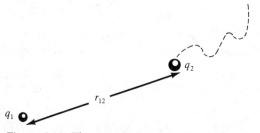

Figure 34.7 The potential energy of a system of point charges is equal to the work done to bring the charges from infinity to their final positions.

34.3 ELECTRIC ENERGY OF SYSTEMS OF CHARGES

That's it. The amount of work done to assemble the charges is equal to the potential energy stored in the system. And that energy comes from the agent that assembled the charges. More complex systems of charges can be treated in the same way. The work done to assemble two charges is equal to the product of the second charge (q_2) and the potential of the first charge at the final position of the second one. When there are more than two charges in a system, the total energy of the system is then obtained by adding all the work required to assemble the charges one at a time. The work done to bring each charge from infinity to its final position is equal to the product of that charge times the potential at that final position due to all the charges that have already been assembled. Example 4 illustrates this idea.

Example 4
Three point charges, $q_1 = -2.0\ \mu C$, $q_2 = 4.0\ \mu C$, and $q_3 = 5.0\ \mu C$, are fixed on a rectangular coordinate system as shown. What is the electric potential energy of the configuration?

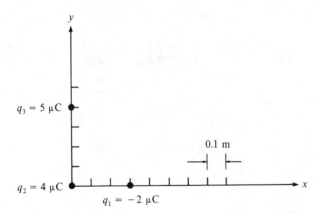

Imagine bringing first q_1, then q_2, and finally q_3 from infinity to their final positions. When q_1 is brought to its final position, no work is done because no other charges are present, so $W_1 = 0$. When q_2 is brought from infinity, work W_2 is done against the electric field created by q_1:

$$W_2 = K_e \frac{q_1 q_2}{r_{12}}.$$

In bringing q_3 from infinity to its final position work must be done against each of the electric fields created by charges q_1 and q_2. That work, W_3, is equal to q_3 times the sum of the potential of these fields,

$$W_3 = q_3 \left(K_e \frac{q_2}{r_{23}} + K_e \frac{q_1}{r_{13}} \right).$$

Thus the potential energy U of the system is the total work done in assembling the charges and is given by

$$U = W_2 + W_3 = K_e \left(\frac{q_1 q_2}{r_{12}} + \frac{q_2 q_3}{r_{23}} + \frac{q_1 q_3}{r_{13}} \right).$$

Inserting numerical values we find

$$U = (9 \times 10^9 \text{ N m}^2/\text{C}^2)[(-2.0 \times 10^{-6} \text{ C})(4.0 \times 10^{-6} \text{ C})/(0.3 \text{ m})$$
$$+ (4.0 \times 10^{-6} \text{ C})(5.0 \times 10^{-6} \text{ C})/(0.4 \text{ m})$$
$$+ (-2.0 \times 10^{-6} \text{ C})(5.0 \times 10^{-6} \text{ C})/(0.5 \text{ m})]$$
$$= 0.03 \text{ J}.$$

Questions

18. A negatively charged particle q_0, initially located 5.0 cm from each of two unequal stationary positive charges, is moved to point P, which is also 5.0 cm from each of the charges, as shown. Which of the following statements is correct about the work done in moving q_0 from its initial position to point P?

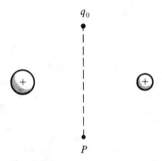

(a) The work depends on the charge of the particle q_0.
(b) The work depends on the stationary charges and the field strengths.
(c) The work depends on the path taken to point P.
(d) The work is zero.

19. Three charges, $q_1 = -6.0$ μC, $q_2 = -2.0$ μC, and $q_3 = 5.0$ μC are placed at the corners of an equilateral triangle of sides 0.05 m. Calculate the energy of the system.

20. In the fission of uranium, the uranium nucleus splits into two fragments, releases several neutrons and energy. Assume that the fission fragments are equally charged nuclei with charge $+46|e|$, where e is the charge of an electron, and that just after fission these nuclei are at rest and separated by twice their radius R, where $R = 1.3 \times 10^{-14}$ m.

(a) Calculate the electric potential energy of the fragments. This is approximately the energy released per fission.

(b) How many fissions are required to produce 1 J of energy?

34.4 CAPACITORS

The Leyden jar was the first capacitor, a device making use of an insulator with a conducting surface on either side. Franklin's investigations showed that the charges on the two sides are equal in magnitude but opposite in sign. The positive and negative charges on the jar are always balanced, and the jar merely maintains a separation of charge. A charged Leyden jar contains the same net charge as one that is not charged. The opening quotation of this chapter reveals that Franklin understood the principle of conservation of charge.

Being a clever experimentalist, Franklin constructed a simple device to demonstrate that the net charge on a Leyden jar is zero. He attached a wire to the outer coating of a charged Leyden jar and bent it upward forming an electrode near the top of the jar, as suggested by Fig. 34.8. He then suspended a conducting cork from a silk thread so it would oscillate back and forth between the end of the wire and a second electrode connected to the inner coating. The cork would be attracted to the positive side, picking up positive charge through conduction, then repelled to the other side, transferring some of its positive charge and thereby becoming negatively charged, then repelled back to the positive side, and so on, oscillating back and

Figure 34.8 Franklin's demonstration that the net charge on a charged Leyden jar is zero.

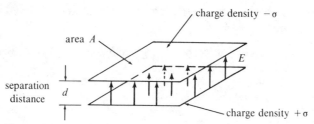

Figure 34.9 A parallel-plate capacitor produces a uniform electric field.

forth between the electrodes until it had completely discharged the Leyden jar. In this manner, Franklin demonstrated that the net charge on the jar is zero.

Franklin also realized that the shape of the jar is not essential. Any device consisting of two conductors insulated from each other will do equally well. He constructed a parallel-plate capacitor consisting of a glass sheet coated with metal on either side, as illustrated in Fig. 34.9, and showed that it has equal and opposite charges on its metal plates. Capacitors are effective because the opposite charges on the plates attract each other but are unable to recombine across the insulator (or gap) between them. They are thus bound firmly in place unless some means such as a conducting wire or Franklin's cork is available to bring them back together.

The charge on the plates of a capacitor creates an electric field between the plates. We've already noticed that the electric field produced by two parallel plates is uniform and has a magnitude given by

$$E = \sigma/\varepsilon_0, \tag{34.8}$$

where σ is the surface charge density on *either* plate. In Section 34.2 we found that the electric potential between two conducting plates increases linearly with the distance from the grounded negative plate. Therefore, if the plates are at a distance d from each other, the magnitude of the potential difference between them is

$$V = \frac{\sigma d}{\varepsilon_0}. \tag{34.9}$$

Now suppose the total charge on a capacitor plate of area A is Q. The surface charge density is simply $\sigma = Q/A$, and Eq. (34.9) can be written as

$$V = \frac{Qd}{\varepsilon_0 A}.$$

In other words, the potential difference between the capacitor plates is proportional to the charge on either plate. The factor $d/(\varepsilon_0 A)$ depends on the geometry of the plates; in this case the separation distance d and the area A.

In general, the charge on the plates of any capacitor is proportional to the potential difference between them:

$$Q = CV. \tag{34.13}$$

The proportionality constant is called the *capacitance* and depends solely on the geometry of the specific capacitor. Capacitance is a measure of the amount of

34.4 CAPACITORS

charge a capacitor holds at a given potential difference; the larger the capacitance, the greater the charge that can be stored for a given potential difference. In Eq. (34.13) V stands for the *magnitude* of the potential difference between the two plates of the capacitor.

If the amount of charge Q and the potential difference of a capacitor are known, then the capacitance can be found from $C = Q/V$. For a parallel-plate capacitor, we immediately see that the capacitance is given by

$$C = \frac{Q}{V} = \frac{Q}{Qd/(\varepsilon_0 A)} = \frac{\varepsilon_0 A}{d}. \tag{34.14}$$

If the capacitance is large, a large amount of charge can be stored at a low potential. To make C large we want a large plate area A and a small separation distance d. In practice, parallel-plate capacitors consist of two parallel sheets of aluminum foil separated only by a thin sheet of plastic. To get a large area in a convenient package, the sheets are usually a few centimeters wide but several meters long. The resulting sandwich is then covered by another sheet of plastic and rolled into a compact cylinder. Examples are shown among those in Fig. 34.10.

The SI units of capacitance are coulombs per volt, C/V, called farads F, in honor of Michael Faraday:

1 C/V = 1 F.

Since 1 C is a huge charge, 1 F is a large capacitance. Most capacitors in common

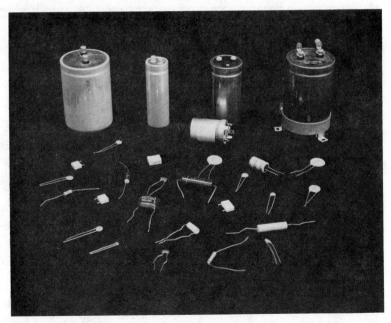

Figure 34.10 Commercial capacitors are produced in a variety of shapes. (Caltech photo by Robert Paz.)

use have capacitances measured in $\mu F = 10^{-6}$ F (microfarads) or pF = 10^{-12} F (picofarads).

Example 5
A parallel-plate capacitor consists of two square plates, each of sides 4.0 cm separated by 2.0 mm.
  (a) Calculate the capacitance.
  (b) How many electrons can be stored on a plate if the capacitor is charged to 50 V?

  (a) We use Eq. (34.14) with $A = (0.04)^2$ m² and work entirely in SI units to get

$$C = (8.85 \times 10^{-12})(0.04)^2/(0.002) = 7.1 \times 10^{-12} \text{ F} = 7.1 \text{ pF}.$$

  (b) The charge $Q = CV$ that can be stored on a plate is

$$Q = (7.1 \times 10^{-12})(50) = 3.5 \times 10^{-10} \text{ C}.$$

The magnitude of the charge of one electron is $|e| = 1.6 \times 10^{-19}$ C, so the number N of electrons that can be stored is

$$N = Q/|e| = (3.5 \times 10^{-10})/(1.6 \times 10^{-19}) = 2.2 \times 10^9.$$

Example 6
A spherical capacitor consists of two thin concentric spherical shells of radii a and b, with $a < b$. A charge Q is placed on the inner shell, and $-Q$ on the outer shell, which is grounded. Calculate the capacitance.

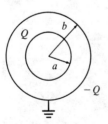

In order to use $C = Q/V$, we need first to calculate the potential difference between the shells, which in turn depends on the field. A concentric sphere of radius r, where $a < r < b$, contains charge Q so, by Gauss's law, the electric field on this sphere has intensity $E = Q/(4\pi\varepsilon_0 r^2)$. Using Eq. (34.3) and substituting for E, we get

$$V(b) - V(a) = -\int_a^b E\, dr = \frac{-Q}{4\pi\varepsilon_0}\int_a^b \frac{dr}{r^2} = \frac{Q}{4\pi\varepsilon_0}\left(\frac{1}{b} - \frac{1}{a}\right).$$

34.4 CAPACITORS

Since $V(b) = 0$, we have

$$V = V(a) = \frac{Q}{4\pi\varepsilon_0}\left(\frac{1}{a} - \frac{1}{b}\right) = \frac{Q(b-a)}{4\pi\varepsilon_0 ab}.$$

Finally, from $C = Q/V$, we get

$$C = \frac{4\pi\varepsilon_0 ab}{b-a}.$$

Questions

21. A popular eighteenth century parlor trick was to have a group of people join hands to form a human chain with the first person touching an electrostatic generator and then have last person touch a metal object. The generator produces a fixed maximum potential difference. Would the shock the last person felt upon touching a metal object depend on (a) the number of people? (b) their size?

22. A parallel-plate capacitor is charged to a potential of 50 V and then wires attached to each plate are brought together, shorting the capacitor and producing a spark. If the plates of the capacitor are moved closer together, and the capacitor is again charged to 50 V, and then shorted, will the second spark contain more charge that the first?

23. A parallel-plate capacitor has a 0.1-mm separation. What must the area of each plate be to produce a capacitance of 2 F? Compare your answer to the area of a window.

24. Suppose a parallel-plate capacitor whose capacitance is 3 pF is attached to a 9-V battery and then disconnected.

    (a) What is the charge on each plate of the capacitor?
    (b) If the distance between the plates is halved while the battery is connected, what is the charge on each plate?
    (c) If the distance between the plates is halved after the battery is disconnected, what is the potential difference between the plates?

25. A cylindrical capacitor consists of two long, thin cylindrical shells of radii a and b and length L, as shown. A charge Q is on the inner shell and $-Q$ on the outer shell. Ignore edge effects and assume the electric field between the cylinders is uniform. Show that the capacitance is given by

$$C = \frac{2\pi\varepsilon_0 L}{\ln(b/a)}.$$

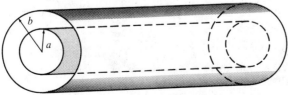

length L

26. The dome of a Van de Graaf generator can be thought of as one plate of a spherical capacitor whose other plate is at infinity.

   (a) Draw on your knowledge of charge to decide how you expect the capacitance of the dome to depend on its radius.
   (b) Calculate the capacitance of an isolated spherical dome of radius R.

34.5 COMBINATIONS OF CAPACITORS

In his studies of the Leyden jar Franklin connected together a bank of parallel-plate capacitors to make what he called an electrical battery. Although the term electric battery is still used today, it means something different from a bank of Leyden jars. However, capacitors are used today in various combinations for applications ranging from electric flashers to computer circuits. Let's now examine how capacitors behave when combined.

In electric circuits the symbol ─┤├─ represents a capacitor whether it be a parallel-plate, cylindrical, or spherical capacitor. Suppose two capacitors of capacitances C_1 and C_2 are *connected in series*, as indicated schematically in Fig. 34.11. When the series combination is connected to an electric generator, charge flows onto one plate, say plate A in Fig. 34.11, making it positively charged. That induces an equal but negative charge on the opposite plate B. The wire connecting the two capacitors is neutral, so plate C of the second capacitor must have a positive charge equal to that on plate B. The positive charge on plate C, in turn, induces an equal but negative charge on D. *For capacitors connected in series the charge on each capacitor is the same*.

When capacitors are connected in series, the combination can be replaced by a single capacitor with an equivalent capacitance, which we will now calculate. Suppose that the potential difference is V_1 across capacitor 1 and V_2 across capacitor 2. The total potential difference across the two capacitors in series is the sum of the potential differences. That's because the total work done is obtained by adding the work done across each capacitor. Therefore, the potential difference across the series combination is

$$V_s = V_1 + V_2 = \frac{Q}{C_1} + \frac{Q}{C_2} = Q\left(\frac{1}{C_1} + \frac{1}{C_2}\right).$$

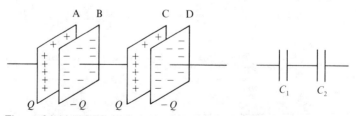

Figure 34.11 When capacitors are connected in series, the charge is the same on each capacitor.

34.5 COMBINATIONS OF CAPACITORS

If we replace the series combination by one capacitor of equivalent capacitance C_s, the capacitance and potential difference V_s are related by $V_s = Q/C_s$, which implies

$$\frac{Q}{C_s} = Q\left(\frac{1}{C_1} + \frac{1}{C_2}\right),$$

hence

$$\frac{1}{C_s} = \frac{1}{C_1} + \frac{1}{C_2}. \tag{34.15}$$

In other words, the reciprocal of the equivalent capacitance of a series combination is the sum of the reciprocals of the individual capacitances. From Eq. (34.15) we see that the equivalent capacitance is less than that of either of the individual capacitors. The generalization to any finite number of capacitors connected in series is given by

$$\boxed{\frac{1}{C_s} = \frac{1}{C_1} + \frac{1}{C_2} + \frac{1}{C_3} + \cdots .} \tag{34.16}$$

Capacitors can also be connected *in parallel*, as shown in Fig. 34.12. When the parallel combination is connected to a generator, charge flows onto each of the capacitors, but not necessarily in equal amounts. The upper plates of the two capacitors are joined by a conducting wire and therefore have the same potential. For the same reason, the lower plates are at the same potential. *Capacitors connected in parallel have the same potential difference across their plates.*

Suppose that the potential difference across each capacitor is V, the capacitances are C_1 and C_2, and the charges on them are Q_1 and Q_2, respectively. The

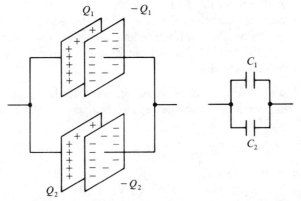

Figure 34.12 Capacitors in parallel have the same potential difference.

total charges stored on the two capacitors is

$$Q = Q_1 + Q_2 = C_1V + C_2V = (C_1 + C_2)V.$$

If the parallel combination of capacitors is replaced by a single capacitor of equivalent capacitance C_p holding charge Q at voltage V, then

$$C_p = \frac{Q}{V} = C_1 + C_2. \qquad (34.17)$$

The equivalent capacitance equals the sum of the individual capacitances. The generalization for any finite number of capacitors connected in parallel is

$$\boxed{C_p = C_1 + C_2 + C_3 + \cdots .} \qquad (34.18)$$

Example 7

In the circuit shown, a potential difference of 2 V exists across the 5-μF capacitor.

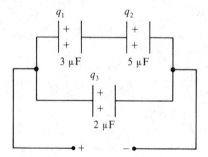

(a) Find the charge on the 3-μF capacitor.
(b) Determine the potential difference across the 3-μF capacitor.
(c) Calculate the potential difference and charge on the 2-μF capacitor.
(d) Determine the total charge put out by a generator that is attached to the positive and negative electrodes.

(a) The 3- and 5-μF capacitors are in series so they have the same charge, that is, $q_1 = q_2$. Because we know both the capacitance and voltage across one of them, we have

$$q_2 = C_2V_2 = (5 \times 10^{-6} \text{ F})(2 \text{ V}) = 10 \text{ μC},$$

which means $q_1 = 10$ μC as well.

(b) Using $V_1 = q_1/C_1$ we get $V_1 = 3.3$ V.

(c) The 2-μF capacitor is in parallel with the series combination of the 3- and 5-μF capacitors. This means that the voltage across the 2-μF capacitor is equal to

34.5 COMBINATIONS OF CAPACITORS

the sum of the voltages across each of the other two: $V_3 = V_1 + V_2 = 5.3$ V. Therefore, $q_3 = C_3 V_3 = (2 \times 10^{-6}$ F$)(5.3$ V$) = 11$ μC.

(d) The total charge put out by the electric generator is equal to the sum of the charges on the 3- and 2-μF capacitors: $Q = q_1 + q_3 = 21$ μC.

Questions

27. A 2- and a 5-μF capacitor are connected in series across a potential difference of 10 V.

    (a) What is the charge on each capacitor?
    (b) What is the potential difference across each?

28. A 4-, a 5-, and a 6-μF capacitor are connected in parallel to a potential difference of 5 V. Find the charge on each capacitor.

29. A 2- and a 5-μF capacitor are first connected in series to a potential difference of 10 V and then disconnected from the source and from each other.

    (a) What is the charge and potential difference on each capacitor?
    (b) If the charged capacitors are then connected to each other as shown in the diagram, what will be the final charge on each capacitor?
    (c) Under the conditions of (b) what will be the final potential difference across each capacitor?

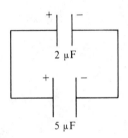

30. Consider the five capacitors shown, where $C_1 = 1$ μF, $C_2 = 2$ μF, $C_3 = 3$ μF, $C_4 = 4$ μF, and $C_5 = 5$ μF. Calculate the capacitance of the equivalent capacitor.

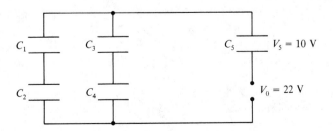

31. For the arrangement of capacitors in Question 30, find (a) the charge on each of the five capacitors and (b) the potential difference across each capacitor.

32. A potential difference V_0 is applied to the circuit shown, which contains four originally uncharged capacitors. A 6-μC charge is on capacitor C and the potential difference across the 8-μF capacitor is 2 V.

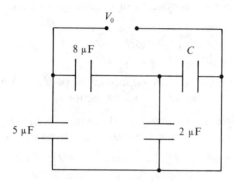

(a) Determine the charge on the 8-μF capacitor.
(b) Find the charge and potential difference for the 2-μF capacitor.
(c) Determine the value of the capacitance of C.
(d) Calculate the potential difference V_0.
(e) Find the total charge supplied by the potential difference V_0.

34.6 ENERGY STORAGE IN A CAPACITOR

The jolting capability of a Leyden jar demonstrates that energy is stored within its confines. Our next task is to determine how much energy is stored and to find where it is located. The energy comes from the work done to charge the capacitor, so first we calculate this work.

Imagine charging a capacitor by transferring charge in small increments from one conductor to the other as illustrated in Fig. 34.13. Suppose that at a certain stage in the process, a positive amount of charge q has been deposited on one conductor, leaving the other with a charge $-q$ and with a potential difference V between the conductors. This potential difference V will depend on the amount of charge q already transferred and we denote it by $V(q)$ to emphasize that it is a function of q. Let ΔW denote the work required to move a small additional positive charge Δq from the negative conductor to the positive one. By Eq. (34.2), this work is given by a line integral,

$$\Delta W = -\Delta q \int_A^B \mathbf{E} \cdot d\mathbf{r},$$

34.6 ENERGY STORAGE IN A CAPACITOR

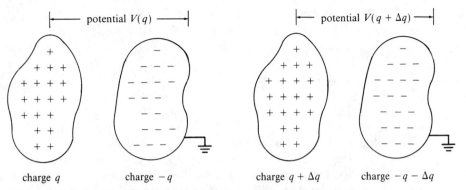

Figure 34.13 Calculating the work done in charging a capacitor.

hence the ratio of work to charge is

$$\frac{\Delta W}{\Delta q} = -\int_A^B \mathbf{E} \cdot d\mathbf{r}. \tag{34.19}$$

But, by Eq. (34.3), the line integral in (34.19) is also equal to the new potential difference $V(q + \Delta q)$ between the conductors, so

$$\frac{\Delta W}{\Delta q} = V(q + \Delta q).$$

Letting Δq shrink to zero, we find

$$\frac{dW}{dq} = V(q).$$

In other words, the derivative of the work with respect to charge is equal to the electric potential. Therefore, to find the total work done in charging a capacitor from 0 to charge Q we simply integrate the potential with respect to charge,

$$W = \int_0^Q V(q)\, dq. \tag{34.20}$$

For a capacitor of any shape, the ratio $q/V(q)$ is a constant C (the capacitance), so $V(q) = q/C$ and the integral in (34.20) now becomes

$$W = \int_0^Q \frac{q}{C}\, dq = \frac{Q^2}{2C}. \tag{34.21}$$

This work is equal to the total energy U stored in the capacitor. If the potential difference across the conductors is V, then $Q/C = V$ and the result in (34.21) can also be written as

$$U = \frac{QV}{2} = \frac{CV^2}{2}. \qquad (34.22)$$

Equation (34.22) holds for a capacitor of any shape.

The energy stored in a capacitor is in the form of electric potential energy. But where is the energy? Does it reside with the charges or is it distributed in the region between the charges? Let's consider the specific example of a parallel-plate capacitor. The result we'll get holds for any capacitor, but by focusing on this simple type we can avoid some of the mathematical complications without losing the essence of the physics.

Consider a parallel-plate capacitor with capacitance C, potential difference V, plate area A, and plate separation distance d. From Eq. (34.14) we know that

$$C = \frac{\varepsilon_0 A}{d},$$

so by (34.22) the energy stored is

$$U = \frac{\varepsilon_0 A V^2}{2d}.$$

By comparing Eq. (34.9) with (34.8) we see that the potential difference $V = Ed$, where E is the field intensity. Hence the energy stored is

$$U = \left(\tfrac{1}{2}\varepsilon_0 E^2\right) Ad. \qquad (34.23)$$

This last result permits us to imagine that the energy is stored in the electric field. The quantity Ad is just the volume of the parallel-plate capacitor. So according to Eq. (34.23) the stored energy is a quantity that depends on the field multiplied by the volume of the region where the energy is stored. The factor $\tfrac{1}{2}\varepsilon_0 E^2$ is the energy per unit volume stored at *each point* between the capacitor plates. This factor is denoted by u and is called the *energy density*,

$$u = \tfrac{1}{2}\varepsilon_0 E^2. \qquad (34.24)$$

It represents the energy per unit volume stored at each point in space where the electric field exists. Although this discussion refers to a parallel-plate capacitor, the concept of energy density is defined by the same equation for capacitors of any shape.

Questions

33. Suppose you have two parallel-plate capacitors having the same areas and voltages, but one has a smaller plate separation.

  (a) Which stores more charge?
  (b) Which stores more energy?

34. How much energy is stored in the electric field between two square plates of 15-cm sides separated by 2.0 mm when the charge on one plate is 200 μC?

35. How much energy is stored in the electric field around an isolated spherical conductor of radius 20 cm charged to 3000 V?

36. A 2- and a 5-μF capacitor are connected in series across a potential difference of 15 V.

  (a) How much energy is stored in each capacitor?
  (b) Suppose that the capacitors are disconnected from the source and reconnected with + to + and − to −. What is the final charge on each capacitor?
  (c) Calculate how much energy is now stored in each capacitor.
  (d) Calculate the total electric potential energy stored in the capacitors for parts (a) and (c). Compare your answers and explain why the two total energies are not equal in the two cases.

34.7 A FINAL WORD

In demonstrating that lightning is electricity, Benjamin Franklin designed two experiments, his famous kite experiment, illustrated in Fig. 34.14, and one in which he placed a lightning rod inside a sentry box. The lightning rod protruded above the center of the box and contained an arrangement that could draw charge from the lightning rod onto a Leyden jar when lightning struck. Franklin understood that this was a very dangerous experiment and prescribed elaborate precautions to protect one's life while performing the experiment.

G. W. Richmann, a German scientist working in St. Petersburg, repeated Franklin's experiment except for one vital ingredient – the precautions. Ball lightning, a pale blue fireball the size of a baseball, descended from the lightning rod and exploded in Richmann's face. He was killed instantly and was found with a burned spot on his forehead and two holes in one of his shoes.

Franklin used caution with his experiments and was able to continue his work in the Colonies. In England he had competitors who claimed to have lightning rods superior to Franklin's. Whereas Franklin thought that the top of a lightning rod should be pointed, his English competitors claimed it should contain a round knob. In the next chapter we shall learn that Franklin had the right idea: At a sharp point the electric field is large and is the preferred configuration for initiating a spark.

Figure 34.14 Benjamin Franklin conducting his kite experiment accompanied by his son. (Courtesy Burndy Library.)

The argument between Franklin and his competitors took place during the War of Independence and was influenced by politics. In an act of solidarity with his countrymen, George III of England issued a royal decree that all lightning rods in England contain round knobs rather than points. He was trying to change the laws of nature by means of a royal decree. As far as we know, he did not succeed.

CHAPTER 35

VOLTAGE, ENERGY, AND FORCE

As frequent mention is made in public papers from Europe of the success of the Philadelphia experiment for drawing the electric fire from clouds by means of pointed rods of iron erected on high buildings, &c. it may be agreeable to the curious to be informed that the same experiment has succeeded in Philadelphia, though made in a different and more easy manner, which is as follows:

Make a small cross of two light strips of cedar, the arms so long as to reach the four corners of a larger thin silk handkerchief when extended; tie the corners of the handkerchief to the extremities of the cross, so you have the body of a kite; which being properly accommodated with a tail, loop, and string, will rise in the air, like those made of paper; but this being made of silk is fitter to bear the wet and wind of a thunder gust without tearing. To the top of the upright stick of the cross is to be fixed a very sharp pointed wire, rising a foot or more above the wood. To the end of the twine, next the hand, is to be tied a silk ribbon, and where the silk and twine join, a key may be fastened. This kite is to be raised when a thunder-gust appears to be coming on, and the person who holds the string must stand within a door or window, or under some cover, so that the silk ribbon may not be wet; and care must be taken that the twine does not touch the frame of the door or window. As soon as any of the thunder clouds come over the kite, the pointed wire will draw the electric fire from them, and the kite, with all the twine, will be electrified, and the loose filaments of the twine will stand out every way, and be attracted by an approaching finger. And when the rain has wetted the kite and twine, so that it can conduct the electric fire freely, you will find it stream out plentifully from the key on the approach of your knuckle...

Ben Franklin in a letter to Peter Collinson, 16 October 1752

35.1 ELECTRIC FIELDS AND POTENTIALS

The various electrical concepts introduced thus far – charges, forces, fields, and potential – help us understand the nature of electricity and its effects. This chapter begins with a further exploration of the relations between these concepts, in particular the relation between fields and potentials.

Recall that electric potential is obtained from the electric field by line integration,

$$V_B - V_A = -\int_A^B \mathbf{E} \cdot d\mathbf{r}, \tag{34.3}$$

the value of the integral being independent of the path joining the points A and B.

We now ask whether the field $\mathbf{E}$ can be recovered from a knowledge of the potential. Specifically, suppose A is a point of zero potential, $V_A = 0$, and suppose we know the potential at every point P, so that

$$V_P = -\int_A^P \mathbf{E} \cdot d\mathbf{r}.$$

We wish to recover $\mathbf{E}$ at the point P. It's not immediately obvious how to do this because $\mathbf{E}$ is a vector quantity whereas the potential V is a scalar. We seek not only the magnitude of $\mathbf{E}$ but its direction as well.

Mathematically, the problem is analogous to recovering a scalar function from a knowledge of its integral. That problem was solved in Chapter 7 by the first fundamental theorem of calculus, which states that

$$f(x) = \frac{d}{dx}\int_a^x f(t)\,dt.$$

In other words, if you know the integral of a function from a fixed point a to a variable point x then the integrand $f(x)$ can be recovered by differentiating the integral with respect to the upper limit x.

We can't apply the first fundamental theorem here because the integral is a line integral and the variable point P is free to move around in space. But we can use the idea behind the first fundamental theorem to solve our problem. Suppose we compare the potential at P with that at a nearby point P'. The potential difference is

$$\Delta V = V_{P'} - V_P = -\int_P^{P'} \mathbf{E} \cdot d\mathbf{r}.$$

If P' is reasonably close to P, the field $\mathbf{E}$ is nearly constant from P to P', so the line integral is nearly equal to $\mathbf{E} \cdot \Delta \mathbf{r}$, where $\mathbf{E}$ is the field at P and $\Delta \mathbf{r}$ is the displacement vector from P to P', as illustrated in Fig. 35.1a. Thus, we have the approximate equation

$$\Delta V = -\mathbf{E} \cdot \Delta \mathbf{r}. \tag{35.1}$$

35.1 ELECTRIC FIELDS AND POTENTIALS

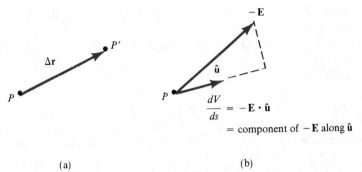

(a)   (b)

Figure 35.1 (a) Difference of potential between two nearby points. (b) The directional derivative of potential related to the field vector.

This gives us the dot product of **E** with a vector $\Delta \mathbf{r}$, but what we really want is **E** itself.

First we note that vector $\Delta \mathbf{r}$ joining P to P' can be written as

$$\Delta \mathbf{r} = \Delta s\, \hat{\mathbf{u}}, \tag{35.2}$$

where $\hat{\mathbf{u}}$ is a unit vector in the direction from P to P', and $\Delta s = |\Delta \mathbf{r}|$ is the distance from P to P'. The approximate equation (35.1) now becomes

$$\Delta V = -\mathbf{E} \cdot (\Delta s\, \hat{\mathbf{u}}), \tag{35.3}$$

with the approximation becoming more exact as Δs becomes smaller. The ratio $\Delta V/\Delta s$ represents the average rate of change of the potential with respect to distance in the $\hat{\mathbf{u}}$ direction. This suggests that we divide by Δs and let Δs shrink to zero to obtain

$$\lim_{\Delta s \to 0} \frac{\Delta V}{\Delta s} = -\mathbf{E} \cdot \hat{\mathbf{u}}. \tag{35.4}$$

The limit on the left of (35.4) is called the *directional derivative* of V in the direction $\hat{\mathbf{u}}$, often denoted by dV/ds. Thus, we can rewrite (35.4) as

$$\frac{dV}{ds} = -\mathbf{E} \cdot \hat{\mathbf{u}}. \tag{35.5}$$

Of course, it is understood that dV/ds depends on the point P as well as the direction $\hat{\mathbf{u}}$, even though the notation dV/ds does not reflect this dependence.

Equation (35.5) shows that the directional derivative of the potential is simply the component of $-\mathbf{E}$ in the direction of $\hat{\mathbf{u}}$, as illustrated in Fig. 35.1b.

If θ is the angle between $-\mathbf{E}$ and the unit vector $\hat{\mathbf{u}}$ we have

$$\frac{dV}{ds} = E \cos \theta \tag{35.6}$$

because $|\hat{\mathbf{u}}| = 1$, so dV/ds has the same sign as $\cos \theta$. From this we see that the directional derivative dV/ds has its largest positive value when $\cos \theta = 1$, that is, when $\hat{\mathbf{u}}$ has the same direction as $-\mathbf{E}$. In other words, at a given point P, the potential undergoes its maximum rate of increase in the direction of $-\mathbf{E}$, this maximum being equal to the field intensity E. Or put another way, the field vector $\mathbf{E}$ points in the direction in which the potential *decreases* most rapidly.

Equation (35.5) can also be used to determine $\mathbf{E}$ explicitly in terms of the potential. We choose a rectangular coordinate system and simply calculate the directional derivative of V in each of the x, y, and z directions. This means that we apply Eq. (35.5) three times, taking $\hat{\mathbf{u}}$ equal to each of the unit vectors $\hat{\mathbf{i}}$, $\hat{\mathbf{j}}$, and $\hat{\mathbf{k}}$. The directional derivative of V in the direction of $\hat{\mathbf{i}}$ is denoted by the symbol $\partial V/\partial x$ and is called the *partial derivative* of V with respect to x. The potential V is a function of three variables x, y, and z (the rectangular coordinates of point P), and the "curly d" in the symbol $\partial V/\partial x$ is used to remind us that we must hold y and z fixed as we differentiate V with respect to x. Therefore, when $\hat{\mathbf{u}} = \hat{\mathbf{i}}$, Eq. (35.5) tells us that

$$E_x = -\frac{\partial V}{\partial x}. \tag{35.7}$$

Similarly, by taking $\hat{\mathbf{u}} = \hat{\mathbf{j}}$ and $\hat{\mathbf{u}} = \hat{\mathbf{k}}$ we find

$$E_y = -\frac{\partial V}{\partial y} \quad \text{and} \quad E_z = -\frac{\partial V}{\partial z}. \tag{35.8}$$

The partial derivative $\partial V/\partial y$ is obtained by holding x and z fixed and differentiating with respect to y, while $\partial V/\partial z$ is obtained by holding x and y fixed and differentiating with respect to z.

Because we know all its components we can write $\mathbf{E}$ itself as

$$\mathbf{E} = -\left(\frac{\partial V}{\partial x}\hat{\mathbf{i}} + \frac{\partial V}{\partial y}\hat{\mathbf{j}} + \frac{\partial V}{\partial z}\hat{\mathbf{k}}\right). \tag{35.9}$$

So this does it. The vector field $\mathbf{E}$ has been recovered from the scalar potential V.

The vector that appears in parentheses on the right of (35.9) is called the *gradient vector* of V and is denoted by the special symbol grad V or by ∇V. (The symbol ∇ is read "del.") Thus, the electric field is the negative of the gradient of the potential,

$$\mathbf{E} = -\text{grad } V \tag{35.10}$$

or

$$\mathbf{E} = -\nabla V.$$

The field intensity E is the length of the gradient vector, given by

$$E = \sqrt{\left(\frac{\partial V}{\partial x}\right)^2 + \left(\frac{\partial V}{\partial y}\right)^2 + \left(\frac{\partial V}{\partial z}\right)^2},$$

whereas the direction of $\mathbf{E}$ is opposite to that of grad V.

Example 1

The electric potential in volts inside a sphere of radius R is given by

$$V(r) = r^2,$$

35.1 ELECTRIC FIELDS AND POTENTIALS

where r is the distance from the center of the sphere. Find the electric field $\mathbf{E}$ at any point inside the sphere.

We describe two methods for determining $\mathbf{E}$, one using Eq. (35.9), and another using Eq. (35.5). To determine grad V from (35.9) we express V in terms of rectangular coordinates and take the partial derivatives to find the components of the gradient vector. If the origin is placed at the center of the sphere, then $V(x, y, z) = x^2 + y^2 + z^2$, so

$$\frac{\partial V}{\partial x} = 2x, \quad \frac{\partial V}{\partial y} = 2y, \quad \text{and} \quad \frac{\partial V}{\partial z} = 2z.$$

Hence,

$$\text{grad } V = 2x\hat{\mathbf{i}} + 2y\hat{\mathbf{j}} + 2z\hat{\mathbf{k}} = 2\mathbf{r},$$

where $\mathbf{r} = x\hat{\mathbf{i}} + y\hat{\mathbf{j}} + z\hat{\mathbf{k}}$ is the radius vector from the origin to the point (x, y, z). This gives the field vector

$$\mathbf{E} = -\text{grad } V = -2\mathbf{r}.$$

For this example there is an alternate way to determine $\mathbf{E}$. By symmetry, the field vector $\mathbf{E}$ must be directed radially, and we also know from Eq. (35.5) that it has the direction in which the potential decreases most rapidly. But $V(r)$ increases as r increases, so $\mathbf{E}$ has the opposite direction to $\mathbf{r}$. The field intensity E is equal to the directional derivative of V in the outward direction, which, in this case, is simply $dV/dr = 2r$. Therefore, $\mathbf{E} = -2\mathbf{r}$, as obtained above.

Example 2

The electric potential at a distance r from a point charge q is given by $V(r) = K_e q/r$. Use this to obtain the electric field.

We could express V in terms of rectangular coordinates as was done in Example 1, but it is simpler to exploit the symmetry. As in Example 1, the electric field must be radially directed in the direction in which $V(r)$ decreases most rapidly. In this example, $V(r)$ decreases as r increases so $\mathbf{E}$ is directed radially outward. Moreover, the directional derivative of V in the outward radial direction is simply dV/dr. This gives us

$$\frac{dV}{dr} = K_e q \frac{d}{dr}\left(\frac{1}{r}\right) = -K_e \frac{q}{r^2}.$$

Therefore,

$$\mathbf{E} = K_e \frac{q}{r^2}\hat{\mathbf{r}},$$

in agreement with Eq. (33.1), which is our previous result for the field of a point charge. The positive sign indicates that the field points radially away from the positive charge q.

Example 3

In Example 4 of Chapter 33 we determined the electric field at an arbitrary point on the axis of a uniformly charged loop of radius a lying in the xy plane and found that

$$\mathbf{E} = K_e \frac{Qz}{(a^2 + z^2)^{3/2}} \hat{\mathbf{k}},$$

where $Q = 2\pi a\lambda$ is the total charge on the loop, with λ the charge density. Use this to determine the potential on the axis.

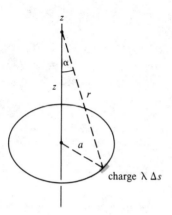

From Eq. (35.5) with $\hat{\mathbf{u}} = \hat{\mathbf{k}}$ we find that the directional derivative of the potential in the positive z direction is

$$\frac{dV}{dz} = -\mathbf{E} \cdot \hat{\mathbf{k}} = -K_e Q \frac{z}{(a^2 + z^2)^{3/2}}.$$

But, as already noted in Example 5 of Chapter 33, the fraction that multiplies $-K_e Q$ on the right is the derivative of the function

$$f(z) = -(a^2 + z^2)^{-1/2}.$$

Thus $V(z)$ and $-K_e Q f(z)$ have the same derivative, so they differ only by a constant. By assigning zero potential to points infinitely far away we see that the constant is zero and hence $V(z) = -K_e Q f(z)$ or

$$V(z) = \frac{K_e Q}{(a^2 + z^2)^{1/2}}.$$

Conversely, if we were given this formula for the potential $V(z)$ we could determine $\mathbf{E} \cdot \hat{\mathbf{k}}$, the component of the field intensity in the positive z direction by taking the negative of the derivative dV/dz.

35.2 EQUIPOTENTIAL SURFACES

Questions

1. Suppose that there is a charge $+Q$ at $x = 0$ and a charge $-Q$ at $x = b$.
   (a) What is the electric field at the point midway between the two charges?
   (b) What is the potential at the midpoint?
   (c) Explain your answer to (b) in light of Eq. (35.5).

2. Must you know the electric potential in a region about a point P in order to calculate the electric field at P, or is it sufficient to know its value at the single point P?

3. A point charge of 5.0 µC is located at $x = 0$.
   (a) Calculate the electric potential at $x = 0.10$ m and at $x = 0.12$ m.
   (b) Compute $-\Delta V/\Delta x$ for the points in (a).
   (c) Find the percentage difference between the answer to part (b) and the value of the electric field at 0.11 m.

4. The electric potential at a distance r from the origin is given by
   $$V(r) = 20r - 50,$$
   where V is in volts and r in meters.
   (a) Where is the potential zero?
   (b) Find the electric field at $r = 0.2$ m.

5. The electric potential inside a sphere of radius R is given by
   $$V(r) = \alpha r^4,$$
   where r is the distance from the center of the sphere and α is a constant that depends on R and on the total charge inside the sphere.
   (a) Find the electric field at any point inside the sphere.
   (b) Is the charge uniformly distributed through the sphere? Justify your answer.

6. The potential in a region of space is given by $V(x, y, z) = 3x^2 - 5y + 2z^3$. Find (a) the potential and electric field at $(0, 2, -1)$; (b) the potential and electric field at $(-2, 1, 0)$; (c) the work done in moving a 5-µC charge from $(0, 2, -1)$ to $(-2, 1, 0)$.

7. Let $\mathbf{r} = x\hat{\mathbf{i}} + y\hat{\mathbf{j}} + z\hat{\mathbf{k}}$ and let $r = |\mathbf{r}|$. If $V(r) = r^n$, where n is an integer (positive or negative), show that the gradient of V is $nr^{n-2}\mathbf{r}$. (This problem can be done in rectangular coordinates, but it is easier to exploit the symmetry.)

35.2 EQUIPOTENTIAL SURFACES

From the fact that the directional derivative of potential in the direction $\hat{\mathbf{u}}$ is given by

$$\frac{dV}{ds} = -\mathbf{E} \cdot \hat{\mathbf{u}}, \tag{35.5}$$

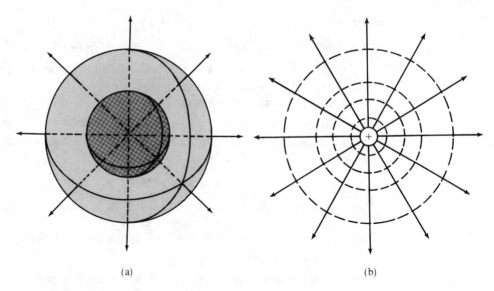

Figure 35.2 (a) Equipotential surfaces for a point charge are concentric spheres. (b) In a two-dimensional cross section they appear as circles.

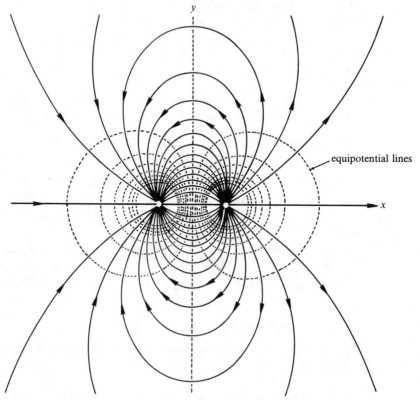

Figure 35.3 Equipotential lines (shown dashed) and field lines for an electric dipole.

35.2 EQUIPOTENTIAL SURFACES

we see that the directional derivative is zero only if **E** is zero or if the unit vector **û** is in a direction perpendicular to **E**. The derivative will also be zero along any portion of space in which the potential is constant. The collection of all points in space at which the potential V has a constant value usually determines a surface, called an *equipotential surface*. The field lines that pierce such a surface will always be perpendicular to it. For example, when the field is generated by an isolated point charge, an equipotential surface satisfies the condition $V = K_e q/r =$ constant, which implies $r =$ constant. The equipotential surfaces in this case are a family of concentric spheres, as illustrated in Fig. 35.2a. The intersection of an equipotential surface and a plane is a curve called an *equipotential line*. The potential is, of course, constant everywhere on an equipotential line so the electric field is perpendicular to the equipotential line at each point. The concentric circles in Fig. 35.2b are equipotential lines obtained by intersecting the equipotential surfaces by a plane through the charge. Figure 35.3 illustrates the equipotential lines for an electric dipole.

In Chapter 33 we found that the electric field is always zero inside a conductor. Therefore the potential is constant everywhere inside a conductor and is equal to the value on the surface. This example shows that the set of points of constant potential can be a solid rather than a surface. The surface of the conductor is an equipotential surface, and also any surface lying inside the conductor is an equipotential surface.

Example 4

(a) What are the equipotential surfaces for a charged parallel-plate capacitor?

(b) What is the distance between an equipotential surface at 30 V and one at 40 V if the potential difference across the capacitor is 100 V and the plates are separated by 2.0 mm?

(a) We already know that inside a parallel-plate capacitor the electric field is uniform, as indicated. The equipotential surfaces are perpendicular to the field lines, so they consist of all the planes parallel to the plates, as indicated in the diagram.

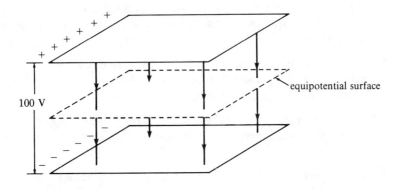

(b) We already know from Eq. (34.9) that the potential inside a capacitor varies as $V(z) = (\sigma/\varepsilon_0)z$, where z is the distance from one plate. When $z = 2.0$ mm and $V = 100$ V, we find $\sigma/\varepsilon_0 = V/z = 5.0 \times 10^4$ V/m. For two equipotential planes a distance Δz apart we also have $\Delta V = (\sigma/\varepsilon_0) \Delta z$, so for $\Delta V = (40 - 30)$ V $= 10$ V we find

$$\Delta z = \Delta V / (\sigma/\varepsilon_0) = (10 \text{ V})/(5 \times 10^4 \text{ V/m}) = 0.2 \text{ mm}.$$

Questions

8. Sketch a few electric field lines and equipotential lines for the case of a charged sphere near a conducting plane as illustrated.

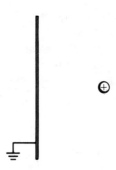

9. Show that equipotential surfaces do not usually intersect. When do exceptions occur?

10. A solid sphere of radius 0.1 m carries a total charge of 4.0 μC distributed uniformly.

  (a) Determine whether or not the potential difference between a concentric sphere of radius 0.2 m and one of radius 0.3 m is the same as that between a sphere of radius 0.3 m and one of radius 0.4 m.

  (b) If the potential difference increases by increments of 10 V on a collection of spherical equipotential surfaces, are the corresponding surfaces equally spaced?

11. An equipotential surface for a charged sphere has a radius of 0.15 m and its potential is 20 V greater than that on another surface with radius 0.05 m larger. What is the charge on the sphere?

12. A sheet of charge has a surface charge density of 5.0 μC/m² and zero potential. Find the distances between equipotential surfaces that have potentials of -100, -200, -300, and -400 V.

13. A very long conducting cylinder of radius 3.0 cm has a potential of 200 V on its surface. If the equipotential surface with a potential of 150 V is a cylinder of radius 8.0 cm, find the radius of the equipotential surface at 100 V.

35.3 VOLTAGES IN THE WORLD

The sparks from a Van de Graaff generator are more spectacular than those from a Wimshurst machine, yet the sparks from a Wimshurst machine can do more damage. The reason for this is the connection between potential energy, charge, and voltage.

We recall that the electrical potential difference ΔV is the work done per unit of positive charge, so the corresponding change in potential energy ΔU of any charge q is simply

$$\Delta U = q \, \Delta V. \tag{34.4}$$

If we know the potential difference ΔV and the size of the charge, then we know how much energy is available for use. This simple relation helps to explain the difference between a Van de Graaff generator and a Wimshurst machine.

Both the Van de Graaff generator and the Wimshurst machine can produce high voltages, but the latter is much more dangerous because it has Leyden jars to store charge. The Van de Graaff generator has no mechanism analogous to Leyden jars for storing charge, aside from the capacitance of its dome with respect to infinity, which is relatively small (see Question 26 in Chapter 34). Therefore, if two machines produce the same voltage, Eq. (34.4) shows that the one that stores the greater charge has greater energy available to do work. If the total charge q is close to zero, the product $q \, \Delta V$ can still be small even if ΔV is 100,000 V.

Why doesn't a common battery produce a shock? A typical battery used in a flashlight or in a portable radio has a meager voltage of 1.5 V. Sparks don't readily jump at that potential and no precautions need be taken in handling ordinary batteries. However, a battery can move an immense amount of charge. Consequently in a battery the amount of energy available for doing electrical work is huge compared to the amount of energy available from either a Van de Graaff generator or a Wimshurst machine. Chapter 36 describes the internal workings of a battery.

To gain more insight into the electrical workings of the world, let's examine typical voltages that we are exposed to in everyday life. One of the most important has to do with the structure of atoms. Imagine an atom as being composed of a positively charged nucleus surrounded by a spherical cloud of electrons. If one electron is missing from the cloud, then the atom becomes a positive ion having a charge $+|e|$.

Now suppose the missing electron is somehow returned to the immediate vicinity of the ion. The electron is attracted by the electric force from the positively charged ion. Remember that the field generated by a spherical negatively charged electron cloud is like that of a single point charge placed at its center, with the potential at a distance r from the point charge being $-K_e|e|/r$. Consequently, it is a simple matter to determine the electric force between the electron and the ion.

By Eq. (34.4), as the electron moves toward the ion it loses potential energy according to $\Delta U = q \, \Delta V$, or

$$\Delta U = (-|e|)\left(K_e \frac{|e|}{r}\right),$$

and so there is a natural potential $\Delta V = \Delta U/(-|e|)$ associated with the atom. It is

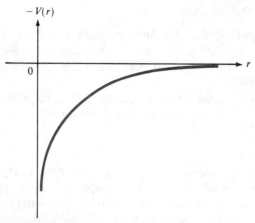

Figure 35.4 Graph of the electric potential of an ion as a function of distance from the nucleus.

extremely important that this potential be relatively large. Why? You, the chair you're sitting on, and all other material objects are made of atoms. If the energy that makes the electron prefer to stay inside the atom weren't large compared to the energy of, say, a Van de Graaff generator, then anytime a Van de Graaff generator were turned on everything nearby would ionize. So the potential in the vicinity of a positive ion is an important voltage.

Although one might expect the natural potential in the vicinity of an isolated ion to be greater than that on a Van de Graaff dome, we will find that quite the contrary is true. To obtain a reasonable estimate of the potential near an ion, we begin with the formula

$$V(r) = K_e \frac{|e|}{r}.$$

Figure 35.4 shows a graph of the negative of the potential as a function of r. (We plot the negative of the potential because we want to emphasize that we are dealing with attractive forces: here the potential is actually positive and the charge of the electrons is negative.)

Although the potential becomes very large at extremely small distance, the electron never reaches the center of the ion. Instead it remains in the electron cloud but not because of the electric force – that force would make the electron orbit collapse to the center of the nucleus. The reasons are quantum mechanical and we'll explore them in Chapter 51. But for now we simply use the fact that the electron remains at a distance from the nucleus on the order of the size of the atom, about 1×10^{-10} m. We can easily calculate the potential at this distance and get

$$V = K_e \frac{|e|}{r} = (9 \times 10^9 \text{ N m}^2/\text{C}^2)(1.6 \times 10^{-19} \text{ C})/(1 \times 10^{-10} \text{ m}) = 14.4 \text{ V}.$$

35.3 VOLTAGES IN THE WORLD

Thus we see that this voltage is not large, as might have been expected, but it is quite small compared to the potential of an electrostatic generator. In fact, it's comparable to that of ten flashlight batteries. Why then aren't electrons torn away from atoms?

The voltages typical of an ion–electron system that constitutes an atom are called *ionization potentials* because they are the potential differences through which an electron must pass in order to gain enough energy to escape from the atom and leave an ion behind. The typical ionization potential for any atom or molecule is of the order of 1 to 10 V. The voltage of a battery is 1.5 V because of the differences in ionization potentials between atoms inside it. In Chapter 36 we'll examine in detail how batteries work. In any case, the ionization potentials are quite small. Then why doesn't an electron separate from an atom and move to the dome of a Van de Graaff machine, which is at much lower potential energy?

The reason is that the energy of the electron is not the only important physical quantity to contend with. The *force* on the electron must be taken into account as well. When the electron is in the atom it feels a force due to the electrical attraction of the ion. Clearly the force between the ion and the electron must be greater than that between the Van de Graaff dome and the electron. This force is determined by the field, so let's examine the relevant fields.

For a spherically symmetric charge distribution we know that the electric field intensity is

$$E = K_e \frac{q}{r^2} \tag{33.1}$$

and the electric potential is

$$V = K_e \frac{q}{r}. \tag{34.6}$$

For this case we see that

$$E = \frac{V}{r}, \tag{35.11}$$

which can be regarded as just a special case of Eq. (35.1). Since we know the potential for both the atom and the Van de Graaff machine as well as the distance r from each to the electron, we can compare the electric fields.

For a typical demonstration Van de Graaff machine the potential is $V = 10^5$ V, and r is the radius of the dome, which is about 25 cm. According to Eq. (35.11), the electric field at the surface of the Van de Graaff dome is about 4×10^5 V/m.

Now for the atom, the potential is 14.4 V and the distance from the ion to the electron is 2×10^{-10} m. According to Eq. (35.11) the electric field due to the atom is on the order of 7×10^{10} V/m. Although the energy of the electron would be lower if it moved to the Van de Graaff dome, the force binding the electron to the atom is 10,000 times greater than the force of the Van de Graaff dome trying to pull the electron away. Electrons simply can't escape easily from the atoms to which they are bound. That's why electrons remain attached to atoms and matter is relatively stable.

Figure 35.5 Potential wells formed by atoms in a metal lattice.

Now let's see what happens to an electron bound to a collection of atoms. Instead of a simple molecule consisting of two atoms, let's consider a molecule formed from a huge number of atoms, all of which share their outermost electrons. Suppose that the atoms are located at fixed positions and form an ordered geometric array – a crystal – that is repeated throughout the material. This picture is an excellent representation of the structure of a piece of ordinary metal. Although the material is electrically neutral, outer electrons are only weakly bound to any one nucleus and may wander from one atom to another as *free electrons*. We can think of these outer electrons as not belonging to any one nucleus but rather to the crystal as a whole.

Each atom creates in its immediate vicinity a potential attractive to electrons, known as a potential well, similar to that in Fig. 35.4, and the sheer number of atoms producing such wells creates a periodic potential like that shown in Fig. 35.5. The outermost electrons from each atom wander freely through this periodic potential. That is our description of a metal.

Although electrons are free to move around inside the metal, at the surface there is a microscopic local field that prevents the electrons from escaping. The size of that local field is just about the same as the field binding an electron to a single atom. So, the electrons can easily respond to fields and move around inside a metal conductor but cannot easily escape from its surface.

The existence of large forces on electrons in an atom raises another question. How are sparks created by a Wimshurst machine? The potential at one of the spherical electrodes of a typical machine is on the order of 50,000 V, and the radius of an electrode is about 2 cm. From Eq. (35.11), the electric field at the surface of the electrode is about 5×10^6 V/m. Even though the potential created by a Wimshurst machine is very large compared to the potential in atoms or in a metal, the electric field is comparatively small. The forces created by the Wimshurst machine are too small to yank electrons out of the metal or to take an atom or molecule in the air surrounding it and pull an electron off of it. Then why does sparking occur when the potential reaches the order of 50,000 V?

There is plenty of energy available to separate atoms from their electrons, that is, to ionize atoms. We discovered that feature earlier. However, the forces are too small to ionize atoms. Ionization occurs through a somewhat different mechanism. In the vicinity of the Wimshurst machine air molecules are flying around and, although most of the molecules are neutral, there are a few stray electrons. Imagine an electron attracted to a positive electrode of a Wimshurst machine. Because the

35.3 VOLTAGES IN THE WORLD

Figure 35.6 Corona plumes from conductors energized at high voltages (Courtesy General Electric High Voltage Transmission Research Facility.)

electron feels a force, it accelerates and gains energy. The chances are good that the electron, as it accelerates toward the positive electrode, will collide with another air molecule and ionize that molecule. If the electron is able to gain an energy of tens of electron volts before the collision, it is quite capable of knocking an electron off the molecule. Consequently, another electron is created and both of them are accelerated and collide with other air molecules, which become ionized, and so on. The whole process cascades, creating an avalanche of electrons that momentarily turns the air into a conductor. For a brief instant a vast number of positive ions are separated from their electrons that are free to move around, and conduct electricity by moving to the positive electrode where they combine with the excess electrons on the electrode of the Wimshurst machine. That's the spark. The light given off is emitted by the air molecules as electrons recombine with them. The air is momentarily a charged gas, known as a plasma, that releases the potential of the Wimshurst machine. If a steady discharge takes place and results in a visible glow or halo, the discharge is known as *corona discharge*. They are sometimes observed around power lines. Figure 35.6 shows corona plumes from conductors energized at high voltages in the laboratory. The plumes occur where there are water droplets protruding from the conductor surface. Each plume is the visual result of many individual streamers that follow at first a straight path (the stem of the plume) and then a random path up to a few inches from the conductor surface.

Example 5

The maximum electric field that dry air can sustain, known as the breakdown field E_{max}, is about 3 MV/m. To what maximum potential can a Van de Graaff dome of radius 0.20 m be raised before a spark is formed in dry air?

From Eq. (35.11) the field and potential at the surface of a conducting sphere are related by $V = Er$, so the maximum voltage that can be sustained is

$$V_{max} = E_{max}r = (3 \times 10^6 \text{ V/m})(0.20 \text{ m}) = 600{,}000 \text{ V}.$$

If a greater charge is placed on the Van de Graaff dome, thereby increasing the potential, there will be a discharge.

Example 6
Why is it more difficult to get a large (juicy) spark from an electrostatic generator on a humid day than on a dry day?

If there's an appreciable number of water molecules in the air, the water ionizes more easily than the nitrogen and oxygen molecules of air. Consequently, the cascade takes place at a somewhat lower voltage. So on a humid day it's more difficult to have an electrostatic generator reach a high voltage; water molecules "drain off" the stored charge on the machine before it becomes large.

Questions

14. After a lightning flash in a thunderstorm there is often a copious amount of rainfall called a cloud burst. In light of your understanding of the formation of sparks, what, if any, is the connection between the lightning and the cloud burst? (See *Scientific American*, July 1985, page 62.)

15. An electrostatic precipitator is a powerful air cleanser that can extract 99% of the ash and dust from gases in chimneys of industrial plants. A precipitator consists of a vertical metal duct that is grounded and a wire running down the center of the duct that is kept at a high voltage, as shown. Dirty gas enters through the lower portion of the duct, and clean gas exits from the upper part. Explain the simple physical principles at work in the precipitator.

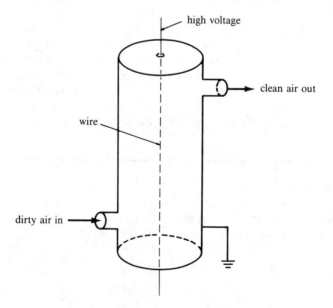

35.4 CHARGE DISTRIBUTION ON CONDUCTORS

16. A popular demonstration device consists of a metal pinwheel with sharp metal points that is made to rotate by connecting it to a high voltage source, as shown. Assume that the voltage source is negative and that the field it creates is not strong enough to strip electrons off of atoms.

(a) Will the pinwheel rotate in a vacuum?
(b) What forces cause it to rotate?

17. How much charge can be placed on a Van de Graaff dome of radius 0.10 m in dry air before it will discharge?

18. Consider n alternating positive and negative charges placed on a line, each of magnitude $|e|$ and each at a distance a from its neighbor, as shown.

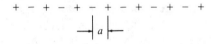

(a) Find the electric potential at the point where the next charge would be placed.
(b) If n is odd, let U_n denote the potential energy of the charge at the center of the configuration. Show that as $n \to \infty$, U_n approaches the value

$$U = -K_e \frac{|e|^2}{a} 2 \ln 2.$$

(*Hint:* The sum of the infinite series $1 - \frac{1}{2} + \frac{1}{3} - \frac{1}{4} + \cdots$ is $\ln 2$.)

(c) If the spacing $a = 2 \times 10^{-10}$ m, determine the value U in eV.

35.4 CHARGE DISTRIBUTION ON CONDUCTORS

Why are the ends of lightning rods pointed? Would blunt rods be just as effective? When you walk across a carpet on a dry day and reach for a doorknob, why does a spark jump from your finger? Why doesn't it jump from the top of your hand? These questions can be answered by considering the dependence of the electric field on the curvature of a surface.

In general, when two charged conductors are brought into contact, the charge redistributes itself so that the electrostatic field is zero inside both conductors. However, the electric field will not be the same everywhere on the surfaces of the

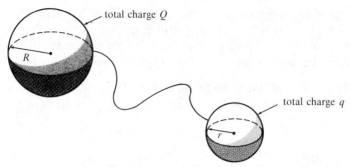

Figure 35.7 A nonspherical conductor formed from two conducting spheres connected by a wire.

conductors. As we discovered in Chapter 33, it depends on the local surface charge density according to

$$E = \frac{\sigma}{\varepsilon_0}. \tag{33.27}$$

The electric potential, though, is the same everywhere on a conductor. Remember that the surface of a conductor is an equipotential surface. This characteristic helps us answer our questions.

To find the effects of curvature on charge density, imagine a nonspherical conductor as being made up of two conducting spheres attached by a wire, as shown in Fig. 35.7. Assume that the two spheres are sufficiently far apart so that the effect of one on the other is negligible. It is difficult to analyze a conductor of general shape, such as your hand, but we can gain some insight by using this simple model.

Suppose the larger sphere has radius R and total charge Q, while the smaller sphere has radius r and total charge q. Let's compare the charge densities on the two spheres. Because the potential is the same on the two spherical surfaces, we have

$$V = K_e \frac{Q}{R} = K_e \frac{q}{r}.$$

This tells us that the ratio of charge to radius is the same on the two spheres: $Q/R = q/r$. But the surface charge densities on the spheres are given by

$$\sigma_R = \frac{Q}{4\pi R^2} \quad \text{and} \quad \sigma_r = \frac{q}{4\pi r^2},$$

and their ratio is

$$\frac{\sigma_r}{\sigma_R} = \frac{q/(4\pi r^2)}{Q/(4\pi R^2)} = \frac{(q/r)}{(Q/R)} \frac{R}{r} = \frac{R}{r}.$$

But $R > r$, so

$$\sigma_r > \sigma_R.$$

35.4 CHARGE DISTRIBUTION ON CONDUCTORS

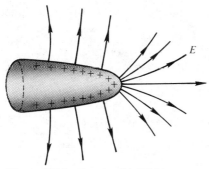

Figure 35.8 The electric field is greater near points of small radius of curvature.

In other words, the surface charge density is greater on the smaller sphere, and so by Eq. (33.27) the electric field is larger at the surface of the smaller sphere. This suggests that a conductor that is blunt at one end and pointed at the other will have a stronger electric field at the pointed end. Figure 35.8 illustrates this feature of electric fields.

Because the electric field is strongest near points of greatest curvature, sparking is more likely to occur at those points. That's why a spark jumps from your finger and not from the palm of your hand. It also explains why Benjamin Franklin was correct in claiming that lightning rods should be pointed: Pointed rods more effectively discharge the potential built up between ground and clouds in a thunderstorm.

Questions

19. A triangular cross section of a charged conductor in the shape of a right circular cone is shown in the accompanying diagram. Sketch a few electric field lines and equipotential lines to indicate the charge distribution on the conductor.

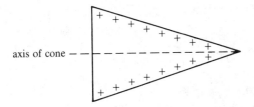

20. It is commonly believed that lightning seeks out oak trees. In fact, a strikingly high proportion of trees destroyed by lightning are trees with rough bark. If the lightning strike occurs early in the rainstorm, a rough-bark tree may be shattered while a nearby smooth-bark tree is unharmed. Why do you suppose this is true?

21. The opening quotation for this chapter describes Franklin's famous kite experiment.

   (a) Why do you think he put a pointed wire on the kite?
   (b) Why did he use a silk ribbon between his hand and the key?
   (c) Why did he use a key?
   (d) Why wasn't Franklin killed?

22. A charge of 30 µC is distributed over two spheres of radius 0.3 and 0.2 m that are connected by a wire.

   (a) Determine the charge on each sphere.
   (b) Find the electric field at the surface of each sphere.

23. A conducting sphere of radius 2.0 cm is given a charge of 3 µC and then connected by a wire to a distant sphere of radius 5.0 cm that is initially uncharged. After the spheres have been connected find (a) the charge on each sphere; (b) the potential of the spheres; and (c) the electric field at the surface of each sphere.

35.5 A FINAL WORD

The theory of electricity is based on a few fundamental facts: There are positive and negative charges. Like charges repel and opposite charges attract. The forces between charges obey the inverse-square law. Yet it took a long time to formulate a theory based on these facts. One of the difficulties in studying electricity is that the basic constituents of the theory, electric charges in a vacuum, are not commonly found in nature. In the real world, electricity always exists as part of material substances and you can't really understand the behavior of electricity without first understanding the behavior of matter. To make things worse, all matter is essentially electrical so you can't understand how matter behaves without first understanding electricity.

At the beginning of the nineteenth century there were two great chemists living in England, John Dalton and Humphrey Davy. Dalton thought that all matter was made up of atoms, with each chemical element having a different kind of atom. That was a very old idea, but Dalton's law of simple and multiple proportions gave the atomic theory a solid scientific basis. Throughout the nineteenth century there were many chemists who accepted the law of simple and multiple proportions but who refused to believe in atoms. They didn't think science should be based on something that couldn't be seen. However, Humphrey Davy had a completely different reason for not believing in Dalton's atoms.

Of the 40 chemical elements known at the time (some of which had been discovered by Davy), 26 were metals. All metals share certain properties, such as surface luster, ductility, the ability to conduct heat, and above all, the ability to conduct electricity. That couldn't be just an accident that happened 26 times to 26 different kinds of atoms. Davy thought there must be some underlying principle of metallization, and that the atoms themselves could not be the ultimate indivisible consitutents of matter.

35.5 A FINAL WORD

It turns out that both Dalton and Davy were right. Matter is made of atoms, one for each element, but they are not indivisible. They are made of internal parts held together by electrical forces. Depending on the details of their structure, they combine to form either electrical insulators or electrical conductors, the conductors being mostly metals.

Long before the structure of atoms was understood, in fact, even before Dalton and Davy began speculating about it, an ingenious Italian named Alessandro Volta invented a number of electrical devices. One of these, the electric battery, transformed the subject of electricity into a valuable tool of science. That story is described in the next chapter.

CHAPTER 36

THE ELECTRIC BATTERY

> After a long silence, which I do not attempt to excuse, I have the pleasure of communicating with you, Sir, and through you to the Royal Society, some striking results to which I have come in carrying out my experiments on electricity excited by the simple mutual contact of metals of different sorts, and even by the contact of other conductors, also different among themselves, whether liquids or containing some liquid, to which property they owe their conducting power. The most important of these results, which includes practically all the others, is the construction of an apparatus which, in the effects which it produces, that is, in the disturbances which it produces in the arms etc., resembles Leyden jars, or better still electric batteries feebly charged, which act unceasingly or so that their charge after each discharge reestablishes itself; which in a word provides an unlimited charge or imposes a perpetual action or impulsion on the electric fluid....
>
> Alessandro Volta, letter of 20 March 1800 to Sir Joseph Banks

36.1 FROG LEGS AND ELECTRICITY

Until the end of the eighteenth century most electricians concentrated their efforts on electrical devices that produced impressive sparks and flashes of light. Instead of breaking new ground, they sought to improve existing devices or searched for even more startling electrical effects to delight audiences. That is, until Luigi Galvani, a professor of anatomy at the University of Bologna, made a leaping advance with frog legs.

Galvani noticed that freshly dissected frog legs were convulsed by muscular contraction when the thigh nerve was touched with a scalpel or an iron rod (but not with glass) while at the same time a spark was drawn from a nearby electrostatic generator. Intrigued by the startling phenomenon, Galvani speculated that the interior and exterior muscle tissue formed a small Leyden jar with the nerve acting as a conductor for the jar. He advanced the idea that animals could be the source of "animal electricity," different from ordinary electricity. Perhaps, Galvani thought, he had uncovered the elusive life force.

In 1791 Galvani published an extensive study on animal electricity. But his theory was undermined by one of his own experiments. Attempting to test whether atmospheric electricity could cause a frog leg to twitch, he pierced a frog leg with a brass hook. As soon as the hook was attached to an iron rack the leg moved without atmospheric electrical disturbance. Spasms also occurred with other combinations of dissimilar metals but not when nerve and muscle were joined by a single metal. This was unlike a Leyden jar, which would be discharged by any metal conductor.

News of Galvani's discovery reached Alessandro Volta, a professor at the University of Pavia, who had great interest in electricity. He had already made several advances in electricity, including a theory of capacitors and the idea of electric tension, which later evolved into electric potential. Searching for a possible source of animal electricity, Volta repeated some of Galvani's experiments on a host of unlucky creatures ranging from headless grasshoppers to bodyless cows. In 1792 he concluded that the source of the electricity was not from the animal itself but from the contact of nerves with dissimilar metals. Volta dismissed Galvani's conclusions and set out to prove that animal electricity was no different from that created by electrostatic machines. His quest was analogous to Franklin's campaign to show that lightning was the same as electricity produced by machines. Volta's search, however, was much more difficult. He did not know that electrostatic machines were high voltage devices, nor that a frog's leg had low voltage. Nonetheless, he discovered that frog legs were by far the most sensitive detectors of charge then available.

To demonstrate the role of the nerve, Volta speculated that the sense of taste could be stimulated by placing two different metals on the tongue. He triumphantly perceived a strong acid taste when he touched a bit of tin to the tip of his tongue

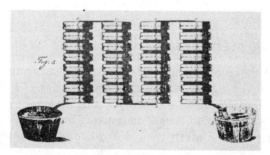

Figure 36.1 The first electric battery, the Voltaic pile.

36.2 THE WORKINGS OF METALS

Figure 36.2 Volta's crown of cups.

while a silver coin rested further back on the tongue. Touching dissimilar metals to other parts of the body could also cause the sensation of light in the eye. Volta recognized that the crucial elements in all these experiments were two dissimilar metals joined by a moist conductor. In his experiments the nerve played the role of the moist conductor. He conjectured that the flow of electric charge was due to an electrical imbalance between the two metals.

Galvani thought Volta's claims were preposterous because they implied that a single scudo (an Italian coin) would contain enough electricity to move the leg of a horse. Undaunted by the attacks of Galvani and his followers, Volta continued his research and in 1800 he announced that he could multiply the electrical effect by placing silver and zinc discs alternately in contact and separated by moistened cardboard (see Fig. 36.1). This arrangement, known as a "Voltaic pile," was the first electric battery. Another arrangement that achieved the same effect was his "crown of cups," a collection of tumblers containing brine and connected to each other by a series of metallic plates of zinc and silver (or copper) alternately in contact, as shown in Fig. 36.2.

36.2 THE WORKINGS OF METALS

Volta's exhaustive research revealed that an operational battery could be made by separating two dissimilar metals by a moist conductor. To understand the physics of the battery we begin by recalling various descriptions of metals and conductors discusssed in earlier chapters.

First, from an electrostatic point of view, a conductor is a region of constant potential. If there are electric charges in the vicinity, or if the conductor itself has a net charge, there will be an electric field, and hence a nonuniform electrostatic potential outside the metal. But inside the metal the electrostatic field is everywhere zero, and the electrostatic potential is uniform, the same everywhere. This occurs because the electrons in the metal can only be at equilibrium if there is no field inside to apply a force to them.

The electrostatic field is theoretically detected by a test charge, a fictitious creation that applies no field of its own. As detected by a test charge, the static field of an electrically neutral piece of metal is zero everywhere, inside and out, and the static potential is the same everywhere, inside and out. However, suppose a real charge, say a negative ion, is brought near an uncharged metal conductor. The ion repels negatively charged electrons in the metal and thereby causes the charges on the surface of the conductor to rearrange themselves, leaving fixed positive ions behind. Having done that, the ion is then attracted by the induced positive charges.

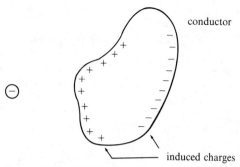

Figure 36.3 An ion brought near a conductor is attracted by the induced positive charges.

In other words, there is an electric field and a force that attracts the ion toward the surface, as illustrated in Fig. 36.3

Example 1

Suppose a large uncharged plane sheet of metal is first connected to ground, and then a negative point charge $-Q$ is brought near the sheet, as shown.
    (a) Determine the potential on the sheet.
    (b) Describe qualitatively the charge distribution on the metal sheet.
    (c) Why is the point charge attracted to the metal?

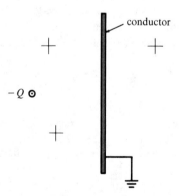

    (a) Because the metal is a conductor, it is a region of constant potential, this constant being zero because the metal is grounded.
    (b) Initially there is no charge on the sheet, but as the negative charge approaches it, negative charges are immediately repelled from the metal sheet to ground (in a time scale of 10^{-8} s) leaving a net positive charge on the sheet. The total positive charge on the sheet is at most $+Q$, but it is not uniformly distributed

over the sheet; more charge is concentrated in regions nearer to the point charge as suggested by the spacing of field lines shown in the diagram.

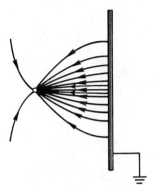

(c) The negative point charge is attracted to the conducting plane by the electric force between it and the induced positive charges. The nearer the point charge is to the plane, the greater the force of attraction.

If the real charge is an *electron*, an additional phenomenon is revealed. While the electron is outside the metal, it is attracted to the surface for the reasons described in Example 1. However, as the electron approaches within atomic distances to the metal's atoms, it is indistinguishable from the electrons in the metal and thus is capable of entering the metal itself and becoming part of the metal's electron system. Only electrons can merge into the sea of free electrons in the metal and give the metal a net charge. If the charge were not an electron but, for example, a tightly bound negative ion, it would not be able to move freely through the lattice.

In Chapter 35 we described a simple model of metals as an array of positive ions held together in a rigid lattice by a cloud of mobile electrons. An equivalent picture is to say that a metal is a giant molecule. The inner electrons of each atom are tightly bound to a single nucleus, forming the positive ions, but the outermost one or two electrons, electrically shielded by the inner ones, don't know which atom they originally came from and are free to wander throughout the metal. These are called conduction electrons. In this electrical soup there is thus a huge quantity of both immobile positive and mobile negative charge. There are powerful electric fields very close to each ion and each electron, but averaged over a few atoms the net field is essentially zero. Otherwise forces would be applied to the mobile electrons making them readjust their positions until the net field becomes zero.

However, at the surface of a metal the situation is different. Here, the powerful field binding an electron to an ion near the surface is not balanced by an equally powerful field drawing it to the outside since there is no nearby ion outside. The situation is illustrated in Fig. 36.4. An electron can easily move from ion 3 to ion 2 because it's equally attracted to both, but it can't move to the left of ion 1 because

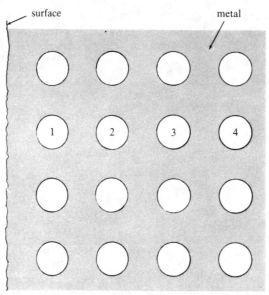

Figure 36.4 Lattice configuration of ions in a metal.

there's no ion there to balance the attraction of ion 1. Thus, electrons are free to move around inside the metal (in response, say, to an external charge as in Fig. 36.3), but they find it very difficult to escape from the metal.

At this point we seem to have a paradox. Consider a test charge and a conduction electron, both inside the metal. The test charge encounters no field and no change in potential as it crosses the surface to the outside of the metal, but for a conduction electron there is a powerful field at the surface preventing its escape. How can that be possible?

One explanation is that, while the test charge detects the effects of all the ions and electrons in the metal combining to produce no field at all, the electron senses the effects of all the ions and all the other electrons, but not its own contribution. So, while a test charge experiences a constant potential everywhere, the electron experiences that same constant value minus the potential due to its own field, wherever it happens to be.

To avoid confusing minus signs due to the negative charge of the electron, the rest of this discussion will be expressed in terms of the energy of an electron (in the convenient units of electron volts, eV) rather than the potential, which is in volts. If the electron is at some arbitrary place inside the metal, we might try to sketch a graph of its potential energy, as shown in Fig. 36.5. Here we imagine that the electron is somehow distributed over the region of one ion, a space of a few Angstroms in diameter. The electron appears to be trapped there, inside a well of potential energy having the $1/r$ form that its own field contributes.

This picture is deceptive, however, because it implies that the electron would encounter a powerful force pushing it back to its present position if it were to try to move. We have already seen that the electron can hop from one ion to the next, but

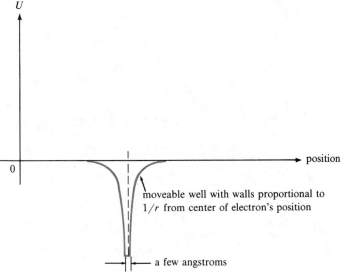

Figure 36.5 The apparent potential energy of one electron in a metal.

if it does, other electrons will be forced to readjust their positions to make the test-charge field zero again. The result is that the well shown in Fig. 36.5 moves with the electron, and, instead of being trapped on one ion, the electron effectively has the same potential energy everywhere in the metal. This is shown in Fig. 36.6. Of course, this situation ends abruptly at the metallic surface. When the electron and its well reach the surface of the metal (at the left side of the figure) there are no

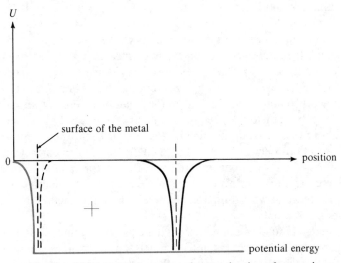

Figure 36.6 The potential energy of a conduction electron in a metal.

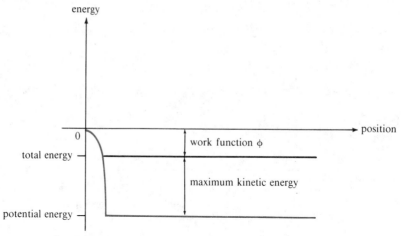

Figure 36.7 The energy of a conduction electron in a metal.

longer any charges available on the left side to rearrange themselves. Thus, the left wall of the well at the surface is a real *fixed* wall that the electron must climb in order to escape from the metal.

Although this picture helps explain the situation, it is not the end of the story. The conduction electrons in a metal don't sit idly on ions waiting for an electric field to come along. Instead, they move in definite orbits of the giant molecule that we call a metal. In other words, they have not only potential energy, but kinetic energy as well. Each conduction electron is in a slightly different orbit, with a slightly different kinetic energy, up to a definite maximum value. Thus we can improve our picture once again, as shown in Fig. 36.7. The most energetic electron in the metal has too little energy to escape from the metal by an amount called the *work function* ϕ.

The work function of a typical metal is of the order of a few eV. At the surface of the metal, the total energy of an electron must change by a few eV over a distance of an Angstrom or so in order to escape. Thus, while the energy needed to escape is not very large (a few eV) the force preventing escape is huge (10^{10} eV/m).

The work function of a metal is the amount of energy an electron loses as it enters from the outside; or, equivalently, it is the amount of energy the electron must gain to escape from the metal. It is similar both in concept and in magnitude to the ionization potential, which is the energy needed to extract an electron from a single atom. The ionization potential is different for each type of atom (hydrogen, helium, etc.) and the work function is different for each metal (copper, zinc, etc.). It is this difference in work functions between dissimilar metals that is responsible for the working of a battery.

To understand this point better, it helps to imagine what happens when two dissimilar metals are brought into contact. To begin with, we'll assume that both conductors are electrically neutral so that, as far as a test charge can tell, there is no field anywhere, and the electrostatic potential is the same everywhere, inside and

36.2 THE WORKINGS OF METALS

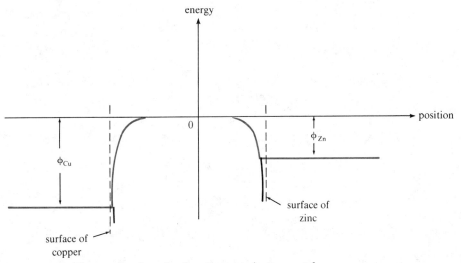

Figure 36.8 The energy of conduction electrons in two metals.

outside both metals. Just before contact, the diagram for the total energy of a conduction electron is as sketched in Fig. 36.8. To be more specific, we assume the conductor on the right is zinc, and that on the left is copper. The work function of copper is larger than that of zinc by about 1 eV:

$$\phi_{Cu} - \phi_{Zn} \approx 1 \text{ eV}.$$

That means there are electrons in the zinc that could lower their energy by that amount if they were able to flow into the copper. However, before contact is made, there is no way for them to overcome the surface force that contains them in the zinc.

Suppose now the metals are brought into contact. At the atomic level, that means there are places at the interface between the two metals where the last zinc ion is adjacent to the first copper ion. The barrier to escape from the zinc has vanished at those places. In fact, there is now a powerful force (a reduction of energy by about 1 eV in a distance of 1 Å) pushing electrons from the zinc into the copper. The result is that a surge of electrons flows from the zinc into the copper at the instant contact is made.

This surge, which occurs whenever two dissimilar metals are brought into electrical contact, was the cause of Galvani's twitching frog's leg.

A short time after contact is made, equilibrium is restored to the system. That happens when electrons are indifferent to whether they are in the copper or in the zinc. The flow of electrons has left a sheet of residual positive charge at the surface of the zinc, and an equal and opposite sheet of negative charge at the surface of the copper, much like the opposing surfaces of a charged capacitor. The energy of a conduction electron is the same everywhere, but the situation as detected by a test charge has changed dramatically. Although the electrostatic potential is constant everywhere inside the zinc, and everywhere inside the copper, the double layer of charge at the interface creates a jump in electrostatic potential, ΔV, given precisely by the formula

$$-e\,\Delta V = \phi_{Cu} - \phi_{Zn}.$$

If the two pieces of metal are now separated, a net charge remains on each, negative on the copper, and positive on the zinc.

Questions

1. The quantity ΔV is known as the *contact potential*. What is the contact potential (in volts) between copper and zinc?

2. If you have a gold filling in your mouth and bite down on a silver spoon, you can feel a tingling sensation. Why? (If you're fortunate to have all natural teeth, try touching a silver spoon to the back of your tongue and a piece of tin to the tip to get the same sensation.)

3. Suppose a negative point charge is brought near a large, uncharged conducting plane that is electrically insulated from its surroundings. If the charge is moved from point A to B in the following figure, describe qualitatively changes in (a) the potential on the plane, (b) the charge on the plane, and (c) the force on the point charge.

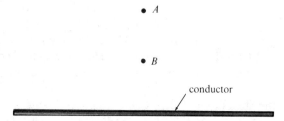

4. Suppose two dissimilar metals are brought together, creating contact potential ΔV, then pulled apart. Does the potential difference between them remain equal to ΔV? For Cu and Zn, make a sketch of V (the potential detected by a test charge) versus position before, during, and after contact between the two metals.

5. In Chapter 35 the surface of any conductor was described as an equipotential surface. Now we've seen that if pieces of zinc and copper are placed in contact the two surfaces will be at different potentials. How can the surface of each metal be an equipotential surface and yet not have identical potentials when placed in contact?

6. Suppose a manufactured metallic object were painted so that it was not possible to tell by eye whether it were made from a single metal or from two or more metals in contact. Would it be possible, in principle, to tell the difference using a test charge outside the object as a probe?

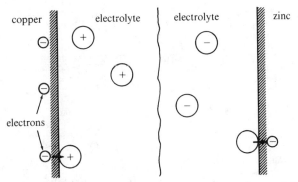

Figure 36.9 Ions transfer electrons at the surfaces of the two electrodes.

36.3 BATTERY BASICS

Volta thought he had found an inexhaustible source of charge – a kind of electric perpetual motion machine that would constantly replenish itself. Through our discussions we have now arrived at an understanding of why charge flows between two dissimilar metals that are placed in contact. However, that flow lasts only for a fleeting moment – just long enough to cause a frog leg to twitch. What is the secret to maintaining the flow over a longer period of time?

The secret lies in chemistry, more precisely, in the chemistry of electrolytes. An *electrolyte* is a fluid in which the molecules dissociate into positive and negative ions in solution that can move about and therefore conduct electricity. Suppose, for example, the pieces of copper and zinc, which we have brought into contact and then separated, are placed in an electrolyte. Initially, the copper has excess electrons, so it tends to attract positive ions from the solution. Whenever a positive ion touches the surface, the surface barrier to electrons momentarily vanishes, and one or more excess electrons are able to escape, neutralizing the ion. At the zinc surface, the reverse phenomenon occurs, negative ions being attracted to the zinc surface, where they are neutralized by surrendering their excess electrons. The situation is sketched in Fig. 36.9.

The net effect of these processes at the two interfaces is to transfer electrons out of the copper and into the zinc, while using up ions from the solution. In a short time, both copper and zinc are restored to their original, electrically neutral condition, and no further activity takes place.

However, because the two metals have been neutralized, they now once again have different internal energies for conduction electrons. If they were removed from the solution and placed in contact again, a new surge of electrons would flow. This is the situation that an electric battery exploits, but without removing the metals from the electrolyte. They remain in the solution, and they are referred to as the *electrodes* of the battery.

To make the physics as simple as possible, suppose a copper bar is connected to the copper electrode, then to the filament of an electric-light bulb, and from there to the zinc electrode, as indicated in Fig. 36.10.

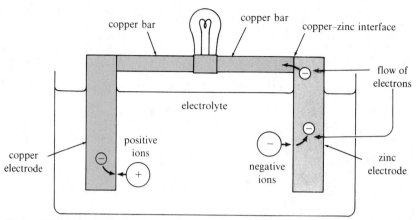

Figure 36.10 A simple battery connected to an electric-light bulb.

At the point where the copper bar touches the zinc electrode, there is contact between dissimilar metals, which has the effect of driving electrons from the zinc into the copper. However, the system does not reach equilibrium, because electrons are being extracted at the same time from the copper electrode at its interface with the electrolyte, and also electrons are being restored to the zinc at its interface with the electrolyte. Thus, there is a steady flow of electrons, through the filament of the light bulb. The source of the energy radiated by the filament is the supply of ions in the electrolyte, which gradually gets used up.

So, the energy of a battery is stored in the chemicals of its electrolyte, but the push that makes current flow is provided by the difference in work functions of its electrodes. The details of the chemistry that takes place in a battery can be very complicated, and depend on both the electrodes and the electrolyte.

A common electrolyte is copper sulphate, $CuSO_4$. When used with copper and zinc electrodes, metallic zinc from the electrode reacts with the copper sulphate – a combination with high potential energy – to produce metallic copper that deposits on the copper electrode, plus zinc sulphate, $ZnSO_4$, with lower potential energy. The chemical reaction can be written as

$$Zn + CuSO_4 \rightarrow Cu + ZnSO_4 + \text{energy}.$$

The energy released in this chemical process appears in electrical form. A potential difference is set up across the circuit, driving the current through the circuit. This particular reaction is reversible. If energy is added to the system by causing electrons to flow out of the copper and onto the zinc, the reaction will proceed in reverse. That's what happens when the battery is recharged.

If the light bulb is disconnected, the open circuit potential difference that exists between the terminals of a battery of this kind depends only on the difference in work function between the two metals:

$$\Delta V = \frac{\phi_{Cu} - \phi_{Zn}}{-e}.$$

36.3 BATTERY BASICS

Because a battery is a chemical source that maintains a constant potential difference, it is often referred to as a *seat of electromotive force*, or *emf*, for short, denoted by $\mathscr{E}$, and measured in volts. (This potential difference was historically identified incorrectly as a force, hence the term electromotive force.) In circuits a battery is denoted by the symbol ⊣⊢ , where the longer line indicates the *positive terminal*, out of which current flows into an external circuit. When two or more batteries are connected so that the positive terminal of one is connected to the negative of another, the batteries are in series, denoted by the symbol ⊣|⊢ , and their emfs combine by addition.

Notice that, in the example we have given, electrons flow from Zn to Cu, but by convention, electric current is said to flow in the opposite direction. (This minor inconvenience is Ben Franklin's doing, although not his fault, of course.) Thus, the copper electrode is considered positive and the zinc electrode negative.

Example 2
Why does touching your tongue to pieces of copper and zinc that are stuck into a lemon produce a tingling sensation?

The lemon juice acts as an electrolyte, and with the copper and zinc it forms a battery. Its voltage is due to the difference in work functions between copper and zinc, and so is the same as any other battery made with copper and zinc electrodes. Because it is a source of potential difference, when hooked to an external electrical circuit, such as your tongue, electrons flow, albeit slowly, from one electrode to the other and stimulate your taste buds.

Example 3
Refer to the circuit shown in Fig. 36.10. Determine the energy of a conduction electron (a) inside the zinc electrode, (b) in the copper bar to the right of the light bulb, and (c) in the copper bar and copper electrode to the left of the light bulb. Also, find the external potential V outside of each of those components, and discuss the behavior of V inside the electrolyte.

First we consider what would happen if the light bulb were unscrewed, leaving a gap in its place. The mobile ions in the electrolyte assure that the two electrodes have the same external potential. Therefore, because of the difference in work functions, a conduction electron has (approximately) 1 eV higher energy in the zinc electrode than in the copper electrode. Electrons in the copper bar on the right have the same energy as those in the zinc electrode because electrons flow easily across the interface between them. It follows that the external potential outside the copper bar is *lower* than that of the electrodes by 1 V (remember, potentials have opposite signs to the energies of electrons).

When the light bulb is put in place, this situation hardly changes, provided ions flow freely through the electrolyte, and electrons flow freely through all the other metal parts. The only significant resistance to flow is in the filament of the light bulb, so that's where electrons lose their energy in flowing from one side of the battery to the other.

The answers to (a), (b) and (c) can be summarized as follows:

	(a) zinc electrode	(b) copper bar	(c) copper electrode
electron energy	$+1$ eV	$+1$ eV	0
external potential	$+1$ V	0	$+1$ V

(The choices of zero energy and potential are arbitrary; only the differences matter.)

Inside the electrolyte, V is nearly uniform. There must, however, be some difference between its values in the solutions outside the copper $(+)$ and zinc $(-)$ electrodes, sufficient to drive enough positive ions toward the copper, and negative ions toward the zinc, to compensate for the rate at which electrons are flowing through the filament of the light bulb.

Questions

7. Why is there such a small current flow out of a lemon battery of the type described in Example 2?

8. Suppose you are an engineer assigned to invent a battery that will make an electric car go 500 km before recharging. Do you have to find better electrodes, or better electrolytes?

9. In the battery discussed in the text, copper from the $CuSO_4$ solution deposits on the copper electrode, while zinc from the zinc electrode dissolves in the solution to form $ZnSO_4$. However, in the lemon battery of Example 2, the electrolyte is lemon juice. Would you conclude that lemon juice contains dissolved copper from the fact that the battery works? Explain.

10. Draw on your knowledge of electric fields and potentials to explain why the total voltage of batteries in series is the sum of the individual voltages.

11. What is the open circuit voltage of a Cu–Zn lemon battery?

12. In Example 3, substitute a lemon for the electrolyte, and a capacitor for the light bulb. Rework the example (a) immediately after the circuit is connected, and (b) a very long time after the circuit is connected.

36.4 REAL BATTERIES

Most batteries in common use today depend on far more complex chemistry than that of the copper–zinc battery discussed in the previous section. Chemists and engineers have devised many types of batteries for different types of applications.

For example, the battery in a quartz crystal wrist watch is small, sturdy, and capable of giving a tiny but constant current at constant voltage for a very long period of time. On the other hand, an automobile battery must provide large bursts of current for relatively short periods of time, it must function over a large range of temperatures, and must be easily rechargeable. The copper–zinc battery is not well suited for either of these applications. It illustrates the basic principles underlying all batteries, but is not very helpful if you need it to start your car on a cold morning.

The modern automative battery got its start in 1859 when Gaston Planté discovered that a battery could be made by placing two lead plates in a dilute solution of sulphuric acid. The battery is charged by passing a current through it in one direction, which has the effect of coating one electrode with an oxide of lead, PbO_2. After this is done, the battery is capable of driving a current in the opposite direction, with the oxide-coated electrode being positive.

Simple as it sounds, this battery involves a series of complicated reactions that took many years to work out. It depends essentially on the chemical reaction

$$Pb + PbO_2 + 2H_2SO_4 \rightleftarrows 2PbSO_4 + 2H_2O + \text{energy}.$$

From the point of view of our earlier discussion we can view this reaction as follows: Oxidizing the surface of one electrode increases its work function ϕ. The work function of a metal is a property of its surface and can be altered by changing the surface. Thus, although the two electrodes start out identical, after the battery is charged they have different work functions. Once this is understood, the explanation of why it works becomes similar to that of the copper–zinc battery.

The search for combinations of materials to make new kinds of batteries is a vigorous field of modern research. A particularly interesting variant of the battery is the *fuel cell*. A fuel cell is a battery in which the electrolyte is continually replaced rather than being used up. In essence, the electrolyte becomes a kind of fuel analogous to gasoline in a car, except that the energy in the chemicals is turned directly into electrical energy rather than being burned to produce heat. For example, hydrogen and oxygen could be combined to form water, the energy of the reaction being converted directly to electrical energy. The advantage of fuel cells is that they are used to drive electric motors, which are much more efficient than internal combustion engines and, in addition, are relatively noiseless and nonpolluting. However, in spite of theoretical superiority, technical problems such as continuous feed of reactants and removal of by-products has prevented universal development of economically viable fuel cell systems. Heat engines are still the principal source of large scale production of electrical energy.

Question

13. Make a list of at least five applications in which batteries are used, and specify, for each application, which properties are most important.

36.5 A FINAL WORD

The unit of electric potential is named for Alessandro Volta, in honor of his pioneering work that led to the development of the battery. Volta was born in 1745 and lived until 1827. His father was one of four brothers, all of whom became priests. Alessandro's father left the priesthood after 11 years to propagate the family name. Among his seven children, three became priests and two became nuns. On the other hand, Alessandro, as he was portrayed by one of his colleagues, understood the electricity of women.

Aside from his research in electricity, Volta worked in other fields, notably the chemistry of gases. His investigations of a flammable gas that formed over the marshes of Lago Maggiore near his home town of Como led to the discovery of methane. He also found that the optimum mixture of oxygen and hydrogen for combustion was in the ratio of 1:2 by volume. In this investigation he found that when oxygen and hydrogen exploded in that ratio, a mist was left on the vessel that contained the gases. He came close to making one of the most important discoveries of the eighteenth century – the composition of water. He didn't make this discovery because the gases were collected in a vessel that also contained water, so he could never tell whether the mist came from the water in the bottom of the vessel or from the explosion. He described his findings to the French chemist Antoine Lavoisier who repeated the experiment using mercury in place of water, and consequently determined the composition of water. Although Volta made a number of additional discoveries in the chemistry of gases, his experimental prowess was most clearly demonstrated in electricity.

In 1796 Napoleon's army invaded and occupied northern Italy. The French replaced the Austrians who had occupied that region before them. In 1801 Volta was sent to Paris by the local administration on an obscure diplomatic mission. The arrival of this provincial Italian in Paris turned into a triumphal tour because Parisians had heard of his discovery of the battery and understood its significance. He discussed his work at the Paris Academy where he captured the imagination of Napoleon. Napoleon himself enthusiastically became Volta's patron, bestowing upon him the title of count and senator of the Kingdom of Italy and awarding him a gold medal and a sizable pension. To ensure the development of more electrical inventions Napoleon instituted a 60,000 franc award for those who would make electrical discoveries comparable to those of Volta and Franklin.

In 1814, at the age of 69, Volta sought retirement but was refused. Napoleon said to him, "A soldier should die on the field of honor." That meant that Volta should return to his laboratory. Fortunately for Volta, the Austrians reoccupied northern Italy in 1819 and he was permitted to retire, which he did, living peacefully as a wealthy man in Como until he died at the age of 82.

Volta excelled as a careful and tireless experimentalist who constantly improved his apparatus to control variables and minimize extraneous effects. He realized the importance of precise measurements in science, saying, "Nothing good can be done in physics unless things are reduced to degrees and numbers." His lively imagination led him to anticipate the inventions of the telegraph and the incandescent gas lamp. But his greatest discovery was that of the electric battery. Not only did it have immediate applications to chemistry, notably to electrolysis, but by providing a source of steady current it led to discoveries connecting electricity and magnetism. These discoveries ultimately transformed civilization.

CHAPTER 37

ELECTRIC CIRCUITS

Viewing the law of the electric circuit from the point at which the labors of Ohm has placed us, there is scarcely any branch of experimental science in which so many and such various phenomena are expressed by formulae of such simplicity and generality.

    Charles Wheatstone, Bakerian Lecture, 15 June 1843

37.1 THE INVENTION OF THE TELEGRAPH

In the early nineteenth century electricity was transformed from an entertaining curiosity to a driving force of civilization. Electric circuits, the subject matter of this chapter, are the basis of virtually all practical applications of electricity. The first application to have a profound effect on everyday life was the telegraph circuit.

    The word "telegraph" was originally used by Claude Chappe in France in 1792 to describe a visual signaling system. By the early 1800s a host of experimenters in

Figure 37.1 An early five-needle telegraph of Cooke and Wheatstone. The needles would point to letters on the board. (Reproduced from *Cooke and Wheatstone: Invention of the Electric Telegraph* by Geoffrey Hubbard, Routledge & Kegan Paul, Ltd. 1965.)

Europe and America discovered a number of schemes for transmitting signals electrically and paved the way for the invention of the electric telegraph.

During a demonstration at the University of Copenhagen in the winter of 1819–20, Hans Christian Oersted found that a compass needle could be deflected by an electric current. Joseph Henry, an American, constructed a signaling apparatus in 1831 that consisted of a horizontal magnetized steel bar supported on a pivot. When a nearby magnet was energized by electric current, one end of the bar was caused to strike a bell, thus conveying a signal. In 1832 a Russian diplomat, Paul Schilling, showed that the deflection of a compass needle could be the basis of a telegraph detector when electric current flowed in a coil of wire around the needle. In 1833 in Germany, Carl Friedrich Gauss and Wilhelm Eduard Weber communicated signals to each other at a distance of 2.3 km with a two-wire single-needle telegraph. In England, at about the same time, William Fothergille Cooke followed up the work of Schilling, and later, with the help of Charles Wheatstone, produced the first practical electric telegraph. One of their early instruments, using five wires and five needles, is shown in Fig. 37.1. By 1838 their system required only two wires and two needles and was widely used in England.

An American, Samuel F. B. Morse, is also associated with the invention of the telegraph, primarily because of the code bearing his name. Morse had the idea of the telegraph as early as 1832, when he returned to the United States after a trip to Europe. He constructed a model in 1835, and applied for a patent in 1837. In 1844 he gave a full-scale demonstration and communicated his famous message "What hath God wrought?" between Washington, D.C. and Baltimore, Maryland. In 1846 the Western Union Telegraph Company was founded, using Morse code to transmit messages. About the same time, public interest in the developments of Cooke and Wheatstone was growing in England and the British Electric Telegraph Company was formed in 1845. By mid-century the simple telegraph circuit had become an essential part of the Industrial Revolution.

37.2 ELECTRICAL CONDUCTION IN WIRES

We turn now to the physics underlying not only the telegraph circuit, but electric circuits in general. In Chapter 33 we learned that under conditions of electrostatic equilibrium the electric field inside a conductor is zero. If any amount of charge is applied to the ends of a conductor, such as a metallic wire, the charges will rapidly move around until equilibrium is again established. For a good conductor, such as copper or silver, this takes place in a fraction of a second.

But if the ends of the wire are connected to a source of electric potential difference, such as an electric generator or the terminals of a battery, then the equilibrium is continually disturbed. Electrons are pushed from one terminal to the other, forming an electric current that continues to flow as long as the source supplies the potential difference. Because of the electric field in the wire, electrons experience a force and are accelerated. In Chapter 35 we described how a metal can be viewed as an array of positive ions held together in a rigid lattice by a cloud of mobile electrons. The motion of an accelerated electron in a metal wire is impeded by collisions with other electrons and with imperfections in the metal lattice. This motion is somewhat analogous to that of a ball in a pinball machine, where gravity provides a constant force on the ball. During its motion, the ball collides with paddles and other bumpers and so wanders about, eventually making its way to the bottom of the machine. Inside a wire, the overall effect of the collisions is a transfer of kinetic energy from the electrons to the vibrational energy of ions in the lattice (heat energy). As a result, the wire becomes warmer, and the velocity of a given electron varies considerably from point to point, as suggested by Fig. 37.2.

Because of complexities due to collisions, the direction of motion of a particular electron cannot be specified, so to analyze the motion, some simplifying assumptions have to be made. For example, for a huge number of electrons there is an average drift velocity at each point in the wire, and it is fruitful to treat the motion as though each electron possesses this average velocity. This makes it possible to

Figure 37.2 Erratic motion of a conduction electron in a wire.

study the dependence of the flow on factors such as the nature of the electric source and the type of wire used as conductor. To determine this dependence quantitatively, certain conventions and definitions have been introduced.

First there is a convention about the direction of flow of charge. In a metallic conductor, such as a copper wire, only electrons are in motion, but in an electrolyte, both positive and negative ions can be in motion. In either case, it is customary to say that the direction of the flow is that of the *positive* charges. Thus, when only electrons are in motion, we pretend that positive charges are flowing in the direction opposite to the actual flow of the electrons.

Next we need a quantitative definition of electric current. Consider a cross section S of the wire. In a given time interval, say from time t to time $t + \Delta t$, a certain quantity of positive charge Δq flows through S. This quantity may depend on t and on Δt, but we assume it does not depend on the location of S nor on the size of S. This is reasonable because no charge is created or destroyed inside the conductor. The ratio $\Delta q / \Delta t$ is the rate of flow of charge through S during that time interval. The current I through S at time t is defined to be the derivative dq/dt, the instantaneous rate of flow through S:

$$I = \frac{dq}{dt} = \lim_{\Delta t \to 0} \frac{\Delta q}{\Delta t}. \tag{37.1}$$

Although there is a direction associated with the flow of positive charge along a wire, the current I is a scalar quantity.

In SI units, current is measured in units of amperes (A), in honor of the nineteenth century French physicist André Marie Ampère. In Chapter 39 we'll find that current is easier to measure than charge. For this reason, the ampere is actually taken as a base SI unit, and the coulomb is defined to be 1 ampere second, 1 C = 1 A s. Recall that a coulomb is the charge carried by a huge number of electrons or protons, 6.2414×10^{18}, so a current of 1 A means that the equivalent of 6.2414×10^{18} positive charges are flowing each second through each cross section.

Now suppose we connect a wire to a battery that maintains a constant potential difference between the ends of the wire, and ask for the current $I = dq/dt$ through an arbitrary cross section S in the wire. At each point of S we attach a small rectangular frame described by an area vector $\mathbf{A}$ whose magnitude A is the area of the frame and whose direction is perpendicular to S, as indicated in Fig. 37.3.

How many equivalent positive charges pass through this frame in a small time interval Δt? Suppose there are, on the average, n positive charges per unit volume, each with average drift velocity $\mathbf{v}$. If the frame is not too large, the drift velocity $\mathbf{v}$ will be nearly constant over the frame. In a small time interval Δt each positive charge moves through a displacement vector $\mathbf{v}\Delta t$, with component $\mathbf{v} \cdot \hat{\mathbf{A}} \Delta t$ in the direction of $\mathbf{A}$, where as usual the notation $\hat{\mathbf{A}}$ denotes a unit vector in the direction of $\mathbf{A}$. A corresponding displacement of the frame traces out a prism with base of area A and edge length $\mathbf{v} \cdot \hat{\mathbf{A}} \Delta t$. The volume of this prism is the area of the base times the altitude, $A(\mathbf{v} \cdot \hat{\mathbf{A}} \Delta t) = \mathbf{v} \cdot \mathbf{A} \Delta t$, so the amount of positive charge Δq in the prism is $n|e|\mathbf{v} \cdot \mathbf{A} \Delta t$. Therefore the average current passing through the frame in

37.2 ELECTRICAL CONDUCTION IN WIRES

Figure 37.3 Charge passing through a small frame of area A.

time Δt, which we denote by $I(A)$, is

$$I(A) = \frac{\Delta q}{\Delta t} = n|e|\mathbf{v} \cdot \mathbf{A}. \tag{37.2}$$

The vector quantity $n|e|\mathbf{v}$ in Eq. (37.2) is called the *vector current density* and is denoted by $\mathbf{J}$. Thus,

$$\mathbf{J} = n|e|\mathbf{v}, \tag{37.3}$$

a vector whose direction is the same as the average drift velocity $\mathbf{v}$ and whose magnitude J is equal to the current per unit area,

$$J = I(A)/A. \tag{37.4}$$

This is the (scalar) current density over the frame of area A. To get the current density at a point we would take the limit of $I(A)/A$ as the area A shrinks to zero.

It might be more suggestive to write $\Delta \mathbf{A}$ instead of $\mathbf{A}$, and ΔI instead of $I(A)$, in which case Eq. (37.2) takes the form

$$\Delta I = \mathbf{J} \cdot \Delta \mathbf{A}.$$

The current flowing through the entire cross section S is just the surface integral

$$I = \iint_S \mathbf{J} \cdot d\mathbf{A}. \tag{37.5}$$

Thus, current is the *flux* associated with the current density vector $\mathbf{J}$, just as flux of another kind is associated with the electric field vector $\mathbf{E}$. Note that the current density is a microscopic quantity, defined at each point of S, whereas current itself is a macroscopic quantity measuring the total flux through S.

When the conductor is a long thin wire with constant area of cross section A, the current density is usually taken to be constant, with the direction of the current density vector parallel to the wire. In this case Eq. (37.5) reduces to

$$I = JA, \tag{37.6}$$

and the current density is simply $J = I/A$, current per unit area. Most of the cases we shall consider involve constant current density.

Example 1
Calculate the average drift speed of electrons in a thin copper wire of radius 0.040 cm carrying a current of 0.5 A, assuming that 8.5×10^{28} electrons/m^3 are available

for conduction in copper.

In this case the current density J is constant and Eq. (37.6) becomes

$$I = JA = n|e|vA,$$

where $A = \pi r^2$ is the area of the circular cross section of the wire. Solving for v and substituting values we find

$$v = \frac{I}{|e|n\pi r^2} = \frac{0.5 \text{ C/s}}{(1.6 \times 10^{-19} \text{ C})(8.5 \times 10^{28} \text{ m}^{-3})\pi(4.0 \times 10^{-4} \text{ m})^2},$$

which yields $v = 7.3 \times 10^{-5}$ m/s. Electrons drift along the wire at less than a snail's pace.

Questions

1. Why is a device to measure electric current analogous to a device for counting automobile traffic?

2. Find an analogy to explain why one would expect the average drift speed for electrons in a wire to be greater in a thin portion than in a thick portion.

3. A current of 1.5 A flows in a wire.

   (a) How much charge flows through a cross section of the wire in 2 min?
   (b) How many electrons does the charge in (a) represent?

4. A tube of radius 3.0 mm transports 1.5×10^{15} electrons each second. Determine (a) the current in the tube and (b) the current density in the tube.

5. A ring of radius a has a uniform linear charge density λ and rotates about an axis through its center and perpendicular to the ring with an angular speed ω. Find the current associated with this flow of charge.

6. A simple model of the hydrogen atom imagines an electron to be orbiting a proton along a circular path of radius r at speed v.

   (a) Determine the current associated with the motion of the electron around the nucleus.
   (b) Calculate the value of the current in (a) if $r = 0.5 \times 10^{-10}$ m and $v = 1.0 \times 10^6$ m/s.

37.3 OHM'S LAW

An important relation between current and potential in electric circuits was discovered by George Simon Ohm in 1826. This relation, now called Ohm's law, states that the current I in a wire is directly proportional to the potential difference V between the ends of the wire. The law is usually written in the form

$$V = IR, \tag{37.7}$$

37.3 OHM'S LAW

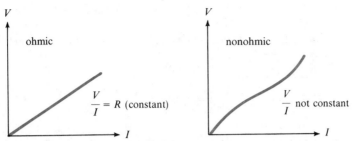

Figure 37.4 Potential difference versus current for ohmic and nonohmic materials.

where the constant of proportionality R, called the *resistance*, depends on the material of the wire. The SI unit of resistance is the ohm, denoted by the capital Greek letter omega,

$$1\ \Omega = 1\ \text{V/A}.$$

Ohm's law is an empirical law, a generalization deduced from experiment. It is not a fundamental law of nature, such as the second law of thermodynamics or Newton's second law, because it fails for some special materials or when large potential differences are present. The resistance of a wire always depends on its length, its cross-sectional area, the type of material, and temperature, but for many materials it does not depend on the current in the wire. Materials that obey Ohm's law, with the constant R independent of I, are called ohmic materials. The graph of potential difference versus current for an ohmic conductor is illustrated in Fig. 37.4. For such materials the resistance between any two points in the material is equal to the potential difference between those points divided by the current, which means that the slope of the potential-versus-current curve is a measure of the resistance—large slope means large resistance. Materials that do not follow Eq. (37.7) with constant R are known as nonohmic materials. A potential-versus-current curve for a nonohmic material is also shown in Fig. 37.4.

Ohm, who was a mathematician by training and only later in life turned to physics, published his findings in a highly abstract way that made the law seem more complicated than it really was. A simpler presentation and further verification of the law appeared in the work of Wheatstone during his development of the telegraph.

A problem that plagued early developers of the telegraph was its short range. The device worked well over short distances, but over long distances the signal became undetectable. Wheatstone dealt with the problem and found that the current became dissipated by the increased resistance of a longer wire. He could maintain the necessary current by increasing the voltage as the length of the wire increased. In 1843, Wheatstone published his findings as experimental verification of Ohm's law.

We've seen that Ohm's law relates the measurable quantities of voltage and current. But what is responsible for the resistance? One answer to this question was given in Section 37.2, where we described the collisions of moving electrons with

Table 37.1 Conductivities and Resistivities for Various Substances at 20°C.

Material	Conductivity σ $(\Omega\ m)^{-1}$	Resistivity ρ $(\Omega\ m)$
Silver	6.2×10^7	1.6×10^{-8}
Copper	5.9×10^7	1.7×10^{-8}
Aluminum	3.6×10^7	2.8×10^{-8}
Lead	4.6×10^6	2.2×10^{-7}
Carbon	2.9×10^4	3.5×10^{-5}
Silicon	6.4×10^2	1.6×10^{-3}
Water	4.0×10^{-6}	2.5×10^5
Amber	2.0×10^{-15}	5.0×10^{14}

other electrons and with imperfections in the lattice structure of a conductor. These collisions act like a resistive or frictional damping force on the motion of the electrons. The average drift velocity of the electrons turns out to be proportional to the force due to the electric field, just as the terminal velocity of a marble falling in a viscous field is directly proportional to the gravitational force causing the motion. Equation (37.3) shows that the current density is proportional to the drift velocity, $\mathbf{J} = n|e|\mathbf{v}$, so the current density is also proportional to the applied electric field:

$$\mathbf{J} = \sigma \mathbf{E}. \tag{37.8}$$

This is called the *microscopic* form of Ohm's law. The factor σ is known as the *conductivity* of the material, and for ohmic materials it reflects an intrinsic property of the material. Often the *resistivity*, ρ, is introduced, being defined as the reciprocal of the conductivity, $\rho = 1/\sigma$. Table 37.1 lists the conductivities and resistivities of certain materials. Note that these numbers vary over some 22 orders of magnitude, an unusually wide range for any physical quantity.

The microscopic form of Ohm's law, Eq. (37.8), can be related to the macroscopic form, Eq. (37.7). Consider a conductor of length L, cross-sectional area A, and conductivity σ, that carries a constant potential difference V across the ends of the conductor, as shown in Fig. 37.5. Inside the conductor the uniform electric field created by the battery is related to the potential difference by Eq. (34.9),

$$E = V/L. \tag{34.9}$$

37.3 OHM'S LAW

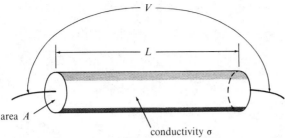

Figure 37.5 Resistance of a conductor of length L and cross-sectional area A.

Substituting into Eq. (37.8), we get $J = \sigma E = \sigma V/L$. Since $J = I/A$ also, we have

$$J = \frac{I}{A} = \frac{\sigma V}{L} = \frac{V}{\rho L},$$

where we replaced σ by $1/\rho$ in the last step. Rearranging terms, we find

$$V = (\rho L/A)I \qquad (37.9)$$

or $V = RI$, where

$$R = \rho L/A. \qquad (37.10)$$

Equation (37.9) is, of course, Ohm's law, and the formula for R in (37.10) shows that the resistance depends on the length, the cross-sectional area, and the material of the conductor. Any object that is specifically fabricated to have resistance is known as a *resistor*.

Example 2

An aluminum wire has a circular cross section of radius 0.10 cm and a length of 5 m. Calculate its resistance.

Using Eq. (37.10) and the entry for aluminum in Table 37.1, we have

$$R = \frac{\rho L}{A} = \frac{(2.8 \times 10^{-8}\ \Omega\ \text{m})(5\ \text{m})}{\pi(1 \times 10^{-3}\ \text{m})^2} = 4.5 \times 10^{-2}\ \Omega.$$

Despite its long length, the resistance of the wire is relatively small. Wires used to connect an appliance to a power source usually have very small resistance compared to the resistance of the appliance itself.

Wheatstone and Ohm surely must have noticed that resistors become heated when a current passes through them. This is known as Joule heating, and there is a simple relation between the current and the heat produced. As it passes through a resistor, a charge loses energy – that is, negative work is done on the charge. The amount of work, $-\Delta W$, is simply the product of the charge Δq and the potential difference:

$$-\Delta W = \Delta q\, V.$$

Then the amount of work done per unit time (the power), is given by the limit

$$P = \lim_{\Delta t \to 0} -\frac{\Delta W}{\Delta t} = \lim_{\Delta t \to 0} \frac{\Delta q}{\Delta t} V = \frac{dq}{dt} V = IV.$$

Using Ohm's law, $V = IR$, we can write this as

$$\boxed{P = I^2 R.} \tag{37.11}$$

This last expression is the rate at which Joule heat is produced by the current in a resistor. It represents the heat produced by collisions of conduction electrons with imperfections and with other electrons in the metal. As heat, this energy escapes from the circuit and cannot be recovered.

Of course, the energy supplied to a resistor must come from some source – a battery or other source of emf. As we learned in Chapter 36, a battery pushes charges through a circuit; in other words, it does work on the charges. In moving a charge Δq through the potential difference $\mathcal{E}$, the battery does an amount of work $\Delta W = \mathcal{E} \Delta q$. An analysis similar to the preceding shows that the power output by the source of emf is

$$P = \mathcal{E} I. \tag{37.12}$$

When a battery is being charged, work is done on charge moving in it, and it absorbs energy. In the following sections we'll see how conservation of energy in electric circuits leads to simple rules for determining currents and voltages in circuits.

Questions

7. Show that the drift velocity of electrons in a wire is related to the electric field by $\mathbf{v} = -\sigma \mathbf{E}/(n|e|)$. The quantity $\sigma/(n|e|)$ is sometimes called the *mobility*.

8. Explain how and why the drift velocity changes with an increase only in (a) the length of a wire, (b) its cross-sectional area, or (c) its conductivity.

9. A current of 2.5 A flows in a silver wire of radius 0.15 mm.
   (a) Find the current density in the wire, assuming that the current is uniformly distributed over the cross-sectional area.
   (b) Determine the electric field in the wire.
   (c) If the wire has a length of 4.0 cm, find the potential difference across its ends.

10. A potential difference of 150 V produces a current of 2 A in a wire. What is the resistance of the wire?

11. A block of carbon is 2.5-cm long and has a square cross section of sides 0.03 cm.
    (a) What is the resistance of the block?
    (b) If a potential difference of 9 V is maintained across its ends, determine the current density in the carbon.

12. A wire of length 1.5 m has a resistance of 0.5 Ω. It is uniformly stretched to a new length of 1.8 m. Determine the new resistance of the wire. (Assume that stretching of the wire doesn't change its resistivity or cross-sectional area. Would you expect it to?)

13. The first trans-Atlantic telegraph messages were sent in 1858 by a cable 3000-km long laid between Newfoundland and Ireland. The cable consisted of seven copper wires, each of 0.73-mm diameter, bundled together and surrounded by an insulating sheath.
    (a) Calculate the resistance of the cable.
    (b) A return path for the current was provided by the ocean itself. Taking the resistivity of sea water to be 0.25 Ω m, and assuming that the equivalent cross-sectional area of current path in the ocean exceeds 1 km^2, show that the resistance of the ocean is much smaller than that of the cable.

14. What is the rate of heat produced by a 15-Ω resistor connected to a potential difference of 18 V?

15. A 25-Ω resistor is connected to a 6-V battery.
    (a) What current flows in the circuit?
    (b) How much heat is produced by the resistance in 1 min?

16. A 50-Ω resistor is connected to a 12-V battery. Determine (a) the current in the resistor, (b) the rate of Joule heat produced in the resistor, and (c) the power supplied by the battery.

17. A resistor R is connected to a battery of emf $\mathscr{E}$. Show that the rate of heat produced by the resistance is equal to the rate of energy supplied by the battery.

37.4 RESISTORS CONNECTED IN SERIES AND IN PARALLEL

In practical applications, a resistor, symbolized by   in a circuit diagram, consists of a long wire wrapped around a spool or a block of material – the familiar cylinders striped with colors. Nowadays practical electric circuits seldom consist of a single resistor connected to a battery or some other source of emf. Rather, they consist of complex networks of resistors and sources of emf. On the other hand, in Wheatstone's day resistance in circuits was just an annoying obstacle to be overcome in the attempt to transmit electrical impulses through wires. As circuitry became more sophisticated, manufactured resistors were developed to produce desired currents in various parts of specific circuits. Through Ohm's law we can describe the relations between the potential drop (the difference in potential between the ends of the resistor), the current, and the resistance of each resistor in a given network, and then combine these results to determine the behavior of the

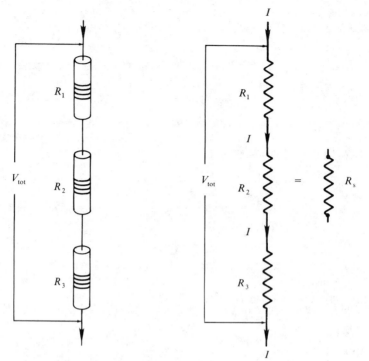

Figure 37.6 Resistors connected in series.

entire network. Let's examine the behavior of resistors in two special but common configurations: series and parallel combinations.

When resistors are connected in series, as shown in Fig. 37.6, the current is the same through each resistor in the circuit, because whatever charge goes into a resistor must also come out. The change in potential across each resistor is given by Ohm's law, $V = IR$. Furthermore, the total change in potential is simply the sum of the changes across each resistor:

$$V_{tot} = V_1 + V_2 + V_3 = IR_1 + IR_2 + IR_3$$
$$= I(R_1 + R_2 + R_3).$$

In other words, if we take three resistors and connect them in series, then connect that combination to a battery, the current through each resistor will be the same as if there were only one resistance with the value R_s, where $R_s = R_1 + R_2 + R_3$. So we have a way of simplifying a circuit element composed of resistors in series. The analysis can be extended to any finite number of resistors connected in series: The equivalent resistance is the sum of the individual resistances:

$$R_s = R_1 + R_2 + R_3 + R_4 + \cdots .$$

(37.13)

37.4 RESISTORS CONNECTED IN SERIES AND IN PARALLEL

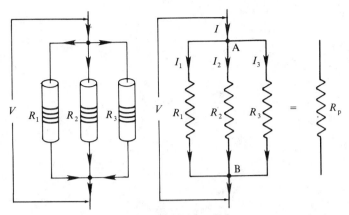

Figure 37.7 Resistors connected in parallel.

Note that this is the same law of combination as capacitors connected not in series, but in parallel.

The result of Eq. (37.13) is also suggested by the formula $R = \rho L/A$ for resistance in terms of the length of a wire. When resistors are in series, the effective length of the combination is the sum of the lengths of the individual resistors, and we expect that the net resistance should be the sum of the individual resistances.

Just as capacitors can be connected in parallel, so too can resistors. An example is shown in Fig. 37.7. Suppose a voltage V is applied and a current I enters the system. The current splits at junction A, and therefore is not the same through each resistor.

According to Ohm's law, the potential drop across each resistance is

$$V_k = R_k I_k, \quad k = 1, 2, 3.$$

The change in potential across each resistor in a parallel combination is always the same as the potential drop between A and B. In other words, we have

$$V = I_1 R_1, \quad V = I_2 R_2, \quad \text{and} \quad V = I_3 R_3.$$

And from this result we can find the current through each resistor:

$$I_1 = V/R_1, \quad I_2 = V/R_2, \quad \text{and} \quad I_3 = V/R_3.$$

Because the same current I that enters the combination at point A also exits at point B, we must have

$$I = I_1 + I_2 + I_3.$$

Substituting for the currents, we get

$$I = \frac{V}{R_1} + \frac{V}{R_2} + \frac{V}{R_3}.$$

Now we consider an equivalent resistance, R_p, connected to a voltage V and having a current I in it, Ohm's law says that $I = V/R_p$, and by equating the last two

expressions for I we find
$$\frac{V}{R_p} = \frac{V}{R_1} + \frac{V}{R_2} + \frac{V}{R_3},$$
which implies that
$$\frac{1}{R_p} = \frac{1}{R_1} + \frac{1}{R_2} + \frac{1}{R_3}.$$

For any finite number of resistors connected in parallel, the equivalent resistance is given by R_p, where

$$\frac{1}{R_p} = \frac{1}{R_1} + \frac{1}{R_2} + \frac{1}{R_3} + \frac{1}{R_4} + \frac{1}{R_5} + \cdots .$$ (37.14)

Resistors in parallel follow the same law of combination as capacitors in series.

Equation (37.14) is also suggested by the formula $R = \rho L / A$ when it is written as $A = \rho L / R$. When resistors are in parallel the effect is to increase the cross-sectional area through which the current flows. Because area is inversely proportional to the corresponding resistance, in a parallel combination the net area is the sum of the cross-sectional areas of the individual resistors and, at the same time is proportional to $1/R_p$, so we expect that the equivalent resistance R_p should be given by Eq. (37.14).

Often a circuit can be analyzed by breaking it down into series and parallel combinations. Once the separate combinations are recognized, they can be replaced by appropriate equivalent resistances. Then new combinations can be recognized and replaced by the equivalent resistance and so on, until a simple circuit results. On the other hand, some circuits cannot be broken down into parallel and series combinations, and more general principles of analysis are needed.

Example 3

Each of the resistors in the circuit shown has a value of 100 Ω. What is the equivalent resistance of the circuit?

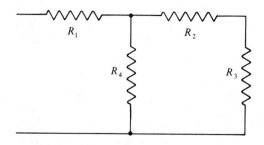

37.4 RESISTORS CONNECTED IN SERIES AND IN PARALLEL

For convenience we'll first label the resistors 1, 2, 3, and 4. We see that R_2 and R_3 are in series, and therefore their equivalent resistance is $R_s = R_2 + R_3 = 200\ \Omega$. Replacing those two resistors by their equivalent resistance, we see that R_4 and R_s are in parallel. The equivalent resistance of this parallel combination is

$$\frac{1}{R_p} = \frac{1}{R_4} + \frac{1}{R_s} = \frac{1}{100\ \Omega} + \frac{1}{200\ \Omega},$$

which gives $R_p = 67\ \Omega$. Finally, R_1 and R_p are in series, and therefore the total resistance of the circuit is $R_{tot} = R_1 + R_p = 167\ \Omega$.

Note. In problems of this type it is suggested that the reader draw intermediate equivalent circuit diagrams at each step of the calculation.

Questions

18. Calculate the equivalent resistance between points A and B of the circuit shown.

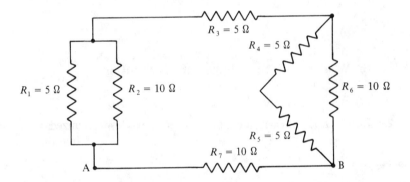

19. For the circuit shown, find (a) the equivalent resistance across the 12-V battery, (b) the current through the battery, and (c) the total power dissipated in all the resistors.

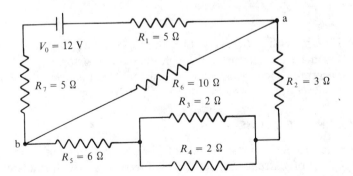

20. A 12-V battery is connected across a 10- and a 40-Ω resistor connected in parallel.

   (a) What current flows through the battery?
   (b) Calculate the ratio of the currents through each resistor.

21. The following questions refer to the circuit shown:

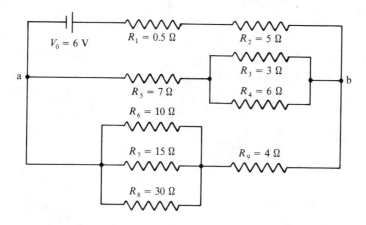

   (a) What is the equivalent resistance of the circuit?
   (b) What is the current through the 5-Ω resistor?
   (c) What is the potential drop across the 7-Ω resistor?
   (d) What is the Joule heating of the 4-Ω resistor?

22. In the circuit shown, the 2.0-V battery has a power output of 2.0 W.

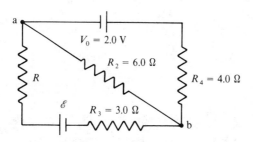

   (a) Determine the current I_4 through the 4.0-Ω resistor.
   (b) Calculate the current I_6 through the 6.0-Ω resistor.
   (c) Determine the resistance of the resistor R that is wound from a 4.0-m length of copper of cross-sectional area 3.4×10^{-8} m^2.

37.5 KIRCHHOFF'S LAWS

In 1845–46, Robert Gustav Kirchhoff began working with Ohm's law during investigations of currents in networks of resistors. He formulated two general principles, now known as Kirchhoff's laws, that extended Ohm's law to currents and

37.5 KIRCHHOFF'S LAWS

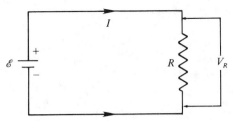

Figure 37.8 An electric circuit consisting of a battery and a resistor.

voltages in complicated circuits. As it turns out, these laws are consequences of two profound laws of nature – conservation of charge and conservation of energy.

To demonstrate how Kirchhoff's theory works, consider a combination of a battery, several resistors, and constant current in a very simple equivalent circuit – a battery connected to a resistor as shown in Fig. 37.8.

A positive charge gets a push – an increase in potential energy – from the emf of the battery. As the charge moves through the resistor (the connecting wires are assumed to have no resistance) it loses energy and there is a decrease in potential V_R. This lost energy comes from the potential energy of the electrons and goes into Joule heat energy that escapes from the circuit. By the time the charge returns to the battery, it is ready for another boost in potential energy to push it through the circuit.

Conservation of energy for the charge demands that the amount of energy dissipated in the resistor be precisely equal to the energy boost given by the battery. That's Kirchhoff's first law:

> The algebraic sum of the increases and decreases in potential around any closed circuit loop must be zero.

By convention, we call potential increases positive, and decreases negative. Kirchhoff's first law states that the algebraic sum of all the potential changes must be zero.

If we apply this law to the simple circuit above, we get

$\mathcal{E} - V_R = 0.$

By Ohm's law, we know that for any resistor,

$V_R = IR,$

so we have

$\mathcal{E} - IR = 0.$

This determines the current that will flow when the resistance R is attached to the battery. Of course, for this simple case we could have invoked Ohm's law to find the current because we have only one battery and one resistor. The power of Kirchhoff's laws is that they apply to *any* closed circuit, regardless of how many batteries, resistors and connections it contains. If the circuit is slightly more complicated, like the one shown in Fig. 37.9, we simply apply the law to each closed path. Analysis of this circuit will give us several equations, which we can solve to find the current in each part of the circuit. But before we actually do that, we need another principle.

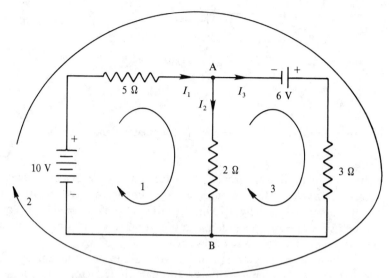

Figure 37.9 Analysis of a circuit using Kirchhoff's laws.

In the simple circuit we first analyzed, we found that as a charge moves around a loop, it gains and loses potential, but that the current remains the same everywhere around the circuit. What happens when there is a junction – a crossing of two or more paths? Imagine current I_1 approaching the junction at A in the circuit in Fig. 37.9. Suddenly it finds a fork in the circuit path. Does it split, and if so, how?

To answer these questions, we invoke the law of conservation of charge, which tells us that no charge can originate or disappear at the junction. So whatever amount of charge per unit time flows into a junction must also flow out of the junction. In other words, all the current that comes into a junction must also exit. This is Kirchhoff's second law:

> At any junction in a circuit, the sum of the currents into the junction equals the sum of the currents out of the junction.

Any circuit path between two junctions is known as a branch. In applying Kirchhoff's laws to circuits, we assign a current to each branch. In circuit diagrams, a junction is sometimes denoted by a large dot.

Example 4

Apply Kirchhoff's laws to find the currents I_1, I_2, and I_3 in the circuit shown in Fig. 37.9.

Applying the second law to junction A, we have

$$I_1 = I_2 + I_3.$$

If we were to apply it to junction B, we would not find anything new. So let's apply the first rule to path 1, using IR for the potential drop in volts across each resistor.

37.5 KIRCHHOFF'S LAWS

This gives us
$$10 - 5I_1 - 2I_2 = 0.$$

We now have two equations but three unknowns, and mathematicians tell us we need three independent equations to find a unique solution for the three unknowns. Let's apply the second rule to loop 2, the outer loop:
$$10 - 5I_1 + 6 - 3I_3 = 0.$$

Now that we've put in the physics, the rest is algebra. The solution in this case turns out to be

$I_1 = 2$ A,
$I_2 = 0$,
$I_3 = 2$ A.

The solution can be checked by applying Kirchhoff's law to loop 3, which gives $3I_3 = 6$ A.

Kirchhoff's laws have great technical application. They tell us how complex electronic circuits function and how they can be designed to produce specific currents and voltages. On the other hand, they stem from two profoundly important principles of nature – conservation of energy and conservation of charge.

Example 5

Find the current through each resistor and the power output of the batteries in the circuit shown.

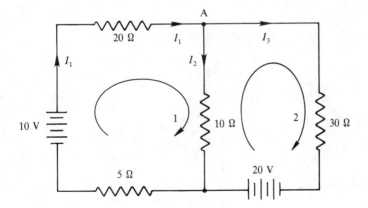

First we assign currents to each branch, as indicated. At junction A we have $I_1 = I_2 + I_3$. Next we apply Kirchhoff's second law to each of the indicated loops: For loop 1, this leads to the condition
$$10 - 20I_1 - 10I_2 - 5I_1 = 0.$$

It is understood that currents are measured in amperes, voltages in volts, and resistances in ohms, so we drop the units in intermediary calculations and insert them in the final answers. Because we ultimately want to know the currents I_1 and I_3 flowing through the batteries, we use the equation $I_2 = I_1 - I_3$ to eliminate I_2 in the above equation. After dividing through by 5 we obtain

$$2 - 7I_1 + 2I_3 = 0. \tag{1}$$

Applying the first rule to the second loop, we get

$$20 + 10I_2 - 30I_3 = 0,$$

and substituting for I_2 (and dividing through by 10), we have

$$2 + I_1 - 4I_3 = 0. \tag{2}$$

We now have two equations (1) and (2) for the two unknowns I_1 and I_3. The solutions are

$$I_1 = 0.46 \text{ A}, \qquad I_3 = 0.62 \text{ A}.$$

To find the rate of power output by each battery, we use Eq. (37.12), $P = I\mathscr{E}$, where I is the current through the respective battery:

$$P_{10} = I_1 \mathscr{E}_{10} = (0.46)(10) = 4.6 \text{ W},$$
$$P_{20} = I_3 \mathscr{E}_{20} = (0.62)(20) = 12 \text{ W}.$$

Ironically, Wheatstone's name is not associated with his important experiment confirming Ohm's law, but with a device known as the "Wheatstone bridge," which was actually invented by a man named Samuel Christie, and later improved by Wheatstone, to measure resistance accurately. A Wheatstone bridge, shown in Fig. 37.10, consists of a resistance R_x to be determined, and three known resistances,

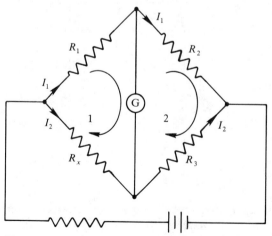

Figure 37.10 A Wheatstone bridge.

37.5 KIRCHHOFF'S LAWS

R_1, R_2, R_3. The object in the center, a galvanometer, detects very small currents. The resistances R_1 and R_2, for example, can be varied until there is no current flowing through the galvanometer; the bridge is then said to be balanced.

Example 6
Determine the unknown resistance R_x when the Wheatstone bridge in Fig. 37.10 is balanced.

Applying Kirchhoff's first law to loop 1 indicated on the circuit in the figure, we get $-I_1R_1 + I_2R_x = 0$, whereas applying the same law to loop 2 we get $-I_1R_2 + I_2R_3 = 0$. The currents can be eliminated from these equations by equating the ratios I_1/I_2. This gives $R_x/R_3 = R_1/R_2$, or

$$R_x = R_3(R_1/R_2).$$

So if R_1, R_2, and R_3 are known resistances, the value of R_x can be found.

Questions

23. For the circuit shown, find (a) the current in each resistor, (b) the power delivered or absorbed by each emf, and (c) the rate of Joule heating in each resistor.

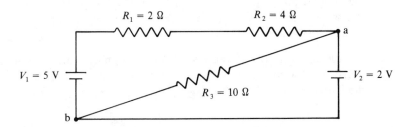

24. In the circuit shown, find (a) the current in each resistor, (b) the potential difference between points A and B, and (c) the power supplied by each battery.

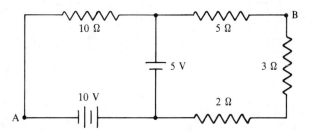

25. For the circuit shown, calculate the current through each resistor.

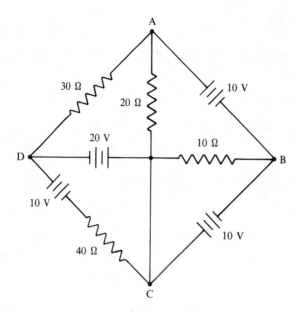

26. In a certain Wheatstone bridge that uses a uniform wire of 1-m length, the resistance R_1 is proportional to the length of the wire from the 0 mark to the balance point and R_2 is proportional to the length from the balance point to the 100-cm mark. If the fixed resistor is $R_3 = 500\ \Omega$, find the unknown resistance if the bridge balances at (a) the 30-cm mark, (b) the 50-cm mark, and (c) at the 90-cm mark.

27. For the circuit in the accompanying illustration, (a) calculate the current in the 100-Ω resistor and (b) use Kirchhoff's laws to find the current through each of the batteries.

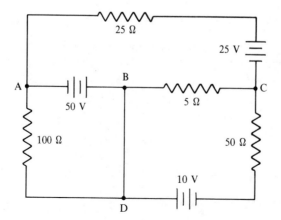

37.6 RC CIRCUITS WITH VARIABLE CURRENTS

28. The following questions refer to the circuit shown. Obtain *numerical* answers for each quantity in the order given.

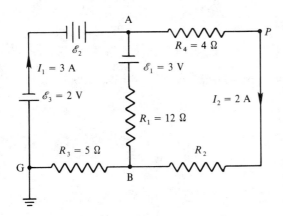

(a) Find the potential of point P with respect to ground.
(b) Calculate the power supplied to the circuit by the emf $\mathscr{E}_1$.
(c) Determine the rate of production of Joule heat in the resistor R_1.
(d) Find the value of the resistance R_2.

37.6 RC CIRCUITS WITH VARIABLE CURRENTS

We now have three elements that can be used in electric circuits: batteries, resistors, and capacitors. We've already explored the physics behind combinations of capacitors and combinations of resistors with batteries. The natural thing to explore next is circuits containing both resistors and capacitors, known as *RC* circuits. As we'll see, the combination of resistance and capacitance determines the rate at which current can increase or decrease.

Consider first the circuit shown in Fig. 37.11, which consists of a charged capacitor C, a resistor R, and a switch S (the symbol ✓• is used to denote an open switch). Suppose that the capacitor has a charge q_0 on it, so that the potential difference across it, given by Eq. (34.13), is $V_0 = q_0/C$. With the switch closed, a closed path is formed allowing positive charges on one capacitor plate to flow through the circuit toward the negative charges on the other plate. At the instant the switch is closed (at time $t = 0$), the charged capacitor acts like a battery of emf $V_0 = q_0/C$ pushing charges through the circuit. By Ohm's law, the current at this instant, I_0, is equal to the potential of the capacitor divided by the resistance: $I_0 = V_0/R$. Eventually the charge on the capacitor is neutralized and the current drops to zero, as illustrated by the graph of current versus time in Fig. 37.12. So we expect the current in the circuit to start out at V_0/R and decrease to zero; similarly the charge starts out at q_0 and decreases to 0, as shown by the graph of charge versus time in Fig. 37.12.

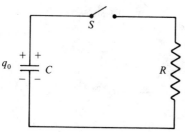

Figure 37.11 A simple *RC* circuit consisting of a capacitor, a resistor, and a switch.

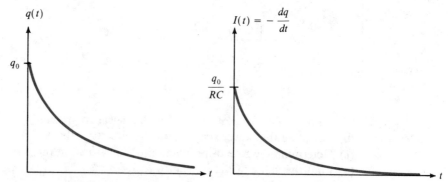

Figure 37.12 Graphs of charge and current versus time for the *RC* circuit of Fig. 37.11 after closing the switch.

Now that we have described qualitatively what happens in an *RC* circuit, let's describe it quantitatively. At time t after the switch is closed, the charge that flows through the wires must come from the capacitor, and the rate of change of charge on the capacitor is just the current that flows through the circuit:

$$I = -\frac{dq}{dt}.$$

The minus sign is needed to make the current positive because dq/dt is negative, resulting from a *decrease* of positive charge on the positively charged capacitor plate. Now apply Kirchhoff's first law. Adding the changes in potential in the direction of the current at any time t, we get

$$q/C - IR = 0.$$

Substituting the derivative of the charge for the current, we obtain the differential equation

$$\frac{q}{C} + R\frac{dq}{dt} = 0,$$

37.6 RC CIRCUITS WITH VARIABLE CURRENTS

or

$$\frac{dq}{dt} = -\frac{1}{RC}q. \tag{37.15}$$

This is a familiar type of differential equation that states the derivative of a quantity (here q) is proportional to the quantity itself. We already know the solution to such an equation:

$$q(t) = q_0 e^{-t/(RC)}, \tag{37.16}$$

where q_0 is the initial charge at time $t = 0$. The quantity RC is called the *time constant* and is denoted by τ. It is characteristic of the circuit and determines how rapidly the capacitor discharges.

Once we have the solution $q(t)$, we can find anything else we want to know about the circuit. For example, the potential difference across the capacitor at any instant, V_C, is given by $V_C = q(t)/C$. The current through the resistor is found from $I = -dq/dt$, and yields

$$I(t) = \frac{q_0}{RC} e^{-t/\tau}. \tag{37.17}$$

From this we find that the potential drop V_R across the resistor is given by $V_R = IR = (q_0/C)e^{-t/\tau}$. The functions $q(t)$ and $I(t)$ are exponentially decreasing functions, plotted in Fig. 37.12.

Now that we understand the effect of capacitance and resistance in the same circuit, let's analyze a slightly more complicated, but immensely more practical circuit – an RC circuit connected to a battery with constant emf $\mathscr{E}$, as shown in Fig. 37.13.

This time we assume that the capacitor is initially uncharged, and that the switch is closed at time $t = 0$. We expect that charge will flow out of the battery and begin to accumulate on the plates of the capacitor. The current in this case deposits positive charge on the positive plate of the capacitor, and hence q increases, dq/dt is positive, and the current is given by

$$I = \frac{dq}{dt}.$$

Applying Kirchhoff's second law we get

$$\mathscr{E} - IR - q/C = 0.$$

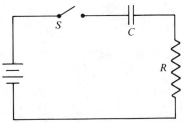

Figure 37.13 An RC circuit with an emf provided by a battery.

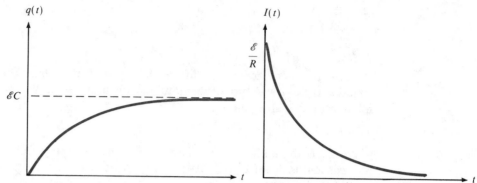

Figure 37.14 Graphs of $q(t)$ and $I(t)$ for the circuit in Fig. 37.13.

Substituting for the current and rearranging terms, we obtain a differential equation that describes the time evolution of the charge on the capacitor:

$$RC\frac{dq}{dt} + q = \mathcal{E}C.$$

We learned how to solve this type of differential equation in Chapter 12. The solution, with the initial condition that $q = 0$ at time $t = 0$, is

$$q(t) = \mathcal{E}C(1 - e^{-t/(RC)}).$$

From this it follows that the current at any instant is given by

$$I(t) = (\mathcal{E}/R)e^{-t/(RC)}.$$

The charge and current as functions of time are plotted in Fig. 37.14. From our analysis and the resulting graphs, we see that initially (at $t = 0$) the current is given by $\mathcal{E}/R$; the capacitor temporarily acts like a *short circuit* or a wire. Only the resistance affects the current. On the other hand, after a very long time ($t = \infty$), the capacitor becomes fully charged and the current becomes zero; at this time the capacitor acts like an *open circuit*, preventing any current flow.

One might wonder what portion of the energy supplied to the circuit is stored in the capacitor, and what portion is dissipated as heat. The energy added to the circuit comes from the battery, and from Eq. (37.12) the rate at which energy is supplied is given by

$$P = I\mathcal{E}. \tag{37.12}$$

The total energy supplied to the circuit is the integral of power over time, from $t = 0$ to $t = \infty$:

$$E_{\text{bat}} = \int_0^\infty P\,dt.$$

Using Eq. (37.12) together with our preceding expression, we get

$$E_{\text{bat}} = (\mathcal{E}^2/R)\int_0^\infty e^{-t/(RC)}\,dt = -\mathcal{E}^2 C e^{-t/(RC)}\big|_0^\infty = \mathcal{E}^2 C.$$

The resistor dissipates energy in the form of heat at a rate given by

$$P_R = I^2 R. \tag{37.11}$$

37.6 RC CIRCUITS WITH VARIABLE CURRENTS

Substituting for the current in this expression and integrating over time to find the total energy dissipated, we get

$$E_R = \frac{\mathscr{E}^2}{R}\int_0^\infty e^{-2t/(RC)}\,dt = -\frac{\mathscr{E}^2 C}{2}e^{-2t/(RC)}\Big|_0^\infty = \frac{\mathscr{E}^2 C}{2}.$$

From Eq. (34.21) we know that the energy stored in a capacitor that has a charge q on it is given by

$$E_C = \frac{q^2}{2C}. \tag{34.21}$$

After an infinite amount of time, the charge on the capacitor is simply $\mathscr{E}C$, so that the energy stored in the capacitor is

$$E_C = \frac{\mathscr{E}^2 C}{2}.$$

Thus we see that

$$E_{\text{bat}} = E_R + E_C.$$

In other words, energy is strictly conserved, as, of course, it must be.

Example 7

The capacitors in the circuit shown are uncharged. After the switch is closed find (a) the initial current through the battery, (b) the final current through the battery, and (c) the final charges on the capacitors.

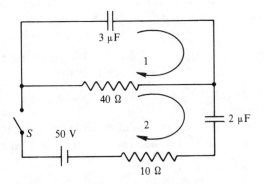

(a) We know that when the switch is closed the capacitors initially act like short circuits; in other words, they act like wires. At that instant no current will flow through the 40-Ω resistor because it is shorted out (the voltage across it is zero) and so the effective resistance of the circuit is only 10 Ω. Applying Kirchhoff's second law to the outer loop obtained by removing the 40-Ω resistor, we get $50 - 10I_0 = 0$, which means that $I_0 = 5$ A.

(b) After a very long time the capacitors act like open circuits – no currents flow through branches containing them. Therefore, the current through the battery becomes zero.

(c) Applying Kirchhoff's second law to loop 1 in the circuit, knowing that the current in all branches is zero, we get $-q_3/C_3 + 40I_{40} = 0$, which gives $q_3 = 0$. Similarly applying Kirchhoff's rule to loop 2, we obtain

$$50 - 40I_{40} - q_2/C_2 - 10I_{10} = 0,$$

and since $I_{40} = I_{10} = 0$, we have $q_2 = 50\,C_2 = 100\ \mu\text{C}$.

Questions

29. An initially charged capacitor of capacitance C is discharged through a resistor R starting at time $t = 0$. Show that at time $t = \tau = RC$ the charge on the capacitor is reduced to 37% of its initial charge.

30. Initially the capacitors in the circuit shown are uncharged and at time $t = 0$ the switch is closed. Determine (a) the initial value of the current through the battery, (b) the current through the battery after a long time, and (c) the charge on each capacitor after a long time.

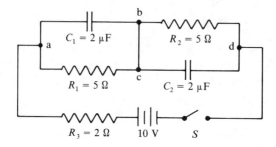

31. In the circuit shown for Question 30, after a very long time the switch is again opened. Find the initial currents through each resistor.

32. In the circuit shown, capacitor C_2 initially has no charge and capacitor C_1 initially has a charge given by $C_1\mathcal{E}/2$. At time $t = 0$ the capacitors are inserted into the circuit as shown. Find (a) the initial currents through each resistor, (b) the final current through the battery, and (c) the final charge on each capacitor.

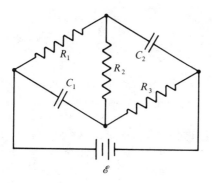

37.7 A FINAL WORD

33. Complete the table for the circuit shown in the diagram in which the switch is closed at time $t = 0$.

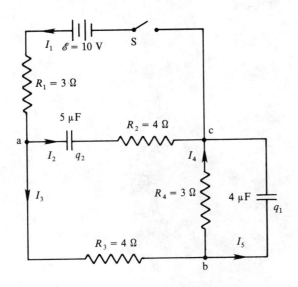

	$t = 0$	$t = \infty$
I_1		
I_2		
I_3		
I_4		
q_1		
q_2		

37.7 A FINAL WORD

While Wheatstone's fame was little or nil, the telegraph flourished and was soon adapted to the ever changing needs of the railroad, the military, and the press. In its earliest stages of commercial use, the telegraph could be found almost exclusively at railway stations – employed there for the routing and scheduling of trains. But even as it becomes a popular subject for newspaper articles, the telegraph soon became one of the primary modes of rapid communication for newspapers themselves, and to suggest that their's had the most up-to-date information, many papers even included the word *telegraph* in their titles, for example, *The Daily Telegraph* in

Figure 37.15 Photograph of Charles Wheatstone. (From *Old Wires and New Waves*: *The History of the Telegraph, Telephone and Wireless* by Alvin F. Harlow. Copyright 1936 by Alvin F. Harlow, renewed 1964 by Mrs. Dora S. Harlow. A Hawthorne Book. Reproduced by permission of E. P. Dutton, a division of New American Library.)

London. And the world still reverberates from the advances in communication initiated by the electric telegraph.

Though he was revered by his colleagues, Wheatstone was not fully appreciated by the public. He was an extremely shy man, morbidly timid before an audience. Legend has it that one evening in 1846, Wheatstone panicked at a lecture at the Royal Institution and fled from the enormous hall filled with colleagues dressed in black tie. Faraday took his place and began speculating on the nature of light, suggesting that light was probably a disturbance of electric, gravitational, and magnetic fields. Although Faraday's ideas were received skeptically at first, they were incorporated into Maxwell's successful theory of electromagnetism 20 years later.

Today, public lectures are still held at the Royal Institution in which famous scientists, dressed in black tie, expound on their latest work. But ever since that day in 1846, it has been customary to lock the lecturer in a little room half an hour before his talk, lest he "do a Wheatstone."

CHAPTER 38

MAGNETISM

In like manner the loadstone has from nature its two poles, a northern and a southern; fixed, definite points in the stone, which are the primary termini of the movements and effects, and the limits and regulators of the several actions and properties. ...Whether its shape is due to design or to chance, and whether it be long, or flat, or four-square, or three-cornered, or polished; whether it be rough, broken-off, or unpolished: the loadstone ever has and ever shows its poles.

William Gilbert, *De Magnete* (1600)

38.1 LODESTONES AND MAGNETIC NEEDLES

Every child who plays with magnets discovers anew what humankind has known for centuries. The ancient Chinese are believed to have observed magnetism, but the ancient Greeks left the earliest written accounts. The Greeks recorded the attractive properties of magnetite, a mineral found in Magnesia, a district of Asia Minor. It had long been known that a bar of magnetite suspended on a thread always points approximately to the north. The rocks containing magnetite became known as *lodestones*, or "leading stones," because they could be used to direct wanderers. By the eleventh century, lodestones were used to indicate directions at sea as well as on land.

In the thirteenth century an important discovery was made by Peter Peregrinus, a Crusader also known as Pierre de Maricourt, who investigated properties of lodestones. By placing a magnetized needle at various positions on a spherical lodestone and tracing out the lines along which the needle pointed, he found that the lines converged at two opposite points on the sphere, which he called *poles*. Peregrinus also discovered that each fragment of a broken magnet is itself a magnet, and that unlike poles attract, while like poles repel. He later designed a compass with a pivoted needle and a graduated scale.

Navigation by magnetic compass gradually replaced navigation by stars, but some seafarers continued to believe that a compass needle was attracted by stars in the Great Bear (the Big Dipper). Although the magnet had a practical use, it was still shrouded in an aura of mystery.

The mystery began to be dispelled when Elizabethan England's most distinguished scientist, William Gilbert, published a celebrated treatise entitled *De Magnete* (On the magnet) in 1600. This was the first comprehensive study of magnetism and took 17 years to complete. Gilbert dedicated it to those who look for knowledge "not only in books but in things themselves." The growing interest in compass navigation may have influenced Gilbert somewhat because he wrote *De Magnete* at the time the English were preparing to meet the Spanish Armada. Gilbert lived from 1544 to 1603, roughly the same period as Johannes Kepler. In 1600, when *De Magnete* was published, Giordano Bruno was burned at the stake in Rome because he believed in the Copernican theory. It was also the year in which Johannes Kepler set out to join Tycho Brahe in Prague.

Gilbert, a student of medicine, received his M.D. at Cambridge University in 1569, and by the mid-1570s was a prominent physician in London. In 1600 he became president of the Royal College of Physicians and was appointed as personal physician to Queen Elizabeth I. When she died in 1603, her only personal legacy was a grant to Gilbert to pursue his hobby, physics, but he had little time to enjoy it because he was a victim of the plague a few months later.

For his work on magnets, Gilbert became known as the "Father of Magnetism." He discovered various methods for producing and strengthening magnets. For example, he found that when a steel rod was stroked by a natural magnet the rod itself became a magnet, and that an iron bar aligned in the magnetic field of the earth for a long period of time gradually developed magnetic properties of its own. In addition, he observed that the magnetism of a piece of material was destroyed when the material was sufficiently heated.

One of Gilbert's major discoveries (he credited himself with 21) was that the earth is a huge magnet, a connection that Peregrinus failed to make. He proved that a compass needle swings north because of the magnetism of the earth itself and not – as some believed – because of a star in the Big Dipper or a mysterious range of iron-capped mountains in the North. Using a spherical magnet and a magnetic needle that was free to rotate in a vertical plane that included the magnetic poles of

38.1 LODESTONES AND MAGNETIC NEEDLES

Figure 38.1 Portrait of William Gilbert. (Courtesy Burndy Library.)

the sphere, as shown in Fig. 38.2, he found that the needle dipped below the horizontal (the tangent plane to the sphere) at different angles, depending on its position on the sphere. Gilbert realized that lines joining points of constant magnetic declination (the angle between the magnetic needle and the horizontal) were also lines of constant latitude on a sphere. Impressed with his discovery, he suggested an application to navigation. Although navigators used compasses at sea, they knew that variations in the earth's magnetism often caused a compass to be unreliable. Gilbert thought circles indicating constant magnetic dip on the earth might be more reliable. However, navigators soon found that the dip along latitude lines varied considerably, and so the idea was abandoned.

Although he is chiefly noted for his work in magnetism, Gilbert made many important contributions to the science of electricity, ranging from the invention of the electroscope to the study of conductors and insulators. To him we owe the term "electricity." He also left a large manuscript devoted to speculative work in general science, which was published posthumously in 1651.

Galileo said *De Magnete* made Gilbert "great to a degree that is enviable." The inscription on Gilbert's tomb is more modest. It reads: "He composed a book,

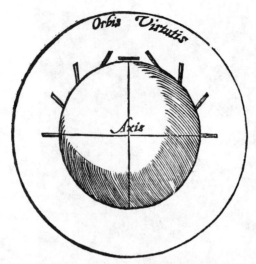

Quò propiores fuerint partes æquinoctiali, eò magis obliquè alliciunt magnetica: at polis viciniores partes magis directè aduocant, in polis directissimè. Eadem etia ratio eft conuersionis magnetu omnium qui funt rotundi & qui funt longi, fed in longis experimentum eft facilius. Nam in quâuis formâ eft verticitas, & funt poli; fed propter malam formam & inæqualem, fæpiùs quibufdam malis impediuntur. Si lapis longus fuerit, vertex verò in finibus, non in lateribus; fortiùs in vertice allicit. Conferunt enim partes vires fortiores in polum rectis lineis, quàm obliquis. Sic lapis, & tellus naturâ confomant motus magneticos.

Figure 38.2 A magnetic needle suspended near a spherical magnet aligns itself at different angles below the horizontal. (From *De Magnete*. Courtesy of the Archives, California Institute of Technology.)

concerning the magnet, celebrated among foreigners and among those engaged in nautical affairs."

38.2 FORCES AND MAGNETIC FIELDS

Anyone who has ever played with magnets is intrigued by their ability to exert force at a distance. How do magnets behave? What factors determine a magnet's force? Quantitative determination of the force between magnets was not made until the eighteenth century when John Michell working in England and Augustin Coulomb in France actually measured the force between two magnetic poles. Individual poles do not exist, but by using long, thin magnetic needles in an attempt to isolate the poles, Michell and Coulomb independently determined that the force between two magnetic poles depends inversely on the square of the distance between the poles, a law similar in form to that for the electric force between charges. If a magnetic pole of strength p_1 is placed in a vacuum at a distance r from another pole of strength

38.2 FORCES AND MAGNETIC FIELDS

p_2, the force between them is described mathematically by the equation

$$\mathbf{F} = K_m \frac{p_1 p_2}{r^2} \hat{\mathbf{r}}. \tag{38.1}$$

Just as two types of electric charge exist, positive and negative, so are there two kinds of magnetic poles: north (positive p) and south (negative p). However, unlike electric charges, positive and negative magnetic poles do not exist independently.

As already mentioned in Chapter 11, pole strengths are measured in SI units of C m/s. The magnetic constant K_m depends on the system of units used; in SI units, it has the defined value

$$K_m = 1.0 \times 10^{-7} \text{ N s}^2/\text{C}^2.$$

Magnets not only attract or repel other magnets, they also attract bits of iron. When iron filings are sprinkled on a piece of cardboard held above a magnet, characteristic patterns are formed, as illustrated in Fig. 38.3. Michael Faraday noticed such patterns and imagined the iron filings to be following lines of force that spread out from the north pole and converged on the south pole. In his *Experimental Researches*, the published account of his investigations, Faraday systematically and elegantly developed the idea of lines of force. He conceived the

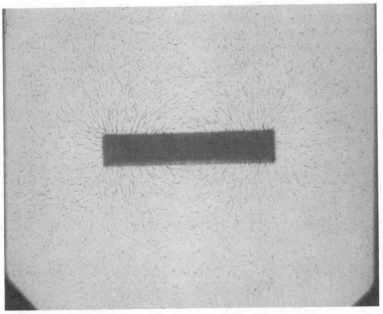

Figure 38.3 Iron filings follow magnetic field lines about a magnet. (Caltech photo by Robert Paz.)

idea of a magnetic field – a collection of lines of force filling the space around a magnet. For Faraday the magnetic field provided a way to describe the behavior of magnets, similar to the way the electric field described the behavior of charges.

The striking resemblance between the force laws for magnetism and electricity suggests that we define the magnetic field mathematically by analogy to the electric field. Even though an isolated magnetic pole does not exist, we can define the magnetic field strength **B** due to an isolated pole p as the force per unit pole:

$$\mathbf{B} = K_m \frac{p}{r^2} \hat{\mathbf{r}}. \tag{38.2}$$

Just as iron filings map out the field of a magnet, the direction and strength of a magnetic field can be mapped out, in principle, by placing a tiny compass at various points in the field and measuring the direction and force on a pole. The direction of the field is the same as the direction of the needle. By measuring the force on a known pole p_0, which is given by

$$\mathbf{F} = p_0 \mathbf{B}, \tag{38.3}$$

the strength of the field can also be determined at each point. The SI unit of the magnetic field is the tesla (T); in Section 38.5 we will find the connection between this derived unit and basic units.

Observations of iron filings and compass needles placed near a magnet suggest a way of visualizing magnetic field lines. The crowding of iron filings near the poles of a magnet suggests that the strength of the field is greater in those regions. By convention, outside the magnet the field points from the north pole to the south pole. Field lines joining the two poles are always drawn to form closed loops, as illustrated in Fig. 38.4, with the lines outside the magnet continuing inside to close the loops. Michell was the first to show that the two poles of a magnet always have equal strength.

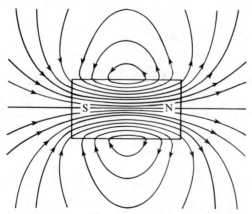

Figure 38.4 Magnetic field lines outside a magnet flow from the north pole to the south pole, forming closed loops.

38.2 FORCES AND MAGNETIC FIELDS

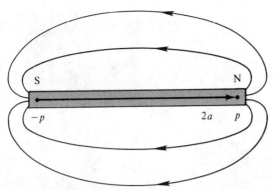

Figure 38.5 A magnetic dipole consists of two equal and opposite poles a small distance apart.

Because magnetic poles always occur in equal and opposite pairs, every magnet can be said to form a *magnetic dipole*. Analogous to electric dipoles, a magnetic dipole consists of two poles separated by a small distance. The *dipole moment* is defined as the product of one of the pole strengths times the distance between poles. If poles of strength p are separated by a displacement vector $2\mathbf{a}$ from south pole to north pole, as indicated in Fig. 38.5, the vector dipole moment is given by

$$\mathbf{m} = 2p\mathbf{a}, \tag{38.4}$$

and the scalar dipole moment is $m = 2pa$.

Magnetic dipoles help us understand the behavior of magnets. Microscopic observations reveal that a magnet consists of distinct regions, called *domains*, about 1 mm in size, that behave as shown in Fig. 38.6. Each domain possesses a north and south pole and behaves like a tiny dipole. In an unmagnetized piece of iron the domains tend to align so as to cancel one another's fields. *Magnetization* refers to the degree of alignment of the domains, which occurs as indicated in Fig. 38.6. As Gilbert noted, a piece of iron can be magnetized by stroking it with a magnet. The magnet's attraction causes the domain boundaries to shift, which allows the better aligned domains to grow at the expense of the poorly aligned ones. When the magnet is removed, the domains remain aligned and produce a net magnetic dipole moment.

The idea of domain explains many other observations about magnets, as, for example, why a magnet can pick up unmagnetized pieces of iron, such as paper clips. The magnet's field causes domains near it to become slightly aligned, thereby making each paper clip a temporary magnet. An opposite pole is formed in the paper clip and is attracted to the nearby pole of the magnet.

Gilbert's observation that heating a magnet can diminish or even destroy its magnetism can be explained in terms of domains. Heating causes an increase in the random thermal motions of the atoms in the domains. This tends to increase the randomness of the domain orientations and hence to decrease the magnetization.

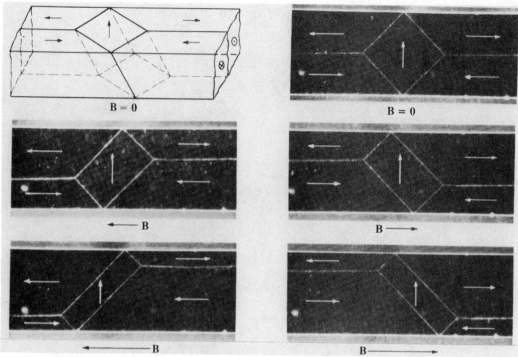

Figure 38.6 Photographs showing the behavior of magnetic domains when an external magnetic field **B** is applied. (From *Introduction to Solid State Physics* by Charles Kittel. Copyright © 1956, Charles Kittel. Reprinted by permission of John Wiley & Sons, Inc. Photo by C. D. Graham.)

As we will see in Chapter 39, a theory that some atoms themselves possess atomic magnetic dipoles leads to a model for the ultimate source of magnetism in magnetic materials.

Example 1

Show that the magnetic field along the axis of a dipole at a point that is a distance r from the center is given approximately by

$$\mathbf{B} = K_m \frac{2\mathbf{m}}{r^3},$$

where $\mathbf{m} = 2p\mathbf{a}$ is the dipole moment.

The dipole consists of two poles of strengths p and $-p$ separated by a distance $2a$. Using Eq. (38.2) we can calculate the field due to each pole at point P in the accompanying diagram. The vector sum of the two fields gives the total field.

38.2 FORCES AND MAGNETIC FIELDS

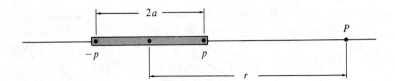

The distance from pole p to point P is $r - a$, so the field due to this pole is simply

$$\mathbf{B}_+ = K_m \frac{p}{(r-a)^2} \hat{\mathbf{r}}.$$

The distance from the pole $-p$ to point P is $r + a$, and the field due to that pole is

$$\mathbf{B}_- = K_m \frac{-p}{(r+a)^2} \hat{\mathbf{r}}.$$

Therefore, the total field at P is the vector sum,

$$\mathbf{B} = \mathbf{B}_+ + \mathbf{B}_- = K_m p \left[\frac{1}{(r-a)^2} - \frac{1}{(r+a)^2} \right] \hat{\mathbf{r}}.$$

Adding the fractions we find

$$\mathbf{B} = K_m p \frac{4ar}{(r^2 - a^2)^2} \hat{\mathbf{r}}.$$

The denominator can be written as

$$(r^2 - a^2)^2 = r^4 (1 - (a/r)^2)^2,$$

and we get

$$\mathbf{B} = K_m \frac{4pa}{r^3 (1 - a^2/r^2)^2} \hat{\mathbf{r}}.$$

This last result is exact, but now we make the approximation that the separation distance of the dipole is small compared to the distance from the dipole to point P. Thus if a/r is small, we can ignore the term a^2/r^2 in the parentheses and obtain the approximation

$$\mathbf{B} \approx K_m \frac{2}{r^3} \mathbf{m}, \tag{38.5}$$

where $\mathbf{m} = 2pa\hat{\mathbf{r}}$ is the vector dipole moment. This approximate field is known as the *dipole field* and is useful in both magnetism and electricity.

Questions

1. Using the idea of magnetic domains, explain qualitatively why a magnet loses some of its magnetism when it is dropped on the floor or is struck sharply with a hammer.

2. Two magnets are positioned near each other as shown. Sketch some magnetic field lines for these magnets.

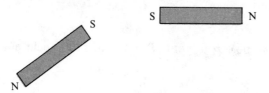

3. The north pole of a long, thin magnet with pole strength 2.0 C m/s, is 20 cm from the south pole of a similar magnet. Determine the force on the north pole.

4. A magnet consists of two poles, each of strength 0.5 C m/s, separated by 1.0 cm. Determine the magnetic field a distance of 9 cm from the north pole and along the axis of the magnet, and compare (a) the exact value with (b) the dipole field approximation.

5. Each of two identical magnets has poles of strength 0.40 C m/s separated by 2.0 cm. The magnets are aligned as shown, with their centers separated by 20.0 cm. Determine the net force on either magnet.

6. A magnet has poles of strength p separated by a distance $2a$. Find the field of the magnet at a point located a distance r along the perpendicular bisector of the magnet, as indicated in the diagram.

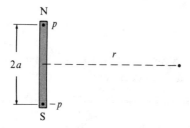

7. Assuming that the magnetic field of the earth is like that of a short dipole buried at the center of the earth, calculate the dipole moment of the earth if the dipole field has a strength of 5.0×10^{-5} T at the north pole. (The radius of the earth is 6.4×10^6 m.)

8. Subatomic particles such as protons possess magnetic dipole moments. If the magnetic dipole moment of a proton is 1.4×10^{-26} C m²/s, find the magnetic field at a distance of 2.0×10^{-10} m along the direction of the magnetic dipole moment vector of a proton.

38.3 MAGNETS AND TORQUES

We are now in a position to explain why a compass needle points north. Think of a compass needle as a magnetic dipole with moment **m** in a uniform magnetic field **B**, as illustrated in Fig. 38.7. A force given by Eq. (38.3),

$$\mathbf{F} = p\mathbf{B}, \tag{38.3}$$

acts on each pole, and although these forces are equal in magnitude and opposite in direction, they tend to rotate the dipole if it is not aligned with the field. So the net force is zero, but the net torque is not zero. Because torque is given by the cross product

$$\boldsymbol{\tau} = \mathbf{r} \times \mathbf{F}, \tag{23.9}$$

the torque about the center due to the north pole is

$$\boldsymbol{\tau}_N = \mathbf{a} \times p\mathbf{B}.$$

The torque about the center due to the south pole $\boldsymbol{\tau}_S$ has the same magnitude, and the right-hand rule reveals it to be in the same direction, so $\boldsymbol{\tau}_S = \boldsymbol{\tau}_N$. Therefore, the net torque on the dipole is given by

$$\boldsymbol{\tau} = 2\mathbf{a} \times p\mathbf{B} = 2p\mathbf{a} \times \mathbf{B},$$

or

$$\boldsymbol{\tau} = \mathbf{m} \times \mathbf{B}, \tag{38.6}$$

where **m** is the dipole moment. The torque is zero when the dipole and the magnetic field are parallel, in other words, when the dipole is aligned in the magnetic field. Consequently, a dipole placed in a uniform magnetic field experiences a torque that

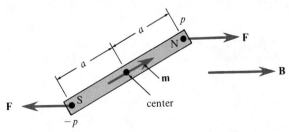

Figure 38.7 Forces acting on a magnetic dipole in a magnetic field create a torque.

tends to rotate it into alignment with the field. For that reason a compass needle aligns itself with the earth's magnetic field and points north.

If a dipole is aligned with a magnetic field, work is required to rotate the dipole against the field. Suppose we rotate the dipole through some angle by applying equal and opposite forces **F** of magnitude pB at the ends of the dipole. The work done by these forces is twice that required to rotate one of the poles, so is given by a line integral

$$W = 2\int \mathbf{F} \cdot d\mathbf{r},$$

where $\mathbf{r} = a\hat{\mathbf{r}}$ is the radius vector that traces out an arc of a circle of radius a, say, subtending the angle of rotation. Using polar coordinates, we have

$$\frac{d\mathbf{r}}{d\theta} = a\frac{d\hat{\mathbf{r}}}{d\theta} = a\hat{\boldsymbol{\theta}},$$

so if the dipole is rotated through an angle α, the work required is

$$W(\alpha) = 2\int_0^\alpha \mathbf{F} \cdot \frac{d\mathbf{r}}{d\theta}\, d\theta = 2a\int_0^\alpha \mathbf{F} \cdot \hat{\boldsymbol{\theta}}\, d\theta.$$

From Fig. 38.8 we see that

$$\mathbf{F} \cdot \hat{\boldsymbol{\theta}} = F\cos(\pi/2 - \theta) = F\sin\theta,$$

and hence the work done is given by

$$W(\alpha) = 2aF\int_0^\alpha \sin\theta\, d\theta = 2apB(1 - \cos\alpha) = mB(1 - \cos\alpha),$$

where $m = 2ap$ is the dipole moment. Note that when $\alpha = 0$ no work is required, and when $\alpha = \pi$ the dipole is reversed end for end, requiring an amount of work $W(\pi) = 2mB$.

The work done on the dipole goes into potential energy stored in the dipole by virtue of its position in the magnetic field. The foregoing calculation shows that the

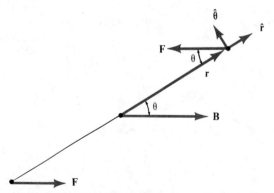

Figure 38.8 Work done in rotating a dipole in a magnetic field appears as potential energy.

change in potential energy is

$$\Delta U = mB(1 - \cos \alpha). \tag{38.7}$$

Example 2
A small magnet is placed in a uniform magnetic field of 0.05 T. The maximum torque on the magnet is measured to be 0.10 N m.
  (a) Determine the dipole moment of the magnet.
  (b) If the length of the magnet is $2a = 10$ cm, find the pole strength.
  (c) Calculate the change in potential energy if the magnet is initially aligned with the field and then rotated 45°.

  (a) According to Eq. (38.6) the maximum torque on the magnet is given by $\tau = mB$. Solving for the dipole moment we get

$$m = \tau/B = (0.10 \text{ N m})/(0.05 \text{ T}) = 2.0 \text{ C m}^2/\text{s}.$$

  (b) Because the dipole moment is given by $m = 2pa$, the pole strength is found to be

$$p = m/(2a) = (2.0 \text{ C m}^2/\text{s})/(0.10 \text{ m}) = 20 \text{ C m/s}.$$

  (c) Using Eq. (38.7) we find that the change in potential energy is

$$\Delta U = mB(1 - \cos 45°) = 0.03 \text{ J}.$$

Questions

9. A small magnet of length 1.5 cm is held at an angle of 30° in a uniform magnetic field of 0.02 T. The torque exerted on it is measured to be 0.30 N m.

  (a) Determine the dipole moment of the magnet.
  (b) Find the pole strength.

10. The dipole moment associated with each atom of iron in an iron bar is 1.8×10^{-23} C m^2/s. Assume that all the atoms in an iron bar of length 5.0 cm and cross-sectional area 2.0 cm^2 have their dipole moments aligned.

  (a) Determine the magnetic moment of the iron bar. (Take the density of iron to be 7.9 g/cm^3.)
  (b) Calculate the torque that must be exerted on the bar to hold it at 60° to the direction of a uniform magnetic field of 1.5 T.

38.4 GAUSS'S LAW FOR MAGNETISM

Another physical quantity associated with magnetism is *magnetic flux*, which can be defined by analogy with electric flux. The magnetic flux passing through a surface S

is defined by the surface integral

$$\phi = \iint_S \mathbf{B} \cdot d\mathbf{A}. \tag{38.8}$$

The SI unit of magnetic flux is called the weber (Wb), which is equal to one tesla meter squared, 1 Wb = 1 T m². In Chapter 33 we found that the electric flux through a closed surface is equal to a constant times the net charge enclosed by the surface. That is Gauss's Law, which quantifies the fact that electric charges are the sources of electric fields.

What are the sources of magnetic fields? This question will be addressed in Chapter 39, but now we pursue the analogy of Gauss's law for magnetism. Gilbert, Michell, and others noted that a magnet always consists of two opposite poles of equal strength, and if a magnet is broken, each piece is itself a magnet with two poles. By repeatedly breaking the pieces we cannot produce a single magnetic pole – a magnetic monopole. Even today, physicists search for magnetic monopoles, but none has yet been found.

We now cast this empirical observation in terms of magnetic flux. Consider a bar magnet like that shown in Fig. 38.9. Because magnetic monopoles do not exist, the magnetic field lines have no "sources" or "sinks" and, consequently must form closed loops, as illustrated in the figure. Now imagine a closed surface S through which magnetic field lines flow, as suggested in Fig. 38.9. The surface S consists of two portions: S_1, where the field lines enter S, and S_2, where the field lines leave S. The flux through S thus consists of two parts, the flux exiting S through S_2, which is positive, and a negative flux of the same magnitude entering S through S_1. The

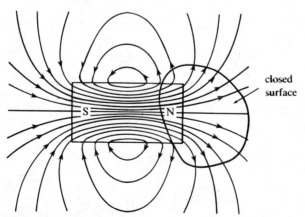

Figure 38.9 The magnetic flux through any closed surface is zero.

38.4 GAUSS'S LAW FOR MAGNETISM

total flux through the closed surface S is therefore zero:

$$\phi = \oiint_S \mathbf{B} \cdot d\mathbf{A} = 0. \tag{38.9}$$

Equation (38.9) is known as *Gauss's law for magnetism*. It summarizes the observation that magnetic monopoles do not exist and that as a consequence magnetic field lines form closed loops.

Example 3

Astrophysicists believe that neutron stars possess extremely large magnetic fields and are formed by the collapse of a massive star. Assuming that a star the size of the sun (radius 7.0×10^8 m), having a magnetic field at its surface of 1.0×10^{-4} T, collapses to the size of the earth (radius 6.4×10^6 m), find the magnetic field at the surface of the collapsed star.

Imagine two concentric spheres about the star, one of which has a radius equal to the original radius of the star, R_0, and the other that has a radius R equal to that of the collapsed star, as illustrated. The same amount of magnetic flux that flows out through the surface of the upper hemisphere of radius R_0 also flows out through the upper hemisphere of radius R. Equating these fluxes we get

$$B(2\pi R^2) = B_0(2\pi R_0^2).$$

Solving for B, the magnetic field at the surface of the collapsed star, we find

$$B = B_0(R_0/R)^2 = (1.0 \times 10^{-4} \text{ T})[(7.0 \times 10^8)/(6.4 \times 10^6)]^2 = 1.2 \text{ T}.$$

Because of flux conservation, the magnetic field of the collapsed star is increased by four orders of magnitude. For a larger collapse the increase in the magnetic field can be even more dramatic.

Questions

11. State some reasons why you would expect the flux of the earth's magnetic field to be greater through the state of Alaska than through Kansas.

12. A certain neutron star is believed to have a magnetic field at its surface of 100 T. If the star originally had a radius 10^5 times larger than its present radius, what was the magnetic field at the surface of the star before it collapsed?

38.5 MAGNETIC FORCE ON MOVING CHARGES

The effects of electric forces on charges were observed since antiquity, just as the magnetic forces on magnets were observed. And since the time of Oersted, magnetic forces on currents have been explored. The advent of electron beams (cathode rays) in the latter part of the nineteenth century vividly demonstrated that moving charged particles are influenced by magnetic fields. For example, a beam of electrons is deflected when a magnet is brought near it, as indicated in Fig. 38.10. The direction of the deflection, and hence the force on the electrons, depends on the direction of the magnetic field; if a south pole is brought near the beam, the deflection is opposite to that caused by a north pole. Surprisingly, the direction of the force is found to be perpendicular to both the velocity vector and to the magnetic field.

If **v** is the velocity of a moving (positive or negative) charge q, and if **B** is the magnetic field vector, experiment shows that the magnitude of the force is given by

$$F = |q|vB \sin \theta, \tag{38.10}$$

where θ is the angle between **v** and **B**. The direction of the force **F** is given by the right-hand rule, as illustrated in Fig. 38.11 for a positive charge q. Because the force **F** is perpendicular to both **v** and **B** it can be expressed very simply as a cross product:

$$\boxed{\mathbf{F} = q\mathbf{v} \times \mathbf{B}.} \tag{38.11}$$

Equation (38.11) holds whether q is positive or negative. We also have $\mathbf{F} \cdot \mathbf{v} = 0$ because **F** and **v** are perpendicular. Because $\mathbf{F} \cdot \mathbf{v}$ is the power, the rate of work done on a moving particle is zero, and thus the kinetic energy is constant.

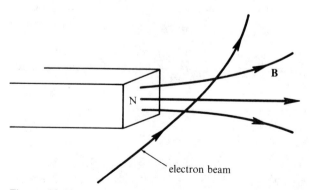

Figure 38.10 An electron beam is deflected by a magnetic field.

38.5 MAGNETIC FORCE ON MOVING CHARGES

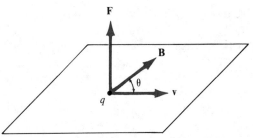

Figure 38.11 The magnetic force on a moving charge depends on its charge and velocity, and on the magnetic field.

In Chapter 33 the electric field was defined by means of the force on a charge at rest. In Section 38.2 we defined the magnetic field in a similar way by examining the force on a magnetic pole. Because magnetic fields also exert forces on *moving* charges, this force can also be used to define the magnetic field in a way that can be easily measured. If a known charge is moving with a known velocity perpendicular to a magnetic field and the force on it is measured, the magnetic field strength can be deduced from

$$B = \frac{F}{qv}.$$

By international agreement, a coulomb is defined in such a way that a charge of 1 C moving at a velocity of 1 m/s perpendicular to a magnetic field of 1 T experiences a force of 1 N. The SI unit of magnetic field intensity, the tesla, is therefore related to other SI units by 1 T = 1 N s/C m. Because a field of 1 T is rather large, magnetic fields are often given in a more common unit, the gauss (G), which is 1/10,000 of a tesla, 1 G = 10^{-4} T.

Suppose that a particle of mass m, charge q, velocity $\mathbf{v}$ is injected into a chamber that has a uniform magnetic field $\mathbf{B}$. What is the subsequent motion of the particle? First, the force $\mathbf{F}$ on the particle is always perpendicular to the velocity. Figure 38.12 is drawn in the plane of $\mathbf{v}$ and $\mathbf{F}$. The magnetic field $\mathbf{B}$ is perpendicular to this plane, and we use the symbol $\otimes$ (indicating the tail of an arrow) if $\mathbf{B}$ points

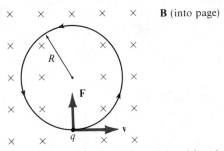

Figure 38.12 A charged particle with velocity perpendicular to a uniform magnetic field moves in a circle.

down into the page, and the symbol ⊙ (indicating the tip of an arrow) if **B** points up out of the page. Although the force cannot change the speed of the particle, it can change its direction. If the magnetic field is constant and also perpendicular to the velocity, the magnetic force acts as a centripetal force and the particle moves in a circle about some center, as illustrated in Fig. 38.12, much like the tension in a string that keeps a twirling rock moving in a circle.

The motion of the particle can be analyzed by applying Newton's second law, $F = ma$, with the centripetal acceleration given by v^2/R, where R is the radius of the circle. Substituting for the force, we get

$|q|vB = mv^2/R.$

The radius R, called the Larmor radius, is given by

$$R = \frac{mv}{|q|B}. \tag{38.12}$$

This shows that R is determined by the momentum and charge of the particle and by the strength of the magnetic field. The angular frequency with which the particle moves around the circle is given by

$$\omega = v/R = |q|B/m. \tag{38.13}$$

Note that this frequency does not depend on the speed of the particle nor on the radius R. Early developers of particle accelerators exploited these facts to accelerate particles in circles in a device known as a cyclotron. For this reason, the frequency ω in Eq. (38.13) is often called the cyclotron frequency.

Figure 38.13 shows the path of an electron in a bubble chamber, a device used to detect charged particles. The chamber, containing liquid hydrogen that is ready to boil, is immersed in a strong magnetic field. When a charged particle moves through the chamber, it ionizes hydrogen atoms and leaves a trail of bubbles along its path. The orbit of the electron is not along a circle but along a spiral, because collisions of the electron with hydrogen atoms decrease the speed of the electron. As the speed decreases, Eq. (38.12) suggests that the radius of the orbit should also decrease, as indeed it does, and the electron spirals inward.

Example 4

A charged particle with a speed of 1.5×10^2 m/s enters a region where there is a magnetic field of 4.5×10^{-3} T and is observed to move in a circle of radius 0.5 m. Determine the minimum mass of the particle.

According to Eq. (38.13), the mass of the particle is given by

$m = |q|B/\omega,$

where $\omega = v/R = 3.0 \times 10^2$ rad/s. To find the minimum mass we use the minimum $|q|$, which is that of an electron, 1.6×10^{-19} C. Substituting values we get

$m = (1.6 \times 10^{-19} \text{ C})(4.5 \times 10^{-3} \text{ T})/(3 \times 10^2 \text{ rad/s}) = 2.4 \times 10^{-24} \text{ kg}.$

38.5 MAGNETIC FORCE ON MOVING CHARGES

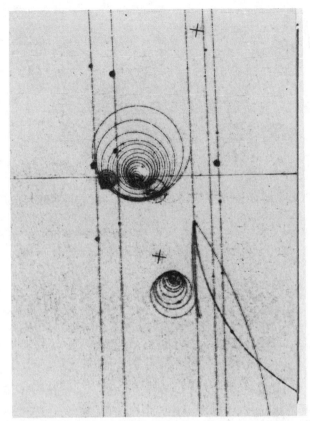

Figure 38.13 Bubble chamber photograph of the spiral path of an electron. (Courtesy of the California Institute of Technology.)

What happens if a charged particle enters a region of a uniform magnetic field with a velocity that is not perpendicular to the field? According to Eq. (38.11) the magnetic force on the particle acts in a direction perpendicular to the field and depends on the component of the velocity perpendicular to the field:

$$F = |q|v_\perp B.$$

No force acts in the direction parallel to the field, and hence the component of the velocity parallel to the field remains constant. The particle undergoes uniform circular motion in a circle of radius

$$R = mv_\perp/(|q|B) \tag{38.12}$$

in a plane perpendicular to the magnetic field and it simultaneously executes uniform straight-line motion in the direction of the field. Consequently, the path of the particle is a helix, as illustrated in Fig. 38.14.

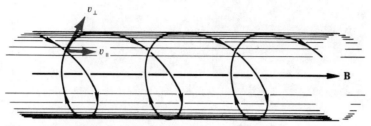

Figure 38.14 Helical motion results when a particle moves in a uniform magnetic field and has a component of velocity parallel to the field.

Although the motion of a charged particle in a nonuniform magnetic field is quite complicated to analyze, we can qualitatively explain a spectacular phenomenon – the Aurora Borealis. Suppose that a positively charged particle is moving in a region where there is a nonuniform magnetic field **B**, as shown in Fig. 38.15. At point 1 the velocity of the particle is into the page and **B** has both vertical and horizontal components. The force on the particle, $q\mathbf{v} \times \mathbf{B}$, has both vertical and horizontal components too, as indicated in Fig. 38.15. The vertical component of force causes the particle to move in a circle, whereas the horizontal component accelerates the particle to the right. However, the speed of the particle cannot be changed by the magnetic field. An increase in the horizontal component of velocity is accompanied by a decrease in the vertical component. Thus, by Eq. (38.12), the particle moves in increasingly larger circles as it accelerates to the right. Because the magnetic field is nonuniform, this motion does not continue forever. At point 2, for example, the magnetic field is horizontal and so the particle experiences no horizontal acceleration at that point. Nonetheless, by its inertia it still moves toward the right. At point 3, the magnetic field has a negative vertical component that causes the particle to accelerate to the left and thereby to slow down. But the magnetic field strength is greater at point 3 than at point 2, so the particle moves in successively smaller circles and eventually acquires a horizontal velocity component

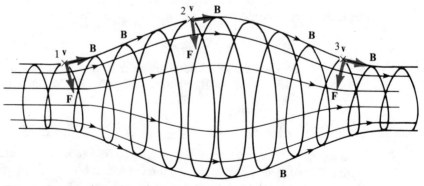

Figure 38.15 A charged particle can become trapped by a nonuniform magnetic field in a region known as a magnetic bottle.

38.5 MAGNETIC FORCE ON MOVING CHARGES

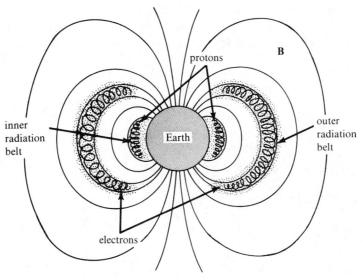

Figure 38.16 The Van Allen radiation belts are regions where charged particles are trapped in the earth's magnetic field.

to the left. Thus, a particle can become trapped by a nonuniform magnetic field in a region known as a magnetic bottle.

The earth's magnetic field forms a huge magnetic bottle. The field is stronger nearer the poles, and charged particles emitted from the sun can become trapped in the field. Regions where protons and electrons are trapped in the magnetic field of the earth are known as Van Allen radiation belts. The trapped particles spiral along the field lines between north and south poles, being reflected in the polar regions, as illustrated in Fig. 38.16. In polar regions these trapped particles penetrate the atmosphere and excite oxygen and nitrogen molecules, which emit light on deexcitation. The light emitted by these molecules forms the Aurora Borealis (and also the Aurora Australis in the southern hemisphere). That's why the "Northern lights" are usually seen in regions near the poles.

If both an electric and a magnetic field exist in a region of space, the total force on a moving charged particle is the superposition of two forces and is given by

$$\boxed{\mathbf{F} = q\mathbf{E} + q\mathbf{v} \times \mathbf{B}.} \tag{38.14}$$

This last equation is known as the *Lorentz force*, named after H. A. Lorentz, who made several advances in understanding the nature of charged particles.

Example 5
A beam of protons moves through a region where both a uniform electric field and a uniform magnetic field exist. The protons move horizontally with a speed of

5.0×10^6 m/s and are observed to be undeflected. If the magnitude of the electric field is 3.0×10^4 V/m and is in the y direction as shown, find the direction and magnitude of the magnetic field.

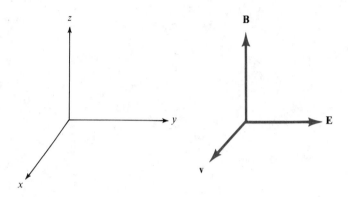

The electric force on the protons will be in the same direction as the electric field, the y direction. Since the beam is undeflected, the magnetic force must be in the $-y$ direction. That implies that the cross product of the velocity and the magnetic field must be in the $-y$ direction. Because the velocity is in the x direction, the magnetic field must be in the z direction. Setting the Lorentz force equal to zero in Eq. (38.14), we obtain $|q|E = |q|vB$, so that

$$B = E/v = (3.0 \times 10^4 \text{ V/m})/(5.0 \times 10^6 \text{ m/s}) = 6.0 \times 10^{-3} \text{ T}.$$

The mass spectrograph represents a useful application of the magnetic force on moving particles. First developed by F. W. Aston and A. J. Dempster in 1919 and later improved by others, the mass spectrograph was designed to measure the masses of isotopes – nuclei of an element that have the same charge but different masses. The basic design of the instrument is illustrated in Fig. 38.17. Ions initially at rest are accelerated through a potential difference V. The loss in potential energy appears as kinetic energy:

$$|q|V = \tfrac{1}{2}mv^2.$$

The ions are then injected into a chamber containing a uniform magnetic field **B** that is perpendicular to the plane of motion of the ions. Because of the magnetic force, the ions move through the magnetic field along a circular path before striking a photographic plate. The distance from the point where the ions enter the chamber to where they strike the plate is twice the radius of the circle, $2R$, where, according to Eq. (38.12),

$$R = \frac{mv}{|q|B}. \tag{38.12}$$

Using this result and energy conservation we can solve for the mass-to-charge ratio

38.5 MAGNETIC FORCE ON MOVING CHARGES

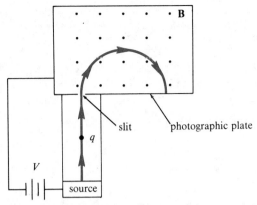

Figure 38.17 Schematic diagram of a mass spectrograph.

of the ions in terms of the known quantities B, R, and V; the result is

$$\frac{m}{|q|} = \frac{B^2 R^2}{2V}.$$

Although this method requires knowledge of the charge of the ions, the charge is usually one or two electron charges. Hence the mass of the ions can be determined.

Example 6

In a modified mass spectrograph a positive ion having a charge-to-mass ratio of 2.0×10^5 C/kg is projected along a line midway between two large plates that are 2.0-cm apart and have a potential difference of 200 V across them. A magnetic field of 0.5 T is perpendicular to the velocity of the ion as shown in the figure.

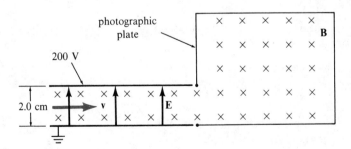

(a) Calculate the speed of the ion if it passes undeflected through the region between the plates.
(b) The ion exits into a region where only the magnetic field exists, as shown in the diagram. Determine where it will strike the photographic plate.
(c) Determine the minimum mass of the ion.

(a) Because the ion passes undeflected through the region containing the electric and magnetic fields, the net force on it must be zero:

$$\mathbf{F} = q(\mathbf{E} + \mathbf{v} \times \mathbf{B}) = 0.$$

This implies that $E = vB$, so that the speed of the ion is $v = E/B$. The uniform electric field is related to the potential difference between the plates by $E = V/d$, where d is the distance between the plates, so we can substitute for E and get

$$v = V/(Bd) = (200 \text{ V})/[(0.5 \text{ T})(2.0 \times 10^{-2} \text{ m})] = 2.0 \times 10^4 \text{ m/s}.$$

(b) According to Eq. (38.12), the radius of the circular orbit in the magnetic field is given by

$$R = \frac{mv}{qB}. \tag{38.12}$$

Substituting for the known quantities we find

$$R = (2.0 \times 10^4 \text{ m/s})/[(0.5 \text{ T})(2.0 \times 10^5 \text{ C/kg})] = 0.2 \text{ m}.$$

(c) The minimum charge on the ion is that of a proton, 1.6×10^{-19} C. Using the given charge-to-mass ratio we find

$$m_{\min} = (1.6 \times 10^{-19} \text{ C})/(2.0 \times 10^5 \text{ C/kg}) = 8.0 \times 10^{-25} \text{ kg}.$$

Questions

13. If a beam of electrons is deflected in passing through a certain region of space, can you be certain that a magnetic field exists there?

14. A beam of electrons passes through a region of space containing both an electric and a magnetic field and is deflected. Will a beam of protons be deflected in the same direction?

15. Explain qualitatively how a charged particle can be trapped in the earth's magnetic field.

16. An electron moves in a circular orbit of radius 0.60 m perpendicular to a magnetic field of 0.20 T.

(a) What is the speed of the electron?
(b) What is its kinetic energy?

17. Compare cyclotron frequencies for a proton and an α particle. An α particle is a helium nucleus consisting of two neutrons and two protons; its mass is approximately four times that of a proton.

18. An electron of kinetic energy 300 keV moves in a circular orbit in a magnetic field of 0.4 T. What is the radius of the orbit?

19. A singly charged ion of magnesium, ^{24}Mg (mass equal to 24 proton masses), inside a mass spectrometer is first accelerated through a potential difference of 4.0 keV and then bent 180° in a magnetic field of 1500 G.

38.6 MAGNETIC FORCE ON CURRENTS

(a) Find the distance from where the ions enter the spectrometer to where they strike the photographic plate.

(b) Calculate the distance on the plate between ^{24}Mg and ^{26}Mg ions.

20. A proton of energy 2.0 keV is projected into a uniform magnetic field of intensity 3.0 T with its velocity at an angle of 80° with respect to the magnetic field. Find the (a) radius, (b) frequency ω, and (c) pitch of the helix that it follows. (The pitch is defined as the distance along the axis of the helix between successive points where the particle has the same position in its circular motion.)

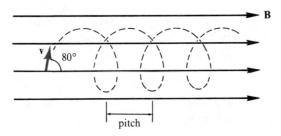

38.6 MAGNETIC FORCE ON CURRENTS

Charged particles moving in magnetic fields experience magnetic forces. Since a current-carrying wire contains charges in motion, we expect that it would likewise experience magnetic forces. Or, because current-carrying wires are neutral, should we not expect magnetic forces? Let's apply our understanding of the magnetic force on moving charges to analyze the effects of magnetic fields on currents. Imagine a section of a current-carrying wire of length Δs and cross-sectional area A in a uniform magnetic field **B**, as shown in Fig. 38.18. The current consists of electrons moving with an average drift velocity **v**. Each electron experiences a force given by Eq. (38.11),

$$\mathbf{F} = q\mathbf{v} \times \mathbf{B}. \tag{38.11}$$

Although the wire is electrically neutral, only the electrons are moving, and hence there is a net force on the wire that is equal to the sum of the forces on the

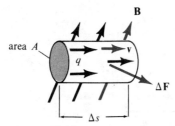

Figure 38.18 A current element in a magnetic field experiences a force.

individual electrons. Because each electron experiences the same force, the force on the segment is simply the number of electrons in the segment times the force on one electron. If there are n electrons per unit volume contributing to conduction, then the number of electrons in the length Δs of wire is $nA(\Delta s)$. Thus the force $\Delta \mathbf{F}$ on the segment is

$$\Delta \mathbf{F} = (q\mathbf{v} \times \mathbf{B})nA(\Delta s).$$

In Chapter 37 we found that the current in the wire is related to the number of electrons per unit volume and drift speed according to

$$I = nqvA. \tag{37.2}$$

If we take the vector $\Delta \mathbf{s}$ to be in the same direction as the drift velocity (and hence also the current) we can define $I\,\Delta \mathbf{s}$ as a current element. Then the force may be written as

$$\Delta \mathbf{F} = I\,\Delta \mathbf{s} \times \mathbf{B}. \tag{38.15}$$

To find the total force on a length of wire, we sum over all the current elements. If the magnetic field is *uniform* along the entire length L of a *straight wire*, the sum becomes

$$\mathbf{F} = I\mathbf{L} \times \mathbf{B}. \tag{38.16}$$

The magnetic force on a current-carrying wire also follows the right-hand rule and is perpendicular to both the current element and the magnetic field, as indicated in Fig. 38.18.

Example 7

Find the force on a semicircular section of wire of radius R carrying a current I in a magnetic field $\mathbf{B}$, as shown.

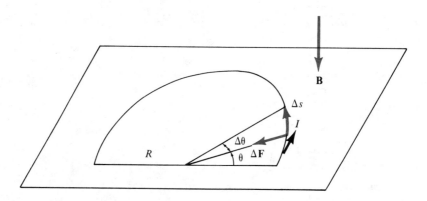

38.6 MAGNETIC FORCE ON CURRENTS

Take a current element on the wire, as shown, and determine the force on it. By Eq. (38.15) the magnitude of the force is $\Delta F = IB\,\Delta s$. But Δs is $R\,\Delta\theta$ so we have

$$\Delta F = IBR\,\Delta\theta,$$

and the direction, given by the right-hand rule, is toward the center of the semicircle. By symmetry only vertical components of the force, $\Delta F \sin\theta$, contribute; the horizontal components cancel. The vertical component on each element is

$$\Delta F \sin\theta = IBR \sin\theta\,\Delta\theta,$$

and the total force is obtained by integration:

$$F = IBR \int_0^\pi \sin\theta\,d\theta = 2IBR.$$

The force turns out, in this case, to be equal to that on a straight wire of length $2R$.

Magnets experience torques when placed in magnetic fields. What happens to a loop of wire in a magnetic field? Consider a rectangular loop of wire lying in the xy plane, as illustrated in Fig. 38.19. Assume that the loop carries a current I and has sides of length L_x and L_y. The loop is in a uniform magnetic field $\mathbf{B}$ in the x direction. What are the forces on the loop? There are no forces on sides 1 and 3 because they are parallel to the magnetic field and so the cross product in Eq. (38.16) is zero. The force on side 2 is simply

$$\mathbf{F}_2 = IL_y B(\hat{\mathbf{j}} \times \hat{\mathbf{i}}) = -IL_y B\hat{\mathbf{k}}.$$

Similarly, the force on side 4 is

$$\mathbf{F}_4 = IL_y B(-\hat{\mathbf{j}} \times \hat{\mathbf{i}}) = IL_y B\hat{\mathbf{k}}.$$

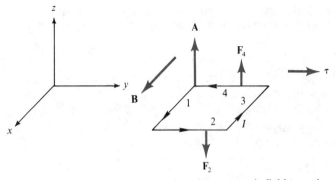

Figure 38.19 A current-carrying loop in a magnetic field experiences a torque.

Although these forces are equal and opposite, they do not act along the same line, and so they create a torque that tends to twist the loop about its center. The torque produced by these forces is

$$\tau = (L_x/2)(\hat{i} \times F_2) + (L_x/2)(-\hat{i} \times F_4)$$
$$= IL_xL_yB/2((\hat{i} \times -\hat{k}) + (-\hat{i} \times \hat{k}))$$
$$= IL_xL_yB\hat{j}.$$

Defining the area vector of the loop as $\mathbf{A} = L_xL_y\hat{k}$, we can write this last result as

$$\tau = I\mathbf{A} \times \mathbf{B}. \tag{38.17}$$

Current loops in magnetic fields behave much like magnetic dipoles: both experience torques from magnetic fields. By analogy we define the magnetic moment of a planar current loop as

$$\mathbf{m} = NI\mathbf{A}, \tag{38.18}$$

where N is the number of turns of wire and A the area of a planar region enclosed by each loop, as indicated in Fig. 38.20. Then the magnetic torque on *any* dipole is given by Eq. (38.6),

$$\tau = \mathbf{m} \times \mathbf{B}. \tag{38.6}$$

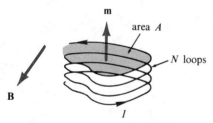

Figure 38.20 A stack of N planar current loops, each of area A carrying current I has a magnetic moment given by $\mathbf{m} = NI\mathbf{A}$.

Although we arrived at Eq. (38.17) for the special case of a rectangular loop, the relation can be generalized to a current loop of any shape by approximating the loop as closely as desired by a number of rectangular loops each having sides parallel and perpendicular to the magnetic field, as illustrated in Fig. 38.21. The torque on a single loop of area $\Delta \mathbf{A}_j$ is

$$\Delta\tau_j = I\Delta\mathbf{A}_j \times \mathbf{B},$$

so that the total torque on the loop is

38.6 MAGNETIC FORCE ON CURRENTS

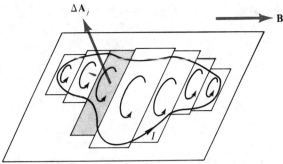

Figure 38.21 A current loop can be approximated by a large number of rectangular loops.

$$\tau = \lim_{\Delta A \to 0} \sum \Delta \tau_j = I \left[\lim_{\Delta A \to 0} \sum \Delta \mathbf{A}_j \right] \times \mathbf{B}$$

or

$$\tau = I\mathbf{A} \times \mathbf{B} = \mathbf{m} \times \mathbf{B}.$$

Questions

21. Of the quantities involved in the force on a current element, which are always at right angles? Which may have any angle between them?

22. Can a circular loop of current-carrying wire be levitated in a uniform magnetic field by using the magnetic force to counter gravity?

23. A straight wire of length 0.2 m carries a current of 1.5 A and is at 60° to a uniform magnetic field of strength 0.03 T. Find the force exerted on the wire.

24. A small circular coil of radius 2.0 cm has 20 turns and carries a current of 0.5 A. Find the torque exerted on it if its area vector is at an angle of 37° from the direction of a uniform magnetic field of strength 0.03 T.

25. Assuming that the earth has a magnetic dipole moment of 6.4×10^{21} C m^2/s, what current would have to be set up in a 1000-turn coil about the earth to cancel the magnetic field of the earth?

26. A wire of length 40 cm and mass 0.02 kg is suspended in a uniform magnetic field of strength 0.02 T by a pair of spring balances. What current should flow in the wire so that the spring balances read zero?

27. A charged particle of mass m, charge q, and speed v moves in a circle in a uniform magnetic field of strength B. Determine the equivalent dipole moment corresponding to the particle's motion in terms of m, q, v, and B.

28. A metal bar of mass m rides without friction along a pair of conducting rails, separated by a distance L, that are perpendicular to a magnetic field **B**, as shown. A source supplies a constant current I through the bar.

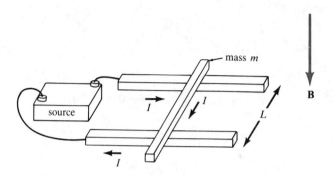

(a) Find the force on the bar.
(b) Assuming that the bar starts from rest at time $t = 0$, find the speed of the bar at time t.

38.7 A FINAL WORD

William Gilbert set out to debunk magical notions of magnetism, yet in building an intellectual bridge between natural philosophy and emerging sciences, he did not completely abandon reference to the occult. For example, he believed that an invisible "orb of virtue" surrounds a magnet and extends in all directions around it. Other magnets and pieces of iron react to this orb of virtue and move or rotate in response. Magnets within the orb are attracted whereas those outside are unaffected. The source of the orb remained a mystery.

In describing the earth as a giant magnet, Gilbert also used a mystical tone in his explanation. The *anima* (soul) of the earth was responsible for its magnetism – Mother Earth guided ships from her soul by magnetic needles. Although his language was that of the natural philosophy of the time, some of his ideas were ahead of his time. His orbs of virtue were a fledgling notion of the idea of fields that would revolutionize physics more than two centuries later.

Gilbert extended his magnetic theories to include the cosmos. He explained the diurnal rotation of the earth, postulated that the fixed stars were not all equidistant from the earth, and envisioned seas on the moon. He described the attraction of the moon to the earth and the tides as being magnetic in origin. Kepler tried to use magnetism to explain the force that swept the planets about the sun, but he required so many ad hoc postulates that others found this theory unacceptable.

Gilbert made other interesting speculations, many of which would be confirmed decades later. For example, he considered comets to be wandering bodies without

38.7 A FINAL WORD

magnetic polarity that could be either within or outside the orbit of the moon. And he imagined a collection of stars in the Milky Way so numerous and so distant as to appear as a mist or cloud – years before the telescope would be invented and the Copernican system confirmed. Although Gilbert discussed the motion of the earth according to Copernicus's model, he neither confirmed nor denied the theory, perhaps fearing the same fiery fate as Giordano Bruno.

In Gilbert's work we find the beginnings of seventeenth century experimental science. He offered data and experiments to confirm observations, he organized his observations into connected phenomena, and he provided a starting point for scientists of the next century. Galileo, the master experimentalist of his age, held Gilbert's work in great esteem and the seventeenth century poet laureate John Dryden eulogized him with the lines:

> Gilbert shall live till the loadstones cease to draw
> Or British fleets the boundless oceans awe.

CHAPTER 39

THE MAGNETIC FIELD

To observe the facts first, to vary their circumstances as much as possible, to accompany this first task with exact measurement so as to deduce from them general laws based solely upon experience, and to deduce from these laws, independently of any hypothesis on the nature of the forces that produce the phenomena, the mathematical value of these forces, that is, the formula that represents them – such is the procedure Newton followed. In general, it has been adopted in France by the savants to whom physics owes the immense progress it has made recently, and it has served as my guide in all my research on electrodynamic phenomena. I have consulted only experience in order to establish the laws of these phenomena, and I have deduced from them the sole formula that can represent the forces to which they are due.

André Marie Ampère (1826)

39.1 THE CONNECTION BETWEEN ELECTRICITY AND MAGNETISM

Hans Christian Oersted's accidental discovery that a current-carrying wire affects a compass needle was the first evidence that electricity and magnetism are related. When word of this discovery reached the French Academy of Sciences in September 1820, most members were skeptical. After all, had not Coulomb shown in the 1780s that there was no interaction between electricity and magnetism? However, the curiosity of one member was keenly aroused, and within a fortnight André Marie

Ampère had not only confirmed Oersted's results but discovered that a coil of wire carrying electric current behaves like an iron magnet. Within a few feverish weeks Ampère created electrodynamics, a new theory describing magnetism as a force between electric currents. In 1826 he summarized his work and that of others in a profound mathematical treatise, *Théorie des Phénomènes Electro-Dynamiques*.

Ampère was born in Lyons, France, in 1775. His father, a devotee of Rousseau's educational theories, exposed young André to a well-stocked library and allowed him to educate himself. Fascinated by numbers and Euclidean geometry, he taught himself Latin in order to read the mathematical works of Euler and Bernoulli. His education was also influenced by a parental conflict that was common in France and Italy – the skeptical philosophical views of his father versus the strict Catholicism of his mother.

The French Revolution broke out when Ampère was 14. His father, who had assumed a post with important police powers, ordered the arrest of the leading Jacobin of Lyons, who was later executed. Ampère's father was guillotined 4 years later when the Republicans took Lyons. This was the beginning of a tragic period in

Figure 39.1 Portrait of André Marie Ampère. (Courtesy of the Archives, California Institute of Technology.)

Ampère's life. His first wife died 4 years after they were married, his second marriage ended in divorce, and he was constantly plagued by monetary problems.

Despite personal setbacks, Ampère managed to carve out a respectable but undistinguished academic career, teaching mathematics first in Lyons and then at the Ecole Polytechnique in Paris. In 1815 he changed fields and became a chemist, coming very close to making some important discoveries. For example, he suspected that chlorine and iodine were elements, but had neither the time nor the resources to substantiate his ideas, and credit for the discovery of these elements went to Humphrey Davy. He independently discovered Avogadro's law, three years after Avogadro had enunciated it. He made a noble attempt to classify the elements according to their properties, but never hit upon Mendeleev's periodic table.

A turning point in his career came at the age of 45 when in the fall of 1820, he began his work on electrodynamics. With the publication of his treatise in 1826, called by some the *Principia* of electrodynamics, Ampère secured his place in the history of science.

39.2 THE LAW OF BIOT AND SAVART

While Ampère was investigating magnetism as a force between electric currents, others followed up Oersted's discovery by investigating the force of currents on magnets. By the end of October 1820, the first quantitative investigations of this force were reported to the French Academy by Jean Baptiste Biot and Felix Savart. They measured the rate of oscillation of a small magnetic dipole suspended at various distances from a long straight wire that carried current, and observed that the magnetic force is directly proportional to the current, and inversely proportional to the distance between the wire and the magnet.

The law of Biot and Savart is usually stated vectorially in terms of magnetic fields, as indicated below in Eq. (39.1). Vector notation and magnetic fields were not yet formulated at the time of Biot and Savart, and their results were expressed in terms of magnitudes of forces. We will derive the vector form of their law from Eq. (38.15), which gives the magnetic force on an element of wire with a current in it:

$$\Delta \mathbf{F} = I \Delta \mathbf{s} \times \mathbf{B}. \tag{38.15}$$

In this formula, $I \Delta \mathbf{s}$ denotes the current element, $\mathbf{B}$ is the magnetic field of the magnet, and $\Delta \mathbf{F}$ is force on the current element. Biot and Savart demonstrated that the current element exerts a force on a magnet. That force is equal and opposite to that of the magnet on the current element, a result to be expected from Newton's third law. In other words, the current element also acts as the source of a magnetic field. We now use this fact to determine this field.

Imagine a current element $I \Delta \mathbf{s}$ located a distance r from the north pole of a magnet, and let $\mathbf{B}'$ denote the magnetic field due to the pole, as illustrated in Fig. 39.2. Then, by Eq. (38.15), the force $\Delta \mathbf{F}'$ exerted on the current element by the pole is given by

$$\Delta \mathbf{F}' = I \Delta \mathbf{s} \times \mathbf{B}'. \tag{38.15}$$

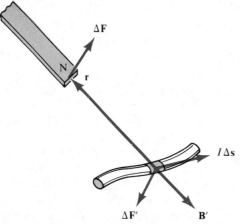

Figure 39.2 According to Newton's third law, a current element exerts a force on a magnetic pole equal and opposite to the force of the pole on the magnet.

The *reaction force* of the current on the pole is

$$\Delta \mathbf{F} = -\Delta \mathbf{F}' = -I\Delta \mathbf{s} \times \mathbf{B}'.$$

The field vector $\mathbf{B}'$ can be expressed in terms of the distance r. If $\mathbf{r}$ denotes the vector from the current element to the pole, as shown in Fig. 39.2, we know from Eq. (38.2) that

$$\mathbf{B}' = -K_m \frac{p}{r^2} \hat{\mathbf{r}}, \tag{38.2}$$

the minus sign coming from the fact that the field is directed from the pole toward the current, in the opposite direction of $\mathbf{r}$. Substituting for $\mathbf{B}'$ in the previous equation we find the force of the current element on the pole,

$$\Delta \mathbf{F} = p \left(K_m \frac{I\Delta \mathbf{s} \times \hat{\mathbf{r}}}{r^2} \right).$$

The quantity multiplying p doesn't depend in any way on the strength of the pole. Following our usual procedure, we define the magnetic field to be the force per unit pole, $\Delta \mathbf{F}/p$. Thus, by definition, the magnetic field $\Delta \mathbf{B}$ due to the current element $I\Delta \mathbf{s}$ is

$$\boxed{\Delta \mathbf{B} = K_m \frac{I\Delta \mathbf{s} \times \hat{\mathbf{r}}}{r^2}.} \tag{39.1}$$

Equation (39.1) is the *elemental* or *local* form of the law of Biot and Savart, and is somewhat analogous to Coulomb's law for a point charge. It identifies the current element as the source of the magnetic field and gives the magnetic field at any point.

39.2 THE LAW OF BIOT AND SAVART

Like the electric field, the magnetic field decreases as the square of the distance from the source. But unlike the electric field, the magnetic field is not in the radial direction. Instead, the magnetic field is perpendicular to both the current element $I\Delta\mathbf{s}$ and the radial direction $\hat{\mathbf{r}}$.

The law of Biot and Savart can also be written in a global form which depends on the geometry of the wire. Because the vectors $\Delta\mathbf{s}$ and $\hat{\mathbf{r}}$ will vary along the wire, we obtain the total field by line integration along the wire,

$$\mathbf{B} = K_m \int_{\text{wire}} \frac{I\,d\mathbf{s} \times \hat{\mathbf{r}}}{r^2}, \qquad (39.2)$$

where $\mathbf{s}$ is the vector function that describes the shape of the wire. The use of the integral is illustrated in the following example.

Example 1
A current flows in a straight wire of finite length. Determine the magnetic field at a point P at distance R from the wire, as shown in the accompanying diagram.

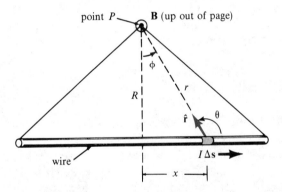

The right-hand rule shows that the field $\mathbf{B}$ will be perpendicular to the plane containing the wire and the point P, and directed up from the page, as indicated in the diagram. To find the magnitude of the field, we place the x axis along the wire with the origin at the nearest point to P. If θ is the angle between the current element $I\Delta\mathbf{s}$ and the unit vector $\hat{\mathbf{r}}$, then the magnitude of the numerator in Eq. (39.1) is

$$|I\Delta\mathbf{s} \times \hat{\mathbf{r}}| = I\Delta x \sin\theta,$$

so the magnitude of $\mathbf{B}$ is given by the integral

$$B = K_m \int_{\text{wire}} \frac{I \sin\theta}{r^2}\,dx. \qquad (39.3)$$

The quantities x, r, and $\sin\theta$ vary over the wire, and to perform the integration we need to express these quantities in terms of a common variable. In this example a convenient variable is the angle ϕ shown in the diagram. First, we have $x = R\tan\phi$, from which we find

$$\frac{dx}{d\phi} = R\sec^2\phi.$$

Next, we have $R = r\cos\phi$, so

$$r^2 = R^2/\cos^2\phi = R^2\sec^2\phi.$$

Finally, $\cos\phi = \sin(\pi - \theta) = \sin\theta$, so the integral in (39.3) simplifies to

$$B = K_m \frac{I}{R} \int_\alpha^\beta \cos\phi \, d\phi,$$

where the angles α and β are determined by the endpoints of the wire. This gives us the result

$$B = K_m \frac{I}{R}(\sin\beta - \sin\alpha). \tag{39.4}$$

Example 1 gives the magnetic field due to any straight segment of wire. For a wire of infinite length, or for a finite wire whose length is so great compared to the distance R that it can be considered infinite, we can take $\beta = \pi/2$ and $\alpha = -\pi/2$ and the result becomes

$$B = K_m \frac{2I}{R}. \tag{39.5}$$

The field strength is directly proportional to I and inversely proportional to R, as originally found experimentally by Biot and Savart.

Iron filings sprinkled on a plane perpendicular to a long straight current-carrying wire form a pattern of circles centered on the wire, as shown in Fig. 39.3a. This pattern suggests that the magnetic field lines are concentric circles about the wire, as shown in Fig. 39.3b, and the analysis in Example 1 supports this visualization. The direction of the field at each point is perpendicular to the wire and also to the vector from the wire to the point. Moreover, as Eq. (39.5) indicates, the magnitude is constant on each circle of a given radius R. To determine the direction of the magnetic field, imagine grasping the wire with your right hand, with the thumb pointing in the direction of the current. Then your fingers curl around the wire in the direction of **B**, as shown in Fig. 39.3b.

The field around a long straight wire can be expressed in a convenient vector form by introducing polar coordinates in the plane of the concentric circles of Fig. 39.3b. Replacing the distance R in Eq. (39.5) by the polar coordinate distance r, and using **r** to denote the radius vector from the wire to an arbitrary point in the

39.2 THE LAW OF BIOT AND SAVART

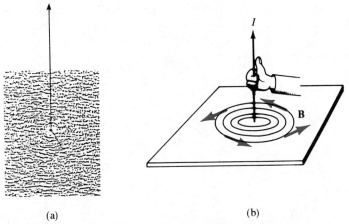

Figure 39.3 (a) Pattern formed by iron filings around a current-carrying wire. (b) Magnetic field lines around a current-carrying wire.

plane of the concentric circles, we see that the field at that point is given by

$$\mathbf{B} = K_m \frac{2I}{r} \hat{\boldsymbol{\theta}}, \qquad (39.6)$$

where $\hat{\boldsymbol{\theta}} = d\hat{\mathbf{r}}/d\theta$ is the usual unit vector perpendicular to $\hat{\mathbf{r}}$.

Example 2
Two long straight wires, each carrying a current of 3.0 A flowing in the same direction, are separated by a distance $d = 0.4$ m. Determine the magnetic field at a point P that is located a distance $d/4$ from one of the wires, as shown.

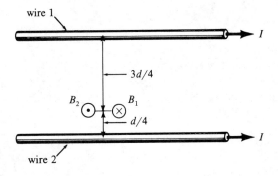

The magnetic field at P is the vector sum of the fields due to each wire. By the right-hand rule, the field due to wire 1 points down into the page whereas the field of wire 2 points up out of the page. Thus the fields are in opposite directions, and the magnitude of the resultant field is given by

$B_P = B_2 - B_1$.

Using Eq. (39.5) to express the fields due to the wires we have

$$B_1 = K_m \frac{2I}{3d/4} \quad \text{and} \quad B_2 = K_m \frac{2I}{d/4}.$$

Therefore, the resultant field has magnitude

$$B_P = K_m \frac{16I}{3d} = (1.0 \times 10^{-7} \text{ N/A}^2)(16 \times 3.0 \text{ A})/(3 \times 0.4 \text{ m})$$
$$= 4.0 \times 10^{-6} \text{ T}$$

and is directed up out of the page.

The next example treats another useful configuration, a circular current-carrying wire.

Example 3

A circular loop of wire of radius R carries a current I, as shown. Determine the magnetic field on the axis of the loop at a distance z above the plane of the wire.

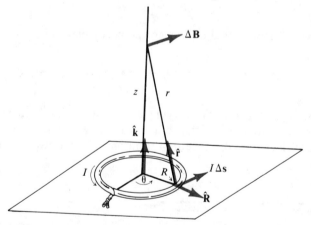

We begin with Eq. (39.1), the elemental form of the law of Biot and Savart. The circular loop is described by the radius vector

$\mathbf{R} = R\hat{\mathbf{R}}$,

the current element is given by

$I\Delta\mathbf{s} = IR\,\Delta\theta\,\hat{\boldsymbol{\theta}}$,

as indicated in the diagram, and $\mathbf{r} = r\hat{\mathbf{r}}$ is the vector from the current element to the

39.2 THE LAW OF BIOT AND SAVART

point at which we are computing the field. Therefore, Eq. (39.1) becomes

$$\Delta \mathbf{B} = K_m \frac{I \Delta \mathbf{s} \times \hat{\mathbf{r}}}{r^2} = K_m \frac{I}{r^2} R \Delta \theta \, \hat{\boldsymbol{\theta}} \times \hat{\mathbf{r}} = K_m \frac{IR}{r^3} \Delta \theta \, \hat{\boldsymbol{\theta}} \times \mathbf{r}. \tag{39.7}$$

From the diagram we see that

$$\mathbf{r} = z\hat{\mathbf{k}} - \mathbf{R} = z\hat{\mathbf{k}} - R\hat{\mathbf{R}},$$

so

$$\hat{\boldsymbol{\theta}} \times \mathbf{r} = z\hat{\boldsymbol{\theta}} \times \hat{\mathbf{k}} - R\hat{\boldsymbol{\theta}} \times \hat{\mathbf{R}}$$
$$= z\hat{\mathbf{R}} + R\hat{\mathbf{k}},$$

and Eq. (39.7) becomes

$$\Delta \mathbf{B} = K_m \frac{IR}{r^3} \Delta \theta \, (z\hat{\mathbf{R}} + R\hat{\mathbf{k}}).$$

The quantities z, R, and r are constant, but $\hat{\mathbf{R}}$ varies with θ. Therefore, integrating over the wire we find

$$\mathbf{B} = K_m \frac{IRz}{r^3} \int_0^{2\pi} \hat{\mathbf{R}} \, d\theta + K_m \frac{IR^2}{r^3} \int_0^{2\pi} d\theta \, \hat{\mathbf{k}}. \tag{39.8}$$

But $\hat{\mathbf{R}} = \cos\theta\, \hat{\mathbf{i}} + \sin\theta\, \hat{\mathbf{j}}$ so $\int_0^{2\pi} \hat{\mathbf{R}} \, d\theta = 0$, which means there is no component of the field in a plane parallel to the plane of the loop. The second integral in (39.8) has the value 2π and we are left with

$$\mathbf{B} = K_m \frac{2\pi I R^2}{r^3} \hat{\mathbf{k}}.$$

The distance z enters through the factor r^3 because $r^2 = z^2 + R^2$, so the result can also be written as

$$\mathbf{B} = K_m \frac{2\pi I R^2}{(z^2 + R^2)^{3/2}} \hat{\mathbf{k}}. \tag{39.9}$$

Two extreme cases of (39.9) are worth noting. First, at the center of the loop we have $z = 0$ and the magnetic field there is given by

$$\mathbf{B}_0 = K_m \frac{2\pi I}{R} \hat{\mathbf{k}}. \tag{39.10}$$

Note that when $z = 0$ the first term on the right of Eq. (39.8) is zero even if the integral is taken over only a portion of the circular segment. Thus, if we take $z = 0$

and integrate over an arc subtending an angle α, then $\int_0^\alpha d\theta = \alpha$ and $r = R$, so the field at the center due to this arc is given by

$$\mathbf{B} = K_m \frac{\alpha I}{R} \hat{\mathbf{k}}.$$

For example, when $\alpha = \pi$ the field at the center of a semicircular loop is just half that of a full circular loop.

Next, if z is very large compared to R, we can neglect R^2 in the denominator on the right of Eq. (39.9) and we obtain the approximation

$$\mathbf{B} \approx K_m \frac{2\pi I R^2}{z^3} \hat{\mathbf{k}}.$$

But πR^2 is the area A of the circular region enclosed by the loop, so we can write this last formula in terms of the dipole moment of a current loop,

$$m = IA, \qquad (38.18)$$

and obtain

$$\mathbf{B} \approx K_m \frac{2m}{z^3} \hat{\mathbf{k}}. \qquad (39.11)$$

This characteristic field with an inverse cube dependence is the same as that obtained in Eq. (38.5) for a magnetic dipole. In other words, a current loop creates a field that, at great distances, is like a dipole field. The field lines traced out by the iron filings in Fig. 39.4 resemble the field lines of a permanent magnet whose dipole moment lies along the axis of the loop.

Next, a remark concerning notation. Formulas like (39.9), which involve a factor π times K_m, are often written in terms of a constant μ_0, called the permeability of free space, defined as $\mu_0 = 4\pi K_m$. In this case, (39.9) becomes

$$\mathbf{B} = \frac{\mu_0}{2} \frac{IR^2}{(z^2 + R^2)^{3/2}} \hat{\mathbf{k}}.$$

Figure 39.4 Iron filings around a circular loop of wire form a pattern that resembles that formed by iron filings around a permanent magnet.

39.2 THE LAW OF BIOT AND SAVART

Figure 39.5 A research magnet consisting of many circular loops. (Caltech photo by Nils Asplund.)

The numerical value of the permeability constant in SI units is

$$\mu_0 = 1.26 \times 10^{-6} \text{ N/A}^2.$$

Because circular coils are easy to manufacture, they are often used to produce magnetic fields for a variety of applications. Figure 39.5 shows a research magnet consisting of two large coils, each containing many circular loops of wire. The fields from the various circular loops add up to create a large field in the gap between the two iron pieces. Iron is used inside the loops to strengthen the field.

Questions

1. The magnetic fields created by wires carrying current to and from a device are often canceled by twisting the wires together. Explain qualitatively why this procedure succeeds.

2. A proton moving with constant speed v is at a distance y from a long, straight wire carrying current I. Find the magnitude and direction of the force on the proton when its velocity is in each of the directions (a), (b), and (c) shown.

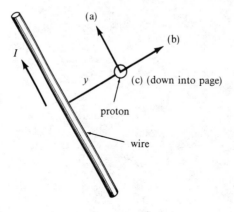

3. A very long wire is bent into an L shape, as shown.

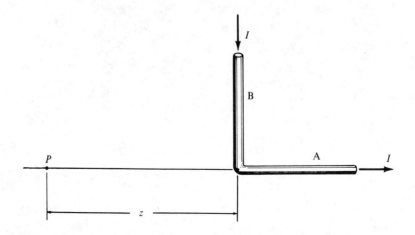

(a) Explain why the contribution to the magnetic field from portion A of the wire is zero at point P.
(b) Find, in terms of z, the magnitude and direction of the magnetic field at point P due to portion B.

4. A long, straight wire carries a current of 0.4 A. If the field measured at a point in space is 1.6×10^{-5} T, find the distance from the wire to that point.

5. Point A is located at a distance of 0.1 m from a long straight wire. Points B and C are at other distances from the wire. The magnetic field measured at points A,

B, and C has the values 0.04, 0.03, and 0.02 T, respectively. Determine the distance from the wire to each of points B and C. Are the points equally spaced?

6. A current of 0.4 A flows in a circular wire loop of radius 0.1 m. Find the magnetic field (a) at the center of the loop, (b) on the axis of the loop at a distance of 0.1 m from the center, and (c) on the axis of the loop at a distance of 0.5 m from the center.

7. A wire carrying a current I is bent into a quarter circle and two straight segments as shown. Determine the magnetic field at point C, the center of the circular arc.

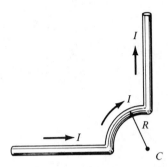

8. A wire loop is made of two concentric semicircular arcs of radii a and b and connecting radial segments as shown. Determine the magnetic field at the center C.

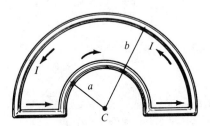

9. Two parallel wires, perpendicular to the page and a distance d apart, carry the same current I, but in opposite directions, as indicated.

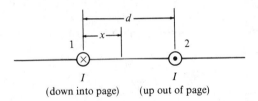

(a) Find the magnetic field at a point between the wires at a distance x to the right of wire 1.
(b) Determine the magnetic field at a distance x to the left of wire 1.
(c) Sketch the graph of the magnetic field as a function of x, with $x < d$, x being positive to the right of wire 1 and negative to the left. The positive direction of the field B is upward and the negative direction is downward.

10. Two long parallel straight wires separated by a distance d carry equal currents in opposite directions as indicated. Determine the magnitude and direction of the magnetic field at point P, which is a distance R from one of the wires, as shown.

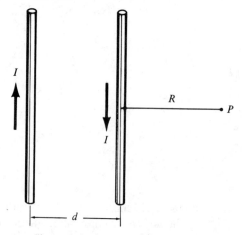

11. Show that the magnitude of the magnetic field at the center of a square current loop of side L carrying current I is given by

$$B = K_m 8\sqrt{2}\, I/L.$$

39.3 FORCE BETWEEN CURRENT-CARRYING WIRES

While Biot and Savart measured the force on a permanent magnet arising from a current element, Ampère examined the force between two current-carrying wires. A short time after he learned of Oersted's discovery, Ampère discovered that there is an attractive force between two parallel wires carrying current in the same direction, and a repulsive force if the currents are in opposite directions. In a series of simple experiments he proved that magnetism was electricity in motion, and deduced the force law between two current-carrying wires.

Ampère's force law can be derived by combining Eq. (38.16),

$$\mathbf{F} = I\mathbf{L} \times \mathbf{B}, \tag{38.16}$$

which gives the magnetic force $\mathbf{F}$ on a current-carrying wire of length L due to a uniform field $\mathbf{B}$, with Eq. (39.6),

$$\mathbf{B} = K_m \frac{2I}{r} \hat{\boldsymbol{\theta}}, \tag{39.6}$$

which gives the field around a long straight wire.

39.3 FORCE BETWEEN CURRENT-CARRYING WIRES

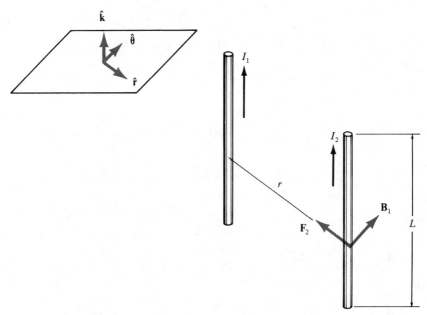

Figure 39.6 Parallel wires carrying current in the same direction attract one another.

Consider two long wires, each of length L and parallel to the z axis, as shown in Fig. 39.6. Suppose that wire 1 carries a current I_1 and wire 2 a current I_2. Wire 1 creates a magnetic field $\mathbf{B}_1$ of constant magnitude along the length of wire 2 that is given by Eq. (39.6),

$$\mathbf{B}_1 = K_m \frac{2I_1}{r} \hat{\boldsymbol{\theta}}, \tag{39.6}$$

where r is the distance between the wires. This field produces a force on wire 2 given by Eq. (38.16),

$$\mathbf{F}_2 = I_2 \mathbf{L} \times \mathbf{B}_1 = I_2 L \hat{\mathbf{k}} \times \mathbf{B}_1.$$

Substituting for $\mathbf{B}_1$ and using $\hat{\mathbf{k}} \times \hat{\boldsymbol{\theta}} = -\hat{\mathbf{r}}$, we obtain

$$\mathbf{F}_2 = -2K_m L \frac{I_2 I_1}{r} \hat{\mathbf{r}}. \tag{39.12}$$

Because $\mathbf{F}_2$ has the direction opposite to $\hat{\mathbf{r}}$, wire 2 is attracted to wire 1, and the magnitude of the force is proportional to the product of the currents and inversely proportional to the distance between the wires. This was one of Ampère's findings. By Newton's third law, we know that wire 1 is attracted to wire 2 by an equal but opposite force.

If the direction of the current is reversed in one of the wires, say in wire 2, then we replace the vector $I_2\mathbf{L}$ by its negative in the foregoing derivation and we obtain Eq. (39.12) with a sign change. This means that the force on wire 2 is repulsive, in the same direction as $\hat{r}$. Wire 1 is also repelled by wire 2 by an equal but opposite force.

Returning to Eq. (39.12) and dividing by L, the length of the wire, we see that the force per unit length of wire is

$$F/L = -2K_m \frac{I_1 I_2}{r} \hat{r}. \tag{39.13}$$

This formula provides the basis for the modern definition of the ampere, the unit of current:

> If two long, straight parallel wires separated by a distance of 1 m carry equal currents, and if the force on each meter of wire is 2×10^{-7} N, then a current of 1 A flows in each wire.

This is the definition alluded to in Chapter 32. The fundamental unit of charge can now be defined in terms of the ampere: A coulomb is the amount of charge that passes through a particular cross section in a wire in 1 s when a steady current of 1 A is flowing in the wire.

Example 4

A 10-cm wire of mass 0.002 kg carrying a current I is suspended by flexible leads 1.0 mm above a long straight wire carrying an equal current in the opposite direction, as shown. The current in the wires is adjusted so that the tension in the leads is zero. Determine the value of the current.

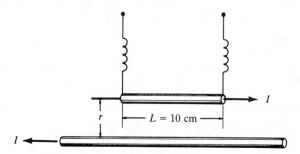

The tension in the leads is zero when the weight of the wire is balanced by the magnetic force acting on it. We know that the magnetic force is given by Eq. (39.12) with the currents equal. Setting this force equal to the weight of the wire we get

$$2K_m L I^2 / r = mg,$$

and solving for I^2 we find

$$I^2 = mgr/(2K_m L)$$

$$= (0.002 \text{ kg})(9.8 \text{ m/s}^2)(0.001 \text{ m}) / [(2 \times 10^{-7} \text{ N/A}^2)(0.1 \text{ m}),$$

39.4 AMPÈRE'S LAW

so

$$I = 31 \text{ A}.$$

Questions

12. Two long, straight wires carrying equal currents attract each other with a force per unit length of 1.8×10^{-3} N/m when they are separated by a distance of 3.6 mm. Find the current in each wire.

13. A rectangular loop of wire has sides $a = 0.2$ m and $b = 0.1$ m and is positioned a distance of 0.05 m from a long straight wire, as shown. If the current in the straight wire is 2.0 A and that in the loop is 3.0 A, find the total force on the loop.

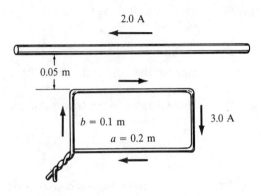

39.4 AMPÈRE'S LAW

In Chapter 33 we found that Gauss's law provides a useful way to determine the electric field of certain charge distributions that possess symmetry. There is a corresponding law, called Ampère's law, which provides a method for determining the magnetic field due to current distributions. Gauss's law is expressed as a surface integral over a closed surface, but Ampère's law is expressed as a line integral over a closed curve.

Ampère's law holds in all cases, but it is most useful in certain applications that involve a great deal of symmetry. Understanding the law in its general form is an important step toward our goal of finding a unifying description of electromagnetic phenomena, which will be described in Chapter 43.

To derive Ampère's law we begin with a special case. Consider a long straight wire carrying a steady current I. Then, by Eq. (39.6), the magnetic field at distance r from the wire is given by

$$\mathbf{B} = \frac{\mu_0 I}{2\pi r} \hat{\theta}, \tag{39.6}$$

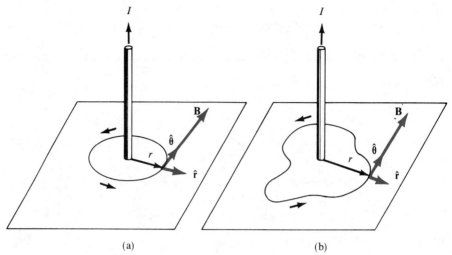

Figure 39.7 The line integral of a magnetic field **B** around a closed path surrounding a wire is proportional to the current.

where now it is convenient to express the result in terms of the permeability constant μ_0 rather than K_m.

In discussing electrostatic fields in Chapter 34 we found that the line integral of the electric field is zero,

$$\oint \mathbf{E} \cdot d\mathbf{r} = 0.$$

This result was a statement that the electrostatic field is conservative. We can ask if the same is true for magnetic fields. To take advantage of the symmetry of the magnetic field in Eq. (39.6) we first integrate along a circular path of radius r in a plane perpendicular to the wire, as shown in Fig. 39.7a. Using polar coordinates in this plane and integrating with respect to θ, we have

$$\oint \mathbf{B} \cdot d\mathbf{r} = \frac{\mu_0 I}{2\pi r} \int_0^{2\pi} \hat{\theta} \cdot \frac{d\mathbf{r}}{d\theta} \, d\theta.$$

But $d\mathbf{r}/d\theta = r\, d\hat{\mathbf{r}}/d\theta = r\hat{\theta}$ because the radius r is constant on the circle. Thus the integrand is $\hat{\theta} \cdot r\hat{\theta} = r$. The constant factor r cancels and the line integral around the circle is simply

$$\oint \mathbf{B} \cdot d\mathbf{r} = \frac{\mu_0 I}{2\pi}(2\pi) = \mu_0 I. \tag{39.14}$$

Unlike the electric field, the line integral is not zero. It is proportional to the current I, and the constant of proportionality is independent of the radius of the path.

Equation (39.14) has a form analogous to that of Gauss's law, which states that the integral of an electric field over a closed surface is proportional to the charge enclosed. It is easy to show that Eq. (39.14) holds not only for a circular path with

39.4 AMPÈRE'S LAW

the wire through its center, but for any closed curve that lies in a plane perpendicular to the wire and loops exactly once around the wire, as shown in Fig. 39.7b. Again, we use polar coordinates and describe the curve by the radius vector **r**. As above, we have $\mathbf{r} = r\hat{\mathbf{r}}$, but now r is not necessarily constant. By the product rule, the derivative $d\mathbf{r}/d\theta$ now involves two terms.

$$\frac{d\mathbf{r}}{d\theta} = r\frac{d\hat{\mathbf{r}}}{d\theta} + \frac{dr}{d\theta}\hat{\mathbf{r}}.$$

The extra term, $(dr/d\theta)\hat{\mathbf{r}}$, makes no contribution to the integral because $\hat{\mathbf{r}}$ is perpendicular to $\hat{\boldsymbol{\theta}}$. Therefore, the dot product $\mathbf{B} \cdot (d\mathbf{r}/d\theta)$ has the same value as above and hence Eq. (39.14) holds for this more general curve. It should be noted that if the curve wraps around the wire N times, the angle θ increases from 0 to $2\pi N$ and the result is

$$\oint \mathbf{B} \cdot d\mathbf{r} = N\mu_0 I. \tag{39.15}$$

This is a general form of Ampère's law. In fact, this equation is valid even if the curve does *not* wrap around the wire. In this case $N = 0$ and the integral around the curve is zero. The reason for this is that the net change in the angle θ is zero if the curve does not wrap around the wire, so the integral $\int d\theta$ has the value zero. Moreover, Eq. (39.15) also holds if the curve is not a plane curve, because portions of a path perpendicular to the field make no contribution to the integral. We omit detailed proofs of these more general statements because they depend on sophisticated mathematical analysis of curves and have nothing to do with physics.

Ampère's law, derived above for current flowing in one straight wire, is applicable to currents flowing in several wires that are not necessarily straight, as illustrated in Fig. 39.8. Suppose there are several currents in a region of space, as shown in Fig. 39.8. Each current I_j generates a magnetic field $\mathbf{B}_j$, and the total field **B** is the vector sum of the individual fields. Therefore, the line integral of **B** around

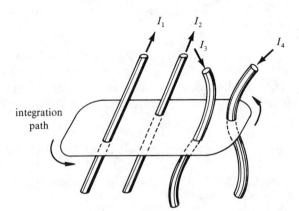

Several currents I_j in a region of space produce a field that is the superposition of the individual fields.

a closed curve is the sum of the line integrals of the individual fields $\mathbf{B}_j$. Since each of these is μ_0 times the current I_j, the sum is μ_0 times the algebraic sum of the currents:

$$\oint \mathbf{B} \cdot d\mathbf{r} = \mu_0 \sum I_j.$$

Thus the line integral of the magnetic field around a closed path is proportional to the algebraic sum of the currents that pierce the region bounded by the path. Taking the algebraic sum means that currents flowing in opposite directions are taken with opposite signs. We use the following convention: Curl the fingers of your right hand in the direction of the path of integration. If current I_j is in the general direction of your thumb, take I_j as positive, otherwise take I_j as negative. Of course, if the path does not enclose the wire containing I_j, there is no contribution to the integral.

Example 5

Two currents, $I_1 = 1$ A and $I_3 = 3$ A, flow down into the page. A third current, $I_2 = 2$ A flows up out of the page. Calculate $\oint \mathbf{B} \cdot d\mathbf{r}$ for each of the closed paths shown.

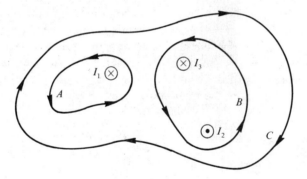

Ampère's law gives us the value of the integral without a knowledge of the field $\mathbf{B}$. For path A, only current I_1 flows through the region bounded by the path. According to the sign convention, I_1 is taken as negative. The line integral along path A is the product of μ_0 and the net current that flows through the region bounded by A:

$$\oint_A \mathbf{B} \cdot d\mathbf{r} = -\mu_0 I_1 = -(4\pi \times 10^{-7} \text{ N/A}^2)(1 \text{ A}) = -4\pi \times 10^{-7} \text{ N/A}.$$

For path B, both I_2 and I_3 flow through the enclosed region. We take I_2 as positive and I_3 as negative and we find

$$\oint_B \mathbf{B} \cdot d\mathbf{r} = \mu_0(I_2 - I_3) = -4\pi \times 10^{-7} \text{ N/A}.$$

39.4 AMPÈRE'S LAW

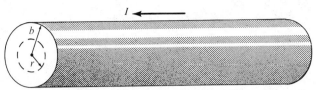

Figure 39.9 A long straight conductor carrying current I uniformly distributed over its cross section.

Finally, all three currents flow through the region bounded by path C. By our sign convention I_1 and I_3 are positive and I_2 is negative. Hence we have

$$\oint_C \mathbf{B} \cdot d\mathbf{r} = \mu_0(I_1 + I_3 - I_2) = 8\pi \times 10^{-7} \text{ N/A}.$$

Ampère's law exhibits a basic relationship that holds between the magnetic field and its source, electric current. It summarizes much of the experimental evidence found in the work of Biot and Savart, Ampère himself, and others. Like Gauss's law for electric fields, Ampère's law can sometimes be used to determine the magnetic field, especially when there is symmetry and a path of integration can be found along which the magnetic field is constant in magnitude and has the same direction as the path. In that case, the line integral is simply the field strength times the length of the path. Next we illustrate the use of Ampère's law in determining the magnetic field inside a conductor.

Imagine a long straight conducting wire with circular cross section of radius b that carries a current I uniformly spread out over its cross section, as illustrated in Fig. 39.9. We know that the field outside the wire is given by Eq. (39.6), with magnitude

$$B = \frac{\mu_0 I}{2\pi r}, \qquad r > b.$$

To find the magnetic field inside the wire we use Ampère's law. The field lines outside the wire are circular, so by symmetry we expect the same to be true inside the wire. Take a circular path of radius $r < b$ inside the wire, as shown in Fig. 39.9. On that path the magnetic field has constant magnitude and is tangent to the circle. Therefore, the line integral of $\mathbf{B}$ around this circle is simply $2\pi r B$, the field strength times the length of the path. By Ampère's law this is also equal to μ_0 times the amount of current flowing through the circular region of radius r. Because the total current I is distributed uniformly over the total cross section of radius b, the portion I_r through the smaller circle of radius r is just I times the ratio of the two areas,

$$I_r = I \frac{\pi r^2}{\pi b^2} = \frac{I r^2}{b^2}.$$

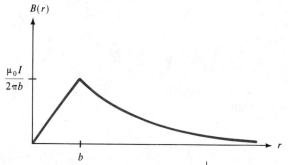

Figure 39.10 Graph of the magnetic field strength of a long straight wire as a function of distance from the center of the wire.

Inserting this into Ampère's law we get

$$2\pi r B = \mu_0 I \frac{r^2}{b^2},$$

and solving for B we find

$$B = \frac{\mu_0 I r}{2\pi b^2}, \qquad r < b. \tag{39.16}$$

This shows that inside the wire the magnetic field strength increases linearly with the distance from the center of the wire, reaching a maximum value at the surface of the wire. The magnetic field strength is plotted as a function of distance from the center in Fig. 39.10.

Another useful case for which the field strength can be determined by Ampère's law is the solenoid, a wire tightly wound into a helix, as shown by the example in Fig. 39.11. The advantage of a solenoid is that it can produce a very strong magnetic

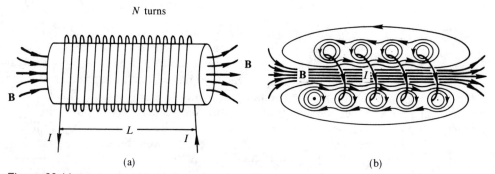

Figure 39.11 (a) A solenoid consists of a wire coiled into a helix. (b) The fields from the separate turns of wire reinforce inside the solenoid and cancel outside.

39.4 AMPÈRE'S LAW

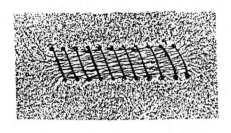 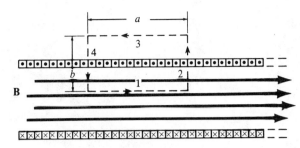

(a) (b)

Figure 39.12 (a) Magnetic field lines of a solenoid. (b) Calculating the magnetic field inside a solenoid.

field in a small region of space. To find the field strength, assume the solenoid is formed from a wire of length L carrying a current I and wound into N circular turns, and apply Ampère's law. Each turn separately would produce a magnetic field about it with circular symmetry. When the turns are brought near to each other to form the solenoid, the fields tend to cancel outside the solenoid and to reinforce each other inside, as illustrated in Fig. 39.11b.

Now we make the approximation that the field is constant inside the solenoid and zero outside of it. This approximation is reasonably good if the length of the solenoid is much greater than its width and the field is measured near the center of the solenoid, as evidenced by the field lines in Fig. 39.12a.

We apply Ampère's law to a rectangular path of sides a and b, as shown in Fig. 39.12b. Along segments 2 and 4 the magnetic field is zero outside the solenoid and perpendicular to the path inside the solenoid, so these segments make no contribution to the line integral of the field. Because $B = 0$ along segment 3, the only nonzero contribution to the integral comes from segment 1, along which the field is constant. So the line integral of the field around the rectangle is simply Ba. Now we calculate the net current that flows through the region bounded by this rectangle. Each turn carries a current I, so we need only count the number of turns that pass through the rectangular region. Because there are N turns altogether and the total length of the solenoid is L, the number passing through the rectangle is simply Na/L, or na, where $n = N/L$ is the number of turns per unit length. (Sometimes N and L are separately specified for a given solenoid, and sometimes n is specified.) Therefore, Ampère's law gives us

$$Ba = \mu_0 Ina,$$

which implies

$$B = \mu_0 nI. \tag{39.17}$$

For example, a solenoid wound with one layer of 0.1-mm wire would have 10^4 turns per meter. A current of 1 A would then produce a field intensity on the order of 0.01 T.

Example 6
A toroid consists of a solenoid that is bent into a torus, as shown. If the toroid has N turns and carries current I, determine the magnetic field strength inside at a distance r from the axis of the toroid.

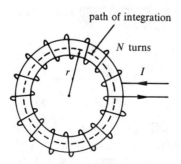

From symmetry we expect that inside the toroid the magnetic field strength is constant at a fixed radius from the axis and that the magnetic field lines are circular. To determine the field strength we apply Ampère's law to a circular field line of radius r, as shown. Because $\mathbf{B}$ has constant magnitude and is tangent to the circle at each point, we have, in polar coordinates, $\mathbf{B} = B\hat{\boldsymbol{\theta}}$ and $d\mathbf{r}/d\theta = r\,d\hat{\mathbf{r}}/d\theta = r\hat{\boldsymbol{\theta}}$, so

$$\oint \mathbf{B} \cdot d\mathbf{r} = Br \int_0^{2\pi} d\theta = B(2\pi r).$$

In other words, the line integral is equal to the product of the field strength B and the length of the path. The path encloses all the windings so the net current flowing through the region bounded by the path if NI. Therefore, Ampère's law states that

$$2\pi rB = \mu_0 NI,$$

which yields

$$B = \frac{\mu_0 NI}{2\pi r}.$$

What do you expect the magnetic field to be outside the toroid?

Questions

14. Determine which of the following wires and paths shown in diagrams (1)–(4) best satisfy the following statements:

(a) $\oint \mathbf{B} \cdot d\mathbf{r} = 0.$

39.5 A FINAL WORD

(b) The magnetic field can be calculated easily using Ampère's law.

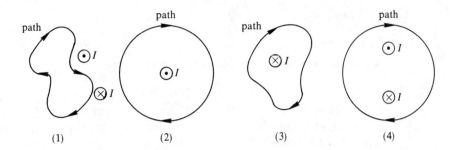

15. A wire of radius 2.0 mm carries a constant current of 4.0 A. Find the magnetic field at (a) 1.5 mm and (b) 5.0 mm from the center of the wire.

16. A 500-turn solenoid has a length of 5.0 cm and carries a current of 6.0 A. Find the magnetic field at the center of the solenoid.

17. A long straight coaxial cable consists of a thin wire of radius a that carries a current I covered by an outer sheath of radius b that carries a current I in the opposite direction. Find the magnetic field (a) inside the cable for $r < a$ and $a < r < b$ and (b) outside the cable.

18. Use Ampère's law to show that the magnetic field outside the toroid of Example 6 is zero. Consider values of r less than the inner radius and also greater than the outer radius.

19. It is sometimes thought that the magnetic field lines between two magnetic poles are straight lines as in the accompanying illustration. By integrating the field around an appropriate path, show that this configuration violates Ampère's law. In reality, the field lines bend outward and "fringe" at the edges.

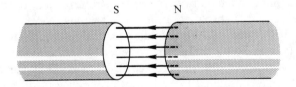

39.5 A FINAL WORD

After Ampère found that he could make a loop of current act like a bar magnet he proposed that electric currents are responsible for magnetism. He thought that a magnetic cylindrical steel bar has electric current flowing inside the bar and suggested that the steel bar is like a stack of loops of current, as illustrated in Fig. 39.13, and should have currents flowing around its surface.

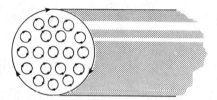

Figure 39.13 Ampère's molecular currents in a piece of iron.

One of Ampère's contemporaries, Augustin Fresnel, immediately criticized this theory, arguing that steel is a poor conductor of electricity and should heat up as current passed through it. But permanent magnets are not unusually warm so there can't be currents constantly flowing inside them. Fresnel also suggested an alternative. Perhaps the currents flow inside the atoms and molecules of the steel without any heating.

Ampère immediately adopted that idea and elaborated on it. In his day it was thought that light was propagated through a medium called the aether. Ampère proposed that the aether consists of a sea of positive and negative electric charge that is usually balanced and neutral. However, a molecule has the ability to split up the positive and negative charge. Positive charge flows around the molecule on one side (say the left side), while negative charge flows around the molecule on the other side (the right side). The result is such that each molecule is a little loop of current that produces a magnetic field. If we take a piece of iron, which is full of this kind of molecule, and line up all the current loops inside, they give magnetic fields pointing in the same direction. The net result is the same as if a sheet of current were flowing around the outside surface of the material. That's how Ampère proposed that a permanent magnet produces a magnetic field.

This is a very curious picture. There are parts of it that we still believe to be true. We believe that currents do flow inside atoms and molecules and, because currents flow in little loops, atoms and molecules have magnetic fields that contribute to the magnetism of a material. That's why a bar magnet behaves the way it does. On the other hand, the picture of the aether splitting up into two charges is an example of thinking that is absolutely brilliant but wrong. Nevertheless, the theory was of some use. It was part of the scaffolding that we're always erecting in science that eventually falls away as the edifice is constructed. In the course of time, professional opinions about the importance of certain events in the history of science change. People considered to be heroes in the history of science come and go, but Ampère's name will always be current.

CHAPTER 40

VECTOR FIELDS AND HYDRODYNAMICS

This method of representing the intensity of a force by the velocity of an imaginary fluid in a tube is applicable to any conceivable system of forces, but it is capable of great simplification in the case in which the forces are such as can be explained by the hypothesis of attractions varying inversely as the square of the distance, such as those observed in electrical and magnetic phenomena.

    James Clerk Maxwell, *Transactions of the Cambridge Philosophical Society*, vol. **X** (1855)

40.1 ACTION-AT-A-DISTANCE REVISITED

When Isaac Newton announced his theory of gravitation in the late seventeenth century, he based it on forces acting at a distance. Somehow the sun exerted a force on the earth through a vast expanse of empty space with no visible connection to transmit that force. This was a new idea that was not readily accepted. Huygens and Leibniz, among others, criticized portions of the *Principia* for not providing a clear, mechanical account of gravitation. Excellent mathematics, they said, but no physics.

After all, common sense dictates that local motion can only be produced by pushing or pulling, and therefore, they concluded, action at a distance implies a perpetual miracle or a spiritual nonmechanical agency, bordering on the occult. In fact, for about 20 years after the publication of the *Principia*, Newton seems to have thought that God was the immediate and omnipresent cause of the mutual attraction of bodies. In the second edition of the *Principia*, published in 1713, Newton hinted at a possible aether mechanism for gravity, but advanced no hypothesis to explain how gravitational force was transmitted across empty space.

Toward the end of the eighteenth century, as the better physics texts placed less emphasis on the search for causes and more on the derivation of one phenomenon from another, attitudes seemed to change. Newton's gravitational theory was recognized as having great utility because of its unifying and predictive power, even though it gave no physical explanation of action at a distance. Forces were viewed as mathematical abstractions, ideas in the mind, not in the real world. Philosophical questions about the causes of action at a distance were raised again in connection with electrical and magnetic forces, and a new point of view was introduced in the nineteenth century, when Michael Faraday proposed that space was filled with lines of force, as described in Chapter 33. In the next generation, James Clerk Maxwell translated some of Faraday's ideas into mathematical language, and another abstraction, the field of force, became part of the literature of physics.

Maxwell, like Newton, constructed a purely mathematical model to describe relations between forces, and made no assumptions about the actual transmission of these forces. He simply noted that no theory was "better established in the minds of men than that of the action of bodies on one another at a distance." He also pointed out that mathematical laws of the same type appear again and again in diverse physical phenomena – gravitation, electricity, magnetism, propagation of light, and heat flow. By focusing on the mathematical laws he could attain generality and precision and "avoid the dangers arising from a premature theory professing to explain the cause of the phenomena."

Whereas attempts to explain action-at-a-distance seemed mystifying and even occult, the mathematical concept of a field was well defined and served as an appropriate subject for scientific inquiry. Moreover, calculations and predictions arising from the mathematics were tested experimentally and found to be in accord with natural phenomena. For example, Maxwell discovered that if an electromagnetic field is disturbed, the disturbance is not transmitted instantaneously, but instead propagates at a definite, finite speed, the speed of light. In fact, light is just a disturbance in the electromagnetic field, its speed inherent in the force laws for magnetism and electricity. We now realize that this discovery was the first great triumph of field theory.

40.2 PROPERTIES OF VECTOR FIELDS

We turn now to a mathematical description of the field concept. The basic ideas used to describe fields originated in *hydrodynamics*, the study of fluid flow, and not in the study of electricity and magnetism. When Maxwell lectured about his theory of electromagnetic fields, he appealed to hydrodynamic analogies to clarify his ideas because physicists of his day were familiar with hydrodynamics and because flowing water is easier to visualize than abstract force fields. We have already appealed to hydrodynamic analogy in discussing electric flux in Chapter 33, and we continue to do so in this chapter.

Mathematically, a *field* is simply a function of position – something that has a definite value at each point in space. If the function values are scalars, the field is called a *scalar field*. The electric potential is an example – the potential at a point (x, y, z) in space is a scalar function of three variables, $V(x, y, z)$. If the function values are vectors, the field is called a *vector field*. It has both a magnitude and a direction at each point under consideration. This chapter is concerned primarily with two examples of vector fields, the electric field **E**, and the magnetic field **B**. To aid in understanding these fields we also study a vector field from hydrodynamics, the *velocity field* **v**, which indicates the speed and direction of flowing water at each point in the fluid.

Before plunging into the mathematics, we summarize some of the results to be obtained in this chapter. For example, we will learn that all three fields store energy in much the same way: The energy in a small portion of space is proportional to the square of the field intensity, E^2, B^2, or v^2. We will also find that water exerts a force similar to the strange magnetic force on current-carrying wires, which we know is perpendicular both to the magnetic field and to the direction of electric current. But the most important result has to do with the character or form of the fields. Electric and magnetic fields are essentially different from one another. Electric fields tend to radiate outward from (or inward toward) point charges, which are their sources. Magnetic fields, on the other hand, never converge to a point. Instead, they circulate around the lines of electric current, which are their sources. Hydrodynamic velocity fields can have either behavior, and learning why will help us visualize and better understand electric and magnetic fields.

The mathematics of vector fields makes use of both derivatives and integrals. The derivatives indicate how fields change locally from point to point, whereas the integrals describe global attributes of the field (effects over regions in the field). This chapter will focus on two global properties of fields called *flux* and *circulation*.

40.3 FLUX OF A VECTOR FIELD

In Chapter 33 we used flowing water to motivate the definition of electric flux. The idea was to place a small rectangular frame of area ΔA in the flow and measure the volume of water that passes through the frame in unit time. Both the area and the orientation of the frame can be described by an area vector $\Delta \mathbf{A}$ whose magnitude ΔA is the area of the frame and whose direction is perpendicular to the frame, as indicated in Fig. 40.1. The dot product $\mathbf{v} \cdot \Delta \mathbf{A}$ is called the flux through the frame and measures the volume of water flowing through the frame in unit time in the direction of $\Delta \mathbf{A}$. Sometimes it is useful to measure the mass of water through the frame per unit time, in which case we simply multiply the volume flux $\mathbf{v} \cdot \Delta \mathbf{A}$ by the density ρ of the water (mass per unit volume) to get the mass flux. Since water is an incompressible fluid, the density ρ is constant, but for fluids other than water (gases, for example), the density ρ could vary from point to point.

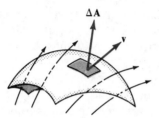

Figure 40.1 Flux through a surface.

To find the total mass flux through a surface S immersed in the flow, we divide S into a large number of small patches with area vectors $\Delta \mathbf{A}_i$ pointing toward the same side of S (which we arbitrarily call the positive side), and take the limiting value of the sum

$$\sum_i \rho \mathbf{v} \cdot \Delta \mathbf{A}_i$$

as the number of patches increases and the size of each patch shrinks to zero. The result is the mass flux through S and is denoted by the surface integral

$$\phi = \iint_S \rho \mathbf{v} \cdot d\mathbf{A}. \tag{40.1}$$

The volume flux is the same integral with ρ replaced by 1.

The electric and magnetic flux were defined earlier in an analogous way:

$$\phi_E = \iint_S \mathbf{E} \cdot d\mathbf{A} \quad \text{(electric flux)}, \tag{33.14}$$

$$\phi_B = \iint_S \mathbf{B} \cdot d\mathbf{A} \quad \text{(magnetic flux)}. \tag{38.8}$$

For a general vector field $\mathbf{V}$, flux can be defined by the same type of integral, with $\mathbf{V}$ replacing $\mathbf{E}$ or $\mathbf{B}$ in the integrand.

A visual description of electric and magnetic flux can be given by using Faraday's lines of force. The magnitude of the field vector ($\mathbf{E}$ or $\mathbf{B}$) can be regarded as a measure of the number of lines per unit area at any point, with the direction of the field vector indicating the direction of the flow. Sometimes the flux ϕ is said to be the "number" of Faraday lines passing through S. This terminology is intended to convey qualitative information only. For example, if one field has a flux 27.9 through S while a second has flux 13.2, this suggests that the first field has roughly twice as many Faraday lines passing through S as the second field. The algebraic sign of ϕ is also significant. Positive ϕ means there is a net flux toward the positive side of S, and negative ϕ means there is a net flux in the opposite direction.

Now consider a closed surface such as a bubble in space. What is the total electric flux through the surface of the bubble? For an electric field, such as that

40.3 FLUX OF A VECTOR FIELD

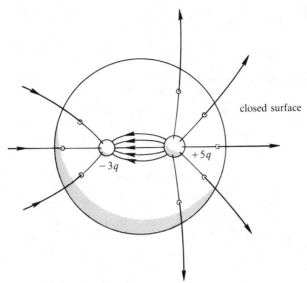

Figure 40.2 The electric flux through a closed surface is nonzero if the surface encloses a net nonzero charge.

shown in Fig. 40.2, the Faraday lines start from positive charges and end on negative ones. If there are more positive than negative charges inside the bubble, it seems reasonable to say there will be more lines coming out than going in. This is consistent with Gauss's law, which states that the flux is proportional to the net charge inside:

$$\phi_E = \oiint_S \mathbf{E} \cdot d\mathbf{A} = \frac{q}{\varepsilon_0}. \tag{33.17}$$

In other words, electric fields have positive flux through a closed surface if there is a net positive charge inside. Positive charges are called *sources*, because field lines radiate away from them, while negative charges are called *sinks*, because field lines flow into them. If the net charge inside S is negative, Gauss's law shows that the flux will also be negative, suggesting that more lines are going in than coming out. In any case the net flux will never be zero if the net charge inside is nonzero. The terms "source" and "sink" originated in hydrodynamics. An electric source is like a nozzle discharging electric field lines, while a sink is like a drain into which field lines flow.

For magnetic fields, the Faraday lines always form closed loops, as illustrated in Fig. 40.3. The lines never end at a point distinct from the one where they began. Gauss's law for magnetism states that the net flux through a closed surface is always zero:

$$\phi_B = \oiint_S \mathbf{B} \cdot d\mathbf{A} = 0. \tag{38.9}$$

Thus, unlike electric fields, magnetic fields can have no sources or sinks. The next section shows how this property of magnetic fields can be visualized by water flow.

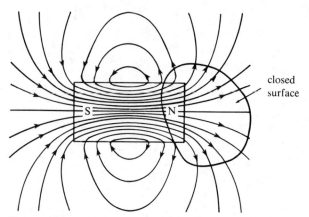

Figure 40.3 The magnetic flux through a closed surface is always zero.

Question

1. A vector field is given by $\mathbf{V}(r, \theta) = (\beta/r)\hat{\theta}$, where β is a constant and r, θ are the polar coordinates of the projection of a point (x, y, z) onto the xy plane. Determine the flux through a spherical surface of radius R with center at the origin. Could this field describe an electric field? Give a reason for your answer.

40.4 CIRCULATION OF A VECTOR FIELD

A connection between fluid flow and magnetic fields is provided by the concept of *circulation*, which is closely related to work, or potential difference. Recall that the work required to move a charge q from point P_1 to point P_2 in an electric field E is

$$W_{12} = -\int_{P_1}^{P_2} q\mathbf{E} \cdot d\mathbf{r} \tag{34.1}$$

and the potential difference is

$$V(P_2) - V(P_1) = -\int_{P_1}^{P_2} \mathbf{E} \cdot d\mathbf{r}. \tag{34.3}$$

If we calculate the work or potential difference for a static electric field around a *closed* path, the result is always zero:

$$\oint \mathbf{E} \cdot d\mathbf{r} = 0.$$

Any vector field with this property is called a *conservative field*.

In Chapter 39 we learned that the magnetic field **B** is not conservative because the line integral around a closed curve need not be zero. Likewise, the velocity field of a fluid flow is not conservative, as we will see presently.

40.4 CIRCULATION OF A VECTOR FIELD

For fluid flow, the line integral of the velocity field taken around a closed curve is called the *circulation*, k:

$$k = \oint \mathbf{v} \cdot d\mathbf{r}. \tag{40.2}$$

The value of k depends on the curve and on the nature of the flow.

As an example of a fluid flow with nonzero circulation, consider a beaker of fluid containing fine aluminum powder that makes the flow lines visible. When the beaker is placed on a turntable and rotated with constant angular speed ω, the flow appears complicated at first, but later settles down to a steady-state condition in which the entire fluid rotates as if it were solid. This steady-state flow is called *solid body rotation*. Every particle of fluid is in uniform circular motion about the central axis, with the same angular speed as the turntable, as indicated in Fig. 40.4.

To calculate the circulation for solid body rotation of a fluid we integrate the velocity field around a closed path. Because the water is flowing in circles, it is convenient to integrate around a circle concentric with the axis of rotation, as shown in Fig. 40.5. In polar coordinates, the velocity of the water at a point a distance r from the axis is

$$\mathbf{v} = \omega r \hat{\theta},$$

and, as we saw in Chapters 26 and 39, we have $d\mathbf{r}/d\theta = r\hat{\theta}$. Substituting into Eq. (40.2), we obtain

$$k = \oint \mathbf{v} \cdot d\mathbf{r} = \int_0^{2\pi} (\omega r \hat{\theta}) \cdot (r\hat{\theta})\, d\theta = \omega r^2 \int_0^{2\pi} d\theta = 2\omega \pi r^2.$$

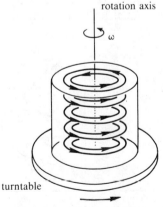

Figure 40.4 In solid body rotation each part of the fluid rotates with the same constant angular speed.

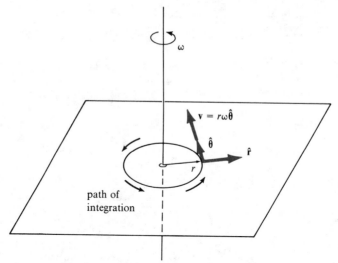

Figure 40.5 The circulation for a fluid in solid body rotation is nonzero.

So in this simple, common flow there is indeed a nonzero circulation. If we calculate the line integral around a different path, the calculation is slightly different but, as we will show presently in Example 1, the result has the same form: $k = 2\omega$ times the area enclosed by the path.

Example 1

Show that for solid body rotation of a fluid with constant angular speed ω, the circulation around any simple closed curve C traversed counterclockwise in a plane perpendicular to the axis of rotation is equal to 2ω times the area enclosed by C.

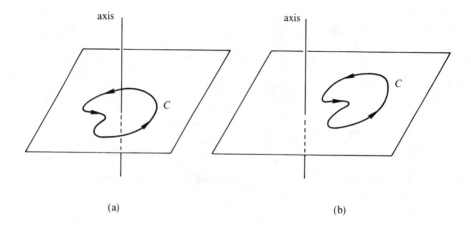

(a)   (b)

40.4 CIRCULATION OF A VECTOR FIELD

First we assume the axis of rotation passes through the interior of the region enclosed by C, as shown in (a). Then we can describe the path in polar coordinates using $\mathbf{r} = r\hat{\mathbf{r}}$, where now r is not necessarily constant but varies as a function of θ. For this curve we have

$$\frac{d\mathbf{r}}{d\theta} = \frac{dr}{d\theta}\hat{\mathbf{r}} + r\frac{d\hat{\mathbf{r}}}{d\theta} = \frac{dr}{d\theta}\hat{\mathbf{r}} + r\hat{\boldsymbol{\theta}},$$

and for the velocity we have $\mathbf{v} = \omega r \hat{\boldsymbol{\theta}}$, so the dot product $\mathbf{v} \cdot d\mathbf{r}/d\theta$ is simply

$$\mathbf{v} \cdot \frac{d\mathbf{r}}{d\theta} = \omega r^2 \hat{\boldsymbol{\theta}} \cdot \hat{\boldsymbol{\theta}} = \omega r^2,$$

because $\hat{\mathbf{r}} \cdot \hat{\boldsymbol{\theta}} = 0$. Therefore, the line integral for the circulation becomes

$$k = \oint_C \mathbf{v} \cdot d\mathbf{r} = \omega \oint_C r^2 \, d\theta = 2\omega \oint_C \left(\frac{1}{2}r^2\right) d\theta = 2\omega \oint_C \left(\frac{1}{2}r^2 \frac{d\theta}{dt}\right) dt.$$

But, from Eq. (26.9) we know that

$$\frac{1}{2}r^2 \frac{d\theta}{dt} = \frac{dA}{dt},$$

the rate at which the radius vector sweeps out area. Therefore,

$$k = 2\omega \oint_C \frac{dA}{dt} \, dt = 2\omega A,$$

where A is the area of the region enclosed by C.

It can be shown that the formula $k = 2\omega A$ also holds even if the axis of rotation does not pass through the region, as shown in diagram (b). An example is given in Question 4.

For the electrostatic field, we found that $\oint \mathbf{E} \cdot d\mathbf{r} = 0$, but for solid body rotation of a fluid we have $\oint \mathbf{v} \cdot d\mathbf{r} \neq 0$. That means that electrostatic field lines never appear like those of solid body rotation. This is a restriction on how the world can be, something that's always important in physics. Is solid body rotation the only forbidden field, or are there others?

To answer this question we analyze how the fluid is set into motion. The rotating wall of the beaker drags the water along by means of its own viscosity. At the wall itself, the water is at rest relative to the wall because molecules that try to move along the wall keep banging into it, losing energy and momentum. A small distance away from the wall there's some flow parallel to the wall, more flow still further away, and so on, as illustrated in Fig. 40.6, where the wall is shown as being straight. In other words, $\mathbf{v}$ is generally in the x direction of Fig. 40.6, but its magnitude increases in the y direction. This is precisely the kind of behavior forbidden to an electric field. In fact, if you calculate the circulation around the rectangular loop shown by the dashed line in Fig. 40.6, you'll find it's nonzero (three

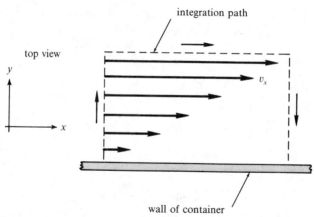

Figure 40.6 Flowing water near the wall of a container moves more slowly than water further away.

sides contribute zero to the integral, but one does not). This type of shearing flow is needed to set up a circulation of any kind. You might think of it as a kind of stirring. Water can be stirred into motion, but electrostatic fields cannot be created by "stirring."

Are there magnetic fields analogous to solid body rotation of water? The field lines for solid body rotation of a fluid,

$$\mathbf{v} = \omega r \hat{\boldsymbol{\theta}},$$

are analogous to the magnetic field lines *inside* a straight wire with current uniformly distributed over its cross section,

$$\mathbf{B} = \frac{\mu_0 I r}{2\pi b^2} \hat{\boldsymbol{\theta}}, \tag{39.16}$$

This is illustrated in Fig. 40.7. Moreover, we know that electric currents are the sources of magnetic fields, as summarized mathematically by Ampère's law,

$$\oint \mathbf{B} \cdot d\mathbf{r} = \mu_0 I. \tag{39.14}$$

Electric current is precisely analogous to hydrodynamic circulation, with $\mu_0 I$ playing the same role as k.

Water can flow out of a nozzle or into a drain. The resulting flow has the same character as the electrostatic field of a positive or negative charge. Magnetic fields, on the other hand, cannot behave that way. Also, water can be stirred into motion, and the resulting flow has the same character as a magnetic field, but electric fields cannot behave that way. The circulation is analogous to electric current – the source of magnetic fields. That's the essential difference between electrostatic and magnetic fields.

There's a simple hydrodynamic example that can exhibit both kinds of behavior. Consider a huge bowl full of water with a plug in the bottom. When the plug is pulled, the water drains out. There's an obvious net flux out of the bowl. Inside, the

40.4 CIRCULATION OF A VECTOR FIELD

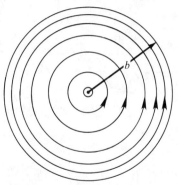

Figure 40.7 Magnetic field lines inside a straight wire of radius b carrying current uniformly distributed over its cross section.

water is flowing in lines toward the hole, like Faraday lines of force converging on a negative charge. This flow is much like an electric field.

But if the water is stirred with a paddle, a vortex quickly develops. The circulation is obvious, but the motion is quite different from solid body rotation. Inward near the vortex, the flow is faster instead of slower as you might expect from the velocity of solid body rotation. As discussed in Chapter 23, each bit of water conserves angular momentum as it spirals inward:

$$L = mvr = \text{constant},$$

where r is the distance from the axis of the vortex. Hence, as the distance from the axis decreases, the velocity increases according to

$$\mathbf{v} = \frac{L}{m}\frac{1}{r}\hat{\boldsymbol{\theta}}.$$

This field has the same mathematical form as the magnetic field of a line current,

$$\mathbf{B} = \frac{\mu_0}{2\pi}\frac{I}{r}\hat{\boldsymbol{\theta}}. \tag{39.6}$$

Any vector field $\mathbf{V}$ of the form

$$\mathbf{V} = \frac{\beta}{r}\hat{\boldsymbol{\theta}}, \tag{40.3}$$

where β is constant, is called a *vortex* field. The field lines are illustrated in Fig. 40.8.

If the circulation of a vortex field is calculated around any path enclosing the axis of the vortex, it is easy to show that the result is always the same, $2\pi\beta$, a number which measures the strength of the vortex. (See Question 8.) This result is similar to that for the magnetic field outside a wire. According to Ampère's law, the line integral of the magnetic field for any path that encloses the current equals a constant times I, the strength of the current. On the other hand, any path that excludes the vortex or the current gives zero. There's circulation around the vortex, but the water outside the vortex core is circulation free. This is *not* true in solid body rotation, which, as shown earlier, has nonzero circulation everywhere. This

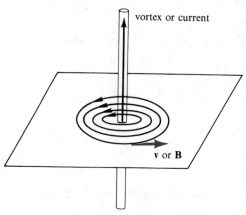

Figure 40.8 Vortex field for spiraling water in a bowl, or magnetic field lines around a current-carrying wire.

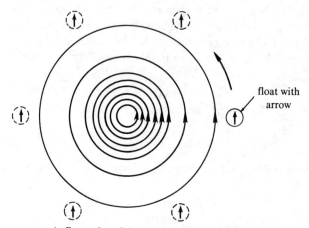

Figure 40.9 A float placed in a vortex flow revolves around a vortex but does not rotate about its own center.

idea is illustrated in Fig. 40.9. If a float is placed in the vortex flow it revolves around the vortex, but it doesn't rotate about its own center.

Example 2

Show that for a vortex field $\mathbf{v} = (\beta/r)\hat{\boldsymbol{\theta}}$, where β is a constant, the circulation is zero for any closed path that does not enclose the vortex.

To calculate the circulation we integrate along a path similar to that used in Example 1. In this case, the factor r cancels in the dot product of $\mathbf{v}$ with $d\mathbf{r}/d\theta$ and

40.4 CIRCULATION OF A VECTOR FIELD

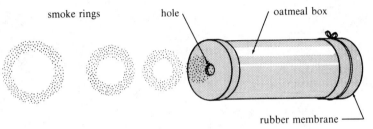

Figure 40.10 An oatmeal box generates smoke rings.

the circulation integral becomes

$$k = \oint \mathbf{v} \cdot d\mathbf{r} = \beta \oint d\theta = 0,$$

because θ returns to its original value if the path does not enclose the vortex.

The core of a vortex doesn't have to be a straight line. It can bend around and even close on itself in a configuration known as a vortex ring. A smoke ring is an example of a vortex ring. The field around a vortex ring is identical to the magnetic field generated by a loop of current; it's what we call a dipole field. Vortex rings in air are very stable, since only the rather weak torque applied via the small viscosity of the air can extract the angular momentum trapped into the structure of the motion. In fact, in a fluid with no viscosity, vortex rings would last forever (but might be impossible to create).

The stability of smoke rings can be illustrated with a simple device made from a cylindrical oatmeal box and a rubber membrane. Smoke rings can be generated by pushing on the membrane. Such a device can be used to blow out candles halfway across a room. Try it.

Questions

2. Which of the following vector fields could describe a magnetic field?

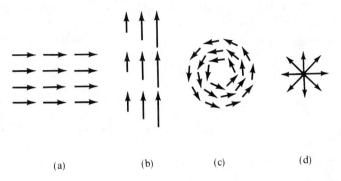

(a)   (b)   (c)   (d)

3. Determine which of the following vector fields could describe a static electric field.

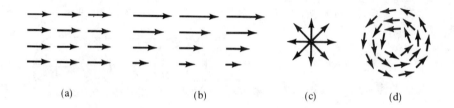

(a)   (b)   (c)   (d)

4. For solid body rotation of a fluid with constant angular speed ω, calculate the circulation around the boundary of the portion of the circular annulus shown and verify that $k = 2\omega$ times the area of the region enclosed.

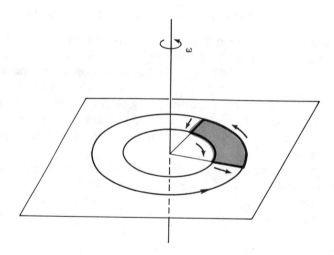

5. Suppose a small paddlewheel is placed in a tank of water undergoing solid body rotation, as shown. Does the paddlewheel rotate about its own axis? Give a physical argument.

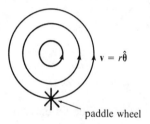

40.4 CIRCULATION OF A VECTOR FIELD

6. Which of the following fields have zero circulation everywhere?

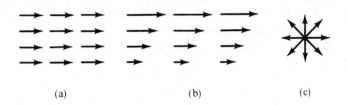

(a)   (b)   (c)

7. Which of the following fields have nonzero circulation somewhere?

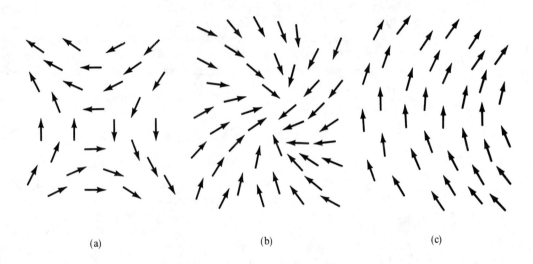

(a)   (b)   (c)

8. Calculate the circulation for the vortex field $\mathbf{v} = (\beta/r)\hat{\boldsymbol{\theta}}$ for a path that encloses the axis of the vortex.

9. A vector field at each point in the plane with polar coordinates (r, θ) is given by $\mathbf{V} = A(r^2 \sin^2\theta \hat{\mathbf{i}} + 2r\cos\theta \hat{\mathbf{j}})$, where A is a nonzero constant. Calculate the circulation around the unit circle described by the vector function $\mathbf{r} = \cos\theta \hat{\mathbf{i}} + \sin\theta \hat{\mathbf{j}}$, and determine whether this field could possibly describe an electric field.

10. A vector field in the plane is described in polar coordinates by $\mathbf{V} = A(3r\cos\theta \hat{\mathbf{i}} + 4r\sin\theta \hat{\mathbf{j}})$. Calculate the circulation around a circle of radius R, and determine whether $\mathbf{V}$ could describe the magnetic field of a current carrying wire at $r = 0$.

11. Given that the vortex field for circulating water is $\mathbf{v} = (\beta/r)\hat{\boldsymbol{\theta}}$, draw on your knowledge of torques to find a semiquantitative explanation why a cork placed in a vortex flow (as in Fig. 40.9) revolves about the vortex but does not rotate about its own center. *Hint.* The cork moves with the water if the water is moving, and circulates with the water if the water is circulating.

40.5 HYDRODYNAMIC ANALOGIES FOR ENERGY AND FORCES

Electric charges create electric fields, and there is energy inherent in the system because it takes work to bring the charges together. In Chapter 34 we adopted the point of view that the energy resides in the field, not in the charges. The field has some energy per unit volume at each point in space. This is the energy density, denoted by u, and is given by

$$u = \tfrac{1}{2}\varepsilon_0 E^2. \tag{34.24}$$

This was obtained by calculating the work in charging a capacitor, but the result holds more generally. Similarly, the energy density in a magnetic field can be calculated from the work needed to produce the flowing current that creates the field; the result turns out to be $B^2/(2\mu_0)$. Aside from constants, the energy per unit volume in both kinds of fields is proportional to the square of the field intensity.

The same is true for the energy stored in water flow. In this case, the energy of the field is the *kinetic energy* of the moving water. The energy of a droplet of mass m is simply $\tfrac{1}{2}mv^2$. So the energy per unit volume is $v^2/2$ times ρ, the mass per unit volume. Thus, we have

$$\text{energy density} = \tfrac{1}{2}\rho v^2 \quad \text{(water)}$$
$$= \tfrac{1}{2}\varepsilon_0 E^2 \quad \text{(electric)}$$
$$= \tfrac{1}{2}B^2/\mu_0 \quad \text{(magnetic)}.$$

In summary, hydrodynamic velocity fields can have the character either of electric fields (sources and sinks) or of magnetic fields (stirred up, circulation). In all cases, the energy density varies as the square of the field intensity.

Finally, what about forces? Remember, there's something peculiar about the magnetic force on a line current: it's perpendicular to both the field and the current at that point, as represented by the cross product

$$\mathbf{F} = I\mathbf{L} \times \mathbf{B}. \tag{38.16}$$

Since we now know that hydrodynamic circulation is analogous to line current, could it be that a similar force acts on a circulation in a velocity field?

The answer can be illustrated by an application of airfoils (cross sections of airplane wings). As an airplane wing is pushed or pulled by the engines, the air flow past it tends to build up pressure below it, and to create a partial vacuum above it, as illustrated in Fig. 40.11. The result of the pressure difference is a net upward force on the airfoil that lifts the plane.

Hydrodynamically, the same situation can be described in a slightly different way. As seen from the airplane, the air is rushing past uniformly, except where it's disturbed by the wing. The wing makes the air slow down below the wing and speed up above it. At each point outside the wing, the flow can be thought of as the sum of two parts. One flow is uniform, as shown by the dashed lines in Fig. 40.12. The other is the effect that opposes the air flow underneath, and adds to it above the wing, as indicated by the curved lines in Fig. 40.12.

This flow has obvious circulation. In other words, the wing can be thought of as acting like a line vortex core, a cause of circulation, which interacts with the steady

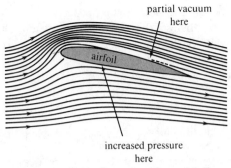

Figure 40.11 Air flow around an airfoil creates lift.

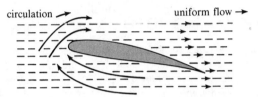

Figure 40.12 Air flow around a stationary airfoil.

flow resulting in an upward force. The air flow is horizontal, the wing (or the direction of the circulation) is perpendicular to the plane of the sketch, and the resulting force (upward) is perpendicular to both. The situation is analogous to a line current in a uniform field resulting in a force perpendicular to both.

40.6 A FINAL WORD

The early Greeks first postulated that all matter was composed of atoms, but the idea remained idle conjecture for centuries. Then, in the eighteenth century, a small step forward in understanding the nature of matter was made by Roger Boscovich, who postulated forces between atoms. His ideas, in turn, influenced scientists of the nineteenth century. One of them, John Dalton, discovered the law of simple and multiple proportions and gave firm scientific foundation to the idea that matter is composed of atoms. However, no one had any idea of what constituted atoms. Then, in 1858, the great German scientist, Heinrich von Helmholtz, made a mathematical discovery about vortex rings. He discovered that a vortex ring in a nonviscous fluid could never be destroyed – it would be conserved forever. Having heard about this, another giant of nineteenth century science, William Thomson (Lord Kelvin), proposed that atoms were vortex rings in the aether that was believed to permeate space and propagate light.

Today we may be inclined to think that nineteenth century physicists were stodgy, conservative, unimaginative people compared to our twentieth century geniuses. However, it would be difficult to imagine anyone being more imaginative than someone who suggests that atoms that make up all matter behave like vortex rings, which, among other peculiarities, have the property of slowing down when a force is applied to them. Nevertheless, that was the suggestion Thomson made – a suggestion taken very seriously by the scientific community. An article on the atom in the famous ninth edition of the *Encyclopaedia Britannica*, published in 1875 at the height of the Victorian era, gives us a glimpse of scientific thought at that time.

After introducing the section "On the Theory of Vortex Atoms," the author states:

> It was reserved for Helmholtz to point out the very remarkable properties of rotational motion in a homogeneous incompressible fluid devoid of all viscosity.... The motion of a fluid is said to be irrotational when it is such that if a spherical portion of the fluid were suddenly solidified, the solid sphere so formed would not be rotating about any axis.

Then there are several pages of mathematical equations detailing the theory.

Further in the article, he writes:

> These properties of vortex rings suggested to Sir William Thomson the possibility of founding on them a new form of atomic theory. The conditions which must be satisfied by an atom are – permanence in magnitude, capability of internal motion or vibration, and a sufficient amount of possible characteristics to account for the difference between atoms of different kinds.

In other words, atoms have to be permanent because we know matter lasts for a long time. They have to have a richness of possible internal properties because in the 1870s it was known that different atoms emitted light of different colors, so atoms had to be capable of vibrating at characteristic frequencies that would produce the observed light. Physicists would later conduct investigations to determine how vortex atoms vibrate.

Further yet in the article, the author states:

> But the greatest recommendation of this theory, from a philosophical point of view, is that its success in explaining phenomena does not depend on the ingenuity with which its contrivers "save appearances," by introducing first one hypothetical force and then another. When the vortex atom is once set into motion, all its properties are absolutely fixed, and determined by the laws of motion of the primitive fluid, which are fully expressed in the fundamental equations. The disciples of Lucretius may cut and carve his solid atoms in the hope of getting them to combine into worlds; and the followers of Boscovich may imagine new laws of force to meet the requirements of each new phenomenon; but he who dares to plant his feet in the path opened up by Helmholtz and Thomson has no such resources. His primitive fluid has no other properties other than inertia, invariable density, and perfect mobility, and the method by which the motion of this fluid is to be traced is pure mathematical analysis. The difficulties of this method are enormous, but the glory of surmounting them would be unique.

Who wrote that? What Brittanica author, writing in 1875, understood the philosophy and history of physics so well and was able to write about it so gracefully? The article is signed J. C. M. – James Clerk Maxwell.

CHAPTER 41

ELECTROMAGNETIC INDUCTION

The experiments described combine to prove that when a piece of metal is passed either before a single pole, or between the opposite poles of a magnet, or near electro-magnetic poles, whether ferruginous or not, electrical currents are produced across the metal transverse to the direction of motion.... If a single wire be moved like the spoke of a wheel near a magnetic pole, a current of electricity is determined through it from one end towards the other.

    Michael Faraday, *Experimental Researches in Electricity* (1831)

41.1 THE INCOMPARABLE EXPERIMENTALIST

After Oersted's 1820 discovery that an electric current produces a magnetic force, scientists searched for a way to reverse the process, to use a magnetic force to produce an electric current. For many years the search proved fruitless, but the struggle continued because it was realized that immortal fame awaited the discoverer.

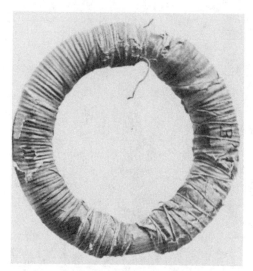

Figure 41.1 Photograph of the historic ring of soft iron used by Faraday in his first successful experiment in electromagnetic induction. (Courtesy of the Archives, California Institute of Technology.)

Twelve years later, in the summer of 1831, the discovery was made in a dimly lit room at the Royal Institution in London. Michael Faraday, a gentle, simple man of 40, worked in a laboratory cluttered with magnets, copper wires, and iron rings. A solitary iron ring enshrouded with two sets of coils lay on a table. (See Fig. 41.1.) A sensitive magnetic needle was attached to one set of coils and a battery to the other. When the battery was connected to the second set of coils, the magnetic needle moved but quickly returned to its original position, and when the connection was broken the needle was again disturbed. Culminating years of agonizing research and frustration, Michael Faraday had completed the circle begun by Oersted – he had undeniably demonstrated that a magnetic field can be used to produce an electric current. His endless perseverance and unerring physical insight pointed to an essential observation others had missed: To produce electric current, the magnetic pole and the electric charge had to be moving or changing with respect to one another. The resulting effect was called electromagnetic induction.

Unlike Oersted's chance discovery, Faraday's success came after a long series of researches in which he explicitly set out to determine if magnetism could give rise to electricity. He first tried to learn if current in one wire could "induce" current in another, just as a charge brought near a conductor induces a separation of charges on the surface of the conductor. In his own words, "These considerations, with their consequence, the hope of obtaining electricity from ordinary magnetism, have stimulated me at various times to investigate experimentally the inductive effect of electric currents. I lately arrived at positive results...which may probably have great influence in some of the most important effects of electric currents."

That was perhaps the understatement of the century. Faraday's momentous discovery, the greatest of his immensely productive life, would not only alter man's view of the universe, but would also alter civilization through new technology. From that time onward, electricity and magnetism were mutually embraced in what would become known as electromagnetism. Soon afterward, colossal steam-powered turbines made of gigantic loops of wire spinning in powerful magnetic fields would generate electricity to light up the world and produce a technological revolution. This chapter focuses on the remarkable phenomenon of electromagnetic induction.

41.2 OBSERVATIONS OF ELECTROMAGNETIC INDUCTION

In 1821 Michael Faraday was asked by the editor of the *Philosophical Magazine* to review a host of reports on electromagnetic experiments that had been submitted to the journal. Faraday's task was to separate real reproducible effects from fantasy, and to determine which theories were consistent with observed facts and which were idle speculation. This undertaking marked a turning point in his career as he set aside his research in chemistry and began investigating electromagnetism.

The results of early experiments on electromagnetic induction in the 1820s appeared somewhat confusing. But once the basic principles were revealed and refined to their essence, electromagnetic induction became comparatively easy to

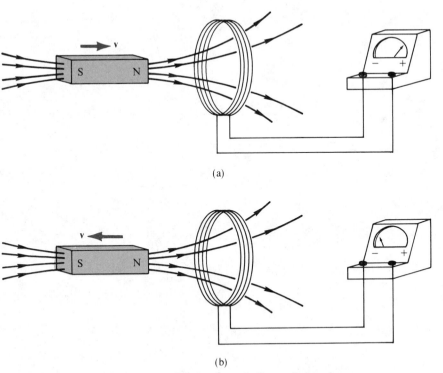

Figure 41.2 The deflection of a galvanometer indicates induced current in one direction when (a) the magnet is brought toward the loop, and in the opposite direction when (b) the magnet is pulled away.

demonstrate and to explain. We begin our analysis by presenting several observations – the same ones Faraday compiled in the historic year 1831.

Faraday showed that currents are induced in a circuit when any of the following conditions prevails:

1. A magnet in motion is brought near the circuit.
2. The circuit is moved in the presence of a magnet or another current-carrying circuit.
3. The current flow in a neighboring circuit is established or interrupted.

Let's examine some simple experiments in which these conditions are satisfied. Suppose a loop of wire is attached to a sensitive galvanometer, as illustrated in Fig. 41.2. The galvanometer detects flow of current in the loop. (Faraday's magnetic needle surrounded by a coil of wire was simply a crude version of a galvanometer.) If the north pole of a magnet is moved toward the loop, the galvanometer deflects in the positive direction, indicating that a current flows in the loop, as shown in Fig. 41.2. But if the magnet is held motionless near the loop, the galvanometer gives no

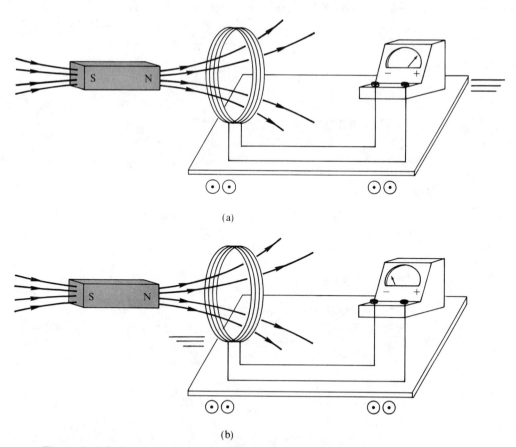

Figure 41.3 Induced current can be created by the motion of a loop of wire (a) toward or (b) away from a fixed magnet.

41.2 OBSERVATIONS OF ELECTROMAGNETIC INDUCTION

deflection, indicating that there is no current. And if the magnet is withdrawn, the galvanometer again deflects, but this time in the opposite direction, as illustrated in Fig. 41.2b, indicating that the induced current flows in the opposite direction. So when a magnet is brought near the loop, an induced current flows in one direction, and when it is withdrawn, the current flows around the loop in the opposite direction.

These observations clearly indicate, as Faraday has observed, that relative motion is the key. With a fixed magnet and a fixed loop there is no induced current. But if the magnet is fixed and the loop is moved toward the magnet, the galvanometer deflects as long as the loop is in motion. For example, if the loop is brought near the north pole of a magnet, as illustrated in Fig. 41.3, the current flows in one direction; if both magnet and loop are at rest, no current flows; and when the loop is moved away from the magnet, the induced current flows in the opposite direction.

The magnetic field near the loop need not be that produced by a permanent magnet. As we learned in Chapter 39, currents are the sources of magnetic fields, so a coil of wire that has a changing current through its windings produces a changing magnetic field. Figure 41.4 shows a switch, a battery, and a coil of wire from a circuit that is near a loop of wire attached to a galvanometer. When the switch is closed, the galvanometer momentarily deflects, indicating a current flow in the loop. With the switch kept closed, no current is induced in the loop because the magnetic field created by the coil is constant. However, when the switch is opened, the magnetic field drops to zero, and that change induces a current in the loop in the opposite direction. This observation reveals that a *changing* magnetic field induces current flow in a stationary loop.

To analyze the forces involved in electromagnetic induction, consider first a loop that is brought near a permanent magnet whose field is coming up out of the page, as in Fig. 41.5. Because the loop is in motion, the electrons in the metal feel a force due to their motion, which is given by

$$\mathbf{F} = q\mathbf{v} \times \mathbf{B}. \tag{38.11}$$

The electrons that are free to move do so in the direction of the force. Because q is negative for an electron, this direction is as indicated in Fig. 41.5. In essence, that's why moving the wire through the magnetic field causes an induced current to flow. When the loop isn't moving, $\mathbf{v} = \mathbf{0}$ and there is no force on the charges.

But there's another way to interpret the phenomenon. Suppose the loop is held fixed and the magnet is moved. Again, a current flows, because it's obviously the same phenomenon, no matter whether the wire or the magnet is in motion. And yet, our simple description of what's happening now fails completely. The force $\mathbf{F} = q\mathbf{v} \times \mathbf{B}$ is now zero, because $\mathbf{v}$, the velocity of the wire relative to some fixed frame of reference, is zero. Therefore, that force cannot be the cause of the current flow.

What *does* cause the current to flow? Initially there are stationary charges inside the wire. When a magnet is moved toward the wire, we observe a current, which means some force is acting on those charges. But whenever we observe a force on a *stationary* electric charge, we attribute it to an *electric field*. That's exactly how we detect electric fields: by placing a stationary charge somewhere in space, and observing whether there's a force on it.

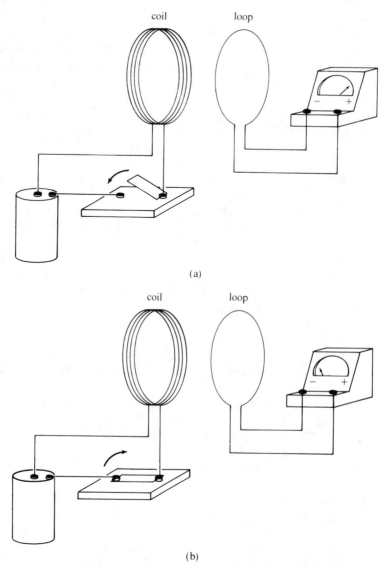

Figure 41.4 (a) When the switch is closed a changing magnetic field is created by the coil on the left and induces a current in the loop on the right. (b) When the switch is opened, a current is induced in the opposite direction.

So, if the magnet is held fixed and the loop is moved, the current is due to the force $\mathbf{F} = q\mathbf{v} \times \mathbf{B}$. However, if the loop is held fixed and the magnet is moved, we discover something new: the moving magnet creates an electric field. This revelation is the real significance of Faraday's discovery.

But there's still a problem to confront. Figure 41.6 shows a closed loop moving through the magnetic field of a permanent magnet, with a current flowing around

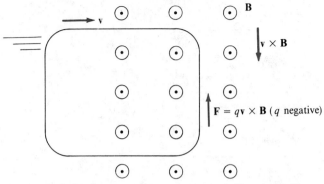

Figure 41.5 The magnetic force on a wire moving through a magnetic field causes electrons inside the wire to flow.

the loop as indicated. Suppose that the magnetic field is uniform and that the loop moves as indicated in Fig. 41.6. In that case the force on the electrons on side 1 is in the positive y direction, and the force on the electrons on side 3 is also in the positive y direction. Because the magnetic field is uniform, these two forces have equal magnitudes but they push current in opposite directions in the wire. This implies that there should be no net current flow at all, which seems to contradict the experimental evidence that current does flow. This apparent contradiction is easily explained. The only way a current will flow is if the magnitude of the magnetic field is not the same on both sides of this loop. Therefore, instead of having a uniform field everywhere, we need a magnetic field whose intensity is larger on one side of the loop than on the other. When that occurs, the force pushing the current one way will be greater than the opposing force trying to push the current the other way, and a net current will flow around the loop.

When the loop is moved near a permanent magnet there is an induced current because the magnetic field due to the permanent magnet is *not* uniform. As the loop

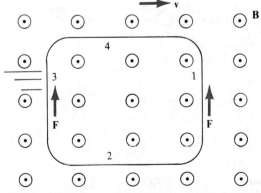

Figure 41.6 Forces acting on charges in a closed loop moving through a uniform magnetic field result in no net current circulation.

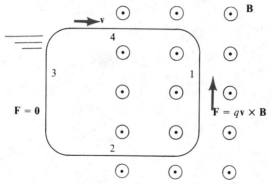

Figure 41.7 Forces of different magnitude acting on charges inside a loop moving through a nonuniform field cause an induced current.

moves nearer to the magnet, it enters a region of greater field intensity. Then the forces on the electrons on opposite sides of the loop are not equal and so an induced current flows (Fig. 41.7).

Faraday's lines can help visualize what's happening. To indicate the magnetic field we draw lines of magnetic flux. Some of these lines thread through the loop. As the loop moves toward the magnet, some lines are left behind, but more new ones enter the loop, and the flux passing through the loop increases with time. That's how Faraday visualized the process: The magnetic flux through the loop changes as the loop moves, and the changing magnetic flux through the loop induces a current in the loop.

We can describe this observation mathematically. The magnetic flux through the loop at any instant is given by the surface integral

$$\phi = \iint_S \mathbf{B} \cdot d\mathbf{A}, \tag{38.8}$$

where the surface S is that enclosed by the loop of wire. Faraday said that the change in flux produces the induced current. To get an idea of how much current is caused to flow, we consider the work done on those charges as the force due to the changing magnetic field pushes them around the wire. Recall that work is defined by a line integral,

$$W = \oint_C \mathbf{F} \cdot d\mathbf{r},$$

where in this case the path of integration C is the closed loop along which the charges are pushed. The important point to note is that this work is not zero:

$$W = \oint_C \mathbf{F} \cdot d\mathbf{r} \neq 0. \tag{41.1}$$

To find the actual value of W, we refer to Fig. 41.8 and note that on side 1 the work done is simply $F_1 h$, because the field (and consequently the force) is constant along that portion of the wire. Along sides 2 and 4, the work done is zero because the

41.2 OBSERVATIONS OF ELECTROMAGNETIC INDUCTION

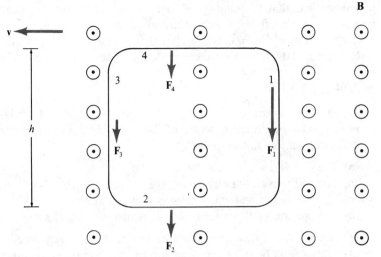

Figure 41.8 Work done by the magnetic force on charges in a closed wire loop moving through a nonuniform magnetic field is nonzero.

force **F** is perpendicular to the path along those sides. Finally, for side 3, the force **F** and the path direction are opposite, so the work done along that side is equal to $-F_3 h$. Therefore, the total work done around the closed path is

$$W = \oint_C \mathbf{F} \cdot d\mathbf{r} = (F_1 - F_3)h, \tag{41.2}$$

which is nonzero because F_1 is greather than F_3. The result is that current flows.

The work done by the force bears a simple relation to the work done by the electric field created by moving the magnet. By definition, $\mathbf{E} = \mathbf{F}/q$, so if we divide both members of Eq. (41.1) by the constant q, we find

$$\oint_C \mathbf{E} \cdot d\mathbf{r} = W/q \neq 0.$$

The work done by the electric field created by the moving magnet is also nonzero.

Earlier in this book we have emphasized repeatedly that the line integral of the electric field around any closed loop is always equal to zero. Now we find that in the case of moving magnets, it is *not* equal to zero. This apparent contradiction is easily explained. When we showed that the line integral around any closed path is zero we were dealing with a *static* electric field due to some distribution of stationary point charges. But now we're dealing with an electric field that's not created by point charges but by a *changing magnetic field*. The line integral around a closed curve of such an electric field is no longer equal to zero.

In the hydrodynamic analogy discussed in Chapter 40, a field whose line integral is not equal to zero is a circulating field of the type created by stirring water. That's exactly what we have here. A moving magnet, or in other words, a magnetic field that's changing in time at a fixed point in space, can stir up a circulating electric field. No distribution of electric charges, no matter how complicated, can ever do that. But a changing magnetic field can create a circulating electric field, just as an electric current can create a circulating magnetic field.

Questions

1. A coil of wire is connected to a sensitive galvanometer and placed near a long bar magnet. Determine under which of the following circumstances there will be a deflection on the galvanometer.

   (a) The coil and magnet are both at rest.
   (b) The coil moves while the magnet is at rest.
   (c) The magnet moves while the coil remains at rest.
   (d) Any motion of the coil or magnet relative to the other.

2. Is the direction of flow of induced current in a loop determined by the direction of the magnetic field, or the direction of the change of the magnetic field, or both? Cite a specific example to support your answer.

3. A magnet is allowed to swing freely as shown. As the magnet approaches the current loop shown in the diagram the direction of the induced current in the loop is as indicated.

   (a) What is the direction of the magnetic field created by the loop? That is, which side of the loop is the north pole and which is the south pole?
   (b) What action does the magnetic field of the loop have on the swinging magnet?

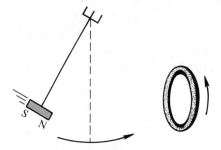

4. A coil of wire is connected to a switch and to a battery as shown in the diagram. Make a sketch of the graph of current versus time for the induced current in the loop if the switch is closed at time $t = 0$, then opened at a later time T.

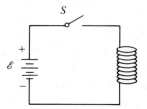

41.3 FARADAY'S LAW

The foregoing section showed that a changing magnetic field creates an electric field. In this section we describe a general principle, first enunciated by Faraday, that expresses a connection between induced electric fields and changing magnetic flux.

For a *static* field, whose line integral around any closed curve is zero, we can define a scalar field V, the potential at each point in space. In Chapter 34 we learned that the difference in potential between any two points in space, $V_B - V_A$, is equal to the negative of the line integral of the electric field from A to B:

$$V_B - V_A = -\int_A^B \mathbf{E} \cdot d\mathbf{r}. \tag{34.3}$$

The difference $V_B - V_A$ depends only on the endpoints A and B and not on the curve joining them because the static field $\mathbf{E}$ is conservative, so the integral is independent of the path and is therefore zero around a closed path. But for an electric field created by a changing magnetic field, the line integral cannot be independent of the path because, as we have just seen, the integral around a closed path can be nonzero. Nevertheless, we can still calculate the integral of $\mathbf{E}$ from A to B and obtain a scalar that depends not only on A and B, but also on the path joining them. This integral, which has the units of potential (work done per unit charge), has a special name – it's called the *induced electromotive force*, or induced emf, and is denoted by $\mathscr{E}_i$. In Chapter 36 we encountered one source of emf – the battery. We now have a second source, the induced emf due to changing magnetic fields.

Example 1
A straight wire of length $h = 0.20$ m moves perpendicular to a uniform magnetic field $\mathbf{B}$ of strength 0.5 T with a constant speed $v = 15$ m/s.
   (a) Describe the motion of electrons in the wire.
   (b) Find the electric field inside the wire.
   (c) Calculate the potential difference between the ends of the wire.

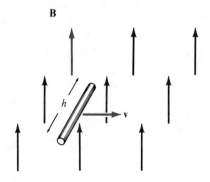

   (a) Each electron in the wire is subject to a magnetic force given by the cross product $\mathbf{F} = q\mathbf{v} \times \mathbf{B}$. By the right-hand rule this force pushes electrons toward one end of the wire, indicated by A in the following diagram. Of course, the free electrons do not continually accelerate under this force. Instead they accumulate rapidly at A and create an electrostatic field that prevents more electrons from accumulating at that end of the wire, as shown in the diagram.

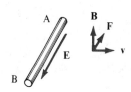

(b) Electrons have negative charge and, because $\mathbf{E} = \mathbf{F}/q$, the equivalent electric field in the wire created by the motion of the wire in the magnetic field is directed from A toward B, as previously indicated. The magnitude of the field is found to be

$$E = F/q = vB = (15 \text{ m/s})(0.5 \text{ T}) = 7.5 \text{ V/m}.$$

(c) By definition, the potential difference is related to the electric field by the integral

$$V_B - V_A = -\int_A^B \mathbf{E} \cdot d\mathbf{r} = -E\int_0^h dr = -Eh = -Bvh. \tag{41.3}$$

Using $h = 0.20$ m and the result of part (a) we find

$$V_B - V_A = -1.5 \text{ V}.$$

If the moving wire were made part of an electric circuit, it would act like a battery with an emf of 1.5 V.

The induced emf tends to drive the current all the way around the loop. It was Faraday's suggestion that the induced emf should be related to the rate of change of magnetic flux through the loop. We can express that relation in mathematical form by considering a square loop of side h that moves with velocity $\mathbf{v}$, as shown in Fig. 41.9. The loop moves in a direction perpendicular to a nonuniform magnetic field that has intensities B_1 and B_2 on opposite sides of the loop at the instant shown in the figure.

Suppose that the loop starts moving at time $t = 0$. After a small time Δt the loop has moved a distance $v\Delta t$. As shown in Fig. 41.9, the leading edge of the loop sweeps out a rectangular region of area $hv\Delta t$ and the loop gains an increment of flux $B_2 hv\Delta t$ through this region. At the same time it loses an increment of flux $B_1 hv\Delta t$ from its trailing edge, so the net change in flux through the loop is

$$\Delta \phi = (B_2 - B_1)hv\Delta t.$$

Dividing this change in flux by Δt and taking the limit as Δt shrinks to zero, we get

41.3 FARADAY'S LAW

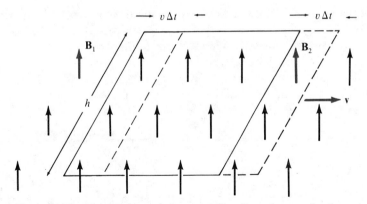

Figure 41.9 The time rate of change of flux through a moving loop.

the time rate of change of flux:

$$\frac{d\phi}{dt} = \lim_{\Delta t \to 0} \frac{\Delta \phi}{\Delta t} = (B_2 - B_1)hv. \tag{41.4}$$

This result can be compared to the induced emf calculated from the work done by the magnetic force moving charges through the loop. This work, the line integral of the force around the loop, was calculated above in Eq. (41.2), which now takes the form

$$\oint \mathbf{F} \cdot d\mathbf{r} = (F_1 - F_2)h.$$

Using $F = qvB$ and dividing by q to get the induced emf $\mathscr{E}_i$ we find

$$\mathscr{E}_i = (B_1 - B_2)hv.$$

Comparing this with (41.4) we obtain *Faraday's law*:

$$\boxed{\mathscr{E}_i = -\frac{d\phi_B}{dt}.} \tag{41.5}$$

This can also be written in terms of the equivalent electric field $\mathbf{E} = \mathbf{F}/q$ and the surface integral defining ϕ_B,

$$\boxed{\oint \mathbf{E} \cdot d\mathbf{r} = \mathscr{E}_i = -\frac{d}{dt} \iint_S \mathbf{B} \cdot d\mathbf{A}.} \tag{41.6}$$

In this last form we are reminded that an induced emf can be created by changing the magnetic field **B**, by changing S (the shape of the loop), by changing the orientation of the loop (as reflected in the dot product), or by a combination of these. (As we remarked in Chapter 33, you will not be asked to do any calculations involving surface integrals. We use them primarily because the notation enables us to express physical laws in simple form.)

Faraday's law specifies the connection between induced emf and the rate at which the magnetic flux linked through a circuit is changing with time. It stands as one of the fundamental laws of electromagnetism.

Example 2
Two metal rods of length h are placed across parallel conducting rails. One rod is kept fixed and the other moves along the rails with a constant velocity **v** perpendicular to a magnetic field **B** as illustrated.

(a) By calculating the electric field, show that the induced emf in the circuit is given by $\mathscr{E}_i = -Bhv$.

(b) Calculate the induced emf by using Faraday's law.

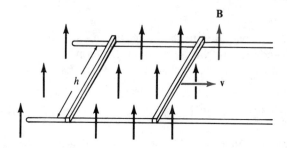

(a) The moving rod is responsible for the induced emf, so by Eq. (41.3) we see immediately that the induced emf is

$$\mathscr{E}_i = -Bhv.$$

(b) As the rod moves, the flux through the circuit increases because the area of the region enclosed by the circuit increases. Let x denote the distance the rod has moved at time t, where we take $x = 0$ at time $t = 0$. Because the wire moves with constant speed v, we have $x = vt$. The area of the region enclosed by the loop at time t is therefore $A = hx = hvt$, and hence the flux through the loop at time t is simply $\phi = BA = Bhvt$. Using Faraday's law in the form of Eq. (41.5), we get

$$\mathscr{E}_i = -\frac{d\phi}{dt} = -Bhv.$$

The two results agree, as expected.

Example 3
A metal rod of length b is pivoted at one end C and rotates with an angular speed ω in a uniform magnetic field **B** directed as shown. The free end of the rod slides

along a circular track and a resistance R is connected between the center and the circular track. Find the current flowing through the resistance.

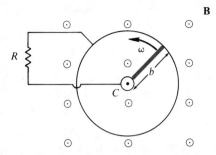

The charges in the moving rod are acted on by a force $\mathbf{F} = q\mathbf{v} \times \mathbf{B}$, which can be ascribed to an equivalent electric field $\mathbf{E} = \mathbf{F}/q = \mathbf{v} \times \mathbf{B}$. The induced emf is given by

$$\mathscr{E}_i = \int_{\text{rod}} \mathbf{E} \cdot d\mathbf{r} = \int_{\text{rod}} \mathbf{v} \times \mathbf{B} \cdot d\mathbf{r},$$

where $\mathbf{r} = r\hat{\mathbf{r}}$, r being the distance from C. Hence

$$\mathscr{E}_i = \int_0^b \mathbf{v} \times \mathbf{B} \cdot \frac{d\mathbf{r}}{dr} dr = \int_0^b \mathbf{v} \times \mathbf{B} \cdot \hat{\mathbf{r}}\, dr.$$

But at any point on the rod we have $\mathbf{B} = B\hat{\mathbf{k}}$ and $\mathbf{v} = \omega r \hat{\boldsymbol{\theta}}$, where $\hat{\boldsymbol{\theta}}$ is a unit vector perpendicular to the rod, so the cross product $\mathbf{v} \times \mathbf{B}$ is equal to

$$\mathbf{v} \times \mathbf{B} = (\omega r \hat{\boldsymbol{\theta}}) \times (B\hat{\mathbf{k}}) = \omega r B \hat{\mathbf{r}}$$

because $\hat{\boldsymbol{\theta}} \times \hat{\mathbf{k}} = \hat{\mathbf{r}}$. Therefore, the integral reduces to

$$\mathscr{E}_i = \omega B \int_0^b r\, dr = \omega B b^2 / 2.$$

Using Ohm's law, we find that the current flowing is $I = \mathscr{E}_i/R = \omega B b^2/(2R)$.

If the rod is replaced by a solid rotating metal disk of radius b, and the resistance is connected between the center and a point on the rim, the induced current is the same. The reason for this is that the disk can be thought of as a collection of radial rods connected in parallel, so the induced emf between the center and the rim is equal to that in any one of the radial segments.

This result can also be obtained from Faraday's law,

$$\mathscr{E}_i = -\frac{d\phi_B}{dt}. \tag{41.5}$$

The flux through a portion of the disk of area A is BA. If this portion is a circular sector subtending an angle θ we have $A = b^2\theta/2$ so

$$\phi_B = \frac{Bb^2\theta}{2}.$$

As the disk rotates, θ changes and $|d\theta/dt| = \omega$, hence the magnitude of $d\phi_B/dt$ is $\omega Bb^2/2$, in agreement with the foregoing result.

Questions

5. An aircraft has a wing span of 20 m and is flying at a speed of 540 km/h perpendicular to the magnetic field of the earth. If the magnetic field strength of the earth is 5.4×10^{-5} T, find the difference in potential between the wing tips of the aircraft.

6. A wire of length 0.20 m moves at 7 m/s through a uniform magnetic field that is perpendicular to it, as shown. If the magnetic field strength is 0.7 T, determine the potential difference between the ends of the wire.

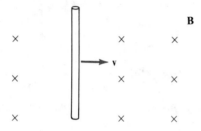

7. A long wire carries a constant current I. A wire of length b moves with velocity v parallel to the wire, as shown. Calculate the emf induced in the moving wire.

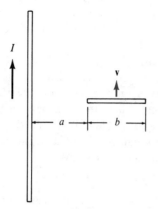

8. A square conducting loop of 20 turns and sides $b = 0.03$ m has a resistance of 100 Ω. It lies in a region where a magnetic field is perpendicular to the loop and changes according to $B(t) = 0.01(5t + 3)$, where t is in seconds and B is teslas.

41.4 LENZ'S LAW

(a) Determine the emf induced in the loop.
(b) Find the current flowing in the loop.

41.4 LENZ'S LAW

Now we can unravel another mystery concerning electromagnetic induction. When a loop of wire moves toward a magnet, an emf and a current are induced in the loop. We learned in Chapter 37 that most materials have resistance, so the flow of current in the loop causes heating. Heat is a form of energy and that energy must come from somewhere. In this case, it comes from the person or mechanism that moves the loop. As we move the loop toward the magnet, we can feel a force resisting the motion, and we must do work to overcome that force.

That force on the loop is created by current flowing in the loop and is in the direction that opposes the change in flux through the loop. The current loop, in turn, creates a magnetic field whose direction is given by the right-hand rule, as illustrated in Fig. 41.10. The magnetic field due to the induced current will be in the opposite direction to the change in the magnetic field that's producing the current. In other words, we have the following principle:

> The induced current always flows to create a magnetic field
> that opposes the change in flux through the circuit.

This observation is called *Lenz's law*, after the Russian physicist Emil Lenz, who pointed out that the direction of the induced current is always such as to oppose the motion that produced it. (See Example 4.) That's reminiscent of mechanical stability. A stable system always reacts to oppose any change in the system. Lenz's law expresses the fact that electromagnetic induction tends to be self-stabilizing.

It's easy to imagine what would happen if Lenz's law were violated. If the induced current in Fig. 41.10 were to move in the direction opposite to that shown, the part of the loop nearer the magnet would act as a south pole and would attract the bar magnet toward the loop. That would increase the flux further, leading to a stronger force, and so on. A slight push would start the magnet accelerating toward the loop, constantly increasing its kinetic energy. The system would be mechanically unstable. Instead, the induced current moves in the direction shown in Fig. 41.10

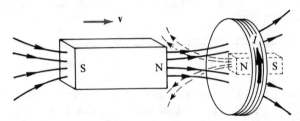

Figure 41.10 An induced current in a loop of wire produces flux in a direction opposite to that of the change in the magnetic field.

and the loop acts as a north pole producing magnetic flux to oppose that of the moving magnet, as indicated in the figure.

Example 4

A permanent magnet approaches a loop of wire as shown in the first diagram. In what direction will the induced current flow?

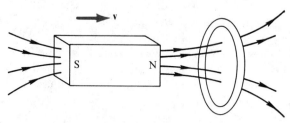

As the north pole of the magnet draws near the loop, the flux is increasing through the loop because the field is more concentrated nearer the poles. The induced current flows to oppose the increase in flux. Hence, it flows counterclockwise when viewed from the magnet. In that way the induced current sets up a magnetic field in the opposite direction of the field from the magnet (see the second diagram). It tries to cancel the increasing flux by keeping the magnetic field constant.

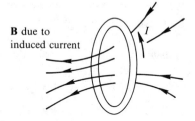

If the magnet is pulled away from the loop, the flux through the loop will be decreasing. Now the induced current will flow clockwise, trying to prevent the decrease in flux through the loop

---

As an application of Lenz's law, consider a pendulum swinging between the poles of an electromagnet consisting of two solenoids with iron cores, as shown in Fig. 41.11. With the electromagnet turned off, the pendulum is caused to swing back and forth. When the electromagnet is turned on, as the pendulum swings into the strong field, currents are induced and flow around in small loops in the copper sheet

41.4 LENZ'S LAW

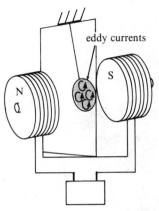

Figure 41.11 Eddy currents in a conducting sheet create a damping force.

at the end of the pendulum, as illustrated in Fig. 41.11. These induced currents, known as *eddy currents*, transform the kinetic energy of the pendulum into heat energy, which warms the copper sheet. The loss of energy causes the pendulum to slow down. This can also be explained in terms of forces. The magnetic force acting on the induced currents in the metal produces a thrust that causes the pendulum to slow down. This effect is known as electromagnetic damping. One application is to create damping in the oscillations of sensitive balances.

Example 5

A conducting sheet is pulled through a region where there is a nonuniform magnetic field, as illustrated. In which direction do eddy currents flow?

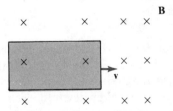

Because the sheet is being pulled into a region where the magnetic field is increasing, the flux through the sheet is also increasing. According to Lenz's law, the induced currents flow to maintain constant flux through the sheet. In this case, the eddy currents flow counterclockwise so as to create a magnetic field in the opposite direction to that of the magnetic field through which the sheet is moving.

Now that we understand the origin and behavior of induced currents, we turn to applications of Faraday's law. Faraday himself succeeded in obtaining an induced

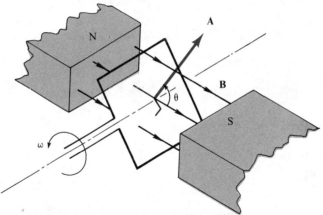

Figure 41.12 The flux through a rotating loop of area A in a uniform magnetic field **B** is $BA \cos \theta$.

current in a coil by rotating it in the earth's magnetic field. At first glance, it appears as if the flux through the loop is not changing, because the earth's magnetic field is (nearly) constant and the area of the loop is also constant. However, flux depends on the angle between the area vector and the magnetic field at each point. Consequently, an induced emf, which depends on the rate of change of flux, can be generated by a change in the orientation of a loop wire.

Suppose that a loop of wire of area A is immersed in a uniform magnetic field **B** and rotates at a constant angular speed ω as shown in Fig. 41.12. The orientation of the loop changes and is given by $\theta = \omega t$. At time t the flux through the loop is simply $\phi = BA \cos \theta = BA \cos \omega t$, where θ is the angle between the area vector of the loop and the vector **B**.

According to Faraday's law, the induced emf is given by

$$\mathcal{E}_i = -\frac{d\phi}{dt} = BA\omega \sin \omega t. \tag{41.7}$$

Thus, a loop of wire rotating uniformly in a magnetic field generates a sinusoidal current. This discovery by Faraday is the basis of the technological revolution brought about by electric generators. In an electric generator, wire loops rotate in a magnetic field, converting mechanical energy into electrical energy. The mechanical energy of the rotating loops, in turn, may have been generated by the chemical energy from the combustion of some fuel, or by the kinetic energy of falling water. In any case, Faraday's law of induction describes the generation of electrical energy.

Questions

9. A conducting sheet is pulled from a region where there is a uniform magnetic field **B**, as illustrated. In what direction will eddy currents flow?

41.4 LENZ'S LAW

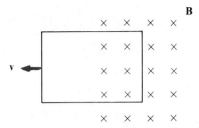

10. A bar magnet is dropped inside a metal pipe that has been evacuated. Despite the absence of air resistance, the magnet reaches a terminal velocity. Explain qualitatively why.

11. A uniform magnetic field of 3.0×10^{-2} T is perpendicular to the plane of a loop of wire of radius 0.04 m and resistance 5.0 Ω. The magnetic field is increasing in magnitude at a rate of 5.0×10^{-3} T/s.

   (a) Find the induced emf in the loop.
   (b) Calculate the rate of Joule heating produced in the loop.

12. A conducting rail of mass m moves without friction along horizontal conducting rails that are separated by a distance h, as shown. The rails are connected to a battery that provides an emf $\mathscr{E}$, and the only resistance of the circuit is R. A uniform magnetic field **B** perpendicular to the plane of the rails exists in the region. If at time $t = 0$ the rail has zero velocity, calculate each of the following:

   (a) The induced emf in the rail when it has speed v. (Ignore the **B** field created by the current $\mathscr{E}/R$.)
   (b) The net force acting on the rail.
   (c) The terminal speed of the rail.

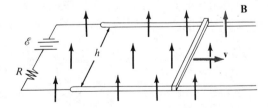

13. A 500-turn coil with resistance 20 Ω and radius 0.04 m is initially perpendicular to a uniform magnetic field of 0.8 T. Determine the amount of charge that flows past a point in the coil if the coil is flipped over.

14. A rectangular loop of wire of sides $a = 0.003$ m, $b = 0.04$ m, and 500 turns is immersed in a uniform magnetic field of 0.02 T. The loop is connected to slip rings, as shown. With what angular speed must the loop rotate so that the maximum potential difference between the two slip rings is 110 V?

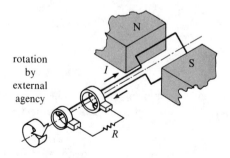

41.5 SELF-INDUCTANCE

In Chapter 39 we found that the magnetic field of a solenoid is nearly uniform inside the solenoid and zero outside it. If a current is suddenly turned on, a magnetic field proportional to the current is established inside the solenoid. The flux inside the solenoid changes from zero to some value that depends on the current and on the solenoid. Because the flux changes, an emf is induced that opposes the change in flux. Any coil of wire that causes a self-induced emf is known as an *inductor*, and can be characterized by a quantity known as *self-inductance*, which we shall define presently.

Consider a solenoid consisting of N turns and length h, as shown in Fig. 41.13. We know from Chapter 39 that if a current I flows through the solenoid the magnetic field inside is given by

$$B = \mu_0 N I / h. \tag{39.17}$$

Once a current starts flowing through the solenoid, the flux inside increases from zero to some final value. At any instant, the flux through the solenoid is due only to the field created by the solenoid itself (if there are no other coils or magnetic fields present) and is given by $\phi_B = NBA$, because B is uniform across the section of each

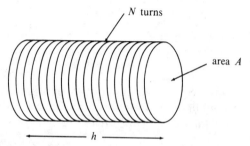

Figure 41.13 A solenoid of length h and N turns of wire used as an inductor.

41.5 SELF-INDUCTANCE

of the N turns of the solenoid. By Faraday's law this change in flux creates an induced emf *in the solenoid itself*. By Eq. (41.5) we have

$$\mathcal{E}_i = -\frac{d\phi_B}{dt} = -\frac{d(NBA)}{dt}.$$

Substituting for B from Eq. (39.17), we get

$$\mathcal{E}_i = -\frac{\mu_0 N^2 A}{h}\frac{dI}{dt}. \tag{41.8}$$

In other words, the induced emf is proportional to the rate of change of the current in the circuit itself. The factor multiplying $-dI/dt$ depends on the geometry of the solenoid and not on the current, so it is constant for a given solenoid. This constant is called the *self-inductance* and is denoted by L. Thus, Eq. (41.8) becomes

$$\boxed{\mathcal{E}_i = -L\frac{dI}{dt}.} \tag{41.9}$$

When this equation is written in the form

$$L = \frac{-\mathcal{E}_i}{dI/dt}, \tag{41.10}$$

it serves as a definition of self-inductance for a coil of any shape or size. This is somewhat analogous to the definition of capacitance given in Chapter 34:

$$C = \frac{Q}{V}. \tag{34.13}$$

The SI unit of inductance is the henry (H), named after an American scientist, Joseph Henry. He was a contemporary of Faraday and independently discovered many features of electromagnetic induction. From Eq. (41.10) we see that 1 H = 1 V s/A.

From the foregoing analysis we find that the self-inductance of a solenoid is given by

$$L = \mu_0 N^2 A/h. \tag{41.11}$$

More generally, the self-inductance is calculated from Eq. (41.10).

Example 6
A toroid is wound with N turns of wire. It has a rectangular cross section with inner radius a, outer radius b, and thickness h, as shown in the first diagram. Determine the self-inductance of the toroid.

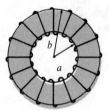

To calculate the self-inductance from Eq. (41.10), we first determine the magnetic field inside the toroid and then the flux. In Chapter 39 we found that the magnetic field inside a toroid of N turns at a distance r from the axis of the toroid is given by

$$B = \frac{\mu_0 NI}{2\pi r}.$$

The flux through a cross section of the toroid can be calculated by dividing the cross section into rectangular strips parallel to the axis of the toroid, as shown in the second diagram. The flux through a strip at a distance r from the axis of the toroid with width Δr and thickness h is given by

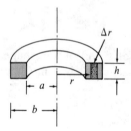

$$\Delta \phi = Bh \, \Delta r = \frac{\mu_0 NIh \, \Delta r}{2\pi r}.$$

The flux through one turn is equal to the integral

$$\frac{\mu_0 NIh}{2\pi} \int_a^b \frac{dr}{r} = \frac{\mu_0 NIh}{2\pi} \ln\left(\frac{b}{a}\right),$$

so the flux through N turns is N times as great,

$$\phi = \frac{\mu_0 N^2 h \ln(b/a)}{2\pi} I.$$

Therefore, according to Faraday's law, the induced emf in the toroid due to a

41.6 MUTUAL INDUCTANCE

changing current through N turns of the toroid is

$$\mathcal{E}_i = -\frac{d\phi}{dt} = -\frac{\mu_0 N^2 h \ln(b/a)}{2\pi}\frac{dI}{dt}.$$

Applying Eq. (41.10) we find that the self-inductance is

$$L = \frac{\mu_0 N^2 h}{2\pi}\ln\left(\frac{b}{a}\right),$$

the coefficient of $-dI/dt$.

Questions

15. A coil with a self-inductance of 1.5 mH carries a current I which is changing at the rate of 150 A/s. Find the induced emf in the coil when $I = 2.0$ A.

16. A solenoid of length 0.80 m and radius 0.02 m contains 2000 turns of wire. Determine the self-inductance of the solenoid.

17. For the solenoid in Question 16, determine the rate of change of current through it when the potential difference across the ends of the solenoid is 12 V.

18. A toroid has a square cross section, an inner radius of 8.0 cm, an outer radius of 10.0 cm and 200 turns. Calculate the self-inductance.

19. Suppose that two inductances, L_1 and L_2, are connected in series and are far enough apart so that the magnetic field from one does not affect the other. Knowing that the total emf induced is equal to the sum of the individual emfs induced by the inductors, show that the inductances may be replaced by an equivalent inductance given by

$$L_{eq} = L_1 + L_2.$$

20. Two inductances, L_1 and L_2, are connected in parallel and are far enough apart to that the magnetic field from one does not affect the other. Show that the inductances may be replaced by an equivalent inductance given by

$$\frac{1}{L_{eq}} = \frac{1}{L_1} + \frac{1}{L_2}.$$

41.6 MUTUAL INDUCTANCE

In the foregoing section we characterized the emf induced in an isolated coil of wire by the self-inductance. Now we characterize the effects of an emf induced in one circuit due to another circuit. Assume that circuit 1, containing a solenoid, is in the vicinity of circuit 2, which for simplicity we take to be a coil of wire, as shown in

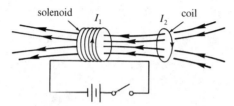

Figure 41.14 A changing magnetic flux in one circuit causes an induced emf in a second circuit.

Fig. 41.14. The magnetic field lines from the solenoid are closed curves so there is magnetic flux through the coil. If the flux changes in the first circuit, then the flux also changes in the second circuit and thereby induces an electromotive force in the second circuit. Let I_1 denote the current in the first circuit and let $\mathscr{E}_{21}$ be the electromotive force in the second circuit.

Let ϕ_{21} denote the flux in the second circuit due to the current in the first circuit. Then the emf produced in the second circuit due to the changing flux ϕ_{21} is given by

$$\mathscr{E}_{21} = -\frac{d}{dt}\phi_{21}.$$

But the flux ϕ_{21} is proportional to the current I_1 flowing in the first circuit; consequently, $\mathscr{E}_{21}$ is proportional to $-dI_1/dt$. The constant of proportionality depends on how the flux of the first circuit is intercepted by the second. The further the circuits are apart, the less flux passes from one through the other. The constant of proportionality is called the *mutual inductance* M_{21} and is defined by the equation

$$\boxed{\mathscr{E}_{21} = -M_{21}\frac{dI_1}{dt}.} \qquad (41.12)$$

The number $-M_{21}$ is the emf in circuit 2 caused by the changing current I_1 in circuit 1, divided by the rate of change of I_1, and is measured in the same units as self-inductance.

Similarly, if circuit 2 carries a changing current I_2, then the emf $\mathscr{E}_{12}$ induced in circuit 1 is proportional to the rate of change of I_2, so there is another coefficient of mutual inductance M_{12} defined by the equation

$$\mathscr{E}_{12} = -M_{12}\frac{dI_2}{dt}.$$

There is a remarkable "reciprocity" theorem that states that for any two circuits the two quantities M_{21} and M_{12} are equal. This is not immediately apparent because, in general, there is no geometrical symmetry between the two circuits. We omit the

41.6 MUTUAL INDUCTANCE

proof of this theorem because it makes use of properties of surface integrals that we have not developed in this text.

The reciprocity theorem allows us to drop the subscripts and speak of *the* mutual inductance M of any two circuits. Of course, the effects of self-inductance are present in each coil and the total emf induced in each circuit is the sum of the emf from self-inductance plus that from mutual inductance:

$$\mathcal{E}_2 = -L_2 \frac{dI_2}{dt} - M \frac{dI_1}{dt},$$

$$\mathcal{E}_1 = -L_1 \frac{dI_1}{dt} - M \frac{dI_2}{dt}.$$

Example 7

A coil consisting of N_2 turns of wire and cross-sectional area A_2 encircles a long solenoid that has N_1 turns, cross-sectional area A_1, and length h. Determine the following:

(a) The flux from the solenoid through the coil.
(b) The mutual inductance of the coil and solenoid.

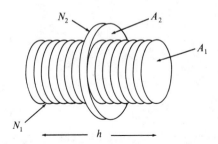

(a) We know that the magnetic field due to the current I_1 flowing in the solenoid is given by

$$B_1 = \mu_0 N_1 I_1 / h.$$

Because we assume the solenoid to be very long, the only flux from it that passes through the loop is due to the magnetic field inside the solenoid. Therefore, the flux through a single turn of wire in the coil is given by

$$\phi_{21} = B_1 A_1 = (\mu_0 N_1 I_1 / h) A_1.$$

(b) The emf induced in the coil is given by

$$\mathcal{E}_{21} = -N_2 \frac{d\phi_{21}}{dt} = -\frac{\mu_0 N_2 N_1 A_1}{h} \frac{dI_1}{dt}.$$

From this and Eq. (41.12) we find that the mutual inductance is given by

$$M = \mu_0 N_2 N_1 A_1/h.$$

The calculation of the mutual inductance is made simpler by treating the current I_1 in the solenoid rather than by treating the current I_2 in the coil. The reason is that the magnetic field due to a coil of wire does not produce a uniform field inside the coil as does a solenoid. Consequently, it is easier to calculate the flux created by the solenoid passing through the coil than vice versa.

Questions

21. A circuit containing a resistor, a battery, and a switch, surrounds a second circuit consisting of a wire and resistor, as shown. Indicate the direction of current flow in each circuit:

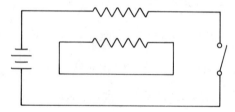

   (a) right after the switch is closed;
   (b) after the switch has been closed for a long time;
   (c) right after the switch is opened.

22. A very small coil with N_2 turns and cross-sectional area A_2 is placed inside a solenoid of cross-sectional area A_1 and length h with the plane of the small coil perpendicular to the axis of the solenoid. Calculate the mutual inductance of the coil and solenoid.

23. A coil consisting of 20 turns is wrapped around a solenoid of self-inductance 8.0 mH and 1000 turns. Determine the mutual inductance between the coil and the solenoid.

24. A rectangular coil with edges a and b is placed a distance c from a very long wire carrying current I, as shown. The wire is assumed long enough that the remaining part of its circuit does not produce any flux through the loop. Show

that the mutual inductance of the wire and loop is given by

$$M = \frac{\mu_0 b}{2\pi} \ln\left(\frac{c+a}{c}\right).$$

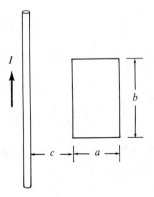

41.7 LR CIRCUITS

Consider a simple loop circuit containing a battery, a resistance (which can be the resistance of the wires in the circuit), and a switch. While the switch is open the current is equal to zero. Some time after the switch is closed, a constant current flows in the circuit, but the current does not reach this constant value instantaneously. As the current begins to flow through the circuit, the flux through the loop changes and consequently there is an induced emf in the circuit that opposes the change in current. To determine exactly how the current changes, the effect of inductance must be taken into account. To analyze the problem mathematically, we consider a simple LR circuit, one that involves both resistance R and inductance L.

In the simple LR circuit shown in Fig. 41.15, a battery provides electromotive force $\mathscr{E}_0$ to an inductor (such as a coil or solenoid) with self-inductance L. The wires in the circuit and the coil itself contribute both resistance and inductance to the circuit. In the diagram all the resistance R in the circuit is indicated by the resistor symbol ⏦, and all the inductance L by an inductor symbol ⏦. When the switch is closed, current flowing in the circuit creates changing magnetic

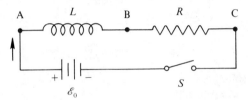

Figure 41.15 A simple inductance–resistance circuit.

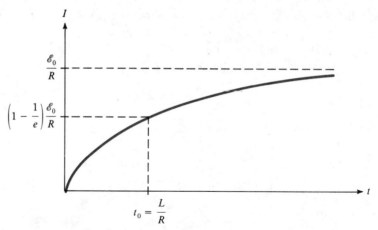

Figure 41.16 Graph of current versus time for a simple LR circuit.

flux in the inductor that induces an emf that, by Lenz's law, opposes the buildup of the magnetic field. With the current flowing in the direction shown in Fig. 41.15, the potential difference between points A and B is $L\,dI/dt$, and the potential difference between points B and C is IR. By Kirchhoff's voltage law the potential difference $\mathscr{E}_0$ between the battery terminals must be the sum of $L\,dI/dt$ and IR, so we have

$$L\frac{dI}{dt} + RI = \mathscr{E}_0. \tag{41.13}$$

This is a differential equation for the current as a function time. It is the same type of differential that we encountered earlier in Chapter 12 and again in Chapter 34 in connection with RC circuits. Its solution is given by

$$I(t) = \frac{\mathscr{E}_0}{R}\left(1 - e^{-(R/L)t}\right). \tag{41.14}$$

Figure 41.16 shows the graph of the current as a function of time. Initially the current starts out at zero, and after a very long time it approaches the steady-state value $I = \mathscr{E}_0/R$, as expected. As I approaches a constant value, dI/dt approaches 0 and the effect of the inductor becomes negligible.

The ratio L/R is called the characteristic time or time constant t_0. At time $t = t_0$ the current reaches the value

$$(1 - e^{-1})\mathscr{E}_0/R,$$

about 63% of its steady-state value.

Now suppose that switch has been closed long enough so that a constant current flows. What happens if the switch is opened? One might expect the current to stop flowing because the circuit is physically broken, which means the current would drop to zero instantaneously. But this does not happen because the inductance in the circuit prevents the current from dropping to zero instantaneously. Instead, the potential difference $L\,dI/dt$ created by the inductor is large enough to allow current to flow across the switch terminals in the form of a spark. You may have noticed such a spark occurring when you pull the plug out of a wall socket for some electrical appliance, such as an iron. In highly inductive circuits, the spark caused by opening the switch can be very impressive.

Example 8

For the LR circuit shown, with $L = 2 \times 10^{-3}$ H, determine (a) the current through each of the two resistors right after the switch is closed, (b) the derivative dI/dt of the current through the 3-Ω resistor right after the switch is closed, (c) the current through each of the resistors after a very long time has elapsed, and (d) the characteristic time for the circuit once the switch is opened after having been closed for a long time.

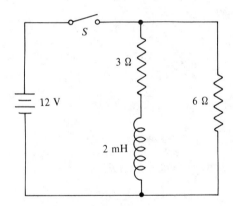

(a) Right after the switch is closed, the inductor acts like an open circuit and the current through the branch containing it is zero. Therefore, the current through the 3-Ω resistor is $I_3 = 0$. Applying Ohm's law to the 6-Ω resistor and knowing that the potential drop across it is 12 V, we have $I_6 = 2$ A.

(b) Because no current flows through the 3-Ω resistor immediately after the switch is closed, the potential drop across the inductor is equal to that across the battery. Therefore, we have $\mathscr{E}_0 = L\,dI_3/dt$, which implies that $dI_3/dt = \mathscr{E}_0/L = (12\text{ V})/(2 \times 10^{-3}\text{ H}) = 6 \times 10^3$ A/s.

(c) After a very long time the inductor acts like a "short" because the potential drop across it, $L\,dI/dt$, is zero. The two resistors then have the same potential drop, which is equal to that of the battery. Therefore,

$$I_3 = (12\text{ V})/(3\text{ }\Omega) = 4\text{ A} \quad \text{and} \quad I_6 = (12\text{ V})/(6\text{ }\Omega) = 2\text{ A}.$$

(d) When the switch is opened, the inductor will oppose a change in the flux through it and cause a current to flow only in the closed circuit, which contains the two resistors and the inductor, as shown.

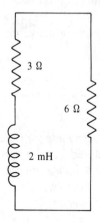

Now we have a simple LR circuit with total resistance $R = 3\,\Omega + 6\,\Omega = 9\,\Omega$. The characteristic time is given by

$$t_0 = L/R = (2 \times 10^{-3}\,\text{H})/(9\,\Omega) = 2.2 \times 10^{-4}\,\text{s}.$$

Why does the current drop to zero and not reach some constant, nonzero value as it did when the switch was closed?

Questions

25. Why do you suppose that house lights dim momentarily when a large appliance, such as a refrigerator or furnace, is turned on?

26. For the circuit shown, determine each of the following:

(a) The potential drop across the resistor right after the switch is closed.
(b) The potential drop across the inductor right after the switch is closed.
(c) The current in the circuit after a long time.
(d) The time constant for the circuit.

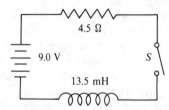

27. For the circuit shown, currents I_2, I_3, and I_6 flow through the 2-, 3-, and 6-Ω resistors, respectively.

41.8 ENERGY AND MAGNETIC FIELDS

(a) Determine the value of each current right after the switch is closed.
(b) Calculate the value of each current after the switch has been closed for a long time.
(c) After the switch is opened, determine the time constant for the circuit.

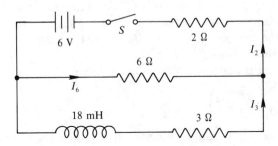

28. Show that the potential drop across the inductor in Figure 41.15 is given by $V_L = \mathcal{E}_0 e^{-tR/L}$ at time t after the switch is closed.

29. A simple circuit has a resistance of 2 Ω, an inductance of 10 mH, a switch, and a 9-V battery, as shown in the diagram. Determine the following:

    (a) The current in the circuit immediately after the switch is closed.
    (b) The value of dI/dt right after the switch is closed.
    (c) The current after the switch has been closed for 5 ms.
    (d) The potential drop across the inductor after the switch has been closed for 10 ms.

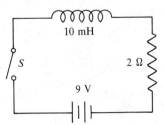

41.8 ENERGY AND MAGNETIC FIELDS

In our study of *RC* circuits in Chapter 37 we found that part of the energy supplied to a circuit by a battery can be stored in the electric field inside a capacitor and part can be transformed into Joule heat through resistance. Now we will find that, with inductance present, energy can be stored in magnetic fields as well.

Consider a circuit consisting of a resistor R, an inductor L, and a battery $\mathcal{E}_0$. Kirchhoff's voltage law gives us the differential equation

$$\mathcal{E}_0 = IR + L\frac{dI}{dt}. \qquad (41.15)$$

Multiplying through by I we obtain

$$\mathscr{E}_0 I = I^2 R + LI\frac{dI}{dt},$$

which we can interpret in terms of rate of change of energy. The term on the left, $\mathscr{E}_0 I$, represents the rate at which energy is supplied to the circuit by the emf $\mathscr{E}_0$. Part of this energy is continuously converted into Joule heating at a rate represented by the $I^2 R$ term, so the remaining term $LI\,dI/dt$ must represent the rate at which energy is stored in the inductor.

If we let U denote the energy stored in the inductor, then

$$\frac{dU}{dt} = LI\frac{dI}{dt} = \frac{d}{dt}\left(\frac{1}{2}LI^2\right).$$

In other words, the energy U and the quantity $LI^2/2$ have the same time derivative so they differ by a constant. But the constant is 0 because $U = 0$ when $I = 0$, and hence

$$\boxed{U = \tfrac{1}{2}LI^2.} \tag{41.16}$$

Equation (41.16) represents the energy stored in any inductor through which a current I is flowing.

We can think of the energy stored in the inductor as energy stored in the magnetic field created by the flowing current. This is analogous to considering the energy stored in a capacitor as energy stored in the electric field. To find the connection between the magnetic field **B** and the energy U in an inductor, we consider a solenoid of length h and cross-sectional area A, which carries a current I through N turns. From Chapter 39 we know that the magnetic field inside the solenoid is uniform and is given by

$$B = \frac{\mu_0 NI}{h}, \tag{39.17}$$

whereas the inductance is

$$L = \frac{\mu_0 N^2 A}{h}. \tag{41.11}$$

Solving Eq. (39.17) for I and substituting into Eq. (41.16) and using (41.11), we find that the energy stored in the magnetic field is given by

$$U = \frac{1}{2}LI^2 = \frac{\mu_0 N^2 A}{2h}\left(\frac{hB}{\mu_0 N}\right)^2 = \frac{B^2}{2\mu_0}hA.$$

The factor hA is the volume of the solenoid and B is the field intensity inside the solenoid, and therefore the coefficient of hA can be regarded as the amount of energy per unit volume stored at each point in the magnetic field. We call this the energy density u_B:

$$u_B = \frac{B^2}{2\mu_0}. \qquad (41.17)$$

Although this formula for energy density was derived for a solenoid, it is true for any magnetic field. It represents the magnetic analogy to energy density for electric fields introduced in Chapter 34:

$$u_E = \frac{\varepsilon_0 E^2}{2}. \qquad (34.24)$$

Questions

30. A circuit consists of a battery, a resistor, and an inductor as shown in the diagram. The switch is closed at time $t = 0$. Find each of the following at time $t = 0.2$ s:

(a) The rate at which the battery supplies energy to the circuit.
(b) The rate of Joule heating.
(c) The rate at which energy is being stored in the magnetic field.

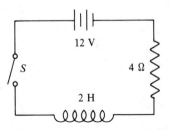

31. For the circuit in Question 30 determine the energy stored in the inductor after the switch has been closed for a long time.

32. At a certain time the potential difference across the ends of a solenoid is 15 V and the current through it is 0.2 A and is changing at the rate of 1500 A/s. Determine the energy stored in the solenoid at that instant.

33. Make a sketch of the graph of the energy stored in the inductor of Question 30 as a function of time.

41.9 A FINAL WORD

Michael Faraday, the son of a blacksmith, was born in 1791 in the outskirts of London. In eighteenth century England, children of the working class did not attend universities. Faraday himself had almost no education at all. He did learn to read and write, but at the age of 13 was apprenticed to a London bookbinder and bookseller, and it was more or less expected that this was to be his life's work. The young Faraday became eager for knowledge and was intent on self-improvement. He began to read some of the books he helped bind, and attended evening lectures, in particular those at the Royal Institution given by the famous chemist, Sir Humphrey Davy. In 1812, after 7 years of apprenticeship, Faraday was so en-

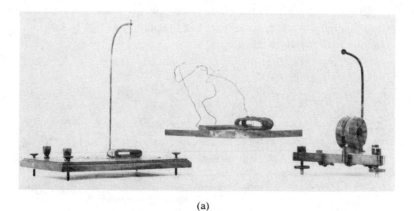

(a)

(b)

Figure 41.17 (a) Early galvanometers made and used by Faraday. (Courtesy of the Archives, California Institute of Technology.) (b) The great electromagnet. (From *Faraday's Diary*, Vol V.)

41.9 A FINAL WORD

thralled by Davy's lectures that he took "the bold and simple step" (as he called it) of writing to Davy, offering his services as an assistant, to work in any capacity whatsoever at the Royal Institution.

In 1813 Davy gave his permission and Faraday became an assistant at the Royal Institution. His services became more and more valuable and, around 1820, Faraday had become an accomplished chemist, although he preferred to call himself a natural philosopher. His mentor Humphrey Davy made many great discoveries in chemistry. For example, he discovered sodium and potassium, identified iodine and chlorine as elements, and invented the hydrogen theory of acidity. But it's often said that Davy's most important discovery by far was Michael Faraday.

The first half of the nineteenth century was the golden age of chemistry. A generation before, chemistry had emerged from the murky depths of alchemy, transformed by Joseph Priestly and Antoine Lavoisier into a science. The next generation, in which Faraday got his start, is that of Humphrey Davy and John Dalton. Then came Dmitri Mendeleev and the periodic table. Chemists at that time were doing exactly what high energy physicists are doing today – trying to discover the nature of the ultimate constituents of matter, and how they combined to form various substances. That was the central problem of chemistry. The problem was finally solved in 1920, at least at the chemical level – when quantum mechanics explained the formation of molecular bonds. That discovery established the foundations of the present science of chemistry.

After Davy's retirement in 1825, Faraday became Director and Professor at the Royal Institution. Photography was introduced in that era, and Faraday was one of the first scientists to be photographed. Figure 41.17a shows some galvanometers that Faraday himself built for his famous experiments, including those on electromagnetism. Figure 41.17b shows the so-called great electromagnet of the Royal

Figure 41.18 Woodcut of Faraday lecturing at the Royal Institution. (From *Faraday's Diary*, Vol. VI.)

Figure 41.19 Photograph of Michael Faraday. (Courtesy of the Archives, California Institute of Technology.)

Institution. In its time it was a national treasure, a scientific resource comparable to the 2-mile-long accelerator that exists today at Stanford, California. The British treasure is a little smaller – it can be placed on the seat of a chair – but the discoveries made with it were certainly no less important than those we make with our national scientific resources today.

Faraday not only became a celebrated scientist, he also became an accomplished lecturer. Figure 41.18 is a woodcut of Faraday delivering one of the evening lectures at the Royal Institution, the same series of lectures that had lured Faraday into science. This particular lecture was considered important enough to deserve a woodcut because the Prince of Wales was in attendance. Today it's worth remembering because the lecturer was Michael Faraday.

CHAPTER 42

ALTERNATING CURRENTS

My personal desire would be to prohibit entirely the use of alternating currents. They are unnecessary as they are dangerous... I can therefore see no justification for the introduction of a system which has no element of permanency and every element of danger to life and property.

I have always consistently opposed high-tension and alternating systems of electric lighting, not only on account of danger, but because of their general unreliability and unsuitability for any general system of distribution.

Thomas A. Edison (1889)

The advent of the alternating-current system of electric power-transmission marks an epoch in the economy of energy available to man from coal... It is in this field as much as in the transmission of energy to great distances that the alternating system, with its ideally simple machinery, is bringing about an industrial revolution...

Nikola Tesla (1900)

42.1 TWO GREAT INVENTORS

The industrial revolution in the latter part of the nineteenth century dramatically changed the face of the world. Steam-powered engines and ships transported people and goods and provided power for factories and printing presses that shaped the future of civilization. In America, the golden land of opportunity, this revolution helped transform a young nation into one of the world's major industrial powers.

Meanwhile, in Europe, Michael Faraday's discovery of electromagnetic induction laid the groundwork for the conversion of electrical energy into mechanical work, and vice versa. Another European, James Clerk Maxwell, cast the laws of electricity and magnetism into a compact mathematical form that we shall explore in Chapter 43. But the technology that put electrical energy into practical use was invented by a native-born American, Thomas Alva Edison (1847–1931), and an immigrant to America, Nikola Tesla (1856–1943), two giants among inventors.

Thomas Edison's mother removed him from school after only 3 months because his teacher thought his constant questions were a sign of stupidity. He studied at home, applying himself particularly to history, essays, and books on science. The young Edison became an entrepreneur at the age of 12 when he boarded trains to sell newspapers. At 15, Edison printed his own newspaper, *The Weekly Herald*, on a small printing press set up in a baggage car. This was the first newspaper ever printed on a moving train. Reading Newton's *Principia* gave Edison a distaste for mathematics, but his innate curiosity and the inspiration he received from Faraday's *Experimental Researches* motivated him to pursue his consuming interest in science and technology. By the age of 23 he had patented an improved stock-ticker system, and with the $40,000 he received from that invention Edison established a laboratory employing 50 people.

Edison's laboratory was a prototype of modern industrial research organizations. He invented by design. When he saw a human need, such as street lighting, he would invent something to fill this need. His approach to every task was careful and methodical, and he did nothing haphazardly. He held over 1000 important patents. Among his inventions were many devices that became commonplace in everyday life, including the incandescent electric-light bulb, the phonograph, and a motion picture camera.

Figure 42.1 Photograph of Thomas Alva Edison. (The Bettmann Archive.)

42.1 TWO GREAT INVENTORS

Figure 42.2 Photograph of Nikola Tesla. (Brown Brothers.)

Nikola Tesla, the other great inventor of the same era, was born in Serbia (now part of Yugoslavia) and was intended for the clergy, but he developed a taste for mathematics and science that led him astray. In 1884 he arrived in the United States with 4 cents in his pocket, a book of poetry, knowledge of a dozen languages, and a desire to work for Edison.

At a time when electric generation and transmission were still in their infancy, Tesla envisioned a second industrial revolution, in which steam power would be supplanted by electric power. Houses and streets would be lighted, factories run, and trains powered by electricity that traveled over miles of copper wire from gigantic electric generators housed in great central stations.

With the technology at hand, that vision seemed unreal. Edison had successfully developed direct current systems for the generation and distribution of electricity, but these systems were small in scale. They depended on local sources of power that could economically supply users less than 1/2 mile from the generating plant. Alternating current systems, which were just beginning to make headway, were handicapped by the lack of an efficient alternating current motor. Nonetheless, Nikola Tesla foresaw that newly evolving devices would overcome these difficulties.

George Westinghouse was one of the few men who had faith in Tesla's ideas. Several patents describing Tesla's inventions were granted in 1888 and bought by

Westinghouse. The first attempts to put them into practice were met with tremendous opposition. But by 1895, 100,000 horsepower of electrical energy was generated at Niagara Falls, an output equaling that of all other generating stations in the United States, and a significant part of it was transmitted and converted to mechanical energy by Tesla's inventions. A year later, part of this energy was transmitted 20 miles to Buffalo to light its streets and homes. The electric revolution was well on its way as old systems were scrapped and replaced by those of Tesla's design. We now turn to a discussion of the physics of alternating currents, the type of current that made Tesla's visions practical.

42.2 ALTERNATING CURRENTS IN SIMPLE CIRCUITS

In earlier chapters we studied the behavior of direct current (dc), the current produced, for example, by connecting a battery to a simple circuit. Tesla's enduring contribution to the age of electronics was the invention of practical, large-scale means to generate and use electricity with alternating current (ac). This chapter examines properties of circuits in which the impressed emfs and the resulting currents vary sinusoidally in time.

An oscillating current produced in a circuit by a voltage that oscillates sinusoidally is called alternating current. In Chapter 41 we learned that a coil of N turns and area A rotating with angular speed ω in a magnetic field **B** produces a sinusoidally induced emf given by

$$\mathscr{E} = \omega NAB \sin \omega t. \tag{41.7}$$

For a given generator, the product ωNAB is a constant, which we denote by $\mathscr{E}_0$ and which represents the amplitude of the voltage. The voltage varies between $+\mathscr{E}_0$ and $-\mathscr{E}_0$. We now investigate the current that results from a sinusoidally varying source of emf, say

$$\mathscr{E} = \mathscr{E}_0 \sin \omega t. \tag{42.1}$$

In circuit diagrams, such an alternating voltage source is represented by the symbol ⊙.

Figure 42.3 shows a simple circuit consisting of a resistor of resistance R and an alternating voltage source. According to Kirchhoff's voltage law, at any instant the algebraic sum of the increases and decreases in potential around the circuit is zero. In this case we have

$$\mathscr{E} - IR = 0,$$

which implies that the current is given by $I = \mathscr{E}/R$, or

$$I = I_0 \sin \omega t, \tag{42.2}$$

where $I_0 = \mathscr{E}_0/R$ is the amplitude of the current. Thus we see that Ohm's law, $\mathscr{E} = IR$, applies to ac as well as dc.

As expected, the current alternates with the same frequency as that of the generator. Moreover, the current and voltage oscillate *in step*, or *in phase*, the peaks of current being reached at the same time as the peaks in voltage, as indicated in Fig. 42.3b.

42.2 ALTERNATING CURRENTS IN SIMPLE CIRCUITS

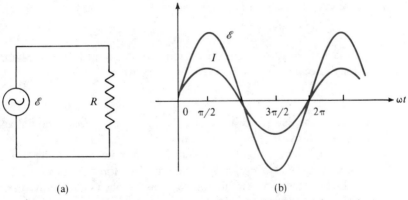

(a)   (b)

Figure 42.3 (a) A circuit consisting of a sinusoidal source and a resistance. (b) Graphs of voltage and current as functions of time.

Now let's modify the circuit by replacing the resistor by a capacitor of capacitance C, as shown in Fig. 42.4a, taking positive current to be in the direction indicated. If Q is the charge at time t on the capacitor plate indicated, the current is the rate of change of charge,

$$I = \frac{dQ}{dt}.$$

But the potential drop across the capacitor is Q/C, so Kirchhoff's voltage law now gives us

$$\mathcal{E} - Q/C = 0,$$

and hence the charge on the capacitor at any instant is given by

$$Q = C\mathcal{E} = C\mathcal{E}_0 \sin \omega t.$$

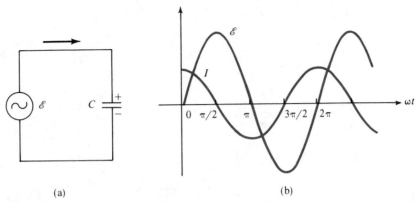

(a)   (b)

Figure 42.4 (a) A circuit consisting of a sinusoidal source and a capacitor. (b) Graphs of the voltage and current in the circuit as functions of time.

Therefore, the current is

$$I = \omega C \mathcal{E}_0 \cos \omega t. \tag{42.3}$$

This last result shows that the current in the circuit is not in phase with the applied voltage. Because $\cos \omega t = \sin(\omega t + \pi/2)$, we can write

$$I = \omega C \mathcal{E}_0 \sin(\omega t + \pi/2),$$

which can be described by saying that the current *leads* the voltage by $\pi/2$ radians or $1/4$ of a cycle. This is illustrated in Fig. 42.4b, where the peaks of the current occur before corresponding peaks of the applied voltage by a quarter of a period. Conversely, we also say that the voltage *lags* the current by $\pi/2$ radians.

In the foregoing circuit the relation between maximum voltage $\mathcal{E}_0$ and maximum current I_0 is

$$I_0 = \omega C \mathcal{E}_0.$$

The ratio $\mathcal{E}_0/I_0$ is called the *capacitive reactance* and is denoted by X_C. Thus we have

$$X_C = \frac{1}{\omega C}$$

and $\mathcal{E}_0 = I_0 X_C$, which is analogous to Ohm's law, $\mathcal{E}_0 = I_0 R$, for a simple circuit with resistance R. The capacitive reactance has the same physical units as resistance. Unlike resistance, the reactance depends on ω, the frequency of the applied voltage.

Next we replace the capacitor by an inductor of inductance L, as shown in Fig. 42.5a. The potential drop across the inductor is given by $-L\,dI/dt$, so Kirchhoff's voltage law now becomes

$$\mathcal{E} - L\frac{dI}{dt} = 0,$$

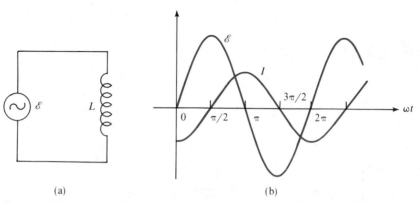

Figure 42.5 (a) A circuit consisting of a sinusoidal source and an inductor. (b) Graphs of the current and voltage in the circuit as functions of time.

42.2 ALTERNATING CURRENTS IN SIMPLE CIRCUITS

which implies

$$\frac{dI}{dt} = \frac{\mathcal{E}_0}{L} \sin \omega t.$$

Integration with respect to t gives us

$$I = -\frac{\mathcal{E}_0}{\omega L} \cos \omega t + \text{constant}.$$

The constant of integration, which depends on the initial conditions, is equal to 0 if the initial current at time $t = 0$ is equal to $-\mathcal{E}_0/(\omega L)$. This gives us

$$I = -\frac{\mathcal{E}_0}{\omega L} \cos \omega t = \frac{\mathcal{E}_0}{\omega L} \sin\left(\omega t - \frac{\pi}{2}\right). \tag{42.4}$$

Again the current and applied voltage are out of phase by 1/4 of a cycle, but in this case the current lags the voltage, as illustrated in Fig. 42.5b. The voltage reaches its maximum value $\mathcal{E}_0$ before the current reaches its maximum, $I_0 = \mathcal{E}_0/(\omega L)$.

For this circuit the ratio of the maximum voltage $\mathcal{E}_0$ to the maximum current I_0 is called the *inductive reactance* and is denoted by X_L. Thus

$$X_L = \omega L,$$

and $\mathcal{E}_0 = I_0 X_L$, which is analogous to Ohm's law. The inductive reactance, like the capacitive reactance, depends on the frequency of the applied emf.

Example 1

An ac generator that produces a maximum voltage of 20.0 V with a frequency of 60.0 Hz is connected first to a 10.0-Ω resistor, then to a 20.0-μF capacitor, and finally to a 5.0-mH inductor. Find the maximum current in each case.

For the resistor we simply have

$$I_0 = \mathcal{E}_0/R = (20.0 \text{ V})/(10.0\ \Omega) = 2.0 \text{ A}.$$

When the source is connected to the capacitor, the maximum current is given by

$$I_0 = \mathcal{E}_0 \omega C = (20.0 \text{ V})(2\pi \times 60.0 \text{ rad/s})(20.0 \times 10^{-6} \text{ F}) = 0.15 \text{ A}.$$

Finally, when the source is connected to the inductor, the maximum current is given by

$$I_0 = \mathcal{E}_0/(\omega L) = (20.0 \text{ V})/[(2\pi \times 60.0 \text{ rad/s})(5.0 \times 10^{-3} \text{ H})] = 11 \text{ A}.$$

Describe how the current would change in each case if the frequency were increased.

Questions

1. A 1.5-μF capacitor is attached to an ac generator of frequency ω. For what frequency ω will the maximum current in the circuit be the same as that in which only a 2.0-Ω resistor is attached to the source?

2. A 1.5-mH inductor is attached to an ac generator of frequency ω. For what frequency ω will the maximum current in the circuit be the same as that in which only a 2.0-Ω resistor is attached to the source?

3. Make a sketch showing how the reactance of a 5.0-μF capacitor varies as a function of the frequency of the applied emf for the range $\omega = 10$ to $\omega = 100$ rad/s. Do the same for the reactance of a 2.0-mH inductor.

4. At what frequency does a 4.0-μF capacitor have the same reactance as a 5.0-mH inductor?

5. A capacitor attached to an ac generator operating at 60 Hz is required to have the same maximum current in the circuit as when a 5.0-Ω resistor is attached to the same source. What is the value of C?

6. An inductor attached to an ac source operating at 1200 Hz draws the same maximum current as when a 50-Ω resistor is connected to the same source. What is the value of L?

42.3 LC CIRCUITS

The circuits in the foregoing section had a resistor or a capacitor or an inductor connected to an ac generator. Now we consider ac circuits with both inductance L and capacitance C, as illustrated in Fig. 42.6.

Kirchhoff's voltage law applied to this circuit now gives us

$$\mathscr{E} - \frac{Q}{C} - L\frac{dI}{dt} = 0.$$

Because $I = dQ/dt$ and $\mathscr{E} = \mathscr{E}_0 \sin \omega t$, this becomes a differential equation for the charge Q on the capacitor:

$$L\frac{d^2Q}{dt^2} + \frac{Q}{C} = \mathscr{E}_0 \sin \omega t. \tag{42.5}$$

We recognize this as the same type of differential equation encountered in Chapter 21 in which the motion of a harmonic oscillator driven by an oscillating

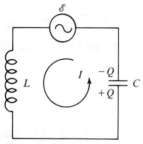

Figure 42.6 An ac generator connected to an inductor and a capacitor.

42.3 LC CIRCUITS

force is described:

$$m\frac{d^2x}{dt^2} + kx = F_0 \sin \omega t. \tag{21.1}$$

In the electric circuit the charge on the capacitor plate (and also the current in the circuit) is oscillating. Comparison of Eqs. (42.5) and (21.1) reveals that the inductance L plays the same role as mass m in a mechanical oscillator. The inductor is a stabilizing influence, which tends to oppose sudden changes in the current due to sudden changes in the voltage, and the inductance L is a measure of the opposition to changing current. Likewise, $1/C$ plays the same role as the spring constant k in a mechanical oscillator. Increasing the capacitance decreases the tendency of the capacitor to oppose changes in charge. Charge, in turn, plays the same role as the displacement of a mechanical oscillator, and current plays the role of velocity.

Because the differential equations for the electric circuit and for a mechanical oscillator have the same form, the phenomenon of resonance also occurs in circuits. The natural resonant frequency of the mechanical oscillator in Eq. (21.1) is $\omega_0 = \sqrt{k/m}$. By comparing equations we see that the corresponding natural resonant frequency of the circuit is given by

$$\omega_0 = \sqrt{\frac{1}{LC}}. \tag{42.6}$$

In mechanical resonance, shattering effects can occur when a small force is applied repeatedly to an oscillating system at the resonant frequency. Similarly, in an ac circuit, a small voltage oscillating near the frequency given by Eq. (42.6) can cause current to flow with large amplitude. As in the case of mechanical oscillators, the amplitude of the charge oscillations is given by a formula analogous to Eq. (21.3):

$$Q_{max} = \frac{\mathcal{E}_0/L}{|\omega^2 - \omega_0^2|}. \tag{42.7}$$

Example 2
Many radio and television stations broadcast simultaneously, each at its own frequency. How does a receiver select one station out of the large number broadcasting simultaneously?

The transmitted radio waves induce tiny voltages at many different frequencies simultaneously in the receiving antenna. To select one station, an *LC* circuit is used. When the inductance or capacitance is adjusted for resonance at just the right frequency, a large current flows at that frequency, and a small current flows at all other frequencies, so the desired station comes through. Turning the dial on a radio changes the area of a variable capacitor and consequently alters the capacitance and hence the frequency to which the circuit is tuned.

Example 3

What inductance attached to a 2.0-mH inductor will cause resonant oscillations at 102.2 MHz?

We know that resonance occurs when

$$\omega_0 = 1/\sqrt{LC}.$$

Solving for C we find

$$C = 1/(\omega_0^2 L) = 1/\left[(102.2 \times 10^6 \text{ rad/s})^2 (2.0 \times 10^{-3} \text{ H})\right] = 4.8 \times 10^{-14} \text{ F}.$$

To change the frequency of oscillations to a lower value, a variable capacitor might be used and the capacitance decreased, say by decreasing the area of the capacitor plates. That's how a radio tuner works.

When a capacitor begins to accumulate charge, it creates an electric field and hence a voltage that opposes the change. In other words, a capacitor opposes the flow of positive and negative charge just as a mechanical spring opposes compression or expansion. On the other hand, when the current in an inductor changes, the changing flux induces a current of its own opposing the change. So, an inductor opposes changes in current, just as the inertia of a mass on a spring opposes changes in velocity. This is another example of how different types of physical problems lead to exactly the same type of differential equation.

Questions

7. An LC circuit contains a 4.0-μF capacitor and a 4.0-mH inductor. Find the period of oscillations when resonance occurs.

8. Two capacitors with capacitances 10.0 and 15.0 pF are connected in an alternating current circuit with a 5.0-mH inductor. Find the resonance frequency if the capacitors are connected (a) in series and (b) in parallel.

9. The inductance of a coil is changed by inserting a magnet into it while the coil is part of an alternating LC circuit with an alternating voltage operating at 60 Hz. If the maximum current occurs when $L = 1.5$ mH, what is the capacitance in the circuit?

10. A variable capacitor and a 6.0-mH inductor are connected in series in a circuit whose source has an amplitude of 20.0 V and oscillates at 120 kHz. Determine the maximum voltage across the capacitor when the capacitance is (a) 7.5 × 10^{-10} F, (b) 1.1 × 10^{-8} F, and (c) 1.2 × 10^{-9} F.

42.4 LCR CIRCUITS

In an oscillating LC circuit, energy passes back and forth between the electric field of the capacitor and the magnetic field of the inductor, just as energy passes from

42.4 LCR CIRCUITS

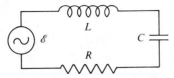

Figure 42.7 An *LCR* circuit with an ac generator.

potential to kinetic in a mechanical oscillator. In a mechanical system, friction or viscosity prevents the amplitude of the oscillations from becoming infinite. In oscillating electric circuits, resistance plays that role. We now add a resistor to the *LC* circuit as shown in Fig. 42.7 and obtain an *LCR* circuit with alternating emf, inductance *L*, capacitance *C*, and resistance *R*. Applying Kirchhoff's voltage law we get

$$\mathcal{E}_0 \sin \omega t - IR - L\frac{dI}{dt} - \frac{Q}{C} = 0.$$

Substituting $I = dQ/dt$ and rearranging terms we obtain the differential equation

$$\frac{d^2Q}{dt^2} + \frac{R}{L}\frac{dQ}{dt} + \frac{Q}{LC} = \frac{\mathcal{E}_0}{L} \sin \omega t. \tag{42.8}$$

We recognize this as the same differential equation encountered in Chapter 21 for a damped simple harmonic oscillator, Eq. (21.4). A detailed analysis of this equation (which is not difficult but somewhat lengthy, and so will be omitted) reveals that if $R^2C < 4L$, the solution Q consists of two parts, a purely sinusoidal term plus a damped sinusoidal term. The damped term has an exponential damping factor that decreases to zero very rapidly for large *t*, so the only significant part of the solution is the purely sinusoidal part. Except for a negligible transient current, the corresponding current that results is also sinusoidal and can be written as

$$I = I_0 \sin(\omega t - \alpha),$$

where the phase angle α satisfies

$$\tan \alpha = \frac{X_L - X_C}{R},$$

and the maximum current I_0 is given by

$$I_0 = \frac{\mathcal{E}_0}{Z},$$

where

$$Z = \sqrt{(X_L - X_C)^2 + R^2}. \tag{42.9}$$

The quantity Z is called the *impedance*. In these formulas, X_L and X_C are the

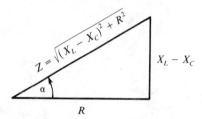

Figure 42.8 Geometrical relationship between the phase angle α and the impedance Z in an LRC circuit.

inductive and capacitive reactances introduced earlier:

$$X_L = \omega L, \qquad X_C = \frac{1}{\omega C}.$$

Figure 42.8 shows a geometrical relationship between the phase angle α and the impedance Z. Both reactances depend on the frequency ω of the applied emf, hence the impedance Z and the resulting maximum current I_0 also depend on ω. The impedance has its smallest value, $Z = R$, when $X_L = X_C$. This occurs when

$$\omega = \frac{1}{\sqrt{LC}} = \omega_0,$$

the natural frequency. At this frequency, also called the resonance frequency, the phase angle α is 0 and the current is in phase with the applied emf.

Example 4

An LCR circuit contains an inductor with $L = 4$ H, a capacitor with $C = 4$ μF, and a 10-Ω resistance. An ac generator with 200-V maximum emf and frequency $\omega = 200$ rad/s is impressed on the circuit. Calculate the resonance frequency ω_0, the maximum current I_0, and the phase angle α.

In this example $LC = (4\text{ H})(4 \times 10^{-6}\text{F})$ so $\omega_0 = 1/\sqrt{LC} = 250$ rad/s. Next we determine the capacitive and inductive reactances:

$$X_C = \frac{1}{\omega C} = \frac{1}{200(4 \times 10^{-6})} = 1250 \ \Omega,$$

$$X_L = \omega L = 200(4) = 800 \ \Omega.$$

Note that $(X_L - X_C)^2 = (450)^2$ is much larger than $R^2 = 100$, a situation that always prevails when the frequency is far from the resonance frequency. Therefore, in calculating the impedance Z we can ignore R^2 and find the approximate value

$$Z = |X_L - X_C| = 450 \ \Omega.$$

The maximum current is

$$I_0 = \mathscr{E}_0/Z = 200 \text{ V}/450 \ \Omega = 0.444 \text{ A}.$$

42.4 LCR CIRCUITS

At the resonance frequency the maximum current would have the much larger value 200 V/(10 Ω) = 20 A. Finally, the phase angle α satisfies $\tan \alpha = (X_L - X_C)/R = -45$, so the phase angle is $\alpha = \tan^{-1}(-45) = -89°$. The current leads the voltage by nearly 90°.

Questions

11. A sinusoidal emf given by $\mathscr{E} = 12 \cos(10^4 t)$ V is applied to an *LCR* circuit. The current has an amplitude of 2.0 A and is out of phase with the applied emf by 37°. Determine (a) the impedance of the circuit, (b) the resistance of the circuit, and (c) the inductance of the circuit if the inductive reactance is 9.0 Ω.

12. Refer to the circuit in Question 11. Determine (a) the capacitive reactance of the circuit and (b) the value of the capacitance in the circuit. (c) If the capacitance is changed so that the circuit is in resonance, what is the new value of the capacitance?

13. An alternating emf given by $\mathscr{E} = 300 \cos(12{,}000 t)$ V supplies current to a circuit containing a resistance R, inductance L, and capacitance C. The inductive reactance is greater than the capacitive reactance. If the current amplitude is 0.2 A and the maximum potential difference across the resistance is 240 V, find (a) the resistance and (b) the phase angle between the current and the applied emf.

14. For the circuit in Question 13, if the capacitance is 5.0 μF, determine the inductance of the circuit.

15. An *LCR* circuit has a resistance of 40.0 kΩ and inductance of 2.0 mH, and is connected to an emf source of amplitude 20 V.

    (a) At what frequency will the reactance of the inductor equal the resistance?
    (b) What is the capacitance in the circuit if resonance occurs at a frequency of 3.18×10^6 Hz?
    (c) What is the maximum current that flows in the circuit at resonance?

16. Show that the impedance Z satisfies the equation

    $$Z^2 = \frac{L^2}{\omega^2}(\omega^2 - \omega_0^2)^2 + R^2,$$

    where ω_0 is the resonant frequency.

17. A radio receiver contains an *LCR* circuit that is tuned by a variable capacitor. It is designed to resonate at frequencies from 500 to 1500 kHz. If $L = 1$ mH, what range of values of C is necessary to cover the range of resonant frequencies?

18. Express the phase angle α in an *LCR* circuit in terms of L, R, ω, and the resonant frequency ω_0. Use your formula to find approximate values for α at very low frequencies and at very high frequencies.

19. The circuit shown contains a battery, an inductor, an initially uncharged capacitor, and three resistors. The switch is closed at time $t = 0$ and the initial current is 1.0 A.

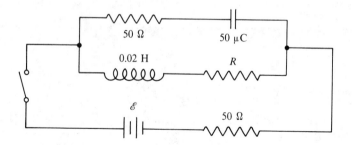

(a) Determine the emf of the battery.
(b) Determine the potential drop across the inductor at time $t = 0$.
(c) After a long time ($t \to \infty$) the current in the circuit is 0.83 A. Determine the value of the resistance R.
(d) If the switch is opened after a long time, explain why there will be no oscillation in the current.

20. In the circuit shown in the diagram, $\mathcal{E} = 24$ V, $C = 2$ μF, and $L = 8$ mH. The currents flow in the directions indicated.

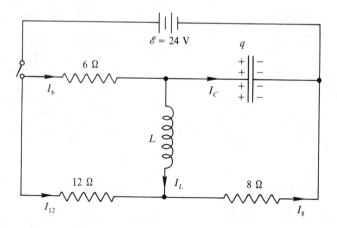

(a) Fill in the values of the various currents in the following table, if the switch is closed at time $t = 0$.

	I_L	I_C	I_6	I_{12}	I_8
$t = 0$					
$t = \infty$					

(b) Calculate dI_L/dt at time $t = 0$.
(c) Find the limiting value of the charge q as $t \to \infty$.

42.4 LCR CIRCUITS

(d) Calculate the energy U_C stored in the capacitor and the energy U_L stored in the inductor at time $t = \infty$.

(e) The switch is opened after a long time. Find a second order differential equation satisfied by the charge q on the capacitor. Determine whether or not the solution will be oscillatory.

21. In the following circuit, $\mathcal{E} = 50$ V, $R_1 = 50$ Ω, $R_2 = 10$ Ω, $C = 5$ μF, and $L = 2.0$ H.

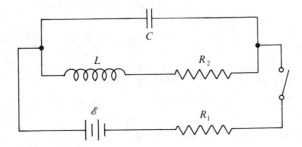

If the switch is closed at time $t = 0$, determine each of the following:

(a) The initial current through the battery.
(b) The limiting value of the current through the battery as $t \to \infty$.
(c) The limiting value of the charge on the capacitor as $t \to \infty$.

22. Refer to the circuit in Question 21. Calculate the limiting values of each of the following as $t \to \infty$:

(a) The magnetic energy stored in the inductor.
(b) The electric energy stored in the capacitor.
(c) The joule heating in the resistor R_1.

23. In the following circuit, $\mathcal{E} = 40$ V, $R_1 = 10$ Ω, $R_2 = 60$ Ω, $R_3 = 30$ Ω, $R_4 = 20$ Ω, $C = 2$ μF, $L_1 = 1.0$ H, $L_2 = 2.0$ H, and $L_3 = 3.0$ H, and the switch is closed at time $t = 0$.

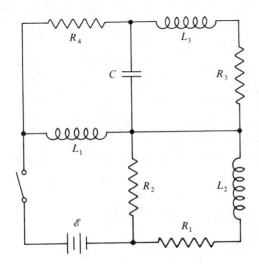

(a) Calculate the initial value of the current through the battery.

Calculate the limiting value as $t \to \infty$ of the current through

(b) the resistor R_2,
(c) the resistor R_1,
(d) the battery.

42.5 POWER IN ac CIRCUITS

To understand why ac is preferable to dc for transmission of electric power over long distances, we perform a simple calculation involving power. Suppose an electric generator produces power P to be transmitted through cables that have resistance R. The longer the transmission cables, the larger R will be, because any cable has electrical resistance proportional to its length.

The power is equal to the current I times the applied voltage V:
$$P = VI. \tag{37.12}$$

The same power can be transmitted at high current and low voltage, or low current and high voltage. Which is better? Passing current through a resistance R causes joule heating given by

$$P_R = I^2 R.$$

This energy escapes from the circuit as wasted heat. Because $I = P/V$, the heat loss is also given by

$$P_R = P^2 R / V^2.$$

For a given power and a given resistance, the higher the voltage, the smaller the loss.

In modern electric power grids, energy is routinely transmitted over thousands of kilometers at hundreds of thousands of volts. However, the consumer wants power at a low, useful voltage. The key to a practical electric power grid is the ability to transform electric energy to high voltage for transport over a long distance, then step it down to low voltage for use in homes. That task is easily achieved with alternating currents by using transformers.

Before discussing how transformers actually work, we first discuss the concept of average power. A resistor in an ac circuit dissipates heat that varies sinusoidally according to the formula

$$P = I^2 R = I_0^2 R \sin^2 \omega t.$$

Figure 42.9 illustrates the power dissipated as a function of time. It shows that the power oscillates rapidly.

Because of rapid oscillations, an indicator on any instrument measuring voltage or current in an ac circuit would have difficulty following the instantaneous values of the voltage or current. In fact, the instrument would tend to average out the fluctuations and record zero for a sinusoidal voltage or current. This is consistent with the mathematical definition of average. For any function $f(t)$, its average over an interval from $t = a$ to $t = b$ is the integral

$$f_{\text{ave}} = \frac{1}{b-a} \int_a^b f(t)\, dt.$$

42.5 POWER IN ac CIRCUITS

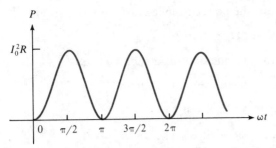

Figure 42.9 Power dissipated through a resistance in an ac circuit as a function of time.

For a sinusoidal function, $f(t) = A \sin \omega t$, this average over a period interval of length $2\pi/\omega$ is easily seen to be zero. (The area of the region above a half-cycle from $t = 0$ to $t = \pi/\omega$ is equal to that under the half-cycle from $t = \pi/\omega$ to $t = 2\pi/\omega$). For that reason, some instruments are designed to measure a different kind of average called the *root mean square* (*rms*). The rms value of any function $f(t)$ over an interval from $t = a$ to $t = b$ is defined to be the positive square root of the quantity

$$\frac{1}{b-a} \int_a^b f(t)^2 \, dt.$$

Example 5
Find the rms value of the sinusoidal current

$$I = I_0 \sin \omega t$$

over one complete cycle, say from $t = 0$ to $t = 2\pi/\omega$.

The rms value of I is the square root of the quantity

$$\frac{\omega}{2\pi} \int_0^{2\pi/\omega} I_0^2 \sin^2 \omega t \, dt = \frac{I_0^2}{2}, \tag{42.10}$$

so

$$I_{\text{rms}} = \frac{I_0}{\sqrt{2}}. \tag{42.11}$$

Similarly, the rms emf of a sinusoidal voltage

$$\mathscr{E} = \mathscr{E}_0 \sin \omega t$$

is given by

$$\mathscr{E}_{\text{rms}} = \frac{\mathscr{E}_0}{\sqrt{2}}. \tag{42.12}$$

In an *RLC* circuit the power supplied at time t is equal to

$$P = \mathcal{E}I = (\mathcal{E}_0 \sin \omega t) I_0 \sin(\omega t - \alpha).$$

Using the formula for the sine of a difference we find

$$\sin(\omega t - \alpha) = \sin \omega t \cos \alpha - \cos \omega t \sin \alpha,$$

so the instantaneous power is given by

$$P = \mathcal{E}_0 I_0 (\cos \alpha \sin^2 \omega t - \sin \alpha \sin \omega t \cos \omega t).$$

Equation (42.10) shows that the average of $\sin^2 \omega t$ over one cycle is $\frac{1}{2}$, and it is easy to verify that the average of $\sin \omega t \cos \omega t$ is 0. Therefore, the average power over one cycle is

$$P_{\text{ave}} = \tfrac{1}{2} \mathcal{E}_0 I_0 \cos \alpha.$$

Because of (42.11) and (42.12) this can also be written as

$$P_{\text{ave}} = \mathcal{E}_{\text{rms}} I_{\text{rms}} \cos \alpha.$$

In a dc circuit the power is the product of the emf and the current. This formula for P_{ave} is analogous, except it contains the rms values of emf and current and is also multiplied by the cosine of the phase angle α. The cosine factor is called the *power factor*. The maximum value of the power factor occurs at resonance, when $\alpha = 0$.

From the right triangle in Fig. 42.8 we see that $\cos \alpha = R/Z$. We also have

$$I_{\text{rms}}/\mathcal{E}_{\text{rms}} = I_0/\mathcal{E}_0 = 1/Z,$$

so the formula for the average power becomes

$$P_{\text{ave}} = \mathcal{E}_{\text{rms}}^2 \frac{R}{Z^2}. \tag{42.13}$$

Questions

24. In an *LCR* circuit a 75-Ω resistor is connected along with a 25-mH inductor and a capacitor to a source of emf alternating at 60 Hz with an amplitude of 10 V. For what values of the capacitance would the average power dissipated in the resistor be (a) a minimum and (b) a maximum?

25. For the circuit in Question 24, find (a) the minimum average power dissipated in the resistor, (b) the phase angle, and (c) the power factor.

26. An electric heater uses an average power of 1200 W when plugged into a 110-V ac outlet. Calculate each of the following:

(a) The maximum instantaneous current and the rms current in the heater.
(b) The maximum and minimum values of the instantaneous power.
(c) The resistance of the heater.

27. (a) Express the power factor $\cos \alpha$ in terms of R, L, ω, and the resonant frequency ω_0.

42.6 TRANSFORMERS

(b) Keep L and C positive and find the limiting value of the power factor as $R \to 0$.

(c) Keep R positive and find the limiting value of the power factor as $L \to 0$.

28. Figure 21.3 shows resonance curves (plotting amplitude versus ω/ω_0) for different amounts of frictional force acting on a mechanical oscillator. Sketch similar curves describing average power, with resistance playing the role of friction.

29. A capacitor of capacitance 2.0 μF has an initial charge of 5.0 μC and is connected across a 2.0-mH inductor as shown in the diagram.

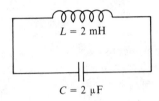

Calculate each of the following for this circuit:

(a) The angular frequency of the oscillations.
(b) The period.
(c) The maximum energy stored in the inductor.

30. A 2.0-μF capacitor with an initial charge of 2.0 μC is connected across a 20-Ω resistor and a 2.0-mH inductor as shown in the following diagram.

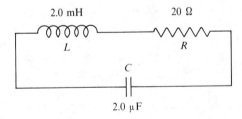

Calculate each of the following:

(a) The angular frequency of oscillations if the resistor were not present.
(b) The actual angular frequency.
(c) The total energy dissipated in the resistor.

42.6 TRANSFORMERS

The first transformer was Faraday's iron ring with two coils wrapped around it. Any two coils of wire properly arranged around a common iron core constitute a transformer. A diagram representing a transformer is shown in Fig. 42.10. When

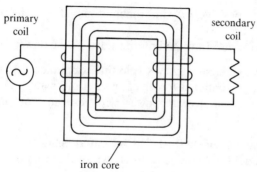

Figure 42.10 Schematic diagram of a transformer.

alternating current is sent through one of the coils (which we call the *primary coil*), it sets up an alternating flux in the core. This varying flux induces an alternating current in the second coil (the *secondary coil*).

As we will show presently, the ratio of the emfs in coil 1 to coil 2 is nearly the same as the ratio of the number of turns in coil 1 to coil 2. For example, if the primary coil has 10 turns and the secondary coil 500 turns, the voltage in the secondary coil will be nearly 50 times greater than that in the primary coil. This is an example of a *step-up* transformer. In a *step-down* transformer the secondary coil has fewer turns than the primary coil.

Although a transformer increases or lowers the voltage of an alternating current, the total power does not change. Increased voltage is compensated for by a decrease in the current. There are, of course, energy losses dissipated as heat in the coils and also in the core. The core losses can be minimized by using a laminated core consisting of many layers of thin steel strips. Thin strips allow the magnetism to change more rapidly and are not heated as much as a solid core.

We consider an ideal transformer, one having no losses and having all the flux linking the primary and secondary coils. Assume an alternating emf is applied to the primary winding, as indicated in Fig. 42.10, where there are N_1 turns in the primary and N_2 in the secondary encircling the core in the same sense. We neglect the resistance of the coils compared to the inductive reactance, so we have an example of two inductive circuits of the type discussed in Fig. 42.5, each with the current lagging the voltage by 1/4 of a cycle. These circuits are diagrammed in Fig. 42.11.

We will show that the induced emf $\mathscr{E}_2$ in the secondary coil is related to the impressed emf $\mathscr{E}_1$ in the primary coil by the equation

$$\mathscr{E}_2 = \mathscr{E}_1 \frac{N_2}{N_1}. \tag{42.14}$$

To prove this we apply Kirchhoff's voltage law to each circuit. In the primary circuit the induced emf across the primary coil must equal the impressed emf $\mathscr{E}_1$. On the other hand, by Faraday's law the induced emf equals the rate of change of flux, and

42.6 TRANSFORMERS

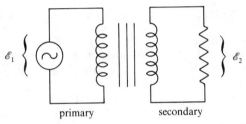

Figure 42.11 Two inductive circuits associated with a transformer.

hence

$$\mathscr{E}_1 = -\frac{d\phi_1}{dt}.$$

Similarly, for the secondary coil we find

$$\mathscr{E}_2 = -\frac{d\phi_2}{dt}.$$

Now if ϕ is the flux through each turn, then $\phi_1 = N_1\phi$ and $\phi_2 = N_2\phi$, and when we eliminate $d\phi/dt$ from the last two equations we immediately obtain Eq. (42.14).

When electric power is transmitted through long distances, the voltage from the generator is transformed to a very high value (in the range from 30,000 to 220,000 V). At substations part of the power is then processed through step-down transformers to neighborhoods at about 4150 V. Another step-down transformer then reduces the voltage further to 115 or 230 V for use in homes. These systems are successful because transformers can transfer power from the primary to the secondary with an efficiency between 90 and 99%.

For an ideal transformer (100% efficient) there are no losses, so the average power into the primary must equal that out of the secondary. Using $P = \mathscr{E}I$, this implies that for an ideal transformer we have

$$\mathscr{E}_{1\,\mathrm{rms}} I_{1\,\mathrm{rms}} = \mathscr{E}_{2\,\mathrm{rms}} I_{2\,\mathrm{rms}}.$$

Questions

31. Can a transformer be used to step up the voltage of a dc generator?

32. A transformer has 500 turns on its primary winding and 50 turns on its secondary. If the rms voltage across the primary is 120 V, find the rms voltage across the secondary, assuming an open circuit for the secondary.

33. A transformer consisting of 2000 turns on the primary is connected to a rms voltage of 220 V and supplies a current of 5 A at 10 V to a secondary winding. Determine (a) the current in the primary and (b) the number of turns in the secondary.

34. A transformer operates at a frequency of 50 Hz. The primary winding has 100 turns and carries a current of 0.5 A, while the secondary has 20 turns and an

emf of 10 V. Determine (a) the emf of the primary and (b) the current in the secondary.

35. Refer to the transformer of Question 34. Determine the self-inductance of the primary winding, assuming that all the impedance is inductive.

42.7 A FINAL WORD

Throughout the latter part of their lives, Tesla and Edison were engaged in a battle of currents. Edison advocated dc as the safest and best type of electricity to power the nation. To point out the dangers of ac to an uninformed public, Edison conducted gruesome experiments. In the neighborhood surrounding Edison's laboratory in West Orange, New Jersey, families noticed that their pets were vanishing. They learned that Edison was paying schoolboys for dogs and cats, which he then electrocuted in crude experiments with alternating current. He intended these experiments to be a warning that society would be "Westinghoused" by ac.

Meanwhile, Tesla was producing bolts of artificial lightning over 100 ft long, and lighting electric lamps at a distance without wires. By 1897, Tesla had built and operated a wireless-controlled boat, laying the foundation for what later became wireless communication. He correctly predicted not only the wireless transmission of ordinary messages but that of pictures and whole pages of newsprint. At a time when Edison fought for direct current, Tesla realized that the future lay in alternating currents.

Tesla's reputation suffered greatly from his battle with the revered Edison. He never received recognition for his achievements in the United States, but in his native Serbia he is a folk hero comparable to Einstein. In 1915 the *New York Times* carried a story reporting that Tesla and Edison were to share the Nobel prize in physics. However, neither Edison nor Tesla ever received the prize. No one knows what really happened. One biographer claims that Tesla declined the honor, not wanting to share it with a mere inventor. Another biographer advanced the theory that it was Edison who objected to sharing the prize and wanted to deprive the impoverished Tesla of the associated monetary award. No real evidence exists to prove that either of them declined the Nobel prize.

CHAPTER 43

MAXWELL'S EQUATIONS

From a long view of the history of mankind – seen from, say ten thousand years from now – there can be little doubt that the most significant event of the 19th century will be judged as Maxwell's discovery of the laws of electrodynamics. The American Civil War will pale into provincial insignificance in comparison with this important scientific event of the same decade.

The Feynman Lectures on Physics, Vol. II, Addison-Wesley Publishing Co. (1964)

43.1 A VICTORIAN GENIUS

Galileo, Kepler, Newton, Faraday – each was endowed with a rare combination of intelligence, curiosity, insight, imagination, and perseverance, and each revolutionized a part of science and altered our perception of the world. The same is true of James Clerk Maxwell, the last great genius of the Victorian era.

Maxwell was born in Edinburgh, Scotland in 1831. He came from a wealthy family, the Clerks of Edinburgh, who had intermarried at various times with descendants of the Maxwells of Middleby who, in turn, had descended from an illegitimate offspring of the eighth Lord of Maxwell. James Clerk Maxwell's father changed his name from John Clerk to John Clerk Maxwell to make it possible to inherit certain property the Clerks were forbidden to acquire because of legal technicalities. If that hadn't happened this chapter would be entitled "Clerk's Equations" instead of "Maxwell's Equations."

Maxwell attended the University of Edinburgh at age 16, and by the time he was 19, when he transferred to Trinity College at Cambridge, he had already published a number of papers in geometry and one in physics. He graduated from Cambridge University summa cum laude in 1854, became a fellow at Trinity College in 1855, and soon was appointed Professor of Natural Philosophy at Marischal College in Aberdeen. From 1860 to 1865 he held a chair with the same title at King's College, London, during which time he published his famous "Dynamical Theory of the Electromagnetic Field." In 1865 at the age of 34 he retired to the family farm in Scotland because of ill health and wrote his celebrated *Treatise on Electricity and Magnetism*. In 1871 he was called back to Cambridge to fill a newly founded Chair of Experimental Physics. There he developed and became first director of the world-famous Cavendish Laboratory. A few years later, at age 48, he died of abdominal cancer.

Although Maxwell secured his place in the history of science by his investigations in electromagnetism and the kinetic theory of gases, he also made fundamental contributions to geometrical optics, color vision, the theory of Saturn's rings, thermodynamics, continuum mechanics, and many other fields, both theoretical and experimental. His contributions appeared in four books and nearly one hundred papers. His *Treatise on Electricity and Magnetism* is a landmark in the history of science, often ranked with Newton's *Principia*.

43.2 THE LINK BETWEEN ELECTRICITY AND MAGNETISM

Today we know that electricity and magnetism are fundamental phenomena, the basic fabric underlying the structure of all matter and perhaps of life itself. But this understanding of the role of electricity and magnetism evolved very slowly. From the time that Thales of Miletus recorded his observations on rubbed amber until Gilbert published his treatise on magnetism 22 centuries later, our knowledge of these matters was primarily lore, a collection of curious facts about the peculiar behavior of amber, lodestones, or mariner's compasses. Only after the invention of devices that could produce and control the flow of electricity did researchers discover that electric currents could generate magnetism. Ampère uncovered the law for calculating magnetic forces due to a current, while Faraday discovered that a changing magnetic field could produce a current in a conductor. These two discoveries provided the fundamental link between electricity and magnetism and began a grand synthesis that was completed by Maxwell when he unified the two kinds of force under a single concept – the electromagnetic field.

Influenced by Faraday's explanations in terms of lines of force, Maxwell translated the descriptions of electric and magnetic phenomena into the language of calculus through a set of four mathematical statements now known as Maxwell's

43.2 THE LINK BETWEEN ELECTRICITY AND MAGNETISM

equations. Three of these equations are merely restatements of results we have encountered in earlier chapters:

1. Gauss's law for electricity:

$$\oint_S \mathbf{E} \cdot d\mathbf{A} = Q/\varepsilon_0. \tag{33.2}$$

2. Gauss's law for magnetism:

$$\oint_S \mathbf{B} \cdot d\mathbf{A} = 0. \tag{38.9}$$

3. Faraday's law for electromagnetic induction, $\mathscr{E}_i = -d\phi_B/dt$, expressed in integral form:

$$\oint \mathbf{E} \cdot d\mathbf{r} = -\frac{d}{dt} \int\int_S \mathbf{B} \cdot d\mathbf{A}. \tag{41.6}$$

Maxwell's fourth equation is a modification of Ampère's law:

$$\oint \mathbf{B} \cdot d\mathbf{r} = \mu_0 I. \tag{39.14}$$

Maxwell realized that this equation had to be modified because, under certain conditions, Ampère's law could not be correct.

Example 1

Show that Ampère's circuit law, as expressed in Eq. (39.14), fails for the circuit shown in Fig. 43.1.

Figure 43.1 shows a circuit containing a capacitor and a closed circular path C surrounding the wire leading to the positive plate. Imagine that at a certain instant a current I is flowing through the wire, so that charge Q is accumulating on the

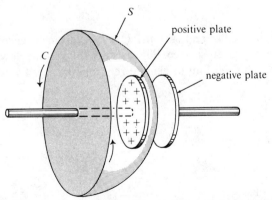

Figure 43.1 An example in which Ampère's law fails.

capacitor. The current generates a magnetic field **B** and, according to Ampère's law, the line integral of **B** around C is proportional to the current.

$$\oint_C \mathbf{B} \cdot d\mathbf{r} = \mu_0 I,$$

where I is the total current passing through the plane circular region enclosed by C. Moreover, Ampère's law states that the same equation holds if the plane circular region is replaced by any surface having C as its boundary, for example, the hemispherical surface S shown in Fig. 43.1. In other words, the line integral of B around C is equal to $\mu_0 I$, where I is the total current passing through S. But the hemispherical surface S passes through the region between the plates of the capacitor and there is no current through S because the charge stops at the capacitor plate inside S. Therefore, Ampère's law cannot hold for this circuit.

Ampère's law fails in the circuit shown in Fig. 43.1 because current is flowing into the dome-shaped region through the circular disk, but none is flowing back out again. On the other hand, precisely because no current is emerging, charge is building up inside, and as a result, electric flux is building up between the capacitor plates, and, therefore, flux lines pass through the hemispherical surface. The electric flux is given by

$$\phi_E = Q/\varepsilon_0.$$

Maxwell argued that this buildup of electric flux could be just what's needed to correct Ampère's equation. The flux is related to the current flowing into the region by the equation

$$I = \frac{dQ}{dt} = \varepsilon_0 \frac{d\phi_E}{dt}.$$

Maxwell thought of the term $\varepsilon_0(d\phi_E/dt)$ as representing a kind of fictitious current to complete the circuit that is interrupted by the capacitor. He called it the *displacement current*,

$$I_d = \varepsilon_0 \frac{d\phi_E}{dt}.$$

Maxwell reasoned that either real current, I, or displacement current, I_d, (or both) might flow through any given surface, but the sum $I + I_d$ that flows into any region must also flow out. This led him to generalize Ampère's law by replacing I in Eq. (39.14) by $I + I_d$:

4. The Maxwell–Ampère law:

$$\oint \mathbf{B} \cdot d\mathbf{r} = \mu_0(I + I_d), \tag{43.1}$$

where

$$I_d = \varepsilon_0 \frac{d\phi_E}{dt}$$

43.2 THE LINK BETWEEN ELECTRICITY AND MAGNETISM

Table 43.1 Maxwell's Four Laws.

	Equation	Meaning
1.	$\oiint_S \mathbf{E} \cdot d\mathbf{A} = Q/\varepsilon_0$	Gauss's law for electricity. Comes directly from Coulomb's law. The electric flux out of a region is proportional to the net charge inside.
2.	$\oiint_S \mathbf{B} \cdot d\mathbf{A} = 0$	Gauss's law for magnetism. Because there are no magnetic monopoles, there is never any net flux out of any region.
3.	$\mathcal{E}_i = -\dfrac{d\phi_B}{dt}$ or $\oint \mathbf{E} \cdot d\mathbf{r} = -\dfrac{d}{dt}\iint_S \mathbf{B} \cdot d\mathbf{A}$	Faraday's law of electromagnetic induction. A changing magnetic flux creates an electromotive force.
4.	$\oint \mathbf{B} \cdot d\mathbf{r} = \mu_0 (I + I_d)$ or $\oint \mathbf{B} \cdot d\mathbf{r} = \mu_0 I + \mu_0 \varepsilon_0 \dfrac{d}{dt} \iint_S \mathbf{E} \cdot d\mathbf{A}$	Maxwell–Ampère law. Circulating magnetic fields are created either by electric currents or by changing electric flux.

and ϕ_E is the flux of the electric field,

$$\phi_E = \iint_S \mathbf{E} \cdot d\mathbf{A}.$$

This small change had a profound effect on the subsequent development of the theory of electromagnetism. It eventually led to the electromagnetic theory of light.

Table 43.1 summarizes Maxwell's four equations, where they came from, and what they mean physically.

Example 2

A constant current I flows into a capacitor whose plates are circular disks of radius R, as shown in Fig. 43.2. Find the magnetic field between the plates.

By symmetry, $\mathbf{B} = B(r)\hat{\boldsymbol{\theta}}$, where $B(r)$ is constant on a circle of radius $r < R$ centered on the axis of the capacitor. Therefore, the line integral of $\mathbf{B}$ around such a circle is the strength of the field times the length of the path:

$$\oint \mathbf{B} \cdot d\mathbf{r} = 2\pi r B(r).$$

The circle is the boundary of a flat disk through which no real current flows, but

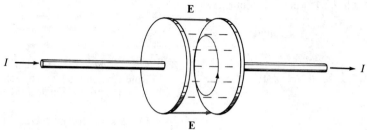

Figure 43.2 Calculation of the magnetic field between capacitor plates.

there is a displacement current given by

$$I_d = \varepsilon_0 \frac{d\phi_E}{dt}.$$

The electric flux through this disk is r^2/R^2 times the total flux through the capacitor, which, in turn, must be equal to I/ε_0. (Why?) Using the Maxwell–Ampère law we find

$$2\pi r B(r) = \mu_0 I_d = \mu_0 I r^2/R^2,$$

hence

$$B(r) = \frac{\mu_0 I}{2\pi R^2} r.$$

Questions

The following questions refer to Fig. 43.3, in which a circular parallel plate capacitor of radius R is being charged by a constant current I. The capacitor is

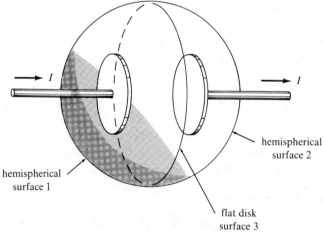

Figure 43.3 Diagram for Questions 1 through 5.

contained within an imaginary spherical surface that is divided into two hemispheres (surfaces 1 and 2) and a flat circular disk (surface 3) that passes between the plates of the capacitor.

Determine the electric flux, the magnetic flux, the real current, and the displacement current for each of the following:

1. Surface 1.
2. Surface 2.
3. Surface 3.
4. The spherical region bounded by surfaces 1 and 2.
5. The hemispherical region bounded by surfaces 1 and 3.

43.3 MAXWELL'S EQUATIONS IN FREE SPACE

The role to be played by the new term in Maxwell's fourth equation is most easily understood if we consider the form of Maxwell's equations in *free space*, by which we mean a region containing no matter. Where there is no matter present, there are no electric charges Q and no electric currents I, so Maxwell's equations in free space become

$$\oint_S \mathbf{E} \cdot d\mathbf{A} = 0, \qquad \oint_S \mathbf{B} \cdot d\mathbf{A} = 0, \tag{43.2}$$

$$\oint \mathbf{E} \cdot d\mathbf{r} = -\frac{d\phi_B}{dt}, \qquad \oint \mathbf{B} \cdot d\mathbf{r} = \mu_0 \varepsilon_0 \frac{d\phi_E}{dt}. \tag{43.3}$$

The two equations in (43.2) can be interpreted as saying that Faraday field lines for both **E** and **B** never end in free space (that is, the density of lines going into any small region of space is the same as that of lines going out). That's because there are no sources (charges or monopoles) for the lines to begin or end on. So, field lines either extend to infinity or end on themselves.

The two equations in (43.3) describe the conditions under which circulating fields can be stirred up. We have already seen, in Chapter 41, that a circulating electric field can be created by a changing magnetic field. That was Faraday's discovery, electromagnetic induction. Maxwell's new term, the right-hand side of the last equation in (43.3), says that the reverse is also true: A changing electric field can stir up a circulating magnetic field. Aside from constants and a minus sign, the fields **E** and **B** are treated symmetrically in Maxwell's equations in free space.

The constants, however, contain an important clue. In Chapter 11 we described Maxwell's stunning discovery that the speed of light c was buried in the forces between charges and magnets,

$$\sqrt{K_e/K_m} = 3 \times 10^8 \text{ m/s} = c.$$

Using the relations $K_e = 1/(4\pi\varepsilon_0)$ and $K_m = \mu_0/(4\pi)$ we find that this equation implies

$$\mu_0 \varepsilon_0 = 1/c^2, \tag{43.4}$$

so the speed of light is explicitly contained in the extra term of the Maxwell–Ampère law.

Questions

6. A cylindrical region of free space is symmetric about the z axis. The electric field in this region is given by $\mathbf{E}(t) = A_0 t \hat{\mathbf{k}}$, where A_0 is a constant and t is time. Find the magnetic field that results from the changing electric flux.

7. According to Maxwell's equations, can a constant (in both space and time) nonzero electric or magnetic field exist in a region where there is no charge and no current? Conversely, if constant electric or magnetic fields do exist in some region, what can be said about the charge and current in that region?

43.4 PLANE WAVES MOVING WITH CONSTANT SPEED

The first great triumph of Maxwell's theory was to predict that disturbances in the electromagnetic field would travel through free space at the speed of light. Before we demonstrate how this happens, we first learn how to describe a disturbance moving at a constant speed, in order to know what we're looking for. We begin with plane waves moving with constant speed.

Consider a curve $y = f(x)$ in the xy plane having the form shown in Fig. 43.4a. The curve has a small peak above the origin, with $f(0) = 4$, but it drops down to zero and has $f(x) = 0$ for $|x| \geq 2$. We wish to find a time dependent function, $F(x, t)$ that will describe the shape of this curve as it moves to the right at a speed $w = 2$. For example, after 1 s the peak will have moved to $x = 2$, after 2 s to $x = 4$, and so on, as suggested by Fig. 43.4b, c, and d.

It's not hard to see that the required function is given by

$$y = F(x, t) = f(x - 2t).$$

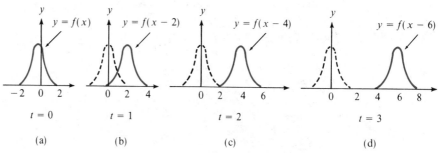

Figure 43.4 A disturbance moving to the right with speed $w = 2$.

Thus, at time $t = 0$, $y = F(x, 0) = f(x)$, the original curve. But when $t = 1$ we have
$$y = F(x, 1) = f(x - 2),$$
which represents a translation to the right of the original function by 2 units of distance. Note that $F(0, 1) = f(-2) = 0$ but $F(2, 1) = f(0) = 4$. Thus the peak is no longer at $x = 0$ but has moved to $x = 2$, and it will continue to move to the right at speed $w = 2$.

More generally, the equation
$$y = F(x, t) = f(x - wt)$$
describes the shape of the curve obtained by moving the original curve $y = f(x)$ to the right at speed w along the x axis. Similarly, the equation
$$y = G(x, t) = f(x + wt)$$
describes the shape that moves to the left at speed w. These remarks are true of any function f, no matter how complicated its shape or mathematical form.

Questions

8. A bell-shaped curve has the equation
$$y = e^{-x^2}.$$
Write an equation $y = F(x, t)$ that represents a disturbance of the same shape that moves to the right at speed $w = 3$.

9. A vertical cross section of the surface of a water wave is described by the equation
$$y = A \sin(kx - \omega t),$$
where A, k, and ω are positive constants.

(a) Show that the speed of the wave is $w = \omega/k$.
(b) Find a function $y = f(x)$ such that $f(x - wt) = A \sin(kx - \omega t)$.

10. Let $y = f(x) = 1 + \cos \pi x$ for $-1 \le x \le 1$, and take $f(x) = 0$ for $|x| \ge 1$. Draw graphs of each of the following functions:

(a) $y = f(x)$.
(b) $y = f(x - 2)$.
(c) $y = f(x + 2)$.
(d) $y = [f(x + 2) + f(x - 2)]/2$.

43.5 THE WAVE EQUATION

Plane waves of the type discussed in the foregoing section are described by an equation of the form
$$y = F(x, t) = f(x - wt), \tag{43.5}$$
where w is a positive constant for waves moving to the right. In mathematical terms,

this means that a function of a single variable, say

$$y = f(z),$$

has been converted to a function of two variables, $y = F(x, t)$, by the substitution

$$z = x - wt.$$

The new function $F(x, t)$ satisfies a simple differential equation relating the space variable x and the time variable t. To find this differential equation we apply the chain rule for derivatives to Eq. (43.5).

Differentiating Eq. (43.5) with respect to t we have

$$\frac{\partial F}{\partial t} = \frac{df}{dz}\frac{\partial z}{\partial t} = -w\frac{df}{dz}, \tag{43.6}$$

because $\partial z/\partial t = -w$. Differentiating (43.5) with respect to x we find

$$\frac{\partial F}{\partial x} = \frac{df}{dz}\frac{\partial z}{\partial x} = \frac{df}{dz}, \tag{43.7}$$

because $\partial z/\partial x = 1$. Eliminating df/dz from these two equations we find

$$\frac{\partial F}{\partial t} = -w\frac{\partial F}{\partial x}. \tag{43.8}$$

It's easy to show that if w is positive this equation describes only disturbances moving to the right with constant speed w. (See Question 11.)

In physical situations we can have disturbances moving to the left as well as to the right. For disturbances moving to the left with speed w the new function of x and t is

$$G(x, t) = f(x + wt),$$

and the same analysis that gave us Eq. (43.8) now leads to the equation

$$\frac{\partial G}{\partial t} = w\frac{\partial G}{\partial x}. \tag{43.9}$$

We can find a single differential equation that describes motion both to the left and to the right by differentiating one more time. Differentiating Eq. (43.6) with respect to t, and using the chain rule again, we find

$$\frac{\partial^2 F}{\partial t^2} = w^2\frac{d^2 f}{dz^2}.$$

In the same way, differentiating Eq. (43.7) with respect to x gives us

$$\frac{\partial^2 F}{\partial x^2} = \frac{d^2 f}{dz^2}.$$

Eliminating $d^2 f/dz^2$ from these two equations we obtain

$$\frac{\partial^2 F}{\partial t^2} = w^2\frac{\partial^2 F}{\partial x^2}. \tag{43.10}$$

This differential equation is called the one-dimensional *wave equation*. It is also

43.5 THE WAVE EQUATION

satisfied by the function $G(x, t) = f(x + wt)$, which represents a displacement moving to the left with speed w. Moreover, it can be shown that any function satisfying the wave equation represents a superposition of displacements moving to the right and to the left with speed w. In other words, if $y(x, t)$ satisfies the wave equation, then

$$y(x, t) = f(x - wt) + g(x + wt)$$

for some choice of functions f and g.

Example 3

Show that the function

$$y(x, t) = A \sin(kx - \omega t) + B \sin(kx + \omega t),$$

where A, B, k, and ω are positive constants, is a solution of the one-dimensional wave equation (43.10). Describe the solution in the case $A = B$.

Differentiating twice with respect to x, and then again with respect to t, we find

$$\frac{\partial^2 y}{\partial x^2} = -k^2 A \sin(kx - \omega t) - k^2 B \sin(kx + \omega t)$$

$$= -k^2 y$$

and

$$\frac{\partial^2 y}{\partial t^2} = -\omega^2 A \sin(kx - \omega t) - \omega^2 B \sin(kx + \omega t)$$

$$= -\omega^2 y.$$

Comparison of these results shows that

$$\frac{\partial^2 y}{\partial t^2} = \frac{\omega^2}{k^2} \frac{\partial^2 y}{\partial x^2}.$$

This is the wave equation with speed w satisfying the condition

$$w^2 = \omega^2/k^2.$$

The same equation is satisfied if either of the sine functions is replaced by a cosine. (See Question 12.)

If $A = B$ the solution is given by

$$y = A\{\sin(kx - \omega t) + \sin(kx + \omega t)\}$$

and represents identical waves traveling in opposite directions. Because of the trigonometric identity

$$\sin(\theta + \alpha) + \sin(\theta - \alpha) = 2 \sin \theta \cos \alpha,$$

the solution becomes

$$y = 2A \sin kx \cos \omega t.$$

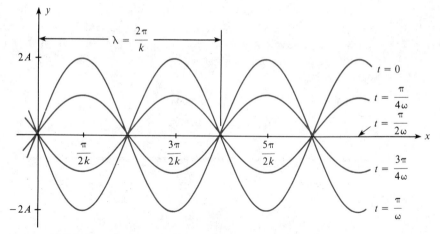

Figure 43.5 A standing wave with equation $y = 2A \sin kx \cos \omega t$ shown for various values of t.

This is called a *standing wave* because the spatial factor, $\sin kx$, does not change with time. The other factor, $\cos \omega t$, makes it appear that the sine curve is oscillating up and down in time. An example is shown in Fig. 43.5.

Solutions of the wave equation of the form sine or cosine of $(kx - \omega t)$ or of $(kx + \omega t)$ are functions that repeat their values when the argument changes by 2π. For a fixed value of t, only x is changing so $\sin(kx - \omega t)$ repeats its values when kx changes by 2π. This means that, as a function of x, $\sin(kx - \omega t)$ is periodic with period $2\pi/k$. This is called the wavelength λ,

$$\lambda = 2\pi/k,$$

and is illustrated in Fig. 43.5. Similarly, for a fixed value of x, the function $\sin(kx - \omega t)$ repeats its values when ωt changes by 2π, so, as a function of time t, $\sin(kx - \omega t)$ is periodic with period $T = 2\pi/\omega$. The reciprocal of T is called the frequency f,

$$f = \omega/(2\pi)$$

(measured in Hz, or cycles per second). The constant ω is the angular frequency (measured in radians per second). The product of frequency and wavelength is $\lambda f = \omega/k$, the wave speed w:

$$w = \lambda f = \omega/k.$$

When a constant factor A multiplies a sine or cosine function, the magnitude of A is called the amplitude of the wave. It represents the maximum value of the wave function.

Questions

11. Show that a disturbance moving to the left with constant speed w does not satisfy the equation $\partial F/\partial t = -w(\partial F/\partial x)$.

12. Show that the function in Example 3 satisfies the wave equation if one or both of the sine functions is replaced by a cosine.

13. In Chapter 22, a number of different kinds of waves were discussed:

  (a) Masses connected by springs.
  (b) Deep water waves.
  (c) Shallow water waves.
  (d) Sound waves in air.

A one-dimensional wave equation is associated with each case. What physical quantity obeys the wave equation in each case? In which cases can we expect solutions of the form

$$y = A \sin(kx - \omega t),$$

where A, k, and ω are positive constants?

14. The vertical displacement of a wave traveling in the x direction is given by the function $y(x, t) = 10 \sin(0.25x - 7.5t)$, where x and y are in centimeters and t is in seconds. Determine the amplitude, the frequency, the wavelength, and the speed of the wave.

15. Give a formula for the vertical displacement of a wave traveling in the negative x direction with a speed of 0.25 cm/s, a wavelength of 4.5 cm, and an amplitude of 5.0 cm.

43.6 ELECTROMAGNETIC WAVES

This section shows that the electric and magnetic fields in free space, as described by Maxwell's equations (43.2) and (43.3), obey the wave equation with wave speed equal to the speed of light. In a sense, this discussion is the culmination of electromagnetic theory, just as the solution of the Kepler problem in Chapter 26 was the culmination of classical mechanics. Its implication is that light, electricity, and magnetism are different aspects of the same phenomenon. That phenomenon encompasses not only visible light but an entire electromagnetic spectrum stretching from radio waves at one end to gamma rays at the other.

To see that Maxwell's equations in free space predict wave motion we begin by assuming that $\mathbf{E}$ and $\mathbf{B}$ depend only on the x coordinate, and on time t. We write

$$\mathbf{E} = E_1(x, t)\hat{\mathbf{i}} + E_2(x, t)\hat{\mathbf{j}} + E_3(x, t)\hat{\mathbf{k}}$$

and

$$\mathbf{B} = B_1(x, t)\hat{\mathbf{i}} + B_2(x, t)\hat{\mathbf{j}} + B_3(x, t)\hat{\mathbf{k}}.$$

First we show that in this case the $\hat{\mathbf{i}}$ components $E_1(x, t)$ and $B_1(x, t)$ do not depend on x.

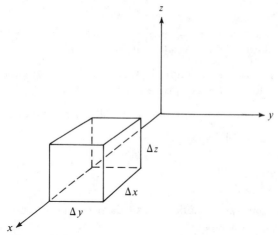

Figure 43.6 The total flux through a small cube is zero.

To show that $B_1(x, t)$ does not depend on x, we apply Gauss's law for magnetism to the surface of a small cube of sides $\Delta x, \Delta y, \Delta z$, shown in Fig. 43.6. Gauss's law says that the total flux emerging from the faces of this cube is zero. Because the components of **B** do not depend on y and z, the flux entering the cube through one of the faces of area $\Delta x \Delta y$ or $\Delta x \Delta z$ must equal the flux leaving the cube through the opposite face. For the faces of area $\Delta y \Delta z$, the flux leaving through the front face is $B_1(x + \Delta x, t) \Delta y \Delta z$, whereas the flux leaving the rear face is $-B_1(x, t) \Delta y \Delta z$, because the outward normal on the rear face is $-\hat{\mathbf{i}}$. By Gauss's law, the sum of all the outward flux must be 0, which implies $B_1(x + \Delta x, t) = B_1(x, t)$. In other words, the first component $B_1(x, t)$ does not depend on x. A similar argument, using Gauss's law for electricity, shows that $E_1(x, t)$ does not depend on x. Therefore, the field vectors can be written as

$$\mathbf{E} = E_1(t)\hat{\mathbf{i}} + E_2(x, t)\hat{\mathbf{j}} + E_3(x, t)\hat{\mathbf{k}}$$

and

$$\mathbf{B} = B_1(t)\hat{\mathbf{i}} + B_2(x, t)\hat{\mathbf{j}} + B_3(x, t)\hat{\mathbf{k}}.$$

This means that at a given instant of time t, as x changes, only the $\hat{\mathbf{j}}$ and $\hat{\mathbf{k}}$ components of **E** and **B** change, hence changes in these vectors occur at right angles to the x axis. Because we have deduced this conclusion from Gauss's law in free space, we know it is a consequence of the absence of charge (and monopoles).

Now apply Faraday's law, Eq. (41.6), integrating **E** around the boundary of the face in the xy plane, traversed in the counterclockwise direction, as shown in Fig. 43.7a. We can write

$$\oint \mathbf{E} \cdot d\mathbf{r} = \int \{ E_1(t) \, dx + E_2(x, t) \, dy + E_3(x, t) \, dz \}$$

$$= \int \{ E_1(t) \, dx + E_2(x, t) \, dy \}$$

because the path of integration is in the xy plane. The contributions from $E_1(t)$

43.6 ELECTROMAGNETIC WAVES

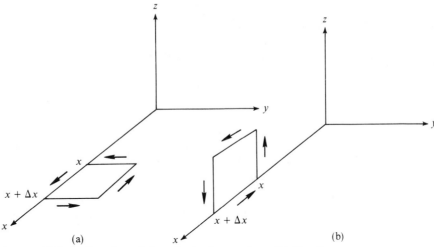

Figure 43.7 (a) One face of the cube in the xy plane. (b) One face in the xz plane.

cancel because E_1 does not depend on x, and the only nonzero contribution to the integral comes from the two edges parallel to the y axis. Therefore, the line integral of **E** around this curve is equal to

$$\oint \mathbf{E} \cdot d\mathbf{r} = \{E_2(x + \Delta x, t) - E_2(x, t)\} \Delta y.$$

If the cube is small, the difference $E_2(x + \Delta x, t) - E_2(x, t)$ is nearly equal to $(\partial E_2/\partial x) \Delta x$, so the last equation becomes

$$\oint \mathbf{E} \cdot d\mathbf{r} = \frac{\partial E_2}{\partial x} \Delta x \Delta y.$$

On the other hand, the right member of Eq. (41.6) (Faraday's law) states that this line integral is equal to

$$-\frac{\partial B_3}{\partial t} \Delta x \Delta y$$

and so, canceling the common factor $\Delta x \Delta y$, we get a relation connecting E_2 and B_3:

$$\frac{\partial E_2}{\partial x} = -\frac{\partial B_3}{\partial t}. \tag{43.11}$$

Next we find another relation connecting E_2 to B_3 by applying the Maxwell–Ampère law, integrating **B** around the boundary of the face of the cube lying in the xz plane, traversed counterclockwise, as shown in Fig. 43.7b. The contribution to the line integral along the horizontal edges of length Δx cancels because $B_1(t)$ is independent of x. The only nonzero contribution is that along the vertical edges, so we have

$$\oint \mathbf{B} \cdot d\mathbf{r} = \{B_3(x + \Delta x, t) - B_3(x, t)\} \Delta z = \frac{\partial B_3}{\partial x} \Delta x \, \Delta z.$$

By the Maxwell–Ampère law, Eq. (43.1), this line integral is also equal to $\mu_0 \varepsilon_0 (\partial \phi_E / \partial t)$ because no conduction current is present, so

$$\frac{\partial B_3}{\partial x} \Delta x \, \Delta z = \mu_0 \varepsilon_0 \frac{\partial \phi_E}{\partial t}.$$

But for a small cube, the flux ϕ_E across the face of area $\Delta x \, \Delta z$ is nearly equal to $E_2(x, t) \Delta x \, \Delta z$ and the last equation becomes

$$\frac{\partial B_3}{\partial x} \Delta x \, \Delta z = \mu_0 \varepsilon_0 \frac{\partial E_2}{\partial t} \Delta x \, \Delta z.$$

Canceling the common factor $\Delta x \, \Delta z$ we find a second relation connecting E_2 and B_3:

$$\frac{\partial B_3}{\partial x} = \mu_0 \varepsilon_0 \frac{\partial E_2}{\partial t}. \tag{43.12}$$

Next we eliminate B_3 from the two relations (43.11) and (43.12). To do this, we differentiate Eq. (43.11) with respect to x and then differentiate Eq. (43.12) with respect to t, and equate the second-order mixed partial derivatives of B_3. This leads to a differential equation for E_2, which states that

$$\frac{\partial^2 E_2}{\partial x^2} = \mu_0 \varepsilon_0 \frac{\partial^2 E_2}{\partial t^2}. \tag{43.13}$$

As promised, Eq. (43.13) is the wave equation, with wave speed

$$w = \frac{1}{\sqrt{\mu_0 \varepsilon_0}},$$

and, of course, we already know from Eq. (43.4) that this particular w is the speed of light.

Example 4

Show that B_3 also satisfies the wave equation (43.13).

Return to Eqs. (43.11) and (43.12) and eliminate E_2 instead of B_3, by differentiating Eq. (43.11) with respect to t and (43.12) with respect to x. The result is

$$\frac{\partial^2 B_2}{\partial x^2} = \mu_0 \varepsilon_0 \frac{\partial^2 B_2}{\partial t^2}.$$

Notice that Eqs. (43.11) and (43.12) were derived, respectively, from Faraday's law and from the Maxwell–Ampère law. The first says that a changing magnetic

43.6 ELECTROMAGNETIC WAVES

field can produce an electric field, while the second says that a changing electric field can produce a magnetic field. Waves arise when these two are combined. In free space there are no electric charges or electric currents to create **E** and **B** fields, but dynamic **E** and **B** fields can be created by each other. This is the essence of the electromagnetic wave.

Example 5
Show that if there is a sinusoidal wave in the electric field, say
$$E_2 = E_0 \sin(kx - \omega t),$$
there must necessarily also be a sinusoidal wave in the magnetic field oscillating in phase with the electric field.

Because the wave speed $\omega/k = c$, we can write E_2 in the form
$$E_2 = E_0 \sin k(x - ct).$$
Substituting this expression for E_2 in Eq. (43.11) we find
$$\frac{\partial B_3}{\partial t} = -\frac{\partial E_2}{\partial x} = -kE_0 \cos k(x - ct).$$
Integrating with respect to t gives us
$$B_3(x, t) = \frac{E_0}{c} \sin k(x - ct) + \text{constant}.$$

In other words, B_3 is equal to a constant plus a sinusoidal wave of amplitude E_0/c, moving in phase with E_2 and with the same speed c. The constant part of B_3 may be zero, but there must be an oscillating component associated with the wave in E_2.

According to Eq. (43.13), disturbances in E_2, the y component of **E**, can move along the x direction at the speed of light. We've also seen that in this same instance, the x component of the field does not vary in space. In other words, these waves – light and the entire electromagnetic spectrum – are purely transverse waves, as distinguished, for example, from sound waves in air, which are purely longitudinal. As Example 5 shows, a wave in the electric field is always associated with a wave in the magnetic field, also transverse, and in phase with the wave in **E**, but perpendicular to it, as shown in Fig. 43.8.

Questions

16. An electromagnetic wave travels in the negative x direction. At a certain point in space the electric field is directed along the positive y axis with an amplitude of 300 V/m.

   (a) Determine the amplitude and direction of the magnetic field associated with this wave.

(b) If the wave varies sinusoidally with a wavelength of 3.0×10^{-7} m and has its maximum oscillation at $t = 0$, find a formula for the electric field at any instant.

17. An electromagnetic wave travels along the positive x direction. When $t = 0$ and $x = 0$, the magnetic field vector of the wave is given by

$$\mathbf{B}_0 = (-2\hat{\mathbf{j}} + 5\hat{\mathbf{k}}) \times 10^{-8} \text{ T},$$

and the wavelength is 4.0×10^{-7} m.

(a) Make a sketch showing the vectors $c\mathbf{B}_0$ and $\mathbf{E}_0$ at $t = 0$ and $x = 0$.
(b) If $\mathbf{E}$ and $\mathbf{B}$ vary sinusoidally in time, write expressions for these fields at any time t and any point x.

43.7 DISTURBANCES CAUSED BY ACCELERATED CHARGES

We have seen that electromagnetic waves can propagate through free space at the speed of light. But how do they get started in the first place? Maxwell found that they would be a consequence of *accelerated charges*. The idea can be understood from the following argument.

Imagine a single negative charge, initially at rest at the origin. Its electric field can be represented by radial Faraday lines, stretching to infinity in all directions. Now suppose that at time $t = 0$ the charge is accelerated up to speed v in a short time interval Δt, after which it continues to move at constant speed v along the positive x axis. We now try to visualize the Faraday field lines at a later time $t \gg \Delta t$.

Because we know that disturbances in the field travel at speed c, the news that the charge has been set in motion cannot have reached those field lines outside of a

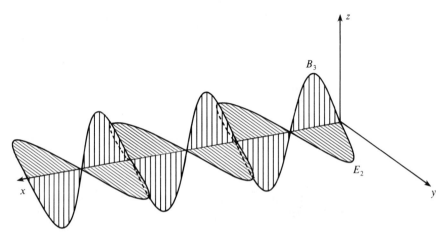

Figure 43.8 Electric and magnetic field oscillations propagating in phase.

43.7 DISTURBANCES CAUSED BY ACCELERATED CHARGES

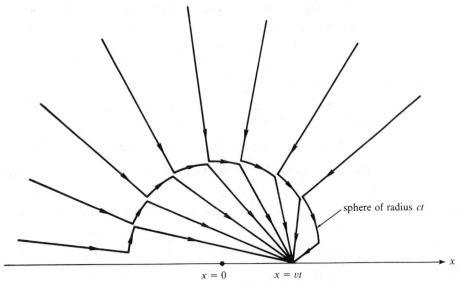

Figure 43.9 Field lines of a moving charge suddenly accelerated.

sphere of radius ct centered at the origin. These lines still point directly inward toward the origin. On the other hand, inside a spherical shell of thickness $c\,\Delta t$, the field lines are pointed directly at the present position of the moving charge, which is at a distance vt from the origin. This is shown in Fig. 43.9. Thus, within the shell, each field line undergoes a sharp jog to connect from its inner form to its outer form.

That shell of thickness $c\,\Delta t$, propagating outward at speed c, contains the burst of electromagnetic radiation caused by accelerating the charge. Within the shell, there is a disturbance in each line, essentially perpendicular to the direction of the line itself. In other words, it is a transverse disturbance, somewhat like the disturbance caused by cracking a whip. We already know that this disturbance in the electric field will create a corresponding pulse in the magnetic field.

Maxwell also found that if charges oscillated sinusoidally, that is, with the acceleration of simple harmonic motion, a succession of wave pulses would radiate away from the wire, again with the speed of light. Bolstered by the new predictions that electromagnetic pulses travel with the speed of light, Maxwell made another bold leap and theorized that light itself is an electromagnetic phenomenon. He then extended his theory to describe the propagation of light in crystals and other media. In essence, that is done by going back to the equations in Table 43.1, where Maxwell's equations include electric charges and currents, and taking into account the charges and currents that can exist in matter. The investigation of Maxwell's equations in various media is a lively field of research today, more than a century after Maxwell's time.

However, Maxwell's theory was not widely accepted by the more conservative physicists during his own lifetime, and he never lived to see the predictions of this theory verified experimentally. It was 10 years after Maxwell's death that electromagnetic waves were actually produced and detected by experimentalist Heinrich Hertz.

Questions

Match each of the following statements with the appropriate Maxwell equation or equations in Table 43.1.

18. Describes sources of electric fields as charges and/or changing magnetic fields.

19. Describes sources of magnetic fields as moving electric charges or changing electric fields.

20. Explains why in electrostatics all the excess charge on a conductor resides on its surface.

21. States that there are no magnetic monopoles.

22. Describes the basic physical principle underlying alternating current generators.

23. Predicts electromagnetic radiation.

43.8 A FINAL WORD

James Clerk Maxwell had a wide grasp of both the history and philosophy of science. Together with T. H. Huxley, he was joint scientific editor of the famous ninth edition of the *Encyclopaedia Brittanica*, and contributed many articles, including one on the atom that was referred to in Chapter 40. His skill as an editor is also revealed in a classic work entitled *Unpublished Electrical Writings of Hon. Henry Cavendish*.

When Maxwell became director of the Cavendish Laboratory he adopted Henry Cavendish as his personal scientific hero. Cavendish had worked mostly by himself and published very little during his lifetime. One celebrated experiment that he did publish was that of determining the universal gravitational constant G (described in Chapter 10 of this text). But Cavendish kept most of his work to himself, with results recorded in laboratory notebooks that he never opened to anyone else. A century after Cavendish's time, Maxwell studied these notebooks, repeated all the experiments as best he could, and reported on Cavendish's work. It turned out that Cavendish had been almost 100 years ahead of his time in electrical science. One revelation that particularly amused Maxwell was that Cavendish, like all eighteenth century electricians, used himself as a galvanometer. He would judge the strength of the electricity in an experiment by grasping the wires and feeling the shock. Maxwell wondered if some people were better galvanometers than others and tested this by asking students and others who came to his laboratory to hold a wire for a moment. Then Maxwell would record the intensity of the person's reaction to the ensuing electric shock. He even did this with the visiting Prime Minister of England. Maxwell loved a joke, and his happy, playful spirit was often expressed in comic verse.

43.8 A FINAL WORD

We conclude this chapter with selected portions of one of his verses entitled "A Problem in Dynamics," dated 19 February 1854.

> An inextensible heavy chain
> Lies on a smooth horizontal plane,
> An impulsive force is applied at A,
> Required the initial motion of K.
> Let ds be the *infinitesimal* link,
> Of which for the present we've only to think;
> Let T be the tension, and $T + dT$
> The same for the end that is nearest to B.
>
> ...
>
> In working the problem the first thing of course is
> To equate the impressed and effectual forces.
>
> ...
>
> Thus managing cause and effect to discriminate,
> The student must fruitlessly try to eliminate,
> And painfully learn that in order to do it, he
> Must find the Equation of Continuity.
>
> ...
>
> If then you reduce to the tangent and normal,
> You will find the equation more neat tho' less formal.
>
> ...
>
> From these two conditions we get three equations,
> Which serve to determine the proper relations
> Between the first impulse and each coefficient
> In the form for the tension, and this is sufficient
> To work the problem and then, if you choose,
> You may turn it and twist it, the Dons to amuse.

CHAPTER 44

OPTICS

> In making some experiments on the fringes of colours accompanying shadows, I have found so simple and so demonstrative a proof of the general law of the interference of two portions of light, which I have already endeavored to establish, that I think it right to lay before the Royal Society a short statement of the facts which appear to me so decisive. The proposition on which I mean to insist, at present, is simply this – that fringes of colours are produced by the interference of two portions of light; and I think it will not be denied by the most prejudiced, that the assertion is proved by the experiments I am about to relate, which may be repeated with great ease whenever the sun shines, and without any other apparatus than is at hand to every one.
>
> Thomas Young, Bakerian Lecture, 24 November 1803

44.1 THE ELECTROMAGNETIC SPECTRUM

Maxwell's spectacular discovery of the laws of electrodynamics was not verified experimentally until nearly a decade after his death. In 1887, the German experimentalist Heinrich Hertz used a remarkably simple method to determine whether electric charges undergoing acceleration radiate electromagnetic waves, as predicted by Maxwell. Two polished metal spheres with an air gap between them were connected to an LC circuit. Large electric charges of opposite sign were built up on

Table 44.1 Wavelengths and Frequencies of Visible Light.

Color	Wavelength (nm)
Red	610–720
Orange	590–610
Yellow	570–590
Green	500–570
Blue	460–500
Indigo	440–460
Violet	400–440

the spheres causing the air in the gap to be ionized and a spark to leap across the gap. Electromagnetic radiation from this circuit induced a current in a similar circuit remote from the first, as observed by sparks across the gap in the receiver circuit. This was a dramatic result, the first successful transmission and detection of nonoptical electromagnetic signals. In 1895, one year after Hertz's death, Marconi produced the first wireless telegraphy system, and by 1901 he transmitted the first wireless signal across the Atlantic.

In vacuum, every electromagnetic wave travels with the speed of light, c, which is equal to the product of its wavelength λ and its frequency f,

$$c = \lambda f. \tag{44.1}$$

Although the speed c is the same for all electromagnetic waves, the wavelength and frequency can vary considerably. They are, of course, inversely proportional to each other – the longer the wavelength, the smaller the frequency. Wavelengths can be as large as a planet (10^5 m) or as small as an atom (10^{-10} m). But visible light occupies only a tiny portion of this range, from a wavelength of 455 nm (violet light) to 780 nm (red light). Table 44.1 summarizes the approximate frequencies and wavelengths of visible light. The first letters of the colors in order of increasing frequency spell out ROY G. BIV – a useful mnemonic.

The collection of all electromagnetic wavelengths (or frequencies) is called the *electromagnetic spectrum*, and it is usually divided into overlapping regions with special names, as indicated in Fig. 44.1. Electromagnetic waves whose wavelengths are much longer than those of visible light are called *radio waves*. These are invisible, of the type first produced and detected by Hertz, and their wavelengths range from centimeters to kilometers, as indicated in Fig. 44.1. These waves carry the death cries of stars from across the galaxy and the love cries of youth from across town. Having such long wavelength, radio waves can be blocked by a metal mesh cage or even by the steel girders on a bridge. Visible light, on the other hand, having much shorter wavelength, easily penetrates a mesh cage. The net result is that, in a mesh cage, a radio can be seen but not heard.

Television signals, short waves, radar, AM, and FM radio signals are special kinds of radio waves. They are generated by electronic circuits that cause charges to oscillate and to radiate energy when accelerated. The frequency of the electronic circuits is in the range of kilohertz to megahertz.

44.1 THE ELECTROMAGNETIC SPECTRUM

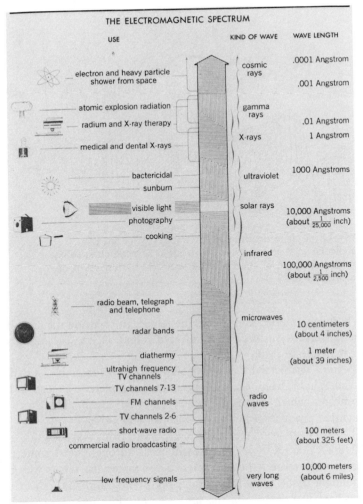

Figure 44.1 The electromagnetic spectrum. (Reprinted from Britannica Junior Encyclopaedia by permission of Encyclopaedia Britannica, Inc.)

Electromagnetic waves with wavelengths shorter than radio waves have found their way into the kitchen. These are microwaves, with wavelengths ranging from a millimeter to tens of centimeters. They are readily absorbed by water molecules in food, and can cause food to be heated rapidly.

Infrared light, IR, is found in the range between microwaves and visible light, and is detected only by the heat it deposits. Snakes and some other creatures are sensitive to infrared light. Glass is opaque to infrared light, which accounts for the so-called greenhouse effect. On a sunny day, visible light passing through the glass of a greenhouse is absorbed by plants and reradiated as infrared light. The infrared

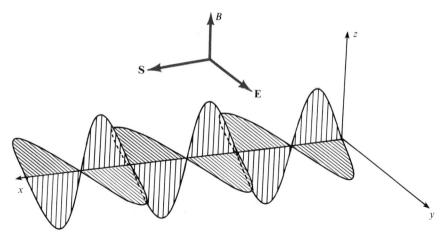

Figure 44.2 A plane electromagnetic wave in space has **E** perpendicular to **B** and travels in a direction perpendicular to both fields. The direction of the Poynting vector is the direction of propagation of the wave.

light becomes trapped and causes the temperature inside the greenhouse to rise. The entire universe is suffused with infrared radiation, the cooled remnants of the Big Bang that started it all.

Beyond visible light is ultraviolet light, UV. Invisible, yet dangerous to life, ultraviolet light from the sun is almost completely absorbed by ozone in the atmosphere, which serves to protect life on Earth. The small portion that is transmitted produces a tan or sunburn for sunbathers.

The next group of electromagnetic waves with wavelengths shorter than ultraviolet are called X rays. They easily pass through most materials. Dense, solid objects absorb more X rays than less dense materials. That's why bones but not the surrounding tissue show up in an X-ray photograph. As indicated in Figure 41.1, X rays have wavelengths of the size of an atom.

Waves in the electromagnetic spectrum with the smallest wavelengths, about the size of the nucleus of an atom, are known as gamma rays and cosmic rays. Gamma rays are produced in nuclear reactions and other extremely energetic processes. Cosmic rays arrive on Earth from all directions in outer space.

Radio waves can cause electrons to oscillate, microwaves can cook food, and gamma rays can cause nuclear reactions. In all cases, electromagnetic radiation transports energy and momentum. That's a characteristic of *all* electromagnetic waves. In Chapter 43 we found that electromagnetic plane waves consist of electric and magnetic fields oscillating in phase, with **E** perpendicular to **B**, and with both vectors perpendicular to the direction of propagation of the wave, as illustrated in Fig. 44.2. The flow of energy transported by an electromagnetic wave can be represented by a quantity known as the *Poynting vector*, named after John Henry Poynting. Defined by the equation

$$\mathbf{S} = \frac{1}{\mu_0} \mathbf{E} \times \mathbf{B}, \tag{44.2}$$

44.1 THE ELECTROMAGNETIC SPECTRUM

the Poynting vector represents the energy per unit time (the power) across a unit area, and in SI units it has units of watts/meter² (W/m²). The Poynting vector is aptly named because, as illustrated in Fig. 44.2, it points in the direction of propagation of the wave, which is, of course, the direction in which the energy is being transmitted. The Poynting vector arises naturally from conservation of energy, but we shall not discuss the derivation of this result.

Example 1

Suppose a plane electromagnetic wave traveling in the z direction is described by the fields

$$\mathbf{E} = E_0 \sin k(z - ct)\hat{\mathbf{i}} \quad \text{and} \quad \mathbf{B} = (E_0/c) \sin k(z - ct)\hat{\mathbf{j}}.$$

Determine the Poynting vector.

By definition, $\mathbf{S} = (1/\mu_0)\,\mathbf{E} \times \mathbf{B}$, so by using the given formulas for the fields we find

$$\mathbf{S} = \frac{E_0^2}{c\mu_0} \sin^2 k(z - ct)\,\hat{\mathbf{i}} \times \hat{\mathbf{j}}.$$

Because $\hat{\mathbf{i}} \times \hat{\mathbf{j}} = \hat{\mathbf{k}}$ and $c = (\mu_0 \varepsilon_0)^{-1/2}$, we obtain

$$\mathbf{S} = \sqrt{\frac{\varepsilon_0}{\mu_0}}\, E_0^2 \sin^2 k(z - ct)\,\hat{\mathbf{k}}.$$

As seen in Example 1, the amplitude of the Poynting vector for a plane wave oscillates according to $\sin^2 k(z - ct)$. For visible light these oscillations occur very rapidly, on the order of gigahertz, and consequently its instantaneous value would be impossible to measure. These rapid oscillations are similar to power oscillations in ac circuits that were discussed in Chapter 42, but now the oscillations can be even more rapid. Any detector, such as an eye or photographic film, will measure the time average of the energy per unit time over the area of the detector for a finite amount of time. Therefore, we proceed as we did in Chapter 42, by taking the time average of the function $\sin^2 k(z - ct)$ over one period. Denoting the time-averaged magnitude of the Poynting vector by $\langle S \rangle$ and writing $\omega = kc$, we find

$$\langle S \rangle = \frac{\omega}{2\pi} \sqrt{\frac{\varepsilon_0}{\mu_0}}\, E_0^2 \int_0^{2\pi/\omega} \sin^2(kz - \omega t)\, dt.$$

The integral has the value π/ω so

$$\langle S \rangle = \sqrt{\frac{\varepsilon_0}{\mu_0}}\, \frac{E_0^2}{2}. \tag{44.3}$$

The time-averaged magnitude of the Poynting vector is proportional to the square of the electric field amplitude and is a measure of the *intensity* of the radiation.

Example 2

A small detector records the intensity of a small light bulb located 4.0 m away to be 0.55 W/m². What is the power output of the lightbulb?

We treat the lightbulb as a point source, and assume the energy moves away from it in spherical waves, so that at every instant the energy is spread out over a sphere. If $\langle P \rangle$ is the time-averaged power output of the source, the energy per unit time per unit area is given by $\langle P \rangle/(4\pi r^2)$, where r is the radius of a sphere centered on the lightbulb. The detector records precisely this quantity as $\langle S \rangle$. Therefore, we have

$$\langle S \rangle = \frac{\langle P \rangle}{4\pi r^2},$$

which reflects the observation that the intensity of light decreases as $1/r^2$. Solving for the time-averaged power, we find

$$\langle P \rangle = 4\pi r^2 \langle S \rangle = 4\pi (4.0 \text{ m})^2 (0.55 \text{ W/m}^2) = 110 \text{ W}.$$

Questions

1. Would it be correct to say that the wavelength of any type of electromagnetic wave is on the order of the size of the system producing it?

2. Why can you get a sunburn on a cool, slightly overcast summer day?

3. The hands of patrons of amusement parks are sometimes marked with ink sensitive to UV light, and their hands are checked when reentering. How do you suppose this works?

4. Yellow light has a wavelength of 570 nm. Calculate the frequency and period of this light.

5. High energy gamma rays have been detected from a source near the center of the galaxy known as Cygnus X-3. If the wavelength of these gamma rays is on the order of 1×10^{-6} nm, what is their frequency?

6. In a particular radio wave the maximum value of the electric field intensity is 8.0 V/m. Determine the maximum value of the magnetic field intensity associated with the wave.

7. Refer to Example 2, and calculate the maximum value of the electric field of the light emitted from the lightbulb.

8. Sunlight strikes the surface of the earth with an intensity of 1.4 kW/m². Knowing that the earth is at a distance 1.5×10^{11} m from the sun, determine the time-averaged power output of the sun.

9. On a warm summer day the intensity of solar radiation is 90 W/m². How close would you have to be to a 1.5-kW electric heater to feel the same intensity of radiation?

44.2 THE NATURE OF LIGHT

Maxwell discovered that light is a wavelike disturbance in the electromagnetic field that travels at constant speed c in vacuum. Before Maxwell's insight shed light on the nature of light, philosophers debated its character, artists gloried in it, astronomers used it, and scientists were intrigued by it. Long before the idea of the electromagnetic field was born, people speculated about the nature and properties of light – the phenomenon that reveals all that we can see. The ancient Greeks thought light was fire, one of the four basic elements. Plato considered light to be streamers or filaments emitted by the eye. The Pythagoreans thought that light traveled from an object to one's eyes in the form of very fine particles, or corpuscles. The Greeks observed that objects cast sharp shadows and concluded that light travels in straight lines. As master geometers, the Greeks were also aware of the fact that the intensity of light decreases inversely as the square of the distance from the source. To them the particle description of light adequately fitted observations.

The particle theory remained essentially unchallenged until the seventeenth century. In 1678 Christian Huygens, a Dutch scientist, formally proposed what many of his predecessors had suspected – that light is a wavelike disturbance. Just as water waves travel from one point to another without water actually being moved from one point to the other, Huygens considered light to be waving in a medium he called the luminiferous aether. Huygens's ideas were not mere speculation. They commanded attention because he could describe many properties of light with the wave theory. The key to Huygens's theory of light was a geometric method of determining the shape of a wavefront at any time from knowledge of the wavefront at some earlier time.

A wavefront is the set of all points of equal phase. The phase of any wave is the argument in the functional form of the wave. For example, in Chapter 43 we learned that plane waves traveling with speed w can be described by an equation of the form

$$y(z, t) = f(z - wt). \tag{43.5}$$

The phase is $z - wt$, and a wavefront is the set of points in the zt plane satisfying $z - wt =$ constant, in this case a family of straight lines. A point source of light produces spherical wavefronts, as indicated in Fig. 44.3a. Sufficiently far from the

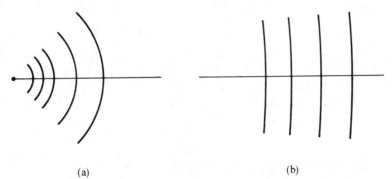

(a)   (b)

(a) A point source of light creates spherical wavefronts.
(b) Far from the source the wavefronts appear as planes.

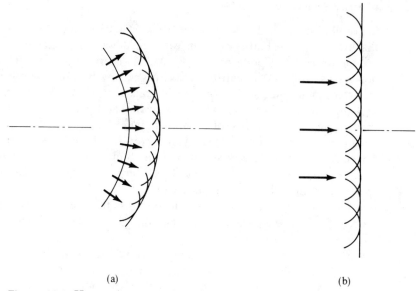

Figure 44.4 Huygens's construction of wavefronts for (a) a spherical wave and (b) a plane wave.

source, the radii of the spherical wavefronts are so large that the waves appear as plane waves, as shown in Fig. 44.3b.

The general problem of following the propagation of electromagnetic waves is extremely complicated. Some of the simpler aspects of the problem are embodied in a fundamental principle enunciated by Huygens:

Huygens's Principle: *Every point on a wavefront can be regarded as a new source of spherical waves that spread out in all directions with the same speed and frequency as the original wave.*

Figure 44.4a and b illustrates this principle for spherical and plane waves, respectively. Any point on a wavefront generates new spherical waves called wavelets. During the time interval equal to one period of the wave, each wavelet travels one wavelength. To construct the new wavefront, first draw wavelets as hemispheres of radius one wavelength from each point on the original wavefront. The surface tangent to all the wavelets represents the new wavefront that has moved one wavelength. Of course, in this construction we ignored wavelets moving backward toward the source, because they are not observed. It was not until 1826 that Fresnel showed that the disturbance did not propagate backward because of interference effects. In 1883 Kirchhoff proved by mathematical analysis that the secondary waves from the individual sources destroy one another by interference except at the wavefront itself, so the wave is propagated only in the direction away from its source.

44.3 REFLECTION AND REFRACTION

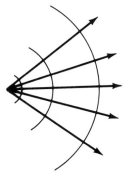

Figure 44.5 Lines perpendicular to the wavefronts are called rays.

The ancients believed that light could be represented by straight lines called *rays*. In Huygens's construction, the rays are along the direction of propagation of the wavelets. They are lines perpendicular to the wavefronts, as illustrated in Fig. 44.5.

Questions

10. Knowing that light travels in straight lines, how should the image of the candle in the accompanying diagram be oriented inside the box? The light passes through a pinhole. This principle was known to Aristotle, and a direct application of it is a box with a pinhole in it, known as a pinhole camera, the predecessor of the photographic camera.

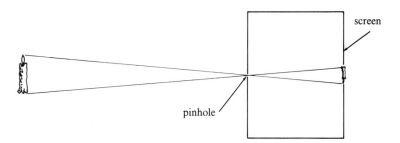

11. Describe some common examples of the behavior of light that indicates light travels in straight lines.

44.3 REFLECTION AND REFRACTION

The Huygens principle is not intuitively obvious, but it gained acceptance because it could be used to explain the laws of reflection and refraction, two phenomena concerning light that had been observed since antiquity. The ancient Greeks

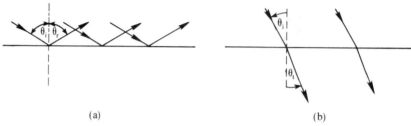

Figure 44.6 (a) Reflection of light. (b) Refraction of light.

observed that when a ray of light is reflected by a mirror, the angle of incidence is equal to the angle of reflection. This is illustrated in Fig. 44.6a, where θ_i denotes the angle of incidence and θ_r the angle of reflection, the angles between the light ray and a line perpendicular to the mirror. The law of reflection states that

$$\theta_i = \theta_r. \tag{44.4}$$

Figure 44.6b illustrates the concept of refraction. When a light ray passes from one transparent medium, such as air, into another of different density, such as water, the ray is bent or *refracted*. When refraction takes place, the angle of incidence θ_i and the angle of refraction θ_t are different. (The subscript t in θ_t suggests transmitted light.) Ptolemy had observed that for a given pair of media, such as air and water, the ratio of the angles θ_i/θ_t is nearly constant. We now know this is true only for small angles. Kepler and others made numerous measurements in an attempt to find a relation that would hold for all angles. That relation, discovered experimentally in 1621 by a Dutch mathematician Willebrord Snell, is called *Snell's law* and states that for a given pair of media the ratio

$$\frac{\sin \theta_i}{\sin \theta_t}$$

is constant.

Although the Huygens principle can be used to explain both these laws, they can be explained more simply by another principle stated by Fermat in 1657. This is called *Fermat's principle of least time* and it states that a ray of light traveling in any combination of media follows the path that minimizes its total travel time. The next two examples show how Fermat's principle implies both the law of reflection and Snell's law of refraction.

Example 3
Deduce the law of reflection from Fermat's principle of least time.

Consider the path of a light ray traveling from point A to a reflecting mirror and then to point B. The following figure shows two broken lines APB and AQB, where P is the point at which the angle of incidence θ_i equals the angle of reflection θ_r, and Q is a point different from P. We are to show that Fermat's principle implies that path APB minimizes the travel time from A to B.

44.3 REFLECTION AND REFRACTION

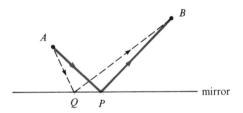

Because the speed of light is the same along both paths APB and AQB, minimizing the travel time is the same as minimizing the total distance traveled. The distance traveled along APB is $AP + PB$, while that traveled along AQB is $AQ + QB$. Now consider the mirror images of the line segments PB and QB, indicated by PB' and QB' in the next figure. The lengths PB and PB' are equal as are the lengths QB and QB'. Therefore, minimizing $AQ + QB$ is the same as minimizing $AQ + QB'$. But the shortest path from A to B' is along a straight line, and this occurs when $Q = P$, so the path APB minimizes the travel time.

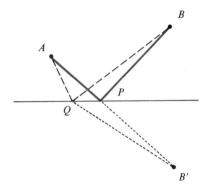

This simple geometric argument was discovered by Heron of Alexandria in the first century A.D.

Example 4
Deduce Snell's law of refraction from Fermat's principle of least time.

The following figure shows a ray of light traveling from point A to point P in one medium (say air) and then from point P to another point B in another medium (say water). The light travels at different speeds in these two media. As it passes from air into water the ray is bent toward the perpendicular at the interface, so the path APB is no longer the shortest path from A to B. But according to Fermat's principle, the total travel time along this path is minimal. We now determine what this path should be.

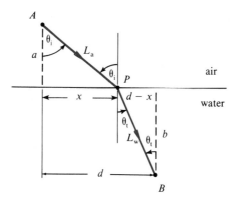

Let v_a denote the constant speed of light in air and v_w the constant speed in water. The total travel time T is the sum of T_a, the time traveling in air, plus T_w, the time traveling in water,

$$T = T_a + T_w.$$

If L_a denotes the length of the path in air and L_w the length of the path in water we have $T_a = L_a/v_a$ and $T_w = L_w/v_w$, so

$$T = \frac{L_a}{v_a} + \frac{L_w}{v_w}. \tag{44.5}$$

Now we express each of the distances L_a and L_w in terms of the distance x shown in the figure. Using the Pythagorean theorem twice we find

$$L_a^2 = x^2 + a^2 \quad \text{and} \quad L_w^2 = (d-x)^2 + b^2. \tag{44.6}$$

This makes T a function of x, and the minimum of T will occur at the point where $dT/dx = 0$. Differentiating Eq. (44.5) we find

$$\frac{dT}{dx} = \frac{1}{v_a}\frac{dL_a}{dx} + \frac{1}{v_w}\frac{dL_w}{dx},$$

so the minimum occurs when the right-hand member is zero, or when

$$v_w \frac{dL_a}{dx} = -v_a \frac{dL_w}{dx}. \tag{44.7}$$

Using Eq. (44.6) to calculate the derivatives of L_a and L_w we obtain

$$2L_a \frac{dL_a}{dx} = 2x \quad \text{and} \quad 2L_w \frac{dL_w}{dx} = -2(d-x).$$

Solving for the derivatives and substituting into Eq. (44.7) we obtain

$$v_w \frac{x}{L_a} = v_a \frac{d-x}{L_w}.$$

But inspection of the figure shows that

$$x/L_a = \sin \theta_i \quad \text{and} \quad (d-x)/L_w = \sin \theta_t,$$

44.3 REFLECTION AND REFRACTION

where θ_i and θ_t are the angles of incidence and refraction, respectively, so the minimum occurs when

$$v_w \sin \theta_i = v_a \sin \theta_t,$$

or when

$$\frac{\sin \theta_i}{\sin \theta_t} = \frac{v_a}{v_w}. \tag{44.8}$$

This implies Snell's law and it reveals that the constant ratio of the sines of the angles is also the ratio of the two speeds v_a/v_w.

A convenient number that characterizes the behavior of light in a medium is the *index of refraction* of the medium, denoted by the symbol n. By definition, the index of refraction is the ratio of the speed of light in vacuum to the speed of light in the medium,

$$n = c/v. \tag{44.9}$$

If light travels in a medium with index of refraction n_i and is refracted in a medium with index of refraction n_t, the same argument that gave us Eq. (44.8) now gives us the general form of Snell's law of refraction

$$\boxed{n_i \sin \theta_i = n_t \sin \theta_t.} \tag{44.10}$$

Table 44.2 lists the refractive indices of several substances. In general, n depends on the wavelength (and therefore the color) of the light. A vivid illustration

Table 44.2 Index of Refraction for Yellow Light (λ = 589 nm).

Substance	Index of refraction
Air at STP	1.00029
Water	1.33
Ice	1.31
Quartz	1.54
Crown glass	1.52
Flint glass	1.66
Diamond	2.42

of this property is the separation of white light into a rainbow of colors by a prism. This property of waves is known as dispersion. The table lists the index of refraction for sodium light. For air at standard temperature and pressure the index is nearly 1. It has very nearly the same value for all visible light.

Example 5

Suppose that light travels from air into a plate of quartz at an angle $\theta_i = 30°$. Determine (a) the speed of the light in the quartz and (b) the angle of refraction θ_t.

(a) According to Eq. (44.9), the speed of light in quartz is

$$v = c/n = (3 \times 10^8 \text{ m/s})/(1.54) = 1.95 \times 10^8 \text{ m/s},$$

which is about two-thirds its speed in vacuum or air.

(b) Using Snell's law to solve for $\sin \theta_t$, we have

$$\sin \theta_t = (n_{\text{air}}/n_{\text{quartz}}) \sin 30° = 0.77,$$

which implies $\theta_t = 19°$. The light bends toward the normal to the surface.

In general, when light passes from smaller to greater n, it bends toward the normal, and when it passes from greater to lesser n, it bends away from the normal.

Figure 44.7 shows a number of rays from a point source in one medium of refractive index n_1 impinging on the surface of another medium of refractive index n_2, where $n_1 > n_2$. Snell's law implies

$$\sin \theta_t = \frac{n_1}{n_2} \sin \theta_i,$$

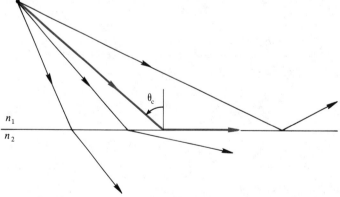

Figure 44.7 The critical angle θ_c is the angle of incidence for which the angle of refraction is 90°. For angles greater than θ_c there is total internal reflection.

44.3 REFLECTION AND REFRACTION

so $\sin \theta_t > \sin \theta_i$ because $n_1/n_2 > 1$. For some incident angle θ_i less than 90° we have $\sin \theta_t = 1$ and hence $\theta_t = 90°$, which means the transmitted light travels parallel to the interface. This value of θ_i is called the *critical angle* θ_c. When $\theta_i > \theta_c$, all the light is reflected, as shown in Fig. 44.7.

Questions

12. When a wave of wavelength λ traveling in vacuum enters a medium of refractive index n, explain why the following quantities change or do not change.

    (a) Wavelength.
    (b) Speed.
    (c) Frequency.

13. The speed of yellow sodium light in a liquid is measured to be 1.88×10^8 m/s. Calculate the index of refraction of this liquid.

14. A laser beam of wavelength 628 nm enters a glass of water at an angle of incidence of 45°. Find the angle of refraction.

15. Determine the critical angle for (a) an air–water surface and (b) an air–diamond surface.

16. A piece of crown glass is adjacent to a piece of flint glass with no air in between, as illustrated in the diagram. If a ray of light strikes the crown glass at an angle of 30°, at what angle does it exit the flint glass?

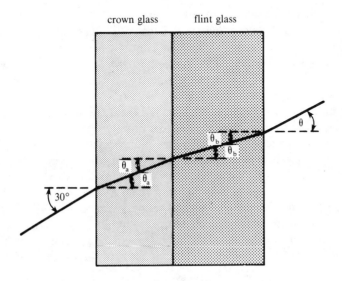

17. Consider two parallel-sided glass plates a and b in vacuum with refractive indices n_a and n_b, placed parallel to each other with a distance D between

them, as shown in the following figure. A ray of monochromatic light enters plate a at the lower left of the figure and emerges from plate b at the upper right. Prove that the entering ray is parallel to the emerging ray, regardless of the thickness of the plates or the distance D.

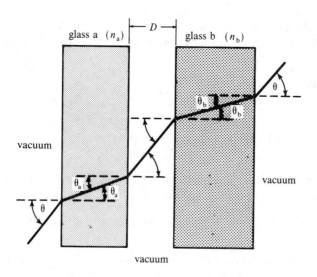

44.4 INTERFERENCE OF LIGHT WAVES

Despite Huygens's success in describing the propagation of light in terms of waves, many scientists advocated a corpuscular theory of light. Most notable among them was Isaac Newton who thought that light was best described as a stream of particles emitted by the light source. He objected to the wave theory because it did not explain why light travels in straight lines. In a lengthy dialogue with Huygens, Newton argued that if a hole is placed between a source of light and a screen, the patch of light on the screen has the same shape as the hole. One would expect this effect if light traveled in straight lines like tiny particles emitted by the source and absorbed by the screen. On the other hand, he argued, if light were truly a wave, it should bend around the hole and not produce a sharp shadow. Newton also asserted that because light travels through an evacuated glass container, light could not be a wave because waves require a medium in which to travel.

In response to Newton's objections, Huygens contended that if the wavelength of light is extremely small—much smaller than the dimensions of the object on which it shines, then bending of the waves would be too minute to observe and the object would cast sharp shadows. As far as light traveling in vacuum, Huygens replied that waves travel in the aether, a medium that permeates all of space and that is not removed when a container is evacuated.

As we shall learn in Chapter 50, the wave-particle controversy continued into the twentieth century when it was shown that light has the dual nature of both particles and waves. The first convincing experimental evidence for the wave nature

44.4 INTERFERENCE OF LIGHT WAVES

of light appeared early in the nineteenth century in the classic investigations on interference by Thomas Young and Augustin Fresnel.

A physician by training, Young was also a self-educated linguist and natural philosopher. He made original studies of the eye while still in medical school, and his doctoral thesis dealt with the physics of sound as it relates to the human voice. In 1802, he found an ingenious way to demonstrate that light exhibits the phenomenon of interference.

When two waves encounter one another, their effects add according to the principle of superposition. If two waves are in phase at some point, they reinforce each other there, with crests adding on crests, thereby creating a stronger wave. This is called *constructive interference*. When two waves are exactly out of phase at some point, with crest arriving on top of trough, they can cancel each other and produce no light intensity at that point. This is called *destructive interference*.

Young demonstrated interference of light with two slits separated by a small distance and illuminated by light from a single source, as shown in Fig. 44.8. He

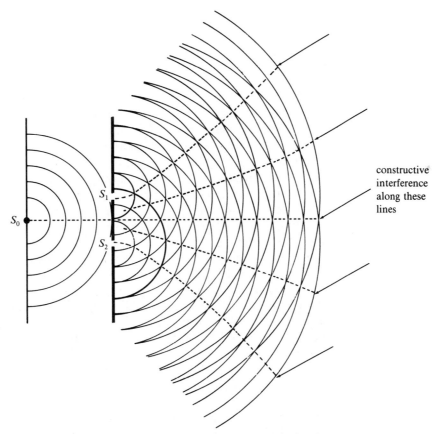

Figure 44.8 Young's double-slit interference experiment. Waves emerging from the slits S_1 and S_2 overlap and produce an interference pattern.

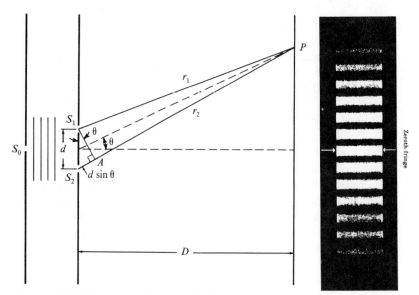

Figure 44.9 Diagram for Young's double-slit experiment.

passed sunlight through first the single slit, S_0, producing a narrow beam of light. He directed the light emerging from the single slit onto two narrow slits, S_1 and S_2, and observed the result on a distant screen. As illustrated in Fig. 44.9, the resulting interference pattern is a series of alternating bright and dark bands.

Young never worked out a detailed mathematical description of interference. Instead, he drew elaborate diagrams of wavefronts, similar to those in Fig. 44.8, to explain his observations. The task of rigorously establishing a theory was taken up later in the century by the French physicist Augustin Fresnel, who synthesized Huygens's wave description with the principle of interference. After his studies in interference, Young ventured to Egypt with Napoleon and deciphered the Rosetta Stone, which advanced the decoding of hieroglyphics. Let's try ourselves to decode the phenomenon of interference.

Suppose P is an arbitrary point on a distant screen located by an angle θ drawn from a line to the midpoint of the slits to P, as illustrated in Fig. 44.9. Let r_1 be the distance light travels from S_1 to P and r_2 the distance from S_2 to P. Now draw a line S_1A from S_1 perpendicular to the line from S_2 to P, as shown in Fig. 44.9. If the slit separation d is much smaller than the distance D to the screen, S_2P and S_1P will be nearly parallel. That means that light from S_2 travels an extra distance S_2A compared to light from S_1 before reaching P. The extra distance is called the path length difference, Δ, and is equal to $d \sin \theta$.

At S_1 and S_2 the light waves are in phase because they originated from the same wavefront emanating from S_0. However, because waves from S_2 travel a greater distance to P than waves from S_1, the waves are not necessarily in phase at P. If the waves are to interfere constructively, the path length difference must be equal to an integral number of wavelengths so that the waves will again be in phase

44.4 INTERFERENCE OF LIGHT WAVES

at P:

$$d \sin \theta = m\lambda, \quad m = 0, 1, 2, 3, \ldots \quad \text{(maxima)}. \quad (44.11)$$

This condition indicates that the bright bands, or maxima, of the interference pattern on the screen will be symmetrically located on either side of the central maximum, which is described by $m = 0$.

The location of the dark bands, or interference minima, can be found by the requirement that the waves interfere purely destructively. In order for the waves to arrive at P out of phase, the path length difference must be half a wavelength or an odd multiple of half a wavelength:

$$d \sin \theta = \left(m + \tfrac{1}{2}\right)\lambda, \quad m = 0, 1, 2, 3, \ldots \quad \text{(minima)}. \quad (44.12)$$

Example 6

A double-slit arrangement is illuminated with light of wavelength 620 nm. The slits are 0.03 mm apart and the screen is a distance of 2.0 m from the slit. What is the distance on the screen from the central interference maximum to the first maximum ($m = 1$)?

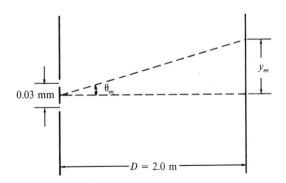

According to Eq. (44.11), the mth maximum is at an angle given by

$$\sin \theta_m = m\lambda/d.$$

The distance y_m of the mth maximum along the screen can be seen from the diagram to be given by

$$y_m = D \tan \theta_m.$$

Using the fact that $\sin \theta_m$ and $\tan \theta_m$ are nearly equal we find

$$y_m = m\lambda D/d.$$

Therefore, for $m = 1$ we have

$$y_1 = (620 \times 10^{-9} \text{ m})(2.0 \text{ m})/(0.03 \times 10^{-3} \text{ m}) = 0.04 \text{ m} = 4 \text{ cm}.$$

The first practical way of measuring the wavelengths of visible light involved measuring the distance between adjacent maxima for double-slit interference.

Will light behave this way if it consists of electromagnetic waves? We can show that electromagnetic waves can interfere to produce the observed interference pattern for a double-slit arrangement by applying the superposition principle. First, we'll make the simplifying assumption that the electric field vectors at a point P on a screen are in the same direction and equal in magnitude. Moreover, we assume that the lines from each slit to point P are parallel. Then we can write the electric fields as

$$E_1 = E_0 \sin \frac{2\pi(z - ct)}{\lambda},$$

$$E_2 = E_0 \sin \frac{2\pi(z - ct - \Delta)}{\lambda},$$

where we have taken into account the path length difference Δ between waves coming from the slits. According to the superposition principle, the resultant electric field at P is given by $E = E_1 + E_2$, so

$$E = E_0 \left\{ \sin \frac{2\pi(z - ct)}{\lambda} + \sin \frac{2\pi(z - ct - \Delta)}{\lambda} \right\}.$$

Using the trigonometric identity

$$\sin A + \sin B = 2 \cos \frac{A - B}{2} \sin \frac{A + B}{2}$$

and writing $\delta = 2\pi\Delta/\lambda$ we find

$$E = 2 E_0 \cos \frac{\delta}{2} \sin \frac{2\pi(z - ct - \Delta/2)}{\lambda}.$$

The constant $\delta = 2\pi\Delta/\lambda$ is known as the phase difference. The resulting time-averaged intensity of the light at P can be found by inserting $2 E_0 \cos(\delta/2)$ in place of E_0 in Eq. (44.3). The result is

$$\langle S \rangle = 2 \sqrt{\varepsilon_0/\mu_0}\, E_0^2 \cos^2(\delta/2). \tag{44.13}$$

If the sources are in phase, then the intensity is the sum of the individual intensities of the waves. For example, the intensity of two flashlights shining on the same spot is the sum of the individual intensities. In that case we have

$$\langle S \rangle = \sqrt{\frac{\varepsilon_0}{\mu_0}}\, E_0^2.$$

The interference given by Eq. (44.13) will result only if, as in Young's experiment, the two beams of light originate from the same source.

The interference of the waves is described by the term $\cos^2(\delta/2)$. A graph of the intensity as a function of δ is plotted in Fig. 44.10. The intensity is maximum

44.4 INTERFERENCE OF LIGHT WAVES

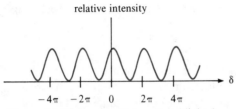

Figure 44.10 Graph of intensity of light from a double-slit arrangement as a function of the phase difference.

whenever $\delta/2 = m\pi$, where $m = 0, 1, 2, 3, \ldots$. Substituting for δ and Δ we find the condition

$$\pi(d \sin \theta)/\lambda = m\pi,$$

or

$$d \sin \theta = m\lambda.$$

That's precisely what we found in Eq. (44.11) by our earlier analysis. Electromagnetic waves exhibit interference in agreement with what is observed.

Interference of light can also be observed with oil films and soap bubbles. In fact, interference is responsible for the multiple colors of soap films, oil slicks, and many insect wings. For example, consider a thin film of refractive index n_2, as illustrated in Fig. 44.11 and assume a light ray arrives on the film nearly perpendicular to the surface. The incident light ray is partially reflected (R_1) and transmitted (T_1) into the film. At the back surface of the film, the light ray is again reflected (R_2) and transmitted (T_2). The reflected ray travels back through the film and is transmitted into the air (T_3). The rays represented by R_1 and T_3 interfere as they travel to an innocent viewer's eye.

The path length difference for the two rays is approximately $2d$ because the incident light is nearly perpendicular to the surface. Before we find the condition for constructive interference, we need to consider two other effects. First, as light travels into any medium, its speed changes according to Eq. (44.9), $v = c/n$. But its

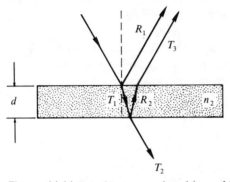

Figure 44.11 Interference produced by a thin film.

frequency f remains the same and $v = \lambda f$, so the wavelength also changes according to

$$\lambda = \lambda_0/n,$$

where λ_0 is the wavelength of light in vacuum.

Second, when light is reflected from a medium with a higher index of refraction than it is traveling in, it undergoes a phase change of 180°. This means that at the interface the reflected light is completely out of phase with the incident light. On the other hand, if the second medium has an index of refraction that is less than the one in which the light is traveling, there is no phase change upon reflection.

Now let's return to the thin film. Assume air is on both sides of the film. Then there is a phase change due to reflection of the ray R_1 but not the ray T_3. Thus even if d were extremely small, the two waves would be out of phase. To compensate for the difference in phase due to reflection, we require that the path length difference be equal to half an odd multiple of wavelengths in the film where the extra half resides. In other words, the condition for constructive interference is

$$2d = (m + \tfrac{1}{2})(\lambda_0/n), \qquad m = 0, 1, 2, 3, \ldots .$$

This explains why a soap film appears black before bursting. As the film becomes thinner, the wavelengths that satisfy the condition for constructive interference become longer and pass outside of the visible spectrum.

On the other hand, if you were observing an oil film on asphalt, for which $n_{oil} < n_{asphalt}$, there would be a phase change for both rays R_1 and T_3. Consequently, the condition for constructive interference would then be

$$2d = m\lambda_0/n_{oil}, \qquad m = 1, 2, 3, \ldots .$$

But films are not usually uniform in thickness, so if white light is incident on the film, different wavelengths interfere constructively at different points depending on the thickness of the film. The result is that the film appears to have different colors at various points, depending on the thickness of the film.

Example 7

An oil film 0.60-μm thick and of refractive index 1.6 floats on water and is illuminated with incident white light perpendicular to its surface. Find which colors in the visible spectrum will be strongly reflected.

First we note that the index of oil is greater than that of water ($n = 1.33$). Consequently, light reflected from the air–oil interface undergoes a phase change of 180°, but light reflected from the oil–water interface does not. Therefore, the condition for constructive interference is

$$2d = (m + \tfrac{1}{2})(\lambda_0/n_{oil}).$$

Solving for λ_0 we find

$$\lambda_0 = \frac{2dn_{oil}}{m + 1/2} = \frac{1920 \text{ nm}}{m + 1/2}.$$

44.4 INTERFERENCE OF LIGHT WAVES

Checking this last result for $m = 0, 1, 2, 3, 4$, we see that for $m = 3$ we obtain $\lambda_0 = 549$ nm, which is the only wavelength in the visible spectrum. Therefore, the film appears greenish.

Interference can also shed light on reflection and refraction at a deeper level. Recall that inside a metal, static electric fields are zero. For the same reason, electrons in a metal respond to oscillating electric fields in such a way as to exclude them from inside the metal. Consequently, any electric field at the surface of a metal must be perpendicular to the surface.

When light is incident on the surface of a metal, the mobile electrons at the surface respond to the force of the oscillating electric field, $\mathbf{F} = q\mathbf{E}$, moving in the opposite direction of $\mathbf{E}$ (due to their negative charge) and thereby preventing the light from entering the interior of the metal. The oscillating electrons reradiate outward as a wave that is completely out of phase with the incident wave. The incident wave and the reradiated wave interfere to cancel exactly the forbidden electric field component along the surface of the metal. This is the phenomenon of reflection.

In other materials, such as glass or water, the electrons are not free enough to reflect the light entirely. The electrons are held to their atoms by a restoring force, like that of a spring, and suffer damping forces due to collisions with other electrons. Consequently, the electrons behave like damped harmonic oscillators. When light is incident on the material, the electrons are forced to oscillate and they respond like forced harmonic oscillators. As we saw in Chapter 18, the oscillations of a damped harmonic oscillator are not completely in phase with the driving force. Likewise, electrons in the material radiate waves that are out of phase with the incident light. Moreover, because the distance between atoms is much smaller than the wavelength of light, the waves from various atoms are slightly out of phase with each other as well. Both the incident wave and the secondary waves travel through the void between atoms at speed c. However, there is one unique direction in which the reradiated light reconstructs into a new wavefront. The net result is to slow down the light and bend it toward the direction perpendicular to the surface. But that's just the phenomenon of refraction.

The same basic phenomena occur not only for visible light but also for the entire electromagnetic spectrum. For example, radio waves reflect from the ionosphere and bend around solid objects. On the other hand, X rays, which have much higher frequencies than visible light, aren't reflected from metals because their frequency is so high that the mobile electrons cannot keep up with the electric field oscillations of the waves.

Because the wavelengths of X rays are comparable to the sizes of atoms or the distances between atoms in solids, when X rays enter solids the effect is much more spectacular – a phenomenon known as X-ray diffraction. Like visible light, X rays are scattered in all directions by each atom of a crystal, and the beam reconstructs not in one direction but in many directions. The precise directions in which these beams emerge depend on the structure and spacing of the atoms. Captured on a

Figure 44.12 Photograph of X-ray diffraction from a sample of silicon. (California Institute of Technology photograph.)

photographic emulsion, the X rays reveal the inner structure of crystals. Figure 44.12 is a photograph of X-ray scattering off a piece of silicon.

Questions

18. The light from two stars very near to each other is focused in a telescope. Will interference fringes be observed? Explain why or why not.

19. Two slits are positioned 0.3 mm apart and placed 1.5 m from a screen. What is the distance between maxima of the interference pattern on the screen if the slits are illuminated with light of wavelength 600 nm?

20. Light of wavelength 470 nm is used to illuminate a double-slit arrangement in which the slits are 0.20 mm apart. How far must a screen be placed so that the bright bands are 1.0 mm apart?

21. A double slit is first illuminated with light of wavelength 600 nm and it is observed that bright bands on a screen 80 cm away are separated by a distance of 4.0 mm. Next the light is replaced by a second source and the bands are observed to be separated by 4.5 mm.

    (a) Determine the separation distance between the slits.
    (b) Find the wavelength of the second source.

22. Suppose that in a double-slit apparatus a very thin, transparent film of refractive index n and thickness t is placed over one slit. Show that the path difference Δ between light emerging from the two slits and traveling to a point on the screen located by the angle θ is given by

$$\Delta = d \sin \theta + t(n - 1).$$

23. Refer to Question 22. Determine the thickness of a thin piece of mica that is placed over one slit of a double-slit apparatus if the refractive index is 1.45, the wavelength of light used is 620 nm, and on a screen 0.6 m away bright bands are 2.0 mm apart.

24. A double-slit apparatus produces maxima on a distant screen that are 3.0 mm apart. If the entire apparatus, slits, source, and screen, is immersed in a gas of refractive index $n = 1.1$, what then will be the distance between maxima?

25. Light of wavelength 600 nm is used in a double-slit experiment in which the slits are 0.2 mm apart and the screen is 1.5 m away. Calculate the ratio of the intensity of light on the screen at an angle of 2.0° from the perpendicular bisector of the slits to that at 2.5°.

26. White light reflected at perpendicular incidence from a soap film has, in the visible spectrum, a single interference maximum for a wavelength of 480 nm. If $n = 1.33$ for the film, determine its thickness.

27. Find the index of refraction of a coating for a glass lens if destructive interference is to occur for light of wavelength 540 nm.

28. Suppose you are scuba diving in a region where an oil film of thickness 500 nm and refractive index 1.2 coats the water. For what wavelengths of visible light would you observe light to be most strongly transmitted?

29. Two rocks 0.5 km off shore obstruct incoming ocean waves. At the beach, interference maxima are observed 25 m apart for waves of wavelength 2.0 m. Calculate the distance between the rocks.

30. A layer of oil of refractive index 1.4 and 0.25 mm thick lies between two glass plates of refractive index 1.6. If the plates are illuminated with incident white light perpendicular to the plates, what wavelengths of visible light will be most strongly reflected?

44.5 A FINAL WORD

Once we know light is a wave, we can understand many of its properties because we know how waves behave. For example, in this chapter we found that the wave nature of light predicts interference. However, there is one aspect of the situation that is deeply troubling. If light is a wave, it is natural to ask, "What is waving"? In water waves, the water is waving, and in sound waves the air density is fluctuating. But what is waving when light from the sun or the stars comes to us through the void of space?

Not knowing the answer, nineteenth century scientists assumed it was the aether, a medium invented by the ancient Greek philosophers who believed it permeated all of space. The wave theorists simply ascribed to the aether the property that it is luminiferous (propagates light). Unfortunately the existence of aether is inconsistent with the principle of relativity, because its existence implies a state of absolute rest: being at rest with respect to the aether. In any other state of motion, one should be able to detect the aether wind – the relative motion of aether blowing by as you move through it.

Throughout the nineteenth century, ingenious attempts were made to detect the motion of the earth through the aether. For one reason or another, all attempts failed. Then, in the 1880s Albert Michelson designed an experiment of exquisite precision, using the interference of light waves as a measuring tool to detect the aether wind. In the next chapter we learn about that experiment and its profound repercussions.

CHAPTER 45

THE MICHELSON–MORLEY EXPERIMENT

"Is there any point to which you would wish to draw my attention?"
"To the curious incident of the dog in the night-time."
"The dog did nothing in the night-time."
"That was the curious incident," remarked Sherlock Holmes.

Arthur Conan Doyle in *The Memoirs of Sherlock Holmes* (1893)

45.1 THE ROOTS OF RELATIVITY

The problem of relativity began when Copernicus wrenched the earth from its fixed position in the cosmos and sent it hurtling around the sun. If the earth is actually moving around the sun, why does it seem to be standing still? Scientists such as Galileo, Kepler, Descartes, and Newton tackled this problem. Galileo argued that by watching a stone fall, you cannot tell whether the earth is at rest or moving. You might wonder whether there is any way to tell.

Figure 45.1 Velocity of a fly relative to the ground is equal to its velocity relative to the car, plus the velocity of the car relative to the ground.

In one sense, all motion is relative. Someone on the moon thinks he sees the earth moving, but a person on Earth thinks the moon is moving. To describe motion you need to specify a frame of reference relative to which the motion is measured. We habitually use nearby walls, buildings, trees, and other objects attached to the solid earth – and therefore at rest with respect to one another – as defining a frame of reference, and then we describe the motion of other objects relative to this frame. But on a boat, we tend to use our local surroundings in the boat – once again, a collection of objects at rest with respect to one another – as defining our frame of reference, even though the boat moves relative to the earth. Kinematically, motion can be described relative to any frame of reference. The choice of frame is merely a matter of convenience.

A frame of reference is simply a conceptual set of coordinate axes relative to which we measure displacements, velocities, and accelerations. When one turns from mere kinematic description of motion to dynamics – the laws of motion – one still has great freedom in choosing the frame of reference, but it becomes necessary to distinguish between different categories of frames. We shall define an *inertial frame* as one in which Galileo's law of inertia holds (a body not acted upon by any net force moves with constant speed along a straight line). We will show that if frame S is inertial, then any frame S' with respect to which S moves with constant velocity is also inertial.

To get an idea why this is so, consider a fly buzzing around inside a moving car, as indicated in Fig. 45.1. Common sense tells us that the velocity of the fly relative to the ground is equal to its velocity relative to the car, plus the velocity of the car relative to the ground.

In general, if frame S' is moving with velocity $\mathbf{v}_0$ relative to frame S, then the velocity $\mathbf{v}$ of a body relative to S is equal to its velocity $\mathbf{v}'$ relative to S', plus $\mathbf{v}_0$:

$$\mathbf{v} = \mathbf{v}' + \mathbf{v}_0.$$

45.2 THE GALILEAN TRANSFORMATION

If $\mathbf{v}$ and $\mathbf{v}_0$ are constant, then so is $\mathbf{v}'$, and therefore the body obeys the law of inertia in frame S' as well as in S, even though its precise location and velocity are different in the two frames. Thus, if S is an inertial frame, so too is S', and the two frames are equivalent for discussing the law of inertia. The equivalence still holds for objects and reference frames moving in different directions as long as the relative velocity of one to the other is constant, as we shall show in Section 45.2. Any inertial frame is suitable to describe the motion; none is preferred over another. In this sense there is no way to tell whether the earth or any other body is in a state of absolute rest. This is the essence of the *Galilean relativity principle*, which can be stated as follows:

> All laws of mechanics observed in one coordinate system are equally valid in any other coordinate system moving with a constant velocity relative to the first.

This statement contains Galileo's principle of inertia and much more. Galileo was earthbound, but Newton, who sought a grand synthesis of the physics of the heavens with that of the earth, extended the principle to the realm of the universe.

In deep space there are no stationary milestones or other markers to recognize rest. Yet Newton needed a frame of reference from which to apply his laws of mechanics. He conceived the ideas of absolute space and absolute time. According to Newton, absolute space was utterly featureless, unchanging, and immovable, a sort of a cosmic scaffolding. In reference to it, Newton could describe any body as being at rest or in relative motion.

Similarly, thought Newton, absolute time existed as an entity in itself, relentlessly flowing without relation to anything external – a cosmic clock marking time for the universe. With absolute space and absolute time, Newton could apply his laws to everything in the universe. However, by introducing absolute space and absolute time, Newton made a sharp distinction between rest and motion. That difference particularly disturbed him because he was well aware of the principle of relativity: Absolute uniform motion in a straight line cannot be detected.

Questions

1. Devise an experiment to tell whether or not you are in an inertial reference frame.

2. An observer on the ground watching a rock dropped from the mast of a moving ship sees the rock follow a parabolic path, whereas an observer on the mast sees the rock follow a straight-line path. Is this difference in the path of the falling rock as seen by two inertial observers a violation of the principle of relativity?

3. Explain why the earth is or is not an inertial reference frame.

45.2 THE GALILEAN TRANSFORMATION

If two inertial observers are in relative motion, measurements of time, position, velocity, and acceleration made by the two observers can be related by a set of

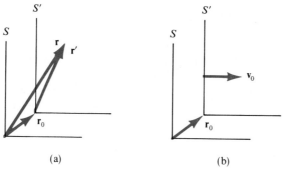

Figure 45.2 Two reference frames with (a) shifted origin and (b) constant relative velocity.

equations known as the *Galilean transformation*. To derive these equations we consider two reference frames called S and S', and assume first that they differ only by a simple displacement of the origin, indicated by the vector $\mathbf{r}_0$ in Fig. 45.2a. Let $\mathbf{r}$ denote the position vector of a body in S, and $\mathbf{r}'$ the position vector of the same body in S', so that

$$\mathbf{r} = \mathbf{r}' + \mathbf{r}_0. \tag{45.1}$$

We assume also that the two observers have the same clock, so if an event takes place in S at time t it also takes place in S' at time $t' = t$. If vector $\mathbf{r}_0$, the shift of origin, is time independent, then velocities and accelerations will be the same in both reference frames because, by differentiating Eq. (45.1), we find

$$\frac{d\mathbf{r}}{dt} = \frac{d\mathbf{r}'}{dt} \tag{45.2}$$

and

$$\frac{d^2\mathbf{r}}{dt^2} = \frac{d^2\mathbf{r}'}{dt^2}. \tag{45.3}$$

Therefore, the two frames are equivalent for discussing Newton's laws of motion, even though the precise location of an object will be different in the two frames.

Next, suppose that the origin is shifted as above and also that frame S' moves at a constant velocity $\mathbf{v}_0$ with respect to S, as indicated in Fig. 45.2b. Then the position of a body in S is related to its position in S' by the equation

$$\mathbf{r} = \mathbf{r}' + \mathbf{r}_0 + \mathbf{v}_0 t. \tag{45.4}$$

Differentiating this equation we find that the velocities $\mathbf{v} = d\mathbf{r}/dt$ and $\mathbf{v}' = d\mathbf{r}'/dt$ differ, but only by a constant amount,

$$\mathbf{v} = \mathbf{v}' + \mathbf{v}_0, \tag{45.5}$$

because $d\mathbf{r}_0/dt = \mathbf{0}$. This means that if Newtons' first law holds in one frame, it will also hold in the other. Differentiating once more, we find that the accelerations

45.2 THE GALILEAN TRANSFORMATION

$\mathbf{a} = d\mathbf{v}/dt$ and $\mathbf{a}' = d\mathbf{v}'/dt$ are the same:

$$\mathbf{a} = \mathbf{a}'. \tag{45.6}$$

Thus, once again the two frames are equivalent for discussing Newton's laws of motion, even though the initial location and velocity of an object will be different in the two frames. This is described by saying that Newton's laws of motion are *invariant* under transformation (45.4). (They are also invariant under (45.1), which is merely a special case of (45.4).) This invariance of the laws of motion is called *Galilean relativity*.

Note that Galilean relativity is built into the form of the fundamental forces of gravity and electricity. The inverse-square laws for the gravitational and electrical forces between objects A and B depend not on the absolute positions $\mathbf{r}_A$ and $\mathbf{r}_B$, which would change in a displaced or moving frame, but on the distance of separation $|\mathbf{r}_A - \mathbf{r}_B|$, which is unchanged by the transformation (45.4) because the terms $\mathbf{r}_0$ and $\mathbf{v}_0 t$ cancel in the subtraction.

It can also be shown that Newton's laws are invariant under a change in the orientation of frame S' relative to S; that is, if S' is rotated as well as translated relative to S. This, too, is part of Galilean relativity: The laws of motion in a tilted frame are the same as in a horizontal one.

If the components of the vectors $\mathbf{r}$, $\mathbf{r}'$, $\mathbf{r}_0$, and $\mathbf{v}_0$ are expressed in rectangular coordinates, the Galilean transformation (45.4) can also be expressed by a corresponding set of scalar equations:

$$x = x' + x_0 + v_{0x}t, \tag{45.4x}$$

$$y = y' + y_0 + v_{0y}t, \tag{45.4y}$$

$$z = z' + z_0 + v_{0z}t. \tag{45.4z}$$

Another equation,

$$t = t',$$

is often appended to this list to indicate that the same clock is being used in both frames. The corresponding scalar equations for velocity components are

$$v_x = v'_x + v_{0x}, \tag{45.5x}$$

$$v_y = v'_y + v_{0y}, \tag{45.5y}$$

$$v_z = v'_z + v_{0z}, \tag{45.5z}$$

while those for acceleration components are

$$a_x = a'_x, \tag{45.6x}$$

$$a_y = a'_y, \tag{45.6y}$$

$$a_z = a'_z. \tag{45.6z}$$

Example 1

A train moves along straight tracks at a constant speed of 80 km/h. Inside the train, a fly sits on a sleeping passenger's head. The train passes an observer at a railroad crossing, and the observer starts a stopwatch when the passenger is directly in front of the observer.

(a) According to the observer, what is the position of the fly 30 s later?
(b) If the fly speeds toward the front of the passenger car at 5 km/h, what is its speed relative to the observer?

(a) Let S denote the reference frame of the ground, with the x axis parallel to the tracks, and attach the S' frame to the sleeping passenger with the origin on the fly and the x' axis parallel to the x axis. At time $t = 0$, we have $x = x' = 0$, and from Eq. (45.4x) we know that at time t, we have $x = x' + v_{0x}t$, where x' is the position of the fly in frame S'. Substituting $x' = 0$, $v_{0x} = 80$ km/h, and $t = 30$ s, we find $x = 0.67$ km.

(b) Our intuition tells us that the speed of the fly relative to the ground is equal to the speed of the train relative to the ground plus the speed of the fly relative to the train, and this is confirmed by Eq. (45.5). From Eq. (45.5x) we find $v_x = v'_x + v_{0x}$, so that

$$v_x = 5 \text{ km/h} + 80 \text{ km/h} = 85 \text{ km/h}.$$

Example 2

A rectangular coordinate frame S is placed on the ground. An observer on the ground finds that the acceleration components of a moving truck are a_x, a_y, a_z. What are the acceleration components according to an observer in a car that is moving parallel to the positive x axis of frame S with a constant speed v_0 relative to the truck?

The reference frame S' of the truck moves with constant velocity $v_0 \hat{\mathbf{i}}$ relative to S, so we have an example of Galilean relativity. Therefore,

$$a'_x = a_x, \qquad a'_y = a_y, \qquad a'_z = a_z.$$

The truck has the same acceleration as measured in both frames S and S'.

The laws of conservation of energy and momentum discussed in Chapters 13 and 19 were derived from Newton's second law. According to the principle of relativity, these laws should be the same in all inertial reference frames. The next two examples verify this for a simple case of elastic collision of two bodies. We apply a conservation law in one frame S, then use the Galilean transformation to find what that law becomes in a frame S' moving with constant velocity $\mathbf{v}_0$ relative to S.

Example 3

Two masses m and M collide elastically, their initial velocities being $\mathbf{u}_0$ and $\mathbf{U}_0$, respectively, in frame S. After the collision, the masses move off with velocities $\mathbf{u}$

45.2 THE GALILEAN TRANSFORMATION

and $\mathbf{U}$ in frame S. Show that if the total momentum of the system in conserved in frame S, then it is also conserved in any frame S' moving with constant velocity $\mathbf{v}_0$ relative to S.

Conservation of momentum in S means that the total momentum of the system before collision equals that after the collision:

$$m\mathbf{u}_0 + M\mathbf{U}_0 = m\mathbf{u} + M\mathbf{U}. \tag{45.7}$$

In frame S' the initial velocities of the masses are $\mathbf{u}'_0$ and $\mathbf{U}'_0$, and the velocities after impact are $\mathbf{u}'$ and $\mathbf{U}'$. These are related to the velocities in S by the equations

$$\mathbf{u}_0 = \mathbf{u}'_0 + \mathbf{v}_0, \tag{45.8}$$
$$\mathbf{U}_0 = \mathbf{U}'_0 + \mathbf{v}_0, \tag{45.9}$$
$$\mathbf{u} = \mathbf{u}' + \mathbf{v}_0, \tag{45.10}$$
$$\mathbf{U} = \mathbf{U}' + \mathbf{v}_0. \tag{45.11}$$

When these quantities are substituted into Eq. (45.7), all the terms involving $\mathbf{v}_0$ cancel and we are left with

$$m\mathbf{u}'_0 + M\mathbf{U}'_0 = m\mathbf{u}' + M\mathbf{U}', \tag{45.12}$$

which shows the total momentum of the system is also conserved in frame S'.

Of course, the explicit value of total momentum is *not* the same in both frames. The actual value of the total momentum in frame S always differs from that in frame S' by the amount $(m + M)\mathbf{v}_0$. But in either frame, the total momentum of the system *before* collision is equal to that *after* collision.

Example 4

Refer to the collision in Example 3 and show that if the total kinetic energy of the system is conserved in frame S, then it is also conserved in frame S'.

This calculation is a bit more complicated because the kinetic energy $\frac{1}{2}mv^2$ involves the square of the speed and not merely the velocity. To simplify the calculation, we express the square of the speed as the dot product of the velocity vector with itself. Conservation of total kinetic energy in frame S states that the kinetic energy before collision equals that after collision. In the notation of Example 3 this gives us

$$\frac{m}{2}\mathbf{u}_0 \cdot \mathbf{u}_0 + \frac{M}{2}\mathbf{U}_0 \cdot \mathbf{U}_0 = \frac{m}{2}\mathbf{u} \cdot \mathbf{u} + \frac{M}{2}\mathbf{U} \cdot \mathbf{U}. \tag{45.13}$$

In this equation, replace all the velocities in frame S by the corresponding expressions in Eqs. (45.8) through (45.11). After the dot multiplication is carried out and the common terms on both sides are canceled, we obtain the equation

$$\frac{m}{2}\mathbf{u}'_0 \cdot \mathbf{u}'_0 + \frac{M}{2}\mathbf{U}'_0 \cdot \mathbf{U}'_0 + \mathbf{v}_0 \cdot (m\mathbf{u}'_0 + M\mathbf{U}'_0)$$
$$= \frac{m}{2}\mathbf{u}' \cdot \mathbf{u}' + \frac{M}{2}\mathbf{U}' \cdot \mathbf{U}' + \mathbf{v}_0 \cdot (m\mathbf{u}' + M\mathbf{U}').$$

Because of conservation of momentum, verified in Example 3, the terms in parentheses multiplying v_0 are equal, and therefore all terms involving v_0 cancel. The equation that remains states that the kinetic energy of the system is also conserved in frame S'.

Questions

In the following problems, assume that the reference frames are inertial frames with time measured so that $t = t'$ in the two frames.

4. A rectangular coordinate system S is attached to the ground with the origin at the base of a lamppost. Another rectangular coordinate system S' with its axes parallel to those of S is attached to a moving car with its origin on the rear bumper. The car moves with a constant speed of 15 m/s in the direction of the positive x axis. In frame S', a radio in the car has coordinates (2 m, 0.5 m, 0). At time $t = 0$, the origin of S' has coordinates (0, 0.3 m, 0) in S.

    (a) Find the coordinates of the radio in frame S when $t = 20$ s.
    (b) A jogger runs along the same street in the same direction as the car with a constant speed of 2 m/s in frame S. Determine the speed of the jogger in frame S'.
    (c) Inside the car, a fly moves from the front of the car directly toward the rear with a speed of 1 m/s in frame S'. Determine the velocity of the fly relative to the ground.

5. A train moves at 10 m/s relative to the earth. Inside the train a runner moves in the same direction as the train with an acceleration of 1.0 m/s² relative to the train. Determine the position of the runner with respect to the ground at time t, if at time $t = 0$ the origins of the two frames coincide and the runner is at the origin with zero speed.

6. A truck moving at 15 m/s along a straight level road carries an aquarium in which a fish swims vertically upward at 3 m/s relative to the truck. What is the speed of the fish relative to the ground?

7. Car A travels east at 35 m/s relative to a station. Car B travels north at 35 m/s relative to the same station. Find the velocity of car A relative to car B.

8. A swimmer can swim 1.5 m/s in an olympic pool. Determine how fast the swimmer can swim relative to the shore in a river that is moving at 1.0 m/s, if he swims

    (a) in the direction of the current;
    (b) in a direction opposite to that of the current;
    (c) perpendicular to the current.

9. In a certain reference frame two objects of masses M and $4M$ move toward each other at velocities v and $-v/4$, respectively, then collide and stick together.

    (a) Determine the velocity of the composite body after collision.
    (b) Calculate the total kinetic energy before and after the collision, and the increase in thermal energy resulting from the collision.

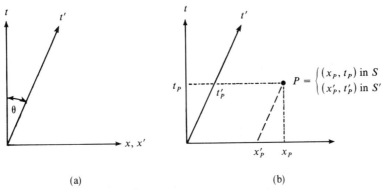

(a)　　　　　　　　　(b)

Figure 45.3 Space–time diagram for two inertial frames.

(c) Repeat the calculations of part (b) in a frame moving at velocity $\mathbf{v}_0$ with respect to the original frame.

45.3 SPACE–TIME DIAGRAMS FOR GALILEAN TRANSFORMATIONS

In Fig. 45.2 we represented two inertial reference frames S and S' by drawing two sets of coordinate axes. There is another way to represent Galilean transformations with a single diagram, called a space–time diagram. For simplicity, we consider only one coordinate axis of S, the x axis, and ignore the y and z axes. To study the behavior of the x coordinate of a moving body as a function of time t we would ordinarily plot x as a function of t, using a horizontal axis for t and a vertical axis for x. However, it has become common practice among most physicists to reverse the axes, using a horizontal x axis and a vertical t axis, a practice introduced by Hermann Minkowski. If a body has coordinate x at time t, we plot the point (x, t) on the space–time diagram. Some authors refer to the point (x, t) as an *event*; Minkowski called it a *world point*. The set of all points (x, t) associated with a given body determine a curve on the space–time diagram giving a visual description of how the x coordinate of the body changes with time. Minkowski called this curve a *world line*. For example, if at time $t = 0$ a body has $x = 0$, then that information is represented by the origin $(0, 0)$. If this body remains at rest in the S frame, its x coordinate will not change as t increases, and the corresponding world line for this body will be a vertical line – the t axis – as shown in Fig. 45.3a.

Now suppose the origin O' of another frame S' moves with constant speed v_0 relative to S and coincides with the origin of S at time $t = 0$. On this same space–time diagram, the world line of O' would be depicted by a straight line. If the axes were oriented in a conventional manner with the same scale along each axis, this line would have slope v_0. But because the axes are reversed, this line is tilted away from the vertical t axis by an angle θ whose tangent is v_0, and because this line represents the path of O' it can be regarded as the t' axis for frame S'. Now we draw the x' axis for frame S' on the same diagram, using the same horizontal axis for x' that we used for x in frame S. The coordinate axes in the S' frame will then be oblique rather than perpendicular, as shown in Fig. 45.3a.

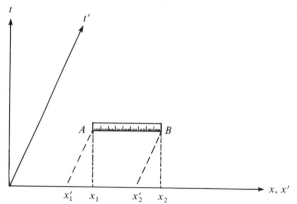

Figure 45.4 Length measurements compared on a space–time diagram.

Next, suppose an event P is described by two observers, one in frame S and the other in frame S'. This event can be described by two pairs of coordinates, (x_P, t_P) in frame S and (x'_P, t'_P) in frame S', as shown in Fig. 45.3b. The geometric relation between these two pairs of coordinates and the respective axes can be described as follows: In frame S, the coordinates (x_P, t_P) are determined geometrically in the usual manner by dropping a perpendicular from P to each of the x and t axes, as shown by the short dashes in Fig. 45.3b. This means that a choice of scale has been made along each of the x and t axes. Now we choose the scale along the x' and t' axes to conform with the Galilean transformation:

$$t' = t,$$
$$x' = x - v_0 t.$$

Geometrically, this means that the horizontal dashed line through P cuts the t' axis at a point representing the time $t'_P = t_P$ in the S' frame, and a line through P parallel to the t' axis (indicated by the longer dashes in Fig. 45.3b) crosses the horizontal axis at a point representing the x' coordinate $x'_P = x_P - v_0 t_P$ in frame S'. In this way, the pair (x'_P, t'_P) represents the coordinates of P in the S' frame, determined by the oblique axes.

Because the scales for measuring distances are different on each of the x and x' axes, the length of an object might be different in the two frames. We will now show that the length of a measuring stick is the same in both frames, provided that the object is observed simultaneously in both frames. Figure 45.4 shows a measuring stick with endpoints A and B. The stick is shown horizontal to indicate that the two endpoints A and B are observed simultaneously in both frames. In frame S, the length of the stick is $x_2 - x_1$. In frame S' its length is $x'_2 - x'_1$. Because the stick is observed at the same time in both frames, we have

$$x'_2 - x'_1 = (x_2 - v_0 t) - (x_1 - v_0 t) = x_2 - x_1,$$

which shows that the measuring stick has the same length in frame S' that it does in frame S.

45.3 SPACE–TIME DIAGRAMS FOR GALILEAN TRANSFORMATIONS

Space–time diagrams will be encountered again in the next chapter in the discussion of special relativity.

Questions

10. Five events A, B, C, D, E are shown on the following space–time diagram:

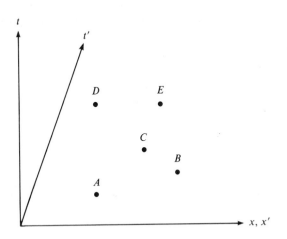

(a) Which events are simultaneous?
(b) Which events occur at the same position in the S frame?
(c) Which events occur at the same position in the S' frame?

11. In the space–time diagram shown, the points A, B, C, D, E, F lie on the world line of a moving particle. Line segment AB is vertical and line segment CD is parallel to the t' axis.

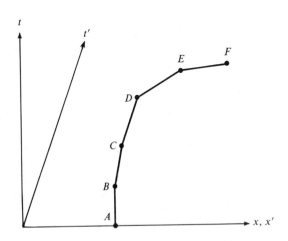

(a) According to an observer in frame S, when is the particle at rest?
(b) According to an observer in frame S', when is the particle at rest?

(c) According to an observer in frame S, when does the particle have its greatest speed?

45.4 RELATIVITY AND THE NATURE OF LIGHT

The Galilean principle of relativity served mechanics well, but fell short in the study of light. The nature of light had been debated for centuries. In the sixth century B.C., Pythagoras believed that light consisted of streams of particles. Newton returned to this idea 22 centuries later. He called the light particles "corpuscles" and used them to explain phenomena such as sharp shadows, straight-line propagation of light, and the ability of light to travel across the vacuum of space. An opposing model developed by Newton's rival, Robert Hooke, and by the Dutch physicist Christian Huygens, proposed that light was a wave. The two theories made different predictions concerning the speed of light.

If light consists of particles, then its speed should depend on the speed of the source, just as the speed of a bullet fired from a moving airplane depends on the speed of the plane. Galilean relativity tells us that the velocity of a bullet is equal to its muzzle velocity relative to the plane plus the velocity of the plane relative to the ground. Consequently, the speed of a bullet depends on the speed of the source, and by analogy, the speed of a light particle should depend on the speed of the light source.

On the other hand, if light is a wavelike phenomenon, its speed does not depend on the source but, instead, depends only on properties of the medium through which it propagates. For example, if you hear a whistle from a fast-moving train, the speed with which the signal reaches your ears has nothing to do with the speed of the train; the signal travels at the speed of sound in air. The pitch or frequency of the sound does depend on the speed of the train (for an approaching train the pitch is higher than for a receding train), but the speed of the sound does not.

The debate raged throughout the seventeenth and eighteenth centuries because no one knew how to measure the speed of light. To make matters worse, other phenomena concerning light could be explained by either of the two theories. As already noted in Chapter 44, Thomas Young performed some remarkable experiments that supported the wave theory. We recapitulate his double-slit experiment, illustrated in Fig. 45.5. Young passed a beam of light through two narrow slits toward a screen. Some of the light passing through one slit strikes the screen at some point P, and some light passing through the other slit strikes the same point P, as indicated by the lines AP and BP in Fig. 45.5a. If the two light waves arriving at P travel the same distance and are in phase, as illustrated by the two sinusoidal curves in Fig. 45.6a, their wave crests reinforce each other and a bright spot, called a fringe, appears on the screen. In this case the waves are said to interfere constructively.

A short distance from P light arrives from two waves that start in phase but don't travel the same distance, as illustrated by lines AQ and BQ in Fig. 45.5b. If distance BQ exceeds AQ by exactly half of a wavelength, then the light arriving at Q from A will be ahead by half of a wavelength relative to that arriving from B. In this case, a wave crest will arrive simultaneously with a wave trough and the waves will cancel each other, leaving a dark fringe on the screen. In this case the waves interfere destructively, as suggested by Fig. 45.6b.

45.4 RELATIVITY AND THE NATURE OF LIGHT

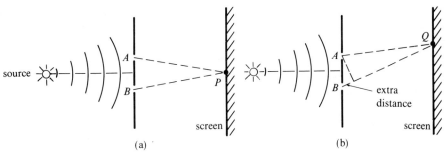

Figure 45.5 Young's double-slit experiment.

Light waves will interfere constructively at any point on the screen where the difference of the travel distances of the two waves is an integer multiple of one wavelength, and they will interfere destructively at any point where this difference is an odd multiple of half a wavelength. The two types of interference should produce alternate bright and dark fringes on the screen. This effect was actually observed in Young's experiments, and because the particle theory could not account for the observed interference patterns, this experiment was taken as conclusive evidence that light is, indeed, a wave.

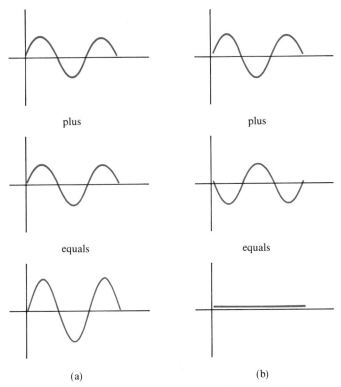

Figure 45.6 (a) Constructive interference. (b) Destructive interference.

Once the wave nature of light was accepted, the next question was to determine what was actually waving. Because sound waves or water waves can only propagate in a medium, it seemed reasonable to assume that light would also require some kind of transmitting medium. To nineteenth century physicists familiar with the mechanical workings of nature, that medium was the *luminiferous aether*. The notion of an aether can be traced back to ancient Greek philosophers, including Aristotle, who believed it to be the medium through which the planets and other heavenly bodies moved. But for nineteenth century physicists, the aether was a colorless, tasteless, odorless medium, rigid enough to propagate light with enormous speed, but tenuous enough to allow the planets to move freely through it. It was soon realized that the aether theory conflicted with the principle of relativity.

According to the principle of relativity, there is no such thing as absolute rest; there is no frame of reference that tells us when things are at rest and when they're not. But if all of space is filled with luminiferous aether that transmits light at a definite speed, then the aether itself could be regarded as a frame at rest, with everything else being in motion relative to it. In particular, one could detect absolute rest or absolute motion simply by measuring the speed of light. It would have one value if you were at rest with respect to the aether, and another value if you were in motion with respect to the aether.

Of course, the stationary aether influenced Maxwell's magnificent theory of electromagnetism, which predicted light to be an electromagnetic wave. Implicit in Maxwell's theory was the prediction that the aether could be detected by the earth's motion through it, but, as we'll see later, most of the effects depend on the square of the ratio of the speed of the earth through the aether to the speed of light, $(v/c)^2$. The ratio v/c was known to be on the order of 10^{-4} so $(v/c)^2$ is on the order of one part in 100 million, and no experimental techniques known in Maxwell's time could attain such sensitivity.

In a sense the success of such an experiment would have been the culmination of the Copernican system, placing the earth in its motion around a fixed sun. Critics might say that although the Copernican theory makes the motions of heavenly bodies mathematically simpler, it's only a clever mathematical device and doesn't correspond to reality because there is no such thing as absolute motion. If the earth could be shown moving through the aether, the Copernican system would have physical reality as well as mathematical convenience. Clearly, the burden of proof was on the experimentalists.

Question

12. Suppose that sodium light of wavelength 5.89×10^{-7} m is used in Young's double-slit experiment. At a particular point on a distant screen, the path length difference between waves arriving from the slits is 4.47×10^{-6} m. Will there be a bright or a dark fringe at that point?

45.5 THE MICHELSON–MORLEY EXPERIMENT

During the nineteenth century various attempts were made to detect the motion of the earth through the aether, and among them was the most significant failure in the

45.5 THE MICHELSON–MORLEY EXPERIMENT

history of physics: the Michelson–Morley experiment. Albert A. Michelson became interested in experiments probing the nature of light when he was a physics instructor at the U.S. Naval Academy. Michelson's forte was careful, precise optical measurements. In 1873 he devised an ingenious experiment that yielded the most accurate value at that time for the speed of light. He undertook such a measurement after reading a letter from the great James Clerk Maxwell to David Peck Todd, a colleague of Michelson's at the Nautical Almanac Office. Michelson was intrigued by Maxwell's statement that no terrestrial method for measuring the speed of light could detect the earth's motion through the aether because the effect would depend on the square of the ratio of the earth's velocity to that of light – an effect too small to be observed.

Maxwell presented a very simple argument. Imagine that the earth is moving at some speed v through the aether. If a beam of light travels at speed c in the rest frame of the aether, then an earthbound observer, moving through the aether at speed v, should measure $c + v$ for the speed of light when the light is traveling against the motion of the earth. Similarly, the observer should measure $c - v$ for the speed of a beam of light moving in the same direction as the earth. By comparing the speed of light in the forward and backward directions with respect to the speed of the earth, one should be able to measure the speed v of the earth through the aether.

The speed of the earth through the aether was taken to be its speed about the sun: 3×10^4 m/s, or 10^{-4} times the speed of light. The difference in the two speeds $c + v$ and $c - v$ is $2v$, which is extremely small compared to c. As we will

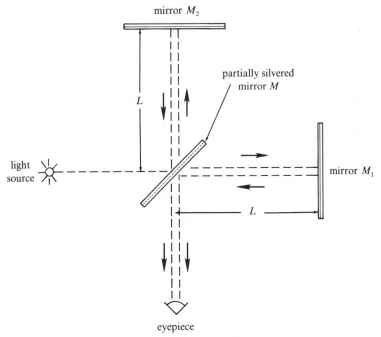

Figure 45.7 Schematic representation of a Michelson interferometer.

show later, the difference in time for light to travel a certain distance down and back depends on $(v/c)^2$, which is even smaller and very difficult to detect by experiment. The challenge of devising an optical instrument sufficiently sensitive to detect the earth's motion led Michelson to study the interference of light.

In 1880 Michelson was granted a leave to study optical techniques in Europe and immediately began working on an apparatus to detect the motion of the earth through the aether. In the laboratory of Hermann von Helmholtz at the University of Berlin, Michelson invented a new instrument of unprecedented sensitivity that has come to be known as the Michelson interferometer.

A diagram representing the principle of a Michelson interferometer is shown in Fig. 45.7. A beam of light is directed at a plate of partially silvered glass that partly transmits and partly reflects light. The silvered surface splits the beam into two components that travel in perpendicular directions, as shown. The transmitted beam travels a distance L to a mirror M_1 and is reflected back along the same path to the silvered glass plate. Similarly, the reflected portion of the beam travels an equal distance L, reflects from mirror M_2, and returns along its original path to the glass plate. At the glass plate both beams recombine as two superposed beams that are directed into a detector, a telescope.

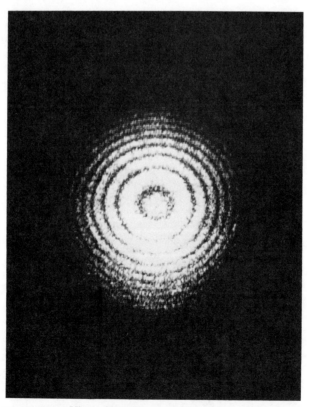

Figure 45.8 View of interference fringes from a Michelson interferometer. (Courtesy of Bob Paz, California Institute of Technology.)

45.5 THE MICHELSON–MORLEY EXPERIMENT

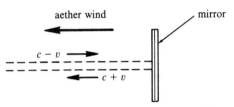

Figure 45.9 Light beam traveling parallel to the aether wind.

If both beams travel the same distance and arrive in phase with one another, they produce a bright fringe from the constructive interference. On the other hand, if the two waves travel different distances and arrive slightly out of phase, destructive interference will result. Consequently a series of bright and dark fringes is seen, similar to those shown in Fig. 45.8. If the length of either arm of the interferometer is changed slightly, the fringe pattern will also change slightly, by an amount that can be detected through the telescope.

Now imagine the whole apparatus moving relative to the aether at speed v in the direction of mirror M_1. From the viewpoint of the laboratory frame, an aether wind is blowing past the apparatus. It is easy to calculate the effect of this aether wind on the travel times for light along the arm of the interferometer parallel to and perpendicular to the wind. If the times are different, the beams will have traveled different distances and that difference will show up as either constructive or destructive interference fringes in the telescope.

When light travels against the wind, as shown in Fig. 45.9, its speed is $c - v$, and when it travels downwind, its speed is $c + v$. Now the time T required for one round trip is the sum of time out plus time back. In each case the time required is simply the distance L divided by the speed for that portion of the trip, so the total time is

$$T = \frac{L}{c - v} + \frac{L}{c + v}.$$

Adding the fractions we obtain

$$T = \frac{2Lc}{c^2 - v^2} = \frac{2L/c}{1 - v^2/c^2}. \tag{45.14}$$

The numerator of this last expression is the time it would have taken if everything were at rest and there were no relative motion to worry about – the total distance $2L$ divided by c, the speed of light. The denominator represents the correction factor due to the aether wind.

To calculate the time it takes light to make the journey up and down perpendicular to the aether wind, it is most useful to describe the process as it would be seen by an observer at rest in the aether. Such an observer would see the interferometer flying by and would be interested in how long light takes to leave the beam splitter and return. As the light travels up and down toward the mirror, as shown in Fig. 45.10, the earth moves a distance vt through the aether, so the light doesn't return to the same place because the mirror is moving.

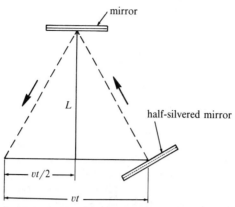
Figure 45.10 Light beam viewed from the aether frame.

The distance the light beam travels can be found from the Pythagorean theorem and is equal to $2\sqrt{(vt/2)^2 + L^2}$. The speed of light in the aether rest frame is simply c, so the time it takes the light beam to travel down and back is this distance divided by c:

$$t = \frac{2}{c}\sqrt{(vt/2)^2 + L^2}.$$

To determine t we square both sides,

$$t^2 = (4/c^2)\left[(vt/2)^2 + L^2\right] = v^2 t^2/c^2 + 4L^2/c^2,$$

then solve for t^2, and take the positive square root to get

$$t = \frac{2L/c}{\sqrt{1 - v^2/c^2}}. \tag{45.15}$$

Comparing Eq. (45.15) with (45.14) we see that the travel times differ by a factor $\sqrt{1 - v^2/c^2}$, and therefore the light beams travel different distances. The actual difference is equal to

$$T - t = \frac{2L/c}{1 - v^2/c^2} - \frac{2L/c}{\sqrt{1 - v^2/c^2}} = \frac{2L/c}{1 - v^2/c^2}\left(1 - \sqrt{1 - \frac{v^2}{c^2}}\right)$$

Multiplying and dividing the expression in parentheses by $1 + \sqrt{1 - v^2/c^2}$ we can rewrite this as

$$T - t = \frac{2L/c}{(1 - v^2/c^2)} \frac{v^2/c^2}{1 + \sqrt{1 - v^2/c^2}}.$$

Now v/c is a small number (on the order of 10^{-4}), and its square $(v/c)^2$ is even smaller (on the order of 10^{-8}), so we can disregard the terms v^2/c^2 in the

45.5 THE MICHELSON–MORLEY EXPERIMENT

denominator and obtain the approximation

$$T - t \approx \frac{2L}{c}\frac{v^2/c^2}{2} = L\frac{v^2}{c^3}. \tag{45.16}$$

This yields an incredibly small time difference $T - t$, and Michelson realized that he couldn't possibly measure the quantity $T - t$ to compare with the predicted value in (45.16). Instead, he realized that by rotating the entire apparatus by 90° he would reverse the roles of the two light paths MM_1M and MM_2M. The time difference between the two waves entering the telescope would also be reversed, thus changing the phase difference between the superposed waves and altering the positions of the interference fringes. He decided to simply rotate the interferometer by 90° and look for a shift of the interference fringes.

As remarked earlier, a bright fringe occurs if the waves arrive in phase and a dark fringe occurs if they arrive half a wavelength $\lambda/2$ out of phase. Therefore, the presence of m bright fringes in a path difference x would mean that m full wavelengths λ fit exactly into the distance x, so $x = m\lambda$ or

$$m = \frac{x}{\lambda}. \tag{45.17}$$

Consequently, a slight change Δx in x would involve a corresponding fringe shift of Δm, where

$$\Delta m = \frac{\Delta x}{\lambda}.$$

The small change in travel distance Δx occurring from rotating the apparatus by 90° would be the product of the corresponding travel time Δt multiplied by the speed of light. The change in travel time Δt is exactly twice the time difference $T - t$ calculated in (45.16); therefore, the corresponding fringe shift caused by rotating the apparatus is

$$\Delta m = \frac{2Lv^2}{\lambda c^2}. \tag{45.18}$$

In Michelson's original apparatus, $L = 120$ cm and $\lambda = 5.7 \times 10^{-5}$ cm, so that a fringe shift of $\Delta m = 0.04$ should have been observed. The sensitivity of his apparatus should have allowed Michelson to measure the fringe shift with an accuracy of about 50%, but he did not observe any shift at all. In 1881 Michelson published an account of his measurements in which he concluded, "The interpretation of these results is that there is no displacement of the interference bands. The result of the hypothesis of a stationary aether is thus shown to be incorrect."

Michelson, together with Edward W. Morley, improved the original apparatus and repeated the experiment at Case School of Science in Cleveland in 1887. By using several mirrors to make the light reflect back and forth through a greater distance, they increased the effective length of the interferometer arms by a factor of 10 over that of the original apparatus, so that a fringe shift $\Delta m = 0.40$ should have been observed. In addition, they mounted the optical parts on a heavy sandstone slab on a wooden float supported by mercury in a trough, as illustrated in

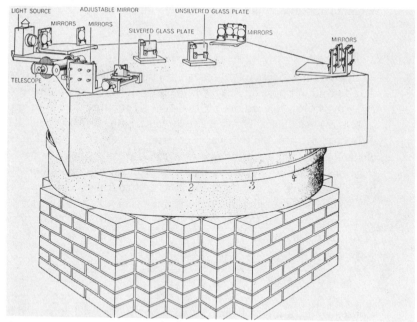

Figure 45.11 Michelson–Morley Experiment by R. S. Shankland. From the Michelson-Morley Experiment by R. S. Shankland. Copyright © 1964 by Scientific American, Inc. All rights reserved.

Fig. 45.11. The arrangement permitted continual observations of the interference fringes as the interferometer was rotated. With this magnificent instrument, Michelson and Morley should have seen a 0.40 fringe displacement with a precision of 0.01. Instead they observed *no displacement whatsoever*. The Michelson–Morley experiment gave a null effect – no effect of the aether was observed.

The experiment was repeated many times under different conditions. Two of the more interesting experiments were done by Miller and by Tomaschek, both in 1924. Miller used sunlight instead of a laboratory light to test whether the effect had anything to do with whether the source of the light was moving with the earth. He used an interferometer arm three times as long as that in the Michelson–Morley interferometer and should have observed a shift of 1.12 fringes with a precision of 0.014. Tomaschek performed the experiment with starlight, wondering if the null result was an effect of the solar system. In this more difficult experiment, Tomaschek used an interferometer with an arm length of 860 cm. He expected a fringe shift of 0.3 and showed it was not more than 0.02. The result of all the trials of the Michelson–Morley experiment, which are summarized in Table 45.1, is that within the precision of each experiment, the effect of the aether is nonexistent.

Questions

13. Michelson and Morley intended to repeat their experiment at intervals of 3 months. Explain why.

45.6 A FINAL WORD

Table 45.1 Repetitions of the Michelson–Morley Experiment

Observer (year)	L (cm)	Δm_{calc}	Precision
Michelson (1881)	120	0.04	0.02
Michelson, Morley (1887)	1100	0.40	0.01
Morley, Miller (1902–04)	3220	1.13	0.015
Miller (1921)	3220	1.12	0.08
Miller (1923–24)	3220	1.12	0.03
Miller (1924)	3220	1.12	0.014
Tomaschek (1924)	860	0.3	0.02
Miller (1925–26)	3200	1.12	0.08
Kennedy (1926)	200	0.07	0.002
Illingworth (1927)	200	0.07	0.0004
Piccard, Stahel (1927)	280	0.13	0.006
Michelson et al. (1929)	2590	0.9	0.01
Joos (1930)	2100	0.75	0.002

14. In the Michelson–Morley experiment of 1887, sodium light of wavelength 5.9×10^{-7} m was used. The experiment would have revealed any fringe shift larger than 0.01 fringes. What upper limit does this result place on the speed of the earth through the aether?

15. Show that if the arms of the Michelson interferometer are of different lengths L_1 and L_2, then the expected fringe shift is given by

$$\Delta m = \frac{(L_1 + L_2)}{\lambda} \frac{v^2}{c^2}.$$

16. In their experiment, Michelson and Morley ignored the effect of the earth's rotation. Discuss how this motion of the earth would affect the experiment.

45.6 A FINAL WORD

The stationary aether was every bit as important to nineteenth century scientists as the stationary Earth was to the Aristotelians before Copernicus. Discarding the aether theory was no small step. According to folklore, when Michelson and Morley

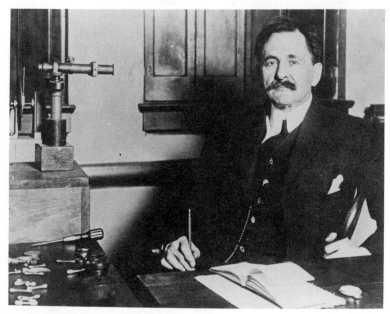

Figure 45.12a Portrait of Michelson. (Courtesy of the University of Chicago Archives.)

Figure 45.12b Portrait of Morley. (Courtesy of Case Western Reserve University.)

performed their experiment they triumphantly expelled the aether theory from the pantheon of physics. However, the actual facts are somewhat different.

Michelson never abandoned the aether theory. Instead, he considered his experiment to have been a failure. In 1907 he won the Nobel prize – the first American to receive this honor – not for showing that the aether theory was wrong, but for inventing the interferometer, which enabled him to make his delicate measurements. Equation (45.17) implies

$$\lambda = \frac{x}{m},$$

a formula that expresses the wavelength λ in terms of the travel distance x and the number of bright fringes m. By measuring m, the Michelson interferometer gives a supremely sensitive way to detect small changes in x by using light of known wavelength.

Michelson never recovered from the shock of the experimental result he didn't want, and he always felt that the only useful purpose served by the entire aether controversy was that it led to the invention of his beautiful device. But to science, his feelings are inconsequential. The importance of his experiment was that it showed that if the aether existed it had no detectable effects. And what was the consequence of this revelation? Many physicists proposed ingenious ideas to account for the experimental results. Some imagined that the earth drags along with it a thin layer of aether as it travels through space. They asserted that the Michelson–Morley experiment showed no effect because it was performed on Earth, where the aether was motionless. Others proposed that light travels at a fixed speed not with respect to the aether, but with respect to its source. Another idea, proposed independently by Lorentz and Fitzgerald, was that the length of the interferometer arm parallel to the aether wind contracted by just the right amount to compensate exactly for the slower times to travel up and downwind as compared to traveling crosswind.

Regardless of the details of these imaginative attempts to reconcile theory with experiment, all attempts were specifically designed to explain away the null result of Michelson and Morley. It is firmly entrenched in the folklore of physics that this experiment motivated Einstein to formulate his theory of relativity. However, when Einstein published his theory in 1905 he made no reference to the Michelson–Morley experiment. Afterward he reported that he had no knowledge of it at that time. We'll return to this point in the next chapter.

CHAPTER 46

THE LORENTZ TRANSFORMATION

> The hypothesis [that the dimensions of solid bodies are slightly altered by their motion through the aether] certainly looks rather startling at first sight, but we can scarcely escape from it, so long as we persist in regarding the aether as immovable. We may, I think, go so far as to say that, on this assumption, Michelson's experiment *proves* the changes of dimensions in question....
>
> H. A. Lorentz, *The Theory of Electrons,* page 196 (Lectures delivered at Columbia University in 1906, published in 1909 and revised in 1915)

46.1 INTERPRETING THE MICHELSON–MORLEY EXPERIMENT

The Michelson–Morley experiment demonstrated that light could travel at the same speed in all directions according to all inertial observers. This seemed to conflict with Maxwell's magnificent theory of electromagnetism, which assumed the existence of an aether to propagate light. Several physicists put forth theories that attempted to preserve the aether and at the same time explain the results of the Michelson–Morley experiment. One of these was George Fitzgerald of Ireland, who proposed in 1889 that all moving bodies were foreshortened in the direction of their motion through the aether.

Specifically, Fitzgerald conjectured that if a linear object like a rod has length L_0 when it is at rest, then its length contracts to $L_0\sqrt{1 - v^2/c^2}$ when it moves with speed v through the aether. This contraction is negligible for ordinary objects moving at speeds that are small compared to c, the speed of light. Even the earth moving rapidly around the sun would only be contracted by an amount equal to the length of a blade of grass. But Fitzgerald's contraction is exactly the amount needed to explain the result of the Michelson–Morley experiment. One arm of the Michelson interferometer would be contracted just enough so that the light beams in the experiment would travel the same time along both arms, and consequently no interference fringe shift would be observed. Despite the ingenuity of the idea, most of Fitzgerald's scientific friends scoffed at it.

In 1892 the Dutch physicist Hendrik Lorentz independently proposed the same idea. Lorentz, who was the world's greatest expert on Maxwell's theory of electromagnetism, had a reputation that commanded attention. He attributed the contraction to interactions of electrons with the aether in which they were imbedded.

Figure 46.1 Photograph of Lorentz and Einstein. (From AIP Neils Bohr Library.)

(Originally the name "electron" applied to both positive and negative charges inside atoms; later the positive charges were found to be different from electrons and became known as protons. Little was known about electrons in the 1890s.) Lorentz's ideas represented a radical departure from traditional dynamics. For example, he proposed that the mass of an electron would vary with its motion through the aether and that its dimensions would contract in the direction of motion because of a modification of molecular forces in that direction. He reasoned that if electrons have these properties and if electric forces bind matter, then macroscopic bodies would exhibit the same properties.

In particular, measuring sticks, used to measure distance, and clocks, used to measure time, would exhibit these properties in their motion through the aether: measuring sticks would contract in the direction of their motion and clocks would run more slowly. To determine precisely how to relate measurements of distance and time so that motion through the aether can't be detected, Lorentz derived a set of equations that relate the measurements of two observers. These equations, known as the *Lorentz transformation*, are the focus of this chapter. They became the basic core of the special theory of relativity later developed by Einstein, even though Lorentz's ideas about electron reactions with the aether did not survive. Lorentz firmly believed in the aether frame and an absolute time scale, but Einstein later argued that those concepts were incorrect, and there was no need to assume the existence of aether.

46.2 THE POSTULATES OF THE SPECIAL THEORY OF RELATIVITY

The French mathematician, Henri Poincaré, objected to the ad hoc explanation of Lorentz and others for the null result of the Michelson–Morley experiment. He believed there must be a central principle to explain the results of the experiment, and proposed the principle of relativity, which states that absolute motion is impossible to detect. In a widely heralded address made at the St. Louis exhibition in 1904, he announced "... the principle of relativity, according to which the laws of physical phenomena should be the same, whether for an observer fixed, or for an observer carried along in a uniform movement of translation; so that we have not and could not have any means of discerning whether or not we are carried along in such a motion."

Poincaré suggested that electromagnetic theory be revised to conform to the principle of relativity. Although he was one of the first to advocate that the relativity of mechanics should be generalized into a universal principle encompassing all physical laws, Poincaré himself did not construct such a theory. On the one hand, he thought that the aether might vanish from the theory of light much like the ideas of caloric and electric fluids vanished into the backwaters of the history of physics. On the other hand, he apparently felt that some kind of medium was necessary for the propagation of light. In any case, Poincaré did not develop the theory of relativity and its radical consequences. That revolution in thought had to await the genius of Albert Einstein.

In 1905 Einstein published three remarkable papers in the same volume of the highly respected physics journal *Annalen der Physik*. The first, entitled "Generation and Transformation of Light," proposed that light can be regarded as a stream of photons. For the results in this paper he was awarded the Nobel prize in 1921. The second, "Motion of Suspended Particles in the Kinetic Theory," was concerned with a phenomenon called Brownian motion and helped to establish the existence of atoms, and in the third, "On the Electrodynamics of Moving Bodies," he introduced a bold new treatment of the theoretical problems that had been plaguing Lorentz, Poincaré, and others. This paper is now regarded as the cornerstone of the theory of special relativity.

Einstein's goal was not to explain the null result of the Michelson–Morley experiment – by Einstein's own account he had no knowledge of that experiment when he wrote his paper. Instead, his goal was to establish relativity as a fundamental universal principle for all of physics. Although he capitalized boldly on its consequences, Einstein did not attempt to account for the principle itself in terms of other hypotheses. He elevated the principle of relativity to the status of a fundamental postulate not requiring further explanation. This is now known as the *first postulate of the special theory of relativity*, which, in Einstein's own words, is stated as follows:

> The same laws of electrodynamics and optics will be valid for all frames of reference for which the equations of mechanics hold good.

Einstein's *second postulate* concerns the speed of light:

> Light is always propagated in empty space with a definite velocity c, which is independent of the state of motion of the emitting body.

Taken together, these postulates imply that the speed of light (one of the "laws" of electrodynamics) must be independent of the state of motion of the observer. Although Einstein did not realize it at the time, the Michelson–Morley experiment provided experimental verification of those postulates. Einstein proceeded to show that these two postulates implied that the fundamental concepts of length and time must be regarded as *relative* quantities rather than *absolute* quantities.

To illustrate the relative nature of time and length we shall describe a thought experiment that is suggested by the Michelson–Morley experiment. Imagine a moving trolley car with an observer in the center of the car, as illustrated in Fig. 46.2a. The observer turns on a light source and measures the time t' it takes for the light to travel vertically upward to a mirror on the ceiling and reflect back down. If the mirror is at height h above the observer, the light travels a total distance $2h$ with constant speed c, hence $2h = ct'$ or

$$t' = \frac{2h}{c}.$$

Now we ask an observer on the ground to make the same observation. He sees things differently because the trolley car is moving past him. He sees the light move along the two legs of an isosceles triangle as shown in Fig. 46.2b, a path whose length is greater than $2h$. But the speed of light is the same for both observers, so,

46.2 THE POSTULATES OF THE SPECIAL THEORY OF RELATIVITY

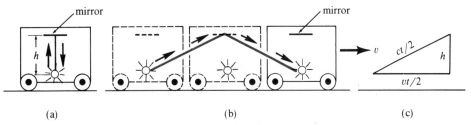

Figure 46.2 A thought experiment with a moving trolley car. (a) Path of light according to an observer in the car. (b) Path according to an observer on the ground. (c) Right triangle used to calculate time t.

relative to the ground observer, it takes a greater time to travel the greater distance. Thus we see that Einstein's postulates imply that time is a *relative* quantity, not an absolute quantity. The observer in the moving car would say that the clock of the observer on the ground runs more slowly than the clock on the train.

It's easy to calculate the exact relation between the two times. Suppose that the car is moving with a horizontal speed v relative to the ground. Then the base of the isosceles triangle has length vt, where t is the time measured by the observer on the ground. Applying the theorem of Pythagoras to the triangle in Fig. 46.2c we find

$$(ct/2)^2 = (vt/2)^2 + h^2.$$

If $v^2 < c^2$ we can solve for t to obtain

$$t = \frac{2h}{\sqrt{c^2 - v^2}} = \frac{2h}{c}\frac{1}{\sqrt{1 - v^2/c^2}} = \frac{t'}{\sqrt{1 - v^2/c^2}}.$$

The factor multiplying t' is usually denoted by γ,

$$\gamma = \frac{1}{\sqrt{1 - v^2/c^2}},$$

and is called the *time dilation* factor. The time taken by the light for the back and forth trip in the moving system appears to the stationary observer to be longer than to the moving observer by the factor γ. It is the same factor we saw in connection with the Michelson–Morley experiment in Chapter 45. Note that $\gamma \geq 1$, with $\gamma = 1$ only if $v = 0$.

The same thought experiment can be used to illustrate that length is also a relative quantity. Suppose the same observers are asked to measure the length of a horizontal rod. Specifically, suppose the rod is on the ground and joins the two points directly below the base of the isosceles triangle in Fig. 46.2b, which is shown again in Fig. 46.3. The observer on the ground measures this distance with a measuring stick and finds it equal to L_0, say. This distance L_0 is also equal to vt, as previously noted. But the observer in the car cannot measure the rod directly because he is in motion relative to the rod. Instead, he calculates the length of the rod by measuring the time it takes to move from one endpoint of the rod to the

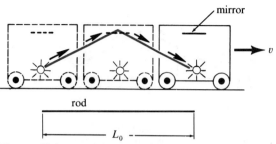

Figure 46.3 A rod measured by an observer on the ground has length L_0. An observer in the moving car determines its length to be L', where $L' < L_0$.

other, and multiplying this by the speed with which he is traveling relative to the rod. As previously seen, the time is t', and the speed is v, so he calculates the length of the rod to be $L' = vt'$. If $v \neq 0$, then $\gamma > 1$ and $t' < t$, so $vt' < vt$ or $L' < L_0$. In other words, the length of the rod is smaller when measured by a moving observer than when measured by a fixed observer. Thus, Einstein's postulates imply that length is a relative quantity, and, in fact, the contraction factor for length is exactly $1/\gamma$, the reciprocal of the time dilation factor. The dilation factor γ will appear again in the Lorentz transformation to be derived in the next section.

Questions

1. Suppose that spacecraft A is moving at $1/3$ the speed of light, $c/3$, relative to a space station. Spacecraft B travels in the opposite direction at $1/2$ the speed of light, $c/2$. If an observer in spacecraft A turns on a flashlight pointed toward the front of the craft, what does an observer in B measure for the speed of light?

2. If you were traveling near the speed of light and looked at yourself in a mirror you were carrying, would you appear "contracted"? Explain your reasoning.

46.3 THE LORENTZ TRANSFORMATION

In the year before Einstein's momentous paper was published, Lorentz presented, as part of his electron theory, a set of relativistic transformation equations that relate the measurements of two inertial observers. Although Lorentz's electron theory was wrong, his transformation equations are correct and they are the basic relations of the special theory of relativity. In this section we shall derive the Lorentz transformation equations from Einstein's two postulates for the special theory of relativity.

Recall that an event is something that happens at some point in space at some instant of time as measured by an observer. In one reference frame S, which we regard as stationary, we label each event with an ordered quadruple of numbers (x, y, z, t), where (x, y, z) represents the point in space and t the time. In another reference frame S', which we regard as moving relative to S, the same event will be

46.3 THE LORENTZ TRANSFORMATION

represented by another quadruple of numbers, say (x', y', z', t'). The Lorentz transformation is a set of equations relating the quantities x, y, z, t with x', y', z', t'.

We consider, as Einstein did in his original paper, a fixed frame S and another frame S' that coincides with S at time $t = t' = 0$ and that moves with constant velocity v parallel to the x axis of S. Because there is no relative motion along the y or z directions, it seems reasonable to assume that

$$y' = y \tag{46.1}$$

and

$$z' = z, \tag{46.2}$$

so we only need to find the relations between x, t and x', t'. We assume throughout that $v \neq 0$.

In his original paper, Einstein claimed that the equations relating x and t with x' and t' must be linear "on account of the properties of homogeneity which we attribute to space and time," but Einstein did not explain what he meant by "homogeneity." Quite possibly he meant that every straight line in the xt plane corresponds to a straight line in the $x't'$ plane, in which case it is known that this does indeed require linear relations. The approach we take here is simply to write down a pair of linear relations,

$$x = Ax' + Bt', \tag{46.3}$$
$$t = Cx' + Dt', \tag{46.4}$$

and show that the coefficients A, B, C, D can be determined in terms of the constant velocity v of the moving frame and the constant velocity c of light. We also want the equations to be solvable for x' and t' in terms of x and t. This requires that $AD - BC$ must be different from zero, or in other words, that $AD \neq BC$. We'll see later that this requirement reduces to $v^2 \neq c^2$.

We now proceed to determine the coefficients A, B, C, D. First we use the fact that the origin of S' is moving at constant velocity v relative to S. The origin in S' corresponds to taking $x' = 0$, so in frame S the origin of S' has coordinates $x = Bt'$ and $t = Dt'$, which means $Dx = Bt$. Differentiating this equation with respect to t and using the fact that the velocity of the moving origin of S' is $dx/dt = v$, we find that

$$Dv = B.$$

This equation determines B in terms of D.

Now the observer in the S' frame thinks he is at rest and that the origin of the S frame is moving relative to S' with constant velocity $-v$. (This is where we use the first postulate of special relativity.) The origin of S corresponds to $x = 0$. Taking $x = 0$ in Eq. (46.3) we find

$$Ax' + Bt' = 0.$$

Differentiating this equation with respect to t' and using

$$-v = dx'/dt',$$

we find $-Av + B = 0$ or

$$B = Av.$$

Comparing this with the previous equation for B we see that $D = A$. Therefore, we can put $B = Av$ and $D = A$ in the transformation equations (46.3) and (46.4) to get

$$x = A(x' + vt'),$$
$$t = Cx' + At'.$$
(46.5)

The last equation can also be written as

$$t = A(Ex' + t'),$$
(46.6)

where $E = C/A$. (Note that $A \neq 0$, otherwise Eq. (46.5) would imply that we would always have $x = 0$, hence $v = dx/dt = 0$, which we have excluded.)

We now have only two coefficients to determine, A and E. The numbers A and E depend only on the two constant velocities v and c and not on the particular type of event being observed. Therefore, to determine A and E we shall examine a very special event.

Consider a light source at rest at the origin O of frame S, and suppose there is a stationary observer in S. At any time t after a light pulse is emitted from O the observer sees a wavefront of light spreading outward from O as a sphere. Because light travels at a constant velocity c, the radius of the sphere is ct. Therefore, any point (x, y, z) on the surface of this light sphere must satisfy the equation

$$x^2 + y^2 + z^2 = (ct)^2,$$
(46.7)

because each side of this equation is the square of the radius of the sphere.

Now consider a second observer in the moving frame S', and assume that frame S' coincides with S at the instant the light pulse is emitted. The second observer sees the light spreading from O' as a different sphere, illustrated in Fig. 46.4. At time t' this sphere has radius ct', so any point (x', y', z') on its surface satisfies the equation

$$x'^2 + y'^2 + z'^2 = (ct')^2.$$
(46.8)

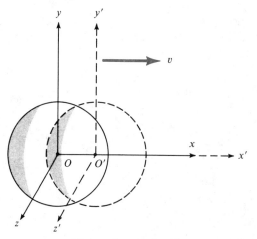

Light spheres spreading outward from sources O and O'.

46.3 THE LORENTZ TRANSFORMATION

Equation (46.7) can be written as

$$y^2 + z^2 = c^2t^2 - x^2,$$

whereas (46.8) can be written as

$$y'^2 + z'^2 = c^2t'^2 - x'^2.$$

But we also have $y = y'$ and $z = z'$, as suggested by Eqs. (45.5) and (45.6), so the last two equations imply

$$c^2t^2 - x^2 = c^2t'^2 - x'^2. \tag{46.9}$$

This means that the general form of the Lorentz transformation in (46.5) and (46.6) must be consistent with (46.9) in the special case of light spheres. It may be of interest to note that the Galilean transformation derived in the previous chapter is *not* consistent with (46.9). In this case, the Galilean transformation becomes

$$x = x' + vt,$$
$$t = t'.$$

Using $t = t'$ in (46.9) we find $x^2 = x'^2$, and this cannot be satisfied if $x = x' + vt$ because $v \neq 0$.

Returning once again to the transformations, we substitute (46.5) and (46.6) into Eq. (46.9) and obtain

$$c^2A^2(Ex' + t')^2 - A^2(x' + vt')^2 = c^2t'^2 - x'^2.$$

When $t' = 0$, this equation must hold for all x', in particular for $x' = 1$, which implies that the coefficients of the x'^2 terms must be equal on both sides, giving us

$$c^2A^2E^2 - A^2 = -1. \tag{46.10}$$

Similarly, the coefficients of the t'^2 terms must be equal,

$$c^2A^2 - A^2v^2 = c^2. \tag{46.11}$$

Equation (46.11) determines A^2 only if $c^2 \neq v^2$, and it gives us

$$A^2 = \frac{c^2}{c^2 - v^2} = \frac{1}{1 - (v/c)^2}. \tag{46.12}$$

When this value of A is inserted in Eq. (46.10) we find that $E = v/c^2$. If we want the positive x axis to correspond to the positive x' axis we take the positive square root and obtain

$$A = \frac{1}{\sqrt{1 - (v/c)^2}}. \tag{46.13}$$

Therefore, we have determined all the coefficients A, B, C, D in terms of v and c. Coefficient A is given by (46.13) and for the others we have

$$B = vA,$$
$$C = EA = (v/c^2)A,$$
$$D = A.$$

The condition $AD \neq BC$, which is needed in order to be able to solve for x' and t' in terms of x and t, now becomes $A^2 \neq (v/c)^2 A^2$ or

$$v^2 \neq c^2,$$

the same condition that allowed us to determine A. In other words, the coefficients in the Lorentz transformation can be determined only if the velocity v of the moving frame is different from the velocity of light. And Eq. (46.13) shows that we need $v^2 < c^2$ to obtain a real value for A.

It is common practice to use the symbol γ for the coefficient A in (46.13):

$$\gamma = \frac{1}{\sqrt{1 - (v/c)^2}}. \tag{46.14}$$

This is the same dilation factor $\gamma > 1$ that we encountered earlier in Section 46.2. The four linear equations relating x, y, z, t and x', y', z', t' now become

$$x = \gamma(x' + vt'), \tag{46.15x}$$
$$y = y', \tag{46.15y}$$
$$z = z', \tag{46.15z}$$
$$t = \gamma(t' + vx'/c^2). \tag{46.15t}$$

This is the *Lorentz transformation*; it relates the coordinates of an event in one inertial frame with those in another inertial frame that is moving with relative velocity v.

The inverse transformation, which gives the coordinates in S' in terms of those in S, is listed in Table 46.1. It can be obtained by solving the foregoing system for x' and t' in terms of x and t. Note that the inverse transformation can also be obtained by simply replacing v with $-v$ and interchanging primed and unprimed quantities.

Table 46.1 Lorentz Transformation for Relative Motion in the x Direction

$x = \gamma(x' + vt')$	$x' = \gamma(x - vt)$
$y = y'$	$y' = y$
$z = z'$	$z' = z$
$t = \gamma(t' + vx'/c^2)$	$t' = \gamma(t - vx/c^2)$

Example 1

In one inertial frame an explosion occurs at the coordinates $(x, y, z, t) = (6 \text{ m}, 0, 0, 2 \times 10^{-8} \text{ s})$. Determine the coordinates of this event according to an observer who is moving along the positive x axis with a relative speed of $0.8c$, assuming that at $t = t' = 0$ the origins coincide.

46.3 THE LORENTZ TRANSFORMATION

To find the coordinates in frame S' we use the Lorentz transformation. A relative speed of $0.8c$ indicates that

$$\gamma = (1 - 0.8^2)^{-1/2} = \tfrac{5}{3}.$$

Therefore,

$$x' = \gamma(x - vt) = (\tfrac{5}{3})(6 - 4.8) = 2 \text{ m},$$
$$t' = \gamma(t - vx/c^2) = (\tfrac{5}{3})(2 \times 10^{-8} - (0.8)6/(3 \times 10^8)),$$
$$t' = 0.67 \times 10^{-8} \text{ s}.$$

Of course, $y' = 0$ and $z' = 0$. Therefore, the coordinates of the event in the S' frame are $(2 \text{ m}, 0, 0, 0.67 \times 10^{-8} \text{ s})$.

In the Lorentz transformation equations the x and t coordinates of an event in one frame become commingled with the relative velocity to give the coordinates x' and t' of the *same* event in the moving frame. In other words, space and time are no longer completely independent of one another. An event that happens at a certain place and time according to one observer happens at different space–time coordinates according to a moving observer. The Lorentz transformation gives us a precise mathematical relation connecting the coordinates of an event measured by two inertial observers. This transformation is the relativistic generalization of the Galilean transformation to which it reduces when $v \to 0$ with c fixed, or when $c \to \infty$ with v fixed (see Question 9).

Example 2

In a certain inertial reference frame, the time between two events occurring in the same position is 10 s. According to an observer moving relative to that frame the events are separated in time by 12 s. What is the relative speed of the two frames?

From the equation for t' in Table 46.1, the time $t'_2 - t'_1$ between two events is given by

$$t'_2 - t'_1 = \gamma\left[(t_2 - t_1) - v(x_2 - x_1)/c^2\right].$$

In the case at hand we have $t'_2 - t'_1 = \Delta t' = 12$ s, $t_2 - t_1 = \Delta t = 10$ s, and $x_2 = x_1$, so $\Delta t' = \gamma \Delta t$, hence $1/\gamma = \Delta t/\Delta t' = 5/6$. Solving for v in the equation $\gamma = 1/\sqrt{1 - v^2/c^2}$, we get

$$v = c\sqrt{1 - (1/\gamma)^2} = 0.55c.$$

Questions

Questions 3 through 6 refer to the following situation. An observer S' is in a train whose rest length is 100 m. A flashbulb is placed at the center of the train.

Clocks located at the front and rear of the train are arranged to begin ticking when light from the flashbulb reaches them. The train is moving at a speed of $0.6c$ relative to an observer S on the ground, who is directly opposite the flashbulb when it flashes. At that instant this observer sets his clock to begin ticking.

3. What is the length of the train according to observer S?

4. Show that the clock of observer S reads $25/c$ s when the light flash reaches the rear of the train.

5. Determine the time on the clock of observer S when the light flash reaches the front of the train.

6. According to S, calculate the amount of time by which the clock at the rear of the train leads the one at the front.

7. If event A occurs before event B in some inertial frame, is there a moving frame in which event B occurs before event A?

8. Two events are simultaneous in one frame and are separated by a distance Δx. Are the events simultaneous in a frame moving with speed v? If not, find the time difference in terms of Δx, v, and c.

9. Show that the Galilean transformation is a limiting case of the Lorentz transformation when (a) $v \to 0$ with c fixed or (b) $c \to \infty$ with v fixed.

10. Show that the Lorentz transformation implies
$$x^2 + y^2 + z^2 - (ct)^2 = x'^2 + y'^2 + z'^2 - (ct')^2.$$
This is described by saying that the quadratic form $x^2 + y^2 + z^2 - (ct)^2$ is an invariant of the Lorentz transformation. It has the same value in any inertial frame.

46.4 LENGTH CONTRACTION

We return once again to the thought experiment described in Section 46.2 in which an observer on a moving trolley car determined the length of a rod on the ground to be smaller than that measured by an observer on the ground. We will show that the Lorentz transformation also describes length contraction.

The observer on the ground is in one frame S. He measures the position of the rear end of the rod at time t_1 and obtaines the value x_1. He measures the position of the forward end of the rod at time t_2 and obtains the value x_2. These measurements can be regarded as labels (x_1, t_1) and (x_2, t_2) for two events in S. The observer in the car is in another frame S' and he describes the same events with labels (x'_1, t'_1) and (x'_2, t'_2). In other words, measuring the length of the rod requires two events to be observed in each frame, and, of course, the labels for these events are related by the Lorentz transformation. By applying the Lorentz transformation for x' and t' twice and subtracting we obtain

$$x'_2 - x'_1 = \gamma(x_2 - x_1) - \gamma v(t_2 - t_1), \tag{46.16}$$

$$t'_2 - t'_1 = \gamma(t_2 - t_1) - \gamma(v/c^2)(x_2 - x_1). \tag{46.17}$$

46.4 LENGTH CONTRACTION

Now if the rod is at rest in the S frame, it doesn't matter when the observer on the ground makes his measurements because the rod is not moving. But in order to get an accurate, honest measurement of its length in the S' frame we must insure that the two ends are marked *simultaneously* in S'. Otherwise, relative to the observer in S', the rod will move between the marking of one end and the marking of the other, and the measured distance will not be correct. This means we require that $t'_2 = t'_1$ in S'. Using this in Eq. (46.17) we determine the time difference $t_2 - t_1$ in terms of $x_2 - x_1$:

$$t_2 - t_1 = \frac{v}{c^2}(x_2 - x_1).$$

But $x_2 - x_1 = L_0$, the rest length of the rod as measured by the observer on the ground, so this last equation gives us

$$t_2 - t_1 = \frac{vL_0}{c^2}.$$

Therefore, if the observer in the S frame makes his measurements at times differing by exactly the amount vL_0/c^2, these events will be simultaneous in frame S'. Substituting this value of $t_2 - t_1$ into Eq. (46.16) we find

$$x'_2 - x'_1 = \gamma\left(L_0 - \frac{v^2 L_0}{c^2}\right) = \gamma L_0\left(1 - \frac{v^2}{c^2}\right) = \frac{L_0}{\gamma}. \qquad (46.18)$$

This shows that the length of the rod as measured in S' is contracted by the factor $1/\gamma$, just as we found by a different argument in Section 46.2.

This contraction in the direction of motion was exactly what Fitzgerald had called for to explain the Michelson–Morley experiment. Of course it should not be surprising that Lorentz's result does the trick; it was specifically designed for that purpose. But Lorentz's purpose was not merely to explain the result of the Michelson–Morley experiment. Instead, he was trying to determine what it would take to make absolute motion undetectable. Although his electron theory was wrong, it had caused him to ask the right question, and he obtained the right answer.

We could, of course, reverse the process by placing a rod in the trolley car in the S' frame and ask an observer in the S frame to measure its length. In this case we use the formulas in the Lorentz transformation that express x and t in terms of x' and t'. Using these formulas twice and subtracting we find

$$x_2 - x_1 = \gamma(x'_2 - x'_1) + \gamma v(t'_2 - t'_1).$$

$$t_2 - t_1 = \gamma(t'_2 - t'_1) + (\gamma v/c^2)(x'_2 - x'_1).$$

Now the rod is at rest in the S' frame so it does not matter when an observer in S' measures its length. But for the observer in S to get an accurate length, he must make his two measurements at the same time, that is, when $t_1 = t_2$. Setting $t_2 - t_1 = 0$ in the second equation we find

$$t'_2 - t'_1 = -(v/c^2)(x'_2 - x'_1).$$

Therefore, if the observer in the S' frame makes his marks at times differing by exactly this amount, the measurements will be simultaneous in frame S. When this is used in the equation for $x_2 - x_1$ we get

$$x_2 - x_1 = \gamma(x_2' - x_1')(1 - v^2/c^2) = (x_2' - x_1')/\gamma.$$

Thus we see that the rod of length $x_2' - x_1'$ in S' appears contracted by the factor $1/\gamma$ in frame S, just as in the earlier discussion it appeared contracted by $1/\gamma$ in frame S'. In other words, if the rod is at rest in one frame, it appears to be contracted by the factor $1/\gamma$ when measured in a frame that is moving relative to the rod. The observations are completely symmetric. The measured length of a rod at rest in a given frame is often called the *proper length*. The shorter length measured in the moving frame is called the *contracted length*.

Example 3

A meter stick is oriented along the y axis as shown. Three observers measure the length of the stick: (A) one is traveling along the positive x axis at a speed $v = 0.8c$, (B) one is traveling along the negative y axis at $0.8c$, and (C) one is traveling at speed $0.8c$ at an angle of $45°$ from the x axis. What lengths are measured by each of these observers?

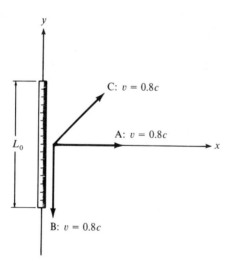

(A) Because the velocity of observer A is perpendicular to the moving meter stick, that observer measures the same length for the meter stick, namely $L_A = 1$ m.

(B) Observer B is traveling parallel to the stick, so according to Eq. (46.18) this observer measures a contracted length given by

$$L_B = L_0\sqrt{1 - v^2/c^2} = (1 \text{ m})\sqrt{1 - 0.8^2} = 0.6 \text{ m}.$$

46.4 LENGTH CONTRACTION

(C) Observer C has a component of velocity parallel to the stick, $v \cos 45°$, so this observer will measure a contracted length given by Eq. (46.18) with $v \cos 45°$ as the relative speed:

$$L_C = L_0\sqrt{1 - (v \cos 45°)^2} = (1 \text{ m})\sqrt{1 - (0.8 \cos 45°)^2} = 0.8 \text{ m}.$$

Questions

11. A rocket ship passes by you at $0.6c$. If you measure its length to be 100 m, what is its proper length?

12. Suppose you decided to travel to a star that is 4 light years away. How fast would you have to travel so that while you are moving the distance would appear to you to be only be 2 light years? (One light year is the distance light travels in 1 year.)

13. Suppose you are in a spaceship when another spacecraft passes you at $0.4c$. An observer in the spacecraft claims that his craft is 120-m long and that yours is 150 m.

    (a) Describe precisely what events are necessary for you to measure the length of the other spacecraft.
    (b) What do you obtain for these measurements?

14. A spaceship is traveling relative to a nearby star at $0.86c$. Observers inside the ship find that it takes 4×10^{-8} s to pass a marker that is at rest relative to the star.

    (a) According to observers in the spaceship, what is the length of their craft?
    (b) According to observers at rest relative to the star, what is the length of the spaceship?

15. A mathematician has a right triangle with angles θ, $90° - \theta$, and $90°$. Another mathematician flies by at relative speed v in a direction parallel to the shortest edge of the triangle.

    (a) Show that the triangle also is measured to be a right triangle by the moving mathematician.
    (b) Find the relation between the angle θ and the corresponding angle θ' measured by the moving mathematician.
    (c) Prove that the hypotenuse r of one triangle is related to the hypotenuse r' of the other by the equation
    $$r' = r\left(1 - (v \cos \theta)^2/c^2\right)^{1/2}.$$

16. Electrons leave an accelerator with a speed of $0.9c$ and travel through an evacuated tube that has a proper length of 2.0 m before reaching their target.

    (a) How long does it take an electron to travel the length of the tube according to an observer in the lab?

(b) How long does it take an electron to travel the length of the tube according to an observer traveling along with the electron?

(c) What is the length of the tube as measured by someone traveling with the electron?

17. Suppose that you are standing near the center of a long horizontal rod, which you perceive to fall with both ends hitting the ground simultaneously. What will a runner who is dashing by at a speed $3c/4$ say about how the rod hits the ground?

46.5 SPACE–TIME DIAGRAMS

The Lorentz transformation gives algebraic relations between the coordinates (x, y, z, t) of an event in one inertial frame S and the coordinates (x', y', z', t') of the same event in another inertial frame S'. Because we assumed that $y = y'$ and $z = z'$ we focus our attention on the two Lorentz equations relating (x, t) and (x', t'):

$$x = \gamma(x' + vt'), \qquad (46.15x)$$

$$t = \gamma(t' + vx'/c^2). \qquad (46.15t)$$

There are various ways to represent these equations geometrically. For example, we could use two separate sets of rectangular coordinate axes, one for frame S and another for frame S', plotting an event as a point (x, t) in the S-frame coordinate system and as a point (x', t') in the S'-frame coordinate system. A collection of events, say the history of a moving particle, would then be represented by a curve C in the xt system and by another curve C' in the $x't'$ system. Such curves are called *world lines*. In this type of representation, geometric relations between the two world lines C and C' due to the special nature of the Lorentz equations are not readily apparent. There are other types of diagrams that are specifically designed to reveal the geometric nature of the Lorentz transformation.

One of these was introduced by Hermann Minkowski in 1908 and is called a space–time diagram, or a Minkowski diagram. We already encountered one type of space–time diagram in Chapter 45 in connection with Galilean relativity. This section describes a similar type of space–time diagram that is suited to special relativity.

Before we explain how to construct Minkowski diagrams, we introduce some notation that gives the Lorentz equations a simpler form. First, we use the variables

$$r = ct \quad \text{and} \quad r' = ct'$$

instead of the variables t and t'. This makes sense from a physical point of view because each of ct and ct' represents a distance, being a product of speed and time, and comparing two distances such as r and x or r' and x' on a graph seems more natural than comparing distance with time. Therefore, we replace t by r/c and t' by r'/c in the Lorentz equations (46.15x) and (46.15t) to obtain

$$x = \gamma(x' + r'v/c),$$

$$r = \gamma(r' + x'v/c).$$

46.5 SPACE – TIME DIAGRAMS

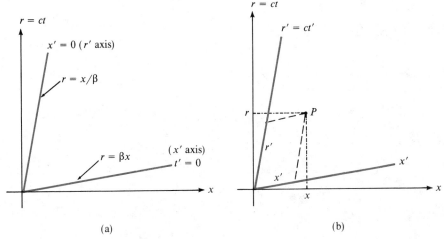

Figure 46.5 Space–time diagram.

The presence of v/c in each equation suggests a further replacement. The quotient v/c is the ratio of the velocity of the moving frame to the velocity of light. We replace this ratio by a new symbol β,

$$\beta = v/c,$$

and the Lorentz equations suddenly appear much simpler:

$$x = \gamma(x' + \beta r'), \quad (46.19x)$$
$$r = \gamma(r' + \beta x'). \quad (46.19r)$$

The constant γ is related to β as follows:

$$1/\gamma = \sqrt{1 - \beta^2},$$

so $\beta^2 = 1 - (1/\gamma)^2$, hence $|\beta| = \sqrt{1 - (1/\gamma)^2}$. The constant β can be positive or negative, but $|\beta| < 1$ because $\gamma > 1$.

The inverse transformation is given by

$$x' = \gamma(x - \beta r), \quad (46.20x')$$
$$r' = \gamma(r - \beta x). \quad (46.20r')$$

Now we draw a set of rectangular coordinate axes in the xr plane, as shown in Fig. 46.5a. Suppose that frame S' moves relative to S at a constant velocity v parallel to the x axis. The origin O' of frame S' corresponds to $x' = 0$, so by Eq. (46.20x') the path of O' according to an observer in S is a straight line of slope $1/\beta$ through the origin:

$$r = x/\beta.$$

If $\beta > 0$ this line has slope greater than 1. The example is shown in Fig. 46.5a is the world line of O' and gives a geometric representation of the r' axis from the point of view of S. Similarly, a geometric representation of the x' axis, from the point of

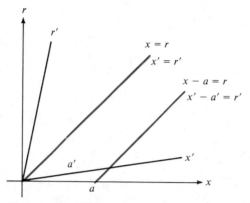

Figure 46.6 World lines of two light signals.

view of S, is obtained by setting $r' = 0$ in Eq. (46.20r'). This gives us a straight line with slope β:

$$r = \beta x.$$

Because the slopes β and $1/\beta$ are reciprocals, the angle between the x and x' axes is equal to that between the r and r' axes.

Minkowski's idea was to use perpendicular axes for the S frame and then to use the lines $r = \beta x$ and $r = x/\beta$ as oblique x' and r' axes for the S' frame on the same diagram, as shown in Fig. 46.5b. Then any given event can be represented by a single point P. From the diagram we can read off its coordinates (x, r) in the S frame and its coordinates (x', r') in the S' frame by dropping lines parallel to the corresponding axes as indicated in the Fig. 46.5b. Once we have numerical values for r and r' it is a simple matter to convert to the values $t = r/c$ and $t' = r'/c$.

A sequence of events (a world line) is represented by a curve on the space–time diagram. That curve serves as the world line for both the xr system and the $x'r'$ system. For example, the world line of a light signal that at time $t = 0$ has $x = 0$ would be represented by the line $r = x$, a line of slope 1. This is because light travels in vacuum with speed c according to any inertial observer, so in time t it travels a distance $x = ct = r$. The world line of light bisects the angle between the x and r axes, as shown in Fig. 46.6.

The very same line is the world line of light in the $x'r'$ system because in time t' it travels a distance $x' = ct' = r'$. If another light beam starts at time $t = 0$ with $x = a$, then it travels a distance $x - a = ct = r$ in time t, so its world line is parallel to the world line through the origin. The world line of any particle moving with speed less than c would be a curve whose tangent line at any point is inclined at an angle with the r axis less than 45°.

In constructing a space–time diagram we could just as well have taken the x' and r' axes perpendicular and used oblique axes for x and r. Putting $x = 0$ and $t = 0$ in Eqs. (46.19x) and (46.19t) we find that the line with equation $r' = -x'/\beta$ represents the r axis, while the line $r' = -\beta x'$ represents the x axis, as shown in Fig. 46.7.

46.5 SPACE–TIME DIAGRAMS

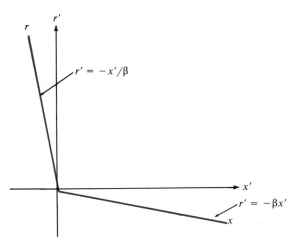

Figure 46.7 Space–time diagram with rectangular $x'r'$ axes.

Now that we understand how space–time diagrams are constructed we can see how they reveal some basic features of special relativity. Figure 46.8a shows two events E_1 and E_2 that occur at the same time $t_1 = t_2$ in frame S. Then $r_1 = ct_1 = ct_2 = r_2$, so in the xr diagram these events lie on a horizontal line $r = r_1$. The x' coordinates of these events are obtained by dropping lines parallel to the r' axis, as shown. The figure suggests that $x'_2 - x'_1 > x_2 - x_1$, and this is readily confirmed by using Eq. (46.20x') twice and subtracting to get

$$x'_2 - x'_1 = \gamma(x_2 - x_1).$$

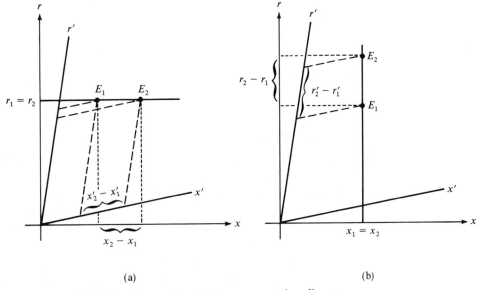

(a)   (b)

Figure 46.8 Relativity of events illustrated on a space–time diagram.

In other words, the difference $x_2' - x_1'$ is greater than $x_2 - x_1$ by the dilation factor γ. A unit distance along the x axis corresponds to a distance γ along the x' axis.

Figure 46.8b shows two events E_1 and E_2 that occur at the same place $x_1 = x_2$ in the S frame but at different times $t_1 < t_2$. These events are on the same vertical line in the xr plane with $r_1 = ct_1 < ct_2 = r_2$. Figure 46.8b suggests that $r_2' - r_1' > r_2 - r_1$, and this is easily confirmed by using Eq. (46.20r') twice and subtracting:

$$r_2' - r_1' = \gamma(r_2 - r_1).$$

This shows that the difference $r_2' - r_1'$ is greater than $r_2 - r_1$ by the dilation factor γ. A unit distance along the r axis corresponds to a distance γ along the r' axis. The magnification factor γ is the same along both the x' and r' axes.

Questions

18. Use a space–time diagram to illustrate that if event A occurs before event B in one inertial frame, A precedes B in any other inertial frame (provided light from A can reach B in the time interval between events).

19. Using a space–time diagram, show that events that are simultaneous in one frame are not simultaneous in another.

20. An inertial frame S' is moving with speed $0.6c$ relative to another inertial frame S.

    (a) Draw a space–time diagram with the xr axes for S perpendicular and with the $x'r'$ axes for S' oblique.
    (b) Plot each of the following events on the diagram of (a): (A) $x = 1$, $r = 1$; (B) $x' = 1$, $r' = 1$; (C) $x' = 3$, $r' = 0$; (D) $x = 0$, $r = 2$.
    (c) Determine the coordinates (x, t) in S and (x', t') in S' for each of the events in (b).
    (d) Draw another space–time diagram with the $x'r'$ axes for S' perpendicular and the xr axes oblique and plot each of the events in (b) on this new diagram.
    (e) Determine the coordinates (x, t) in S and (x', t') in S' for each of the events in (b).

46.6 A FINAL WORD

Poincaré enunciated the principle of relativity and Lorentz published his transformation equations before Einstein wrote his 1905 paper. In the second volume of Sir Edmund Whittaker's *The History of Theories of Aether and Electricity*, published in 1953, there is a chapter on relativity, pointedly entitled, "The Relativity Theory of Poincaré and Lorentz." Einstein is mentioned for the first time in a paragraph on the thirteenth page, which is famous for what it says and what it doesn't say:

> In the autumn of the same year (1905) Einstein published a paper which set forth the relativity theory of Poincaré and Lorentz with some amplification, and which attracted much attention.

46.6 A FINAL WORD

Whittaker continues,

> He asserted as a fundamental principle the constancy of the velocity of light, i.e., that the velocity of light in a vacuum is the same in all systems of reference which are moving relatively to each other, an assertion which at the time was widely accepted. In this paper Einstein gave modifications which must now be introduced into the formulae for aberration and for the Doppler effect.

Aberration has to do with the direction in which you point a telescope to see a star if you're on a moving platform. The earth is such a moving platform and the aberration of a star is slightly displaced because of relativistic effects. The Doppler effect concerns the observation that light seen from a moving source has its color changed (because of a frequency change) although the velocity is not changed. So Whittaker is willing to concede only these two minor points as Einstein's contribution to relativity theory.

Is it true that Einstein made only minor contributions to the theory of relativity? It is certainly true that he made only minor contributions to the equations of relativity, because Lorentz had derived his transformation equations before Einstein published his 1905 paper.

If we think of the theory of relativity as being only a description of the behavior of moving rods and clocks, then Einstein had little to do with that theory – all the key results were worked out before 1905. But these earlier results were developed with one purpose in mind – to explain the embarrassing Michelson–Morley experiment. The motivation for Einstein's work was something completely different. He decided that the principle of relativity and the constancy of the speed of light in all frames of reference were fundamental laws of the world, and these became his two postulates. He didn't assert that these were needed to explain the Michelson–Morley experiment. Instead, for philosophical reasons that are very deep, he said that the world must satisfy these postulates.

To understand the significance of his ideas it's useful to think of a simpler analogy – the principle of inertia. Suppose for a moment that you are Galileo and you have just discovered that a body in motion tends to stay in motion in a straight line. There are various things you could do once you've made this discovery. For example, you could try to find a mechanism that explains it. Do bodies have little motors inside them that keep them moving? What keeps them going so they don't come to rest? You would soon find that this type of scientific inquiry leads you nowhere, because you are asking the wrong questions. Moreover, you are speculating on matters that are impossible to test by experiment. A more fruitful approach is to accept inertia as a fundamental principle governing the way the universe works, and try to explain other phenomena using this principle. This is exactly what Einstein did. He said that the universe is governed by the principle of relativity and the constancy of the speed of light. Everything else comes about as a logical consequence of these principles.

Poincaré had made an attempt in the same direction, believing that the relativity principle should be the fundamental principle from which other things are derived. But Einstein understood that by adopting these principles we would have to change our intuitive notions of space and time. When he explained that in his paper, he did so in the simplest and most direct way you can imagine. He said,

> If we wish to describe the motion of a material point, we give the values of its coordinates as functions of time. We must bear carefully in mind that a mathematical description of this kind acquires no physical meaning unless we are quite clear as to what we understand by "time." We have to take into account that all our judgments in which time plays a part are only judgments of simultaneous events. If, for instance, I say, that a train arrives here at 7 o'clock, I mean something like this. The pointing of the small hand of my watch to seven and the arrival of the train are simultaneous events.

That's part of the definition of time. Next he goes into a discussion of how to establish whether two events are simultaneous if they don't take place at the same position in space. Einstein shows that the only way one can establish that is to make some way of synchronizing clocks by exchanging light signals back and forth. But once you exchange light signals, you're stuck with the constancy of the speed of light as seen by all observers and it necessarily means that observers in different frames moving at different velocities will make different decisions about whether events are simultaneous or not. What Einstein did was to yank us loose from our absolute notions of the meaning of space and time. It was an act as profound and revolutionary as Copernicus yanking the earth loose from its stationary position at the center of the universe.

CHAPTER 47

VELOCITY AND TIME

So we see that we cannot attach any *absolute* signification to the concept of simultaneity, but that two events which, viewed from a system of coordinates, are simultaneous, can no longer be looked upon as simultaneous events when envisaged from a system which is in motion relatively to that system.

Albert Einstein, "On the Electrodynamics of Moving Bodies" (1905)

47.1 PROPER LENGTH AND PROPER TIME

The special theory of relativity describes consequences of two assumptions: For all inertial observers the laws of nature are the same, and the speed of light is constant. These properties are inherent in the Lorentz transformation, which connects measurements of space and time coordinates in one frame with those in another. This chapter explores further consequences of these equations as they apply to measurements of velocity and time. First we recapitulate some of the ideas concerning length and time in special relativity that were described in Chapter 46.

As usual we consider two reference frames S and S', where S' is regarded as moving with constant velocity v parallel to the positive x axis of S. If the origins of S and S' coincide at time $t = t' = 0$ the Lorentz transformation takes the form

$$x = \gamma(x' + vt'), \quad (46.15\text{x})$$
$$y = y', \quad (46.15\text{y})$$
$$z = z', \quad (46.15\text{z})$$
$$t = \gamma(t' + vx'/c^2), \quad (46.15\text{t})$$

while the inverse transformation is given by

$$x' = \gamma(x - vt), \quad (47.1\text{x}')$$
$$y' = y, \quad (47.1\text{y}')$$
$$z' = z, \quad (47.1\text{z}')$$
$$t' = \gamma(t - vx/c^2), \quad (47.1\text{t}')$$

where γ is the dilation factor, a number greater than 1 given by

$$\gamma = \frac{1}{\sqrt{1 - v^2/c^2}}.$$

In the last chapter we learned that if a rod at rest along the x axis in one frame has length L_0 when measured by an observer in that frame, then an observer in the other frame determines the length of the rod to be contracted by the factor $1/\gamma$. The length L_0 is called the *proper length*, whereas L_0/γ is called the *contracted length*.

The Lorentz equations also imply time dilation. If T_0 is the time between two events measured on a clock that is at rest in one frame, then an observer in the other frame determines that the time between the same two events is greater than T_0 by a factor γ. For example, if two events E_1 and E_2 occur at the same place $x_1 = x_2$ in the S frame but at different times $t_1 < t_2$, then by applying Eq. (47.1t') twice and subtracting we find

$$t'_2 - t'_1 = \gamma(t_2 - t_1).$$

In other words, if the observer in S thinks that the time between a tick and tock is 1 s, the observer in S' measures this time to be γ s. Because $\gamma > 1$ the observer in S thinks that the S' clock is running slow.

If we call T_0 the *proper time* – the time between two events measured on a clock where the events occur at the same place – and T the time between the events measured by an observer moving with speed v, the foregoing relation becomes $T = \gamma T_0$ or

$$T = \frac{T_0}{\sqrt{1 - v^2/c^2}}.$$

Because the two events occur at the same position in the rest frame, one clock suffices to measure the proper time. In the moving frame, the events occur at different positions, and hence to measure the time between the two events, two clocks would be used, one at each position. A useful mnemonic is, "more clocks,

47.1 PROPER LENGTH AND PROPER TIME

greater time." Whenever we use a familiar phrase such as "the stationary observer thinks the moving clock is running slow," we are really envisaging a series of operations related to distinct events involving as many clocks and signals as necessary.

The time dilation factor γ and the length contraction factor $1/\gamma$ are reciprocals. When v^2 is small compared to c^2, γ is nearly 1 and the relativistic effects become negligible. On the other hand, as $|v|$ gets closer to the speed of light, γ becomes very large and the relativistic effects become more significant. In the limiting case when $v^2 \to c^2$ we find $\gamma \to \infty$ and $1/\gamma \to 0$, which means that a meter stick shrinks to a point and a clock stops completely. We also know that the inequality $v^2 > c^2$ would imply that γ is the square root of a negative number, and this makes no physical sense. In other words, no speed can exceed that of light. That's what Poincaré had in mind when he made his famous speech in St. Louis in 1904 and remarked that in the new mechanics nothing will be able to travel faster than the speed of light.

Example 1

An extremely fast train moving at a speed of $0.6c$ passes two fixed posts on the ground that are separated by 80 m according to observers on the ground. These observers note that at some given time the position of the rear end of the train coincides with that of one of the posts, and at the same instant the position of the front end coincides with that of the other post. Calculate the length of the train according to (a) an observer on the ground and (b) an observer on the train. Calculate the amount of time it takes for the full length of the train to pass one post according to (c) an observer on the ground and (d) an observer on the train. (e) Calculate the distance between posts according to an observer on the train.

(a) Observers on the ground measure the length of the train to be $L = 80$ m, the same as the distance they measured between the posts.

(b) Let L_0 denote the length of the train measured by an observer on the train. The length L measured in (a) is a contraction of L_0 by the factor $1/\gamma$. Now

$$1/\gamma = \sqrt{1 - (0.6)^2} = 0.8,$$

so

$$L_0 = L\gamma = L/0.8 = (80 \text{ m})/0.8 = 100 \text{ m}.$$

(c) According to an observer on the ground, the train must travel its full length L at speed v to pass one post. Therefore, the time it takes is

$$T = L/v = (80 \text{ m})/[(0.6)(3 \times 10^8 \text{ m/s})] = 0.44 \times 10^{-6} \text{ s}.$$

(d) An observer on the train sees the posts move past him. With his clock he measures T_0 for the time between the first and second post passing him. The ground-based observer measures the time interval between these same events to be

the dilated time T of part (c). Therefore, $T = \gamma T_0$, so
$$T_0 = T/\gamma = (0.44 \times 10^{-6} \text{ s})(0.8) = 0.35 \times 10^{-6} \text{ s}.$$

(e) According to the observer on the train, the posts move past him at speed $0.6c$ in time T_0. Therefore, the distance between the posts is simply $d = vT_0 = 0.6(3 \times 10^8 \text{ m/s})(0.35 \times 10^{-6} \text{ s})$, or $d = 64$ m. The same result can also be obtained by length contraction, $d = 80 \text{ m}/\gamma = (0.8)80 \text{ m} = 64$ m.

Questions

Questions 1 through 4 refer to the following hypothetical situation: A class of physics students is given a quiz to be completed in 10 min according to the instructor's clock. Relative to the students, the instructor moves at a speed of $0.8c$ and he sends a light signal to the class when his clock reads 10 min. The students stop writing when the light signal reaches them.

1. According to the class, what is the instructor's position when the instructor's clock reads 10 min? (Assume that the two coordinate systems coincide at $t = t' = 0$.)

2. According to the class, what is the time when the instructor's clock reads 10 min?

3. How long does it take the light signal to reach the class?

4. According to the students, how much time did they have to work the quiz?

5. Let T_1 be the lifetime of a meson (a type of radioactive particle) as measured in an inertial frame S_1 in which the meson is at rest. In a different inertial frame, S_2, moving with velocity v_{12} relative to S_1, this lifetime is measured to be T_2. Let T_3 be the lifetime measured in a third frame, S_3, which has a velocity v_{13} relative to S_1 and v_{23} relative to S_2. Among the following equations relating measurements of lifetimes, determine those that are correct, and correct those that are incorrect.

   (a) $T_1 = T_3\sqrt{1 - v_{13}^2/c^2}$.
   (b) $T_2 = T_1\sqrt{1 - v_{12}^2/c^2}$.
   (c) $T_3 = T_2/\sqrt{1 - v_{23}^2/c^2}$.
   (d) $T_3 = T_1/\sqrt{1 - v_{23}^2/c^2}$.

6. A spaceship must travel 8.0×10^{12} m in order to reach its home base. If the life support systems can function for 48 h, what is the minimum speed that the ship must have if the crew is to survive?

47.2 COMBINATIONS OF VELOCITIES IN SPECIAL RELATIVITY

In Galilean relativity, studied in Chapter 45, we learned that if frame S' moves with constant velocity **v** with respect to frame S and if the position **r** of a body in S is

47.2 COMBINATIONS OF VELOCITIES IN SPECIAL RELATIVITY

related to its position $\mathbf{r}'$ in S' by the equation

$$\mathbf{r} = \mathbf{r}' + \mathbf{r}_0 + \mathbf{v}t, \tag{45.4}$$

then the velocities $\mathbf{u} = d\mathbf{r}/dt$ and $\mathbf{u}' = d\mathbf{r}'/dt$ differ only by a constant amount,

$$\mathbf{u} = \mathbf{u}' + \mathbf{v}. \tag{45.5}$$

In particular, if the relative motion is entirely in the x direction, so that $y = y'$ and $z = z'$, the vector equation (45.5) gives three scalar equations for the velocity components:

$$u_x = u'_x + v, \qquad u_y = u'_y, \qquad u_z = u'_z,$$

where now v is the velocity of frame S' relative to S. A moment's reflection tells us that the first of these equations cannot possibly hold in special relativity. For example, if v is close to the speed of light, say $v = 0.8c$, and if a particle is moving at great speed in S', say $u'_x = 0.7c$, then the formula for u_x gives $1.5c$ for the speed of the particle in frame S, a speed greater than the speed of light, which is impossible. This means that in special relativity velocities do not combine by simple vector addition.

To find exactly how velocities do combine in special relativity, we refer to the Lorentz equations. Differentiating Eq. (46.15x) with respect to t we find

$$\frac{dx}{dt} = \gamma \left(\frac{dx'}{dt} + v \frac{dt'}{dt} \right). \tag{47.2}$$

By the chain rule we have

$$\frac{dx'}{dt} = \frac{dx'}{dt'} \frac{dt'}{dt},$$

so Eq. (47.2) becomes

$$\frac{dx}{dt} = \gamma \left(\frac{dx'}{dt'} + v \right) \frac{dt'}{dt}. \tag{47.3}$$

With the notation

$$u_x = \frac{dx}{dt} \quad \text{and} \quad u'_x = \frac{dx'}{dt'},$$

Eq. (47.3) becomes

$$u_x = \gamma (u'_x + v) \frac{dt'}{dt}. \tag{47.4}$$

By differentiating Eq. (46.15t) with respect to t' we find

$$\frac{dt}{dt'} = \gamma \left(1 + vu'_x/c^2 \right). \tag{47.5}$$

In Example 2 we show that $1 + vu'_x/c^2 \neq 0$, so we can take reciprocals in Eq. (47.5) and substitute into (47.4). The factor γ cancels and we obtain

$$u_x = \frac{u'_x + v}{1 + vu'_x/c^2}.\tag{47.6x}$$

This is the law of combination of velocities in special relativity. The sum of velocities in the numerator, $u'_x + v$, is what Galilean relativity gives for u_x, but in special relativity this sum must be divided by the quantity $1 + vu'_x/c^2$. If both v and u'_x have the same sign this quantity is greater than 1, and hence u_x is smaller than what it would be in Galilean relativity.

For example, if $v = 0.8c$ and $u'_x = 0.7c$, as in the previous example, we find

$$u_x = \frac{0.7c + 0.8c}{1 + (0.8)(0.7)} = \frac{1.5c}{1.56} = 0.96c,$$

a speed less than that of light. In fact, even in extreme cases the velocities cannot combine to exceed the speed of light. For example, if frame S' is traveling at velocity v relative to S and if an observer in S' sees a beam of light traveling in the same direction at speed c, then $u'_x = c$ and Eq. (47.6x) tells us that an observer in S sees the same beam traveling at velocity u_x where

$$u_x = \frac{c + v}{1 + cv/c^2} = \frac{c(c + v)}{c + v} = c.$$

In other words, the speed of light is the same for all inertial observers, as it should be according to Einstein's second postulate.

Example 2
Show that we always have $1 + vu'_x/c^2 \neq 0$.
The only way we could have $1 + vu'_x/c^2 = 0$ is if

$$\left(\frac{v}{c}\right)\left(\frac{u'_x}{c}\right) = -1.$$

But in deriving the Lorentz equations in Chapter 46 we found that $v^2 < c^2$, so the first factor on the left has absolute value < 1, and the only way the product could be -1 is for the other factor to have absolute value > 1. This would mean $|u'_x| > c$, which is impossible because nothing travels faster than the speed of light. Therefore, $1 + vu'_x/c^2 \neq 0$.

Equation (47.6x) gives us the x component of velocity in the S frame in terms of the x' component of velocity in the S' frame. What about the velocity components in the y and z directions? You might think that $u_y = u'_y$ and $u_z = u'_z$ because $y = y'$ and $z = z'$. We'll see in a moment that these equations do not hold. The physical reason is that a velocity component is equal to a length divided by a time. The observers will agree on the distance traveled in the y and z directions, but they

47.2 COMBINATIONS OF VELOCITIES IN SPECIAL RELATIVITY

will not agree on the time it took. So the velocity components should be different. To find the exact velocity relations we proceed as we did in obtaining Eq. (47.6x). Differentiating Eq. (46.15y) with respect to t we get

$$\frac{dy}{dt} = \frac{dy'}{dt} = \frac{dy'}{dt'}\frac{dt'}{dt}.$$

With the notation $u_y = dy/dt$ and $u'_y = dy'/dt'$ this becomes

$$u_y = u'_y \frac{dt'}{dt}.$$

In Galilean relativity we have $t = t'$, hence $dt'/dt = 1$ and $u_y = u'_y$. But in special relativity the times are related by Eq. (46.15t), and dt'/dt is determined by taking reciprocals in Eq. (47.5). This gives us

$$u_y = \frac{u'_y}{\gamma(1 + vu'_x/c^2)}. \tag{47.6y}$$

In the same way we obtain the corresponding result for $u_z = dz/dt$. In summary, the transformation equations for velocities in special relativity are given by

$$u_x = \frac{u'_x + v}{1 + vu'_x/c^2}, \tag{47.6x}$$

$$u_y = \frac{u'_y}{\gamma(1 + vu'_x/c^2)}, \tag{47.6y}$$

$$u_z = \frac{u'_z}{\gamma(1 + vu'_x/c^2)}. \tag{47.6z}$$

These equations are a bit more complicated than their Galilean counterparts because of the extra factors in the denominators. Note that an extra factor γ appears in the denominator of the last two equations. Despite appearances, these equations are symmetric in the primed and unprimed quantities. The corresponding equations for expressing the velocity components u'_x, u'_y, and u'_z in terms of u_x, u_y, and u_z can be obtained by simply replacing v by $-v$ and interchanging primed and unprimed quantities. Verification of this fact is requested in Question 13.

Example 3
Assume frame S' moves with velocity v in a direction parallel to the positive x axis of frame S. Consider a particle moving parallel to the positive y axis of frame S with velocity u_y. Determine the velocity components u'_x, u'_y, and u'_z of this particle in frame S'.

Because the particle is moving parallel to the y axis in frame S we have $u_x = u_z = 0$. Therefore, from Eqs. (47.6x) and (47.6z) we find

$$u'_x = -v \quad \text{and} \quad u'_z = 0.$$

Equation (47.6y) now gives us

$$u'_y = \gamma u_y(1 + vu'_x/c^2) = \gamma u_y(1 - v^2/c^2) = u_y\sqrt{1 - v^2/c^2}.$$

The geometric relation between u_y and u'_y is shown in the accompanying figure.

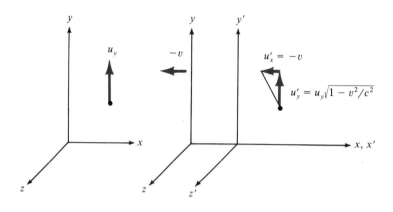

Example 4

A fast train moves at a speed of $0.6c$ relative to the ground. Inside the train a runner moves at a speed $0.6c$ relative to the train. Calculate the velocity components of the runner relative to the ground if he is running (a) in the same direction as the train and (b) in the opposite direction.

Let S denote the ground frame and S' that of the train. We are given that the relative speed of S' to S is $v = 0.6c$, whereas $u'_x = 0.6c$ in part (a) and $u'_x = -0.6c$ in part (b). According to the velocity transformation, we have

$$u_x = \frac{u'_x + v}{1 + vu'_x/c^2}. \tag{47.6x}$$

Substituting values we find for (a) $u'_x = 0.6c$ implies $u_x = (40/41)c$, and for (b) $u'_x = -0.6c$ implies $u_x = 0$. Similarly, we find $u'_y = u'_z = 0$ so $u_y = u_z = 0$ for both (a) and (b).

Questions

Questions 7 through 9 refer to the following situation: A spacecraft of proper length 100 m moves at a speed $3c/5$ with respect to the earth and contains an alien

47.3 THE FIZEAU EXPERIMENT

at the rear of the craft. The alien fires a bullet toward the front of the spacecraft at a relative speed of $3c/5$.

7. Determine the speed of the bullet relative to the earth.

8. Calculate the length of the spacecraft as measured by observers moving along with (a) the alien, (b) the earth, and (c) the bullet.

9. Determine how long the bullet takes to reach the front of the spacecraft as measured by (a) an observer on the earth, (b) the alien, and (c) someone moving along with the bullet.

10. Train A travels north at a speed $4c/5$ relative to a certain station whereas train B travels east at speed $3c/5$ relative to the same station. Find the velocity components of train A relative to an observer on train B.

11. An atom is moving with respect to a lab at a speed of $0.3c$ along the positive x direction, and emits an electron having a speed of $0.8c$ in the positive y direction in the rest frame of the atom.

    (a) Find the components of the electron's velocity according to an observer in the lab.

    (b) Calculate the magnitude and direction of the electron's velocity in the lab frame.

12. Two spaceships traveling in the same direction pass the earth. According to an Earth-based observer, spaceship A has a speed of $0.7c$ and spaceship B has a speed of $0.9c$.

    (a) What is the speed of spaceship B according to an observer on A?

    (b) The commander of spaceship B, not wanting to be called an unidentified flying object, flashes a signal to Earth. The observer on Earth measures the duration of the signal to be 10 s. What is the duration of the signal as measured on spaceship B?

    (c) What does an observer on spaceship A measure for the duration of the signal?

13. Use the method in the text to express the velocity components u'_x, u'_y, and u'_z in terms of u_x, u_y, and u_z. Your results should agree with those obtained by replacing v by $-v$ and interchanging primed and unprimed quantities in Eqs. (47.6x, y, z).

47.3 THE FIZEAU EXPERIMENT

Einstein's postulate about the speed of light being constant refers to the speed of light in a vacuum. Long before Einstein, many investigators tried to determine whether the speed of light was finite or infinite. The first experimental evidence for the finiteness of the speed of light was obtained in the seventeenth century by a Danish astronomer Ole Roemer as a result of observations on the eclipses of

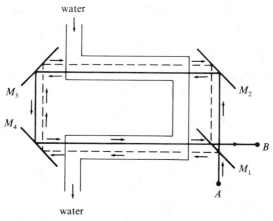

Figure 47.1 Diagram for the Fizeau experiment.

Jupiter's satellites. Later, an English astronomer James Bradley explained an apparent seasonal change in the position of a star called γ Draconis by the finiteness of the speed of light.

In 1849, a French scientist, Armand Fizeau, made the first terrestrial determination of the speed of light, that is, the first determination that did not involve astronomical constants. Using a rotating toothed wheel as a light switch, Fizeau arrived at the value 3.13×10^8 m/s for the speed of light in air. This was not very accurate but it paved the way for more precise measurements. Leon Foucault, who had worked with Fizeau, improved the apparatus and obtained the value 2.98×10^8 m/s in 1862. Much later, Albert A. Michelson modified the Foucault method on a spectacular scale using a rotating mirror apparatus, and in 1927 he announced the highly accurate value $(2.99796 \pm 0.00004) \times 10^8$ m/s. The accepted value today is very close to 2.99776×10^8 m/s.

Fizeau carried out another experiment in 1851 that can be cited as physical evidence for the formulas we obtained on relativistic velocities. Fizeau made light travel through water that, in turn, was flowing with a speed v relative to his laboratory. A diagram of the experimental arrangement is shown in Fig. 47.1. Water flows in a U-shaped tube at known speed v. Light from source A hits a half-silvered mirror M_1 and is split into two beams, indicated by solid and dashed lines in Fig. 47.1, that reflect off mirrors M_2, M_3, and M_4 as shown, then recombine at M_1 and proceed to B. The relative time delay between the two circuits can be measured by interference methods.

When light passes through a transparent medium such as glass, water, or air, its speed is not c but c/n, where n is the index of refraction, introduced in Chapter 44. For example, the index of refraction of crown glass is about 1.52, that of water is about 1.333, while that of air is about 1.0003. Fizeau determined that the speed of the light parallel to the stream of water was given by

$$u = \frac{c}{n} + \alpha v, \tag{47.7}$$

where n is the index of refraction of water and α is a constant called the "drag coefficient." Fizeau's experiment indicated that α is about 0.434. Michelson and Morley repeated Fizeau's experiment in 1886 with improved accuracy and obtained the same value for α.

We can examine these findings in light of Eq. (47.6x). If c/n is the speed of light in water, as measured by an observer at rest relative to the water, and if v is the speed of the water relative to the laboratory observer, then the speed of light u relative to the laboratory observer is given by the composition law in Eq. (47.6x),

$$u = \frac{(c/n) + v}{1 + [v/(nc)]}. \tag{47.8}$$

Now if $x \ll 1$ we have the approximate formula

$$\frac{1}{1 + x} \approx 1 - x, \tag{47.9}$$

which is highly accurate if x^2 is negligible compared to x. To see where this comes from we use the algebraic identity

$$1 - x^2 = (1 + x)(1 - x).$$

If x^2 is very small the left member of this identity is nearly 1 so $(1 + x)$ and $(1 - x)$ are nearly reciprocals of each other, and we get (47.9). Taking $x = v/(nc)$ and noting that v is very small compared to c, we conclude that x^2 is negligible compared to x. Therefore, using the approximation (47.9) in (47.8), we obtain

$$u \approx \left(\frac{c}{n} + v\right)\left(1 - \frac{v}{nc}\right) = \frac{c}{n} + v\left(1 - \frac{1}{n^2}\right) - \frac{v^2}{nc} \approx \frac{c}{n} + v\left(1 - \frac{1}{n^2}\right),$$

where in the last step we neglected $v^2/(nc)$. Comparing this with Fizeau's formula (47.7) we find that

$$\alpha = 1 - \frac{1}{n^2} = 1 - \frac{1}{(1.333)^2} = 0.437,$$

which agrees well with Fizeau's experimentally determined value 0.434.

47.4 THE MUON EXPERIMENT

According to special relativity, moving clocks and meter sticks behave in ways that are contrary to our intuition. A clock in motion appears to run more slowly than one at rest – an effect that has nothing to do with the inner workings of the clock, but rather with the nature of time itself. In ordinary experience this is not noticeable because the time dilation factor γ is very close to 1 for any commonplace observation. In order to demonstrate this effect experimentally one would have to observe a clock moving at a speed close to that of light.

Such clocks actually exist. They are subatomic particles that were first found in the upper atmosphere and are now created routinely by powerful particle accelerators. Observations of these particles led to one of the first experimental verifications of time dilation.

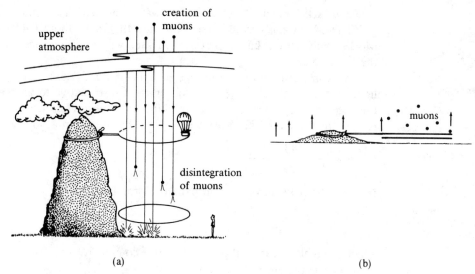

Figure 47.2 The muon experiment, according to observers (a) on Earth and (b) traveling with the muons.

The earth's atmosphere is constantly bombarded by cosmic rays, particles of very high energy that originate from outer space. When these particles enter the outer layers of the atmosphere about 15 km up, they collide with other particles and produce many kinds of subatomic, subnuclear fragments. One of the products is the muon, or mu meson, an unstable, subnuclear particle that has a mean lifetime of 2.2 μs in its own reference frame. Not every muon will decay in exactly that time, but in a large population of muons the average decay time will be 2.2 μs.

As a result of the collisions that occur at the top of the atmosphere, the muons are propelled downward at high speeds very close to the speed of light. Because we know how long they live, we can calculate how far they travel before they decay. On the average, this distance is the product of their speed c and the decay time, 2.2×10^{-6} s, which is about 700 m. Figure 47.2 illustrates this distance.

Most of the muons are created about 15 km above the earth's surface and, because they are expected to decay after traveling less than 1 km, they should never reach the earth. It's true that some muons live longer than others, but not long enough to travel 20 times the distance they're supposed to. On the other hand, observations reveal that large quantities of muons *do* reach the earth. The explanation for this is time dilation. Because the muon is moving at close to the speed of light, its clock is running slow – that is, it has an internal clock that makes its decay time seem longer than 2.2 μs according to an observer on Earth. We haven't the foggiest notion of the nature of the internal clock, and we don't need to know how it works. But we do know that it will obey the laws of special relativity and exhibit time dilation. In this extra time, the muon travels a much longer distance through the atmosphere than it's supposed to, and that's why it makes it to Earth.

47.4 THE MUON EXPERIMENT

In 1941 B. Rossi and D. B. Hall made measurements comparing the flux of muons at the top of a mountain and at sea level to see how fast the muons decay along the way. The observations indicate that muons survive about 9 times as long as they would if they were at rest with respect to the earth.

From the point of view of the muon itself, time dilation does not occur. The muon thinks it is at rest and that the earth is moving toward the muon at nearly the speed of light. The muon's internal clock tells it to decay right on schedule, 2.2 μs after it is created. But, according to the muon, the earth and its atmosphere are approaching at such great speed that the thickness of the atmosphere undergoes Lorentz contraction. The atmosphere appears so thin that the muon penetrates it and strikes the earth before 2.2 μs have elapsed.

Example 5

An experiment shows that muons travel 2000 m in 6.71×10^{-6} s from the time they are created at the top of a mountain to the time they reach sea level.

(a) Calculate the speed of the muons relative to the earth.
(b) Determine the time dilation factor γ for the muons moving at this speed.
(c) In the reference frame of the muons, what is the altitude of the mountain above sea level?

(a) According to an observer on Earth, the muons travel a distance $L_0 = 2000$ m in a time $T = 6.71 \times 10^{-6}$ s. Therefore, their speed is given by

$$v = L_0/T = (1000 \text{ m})/(6.71 \times 10^{-6} \text{ s}) = 2.98 \times 10^{-8} \text{ m/s} = 0.994c.$$

(b) Using $v/c = 0.994$ we find $1 - (v/c)^2 = 0.012$ so

$$\gamma = 1/\sqrt{0.012} = 9.13.$$

(c) The moving muons see a contracted length for the height of the mountain. According to length contraction, $L = L_0/\gamma$, so substituting for L_0 and γ, we find $L = 219$ m. Thus, observers on Earth see muons travel a distance of 2000 m in 6.71×10^{-6} s at a speed of $0.994c$, whereas the muon sees the earth move toward it a distance of 219 m in 0.73×10^{-6} s.

Questions

14. Point out the error in the following argument: A clock, at rest with respect to an inertial frame S, records a time T_0 for an egg to hatch in that frame. In frame S' moving with respect to the S frame, a clock records a longer time T for that egg to hatch. Therefore, it is possible to distinguish between the two inertial frames, thus violating Einstein's principle of relativity.

15. A spaceship travels at constant speed relative to Earth and reaches the star system of Procyon, a distance of 10 light years. Suppose the trip takes 18 yr as measured by clocks aboard the spaceship.

  (a) What is the speed of the spaceship relative to Earth?
  (b) According to observers on Earth, how long does the trip take?
  (c) According to the space travelers, how far is Earth from Procyon?

16. In an inertial frame S a child shines a flashlight at an angle of 30° from the x axis. Another child in a frame S' passing by at a speed of $0.8c$ parallel to the x axis observes the beam of light.

  (a) What are the components of the velocity of light according to the child in the S' frame?
  (b) What is the speed and direction of the light in the S' frame?

Questions 17 through 21 refer to a certain type of radioactive particle, called a charged pion, which has an average lifetime of 2.6×10^{-8} s when observed at rest. Assume that charged pions emerging from an accelerator have a speed of $0.8c$ relative to the lab.

17. According to an observer in the lab, what is the average lifetime of a moving pion?

18. What is the average distance traveled by a pion in the lab before it decays?

19. According to an observer moving along with the pion, what is the average distance from the point where the pion emerges from the accelerator to the point of decay?

20. A fervent superphysicist runs after an emerging pion with a speed of $0.6c$ relative to the lab. According to the superphysicist, how fast is the pion traveling?

21. According to the superphysicist, what is the average lifetime of a pion?

22. Two scientists traveling in separate spaceships observe the formation and decay of a beam of muons. The clock on ship A measures an average lifetime of 2.2 μs, while the clock on ship B measures an average lifetime of 4.4 μs. The scientist on one of the ships notices that the muons are not moving relative to her.

  (a) Which scientist sees the muons at rest with respect to her spaceship?
  (b) Calculate the relative speed of the two spaceships.

47.5 THE TWIN PARADOX

The idea that time does not run at the same pace for every observer has given rise to a number of paradoxes that puzzled and stimulated every generation of physics students in this century. All these paradoxes arise from incomplete understanding and all have been satisfactorily explained. They appear to be paradoxes only

47.5 THE TWIN PARADOX

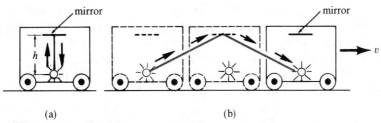

(a) (b)
Figure 47.3 The trolley car experiment.

because some of the implications of special relativity seem to conflict with intuitive ideas of distance and time. One of the most celebrated of these, called the twin paradox, will be discussed in this section.

Before introducing the twin paradox, let's recall the trolley car experiment discussed in Section 46.1 and illustrated in Fig. 47.3. If you are in the car, an observer on the ground sees the light take a longer path than you do, as indicated in Fig. 47.3, so he thinks your clock is running slow. But if the ground observer performs an experiment in his laboratory to measure the speed of light, the conditions are reversed. To you in the car, the laboratory seems to be moving and the path of the light in the laboratory seems longer to you, and therefore you conclude that the other person's clock is running slow. The two situations are symmetrical, each of you thinking that the other's clock is slow.

Nevertheless, it seems somewhat bizarre that when 2 s have passed on your clock, you may think only 1 s has passed on his, and when 2 s have passed on his clock, he may think only 1 s has passed on yours. The explanation for this is that you and the other person can meet and set your clocks together once and only once. After that you never meet again because each of you is moving at a constant speed in a straight line relative to the other. So if you want to compare clock readings after you separate, you can do so only by sending signals to each other. One way or another, any such signal is equivalent to a beam of light traveling at speed c. Each of you judges it to have traveled along a path of different length and therefore each thinks it has taken a different time.

The twin paradox involves a pair of twins, a sister and brother, A and B, who separate and later meet again at the same place. They were born at the same time so their biological clocks are synchronized at birth. They are able to compare their clocks at the beginning of the trip and at the end. Brother B takes up a career on Earth as a light-clock repairman, but his more adventurous sister A becomes a space trucker, making the milk run to a nearby solar system, 10 light years away, in a freight rig that manages $\gamma = 10$. This means her speed v is very nearly c, so brother B sees his sister travel 10 light years away, and 10 more back in just slightly more than 20 yr. On the other hand, sister A sees the distance to her destination Lorentz contracted by a factor $1/\gamma = 1/10$, so by traveling at nearly the speed of light she makes the journey in 1 yr each way. Thus, each time she returns to Earth from one of her journeys, she is 2 yr older, but her brother has aged by 20 yr.

This scenario is indeed what the special theory of relativity predicts. The paradox is that we expect each observer to think the other's clock to be running slow (any clock, including the human metabolism). Doesn't sister A think she is standing still while brother B races away on spaceship Earth?

Indeed, as long as the twins are in uniform relative motion, each would judge the other's clock to be running slow. But that has no practical consequence as long as they are far apart (and continue to get farther apart), unable to compare their biological clocks directly. The only way they can get back together again is for one of them to switch inertial frames. That switch, from going away to coming back, by only one of the twins, destroys the symmetry between their points of view. That is why there is no paradox.

Questions

23. An astronaut travels at a speed of $0.5c$ relative to the earth. Does she detect her heartbeat to be slower or the spaceship to be shorter than when she was at rest? Explain your answer.

24. Two observers agree to test time dilation. One in the S' frame moves at a speed $0.6c$ relative to the other in the S frame. When their origins coincide they start their clocks. They agree to send a signal when their clocks read 60 min and send a confirmation signal when each receives the other signal.

  (a) When does the observer in the S frame receive the first signal from the observer in S'?
  (b) When does the observer in S receive the confirmation signal?
  (c) Make a table showing the times in S when the observer sent the first signal, received the signal, and received the confirmation signal. How does this compare with a table that the observer in S' would construct?

25. Astronaut A leaves her twin brother B on Earth and departs on a trip to Alpha Centauri, a distance of 4 light years, and back. She travels at a speed of $0.6c$ relative to Earth both ways, and sends a light signal every 0.01 yr according to her calendar. Her twin B sends a light signal every 0.01 yr according to his own calendar.

  (a) How many signals does B receive before A turns around?
  (b) How many signals does A receive before she turns around?
  (c) What is the total number of signals each twin receives from the other?
  (d) By how much younger is twin A than B when she returns? Do both twins agree on this result?

26. For the situation in Question 25, make a sketch on a space–time diagram, showing the trip and a few representative light signals being sent and received on each part of the trip.

47.6 A FINAL WORD

You might wonder why the twins have different ages after one makes a relativistic trip. What caused the astronaut to come back younger than her brother? Was it the acceleration that did it? The change of direction? How did it happen? As far as we know there is no way to answer these questions. It's like asking why bodies in motion remain in motion according to the law of inertia. That's the way the world is. Objects just follow the law of inertia, and we deduce features of motion from that. Similarly, the fundamental nature of time and space produces the result that the astronaut returns younger than her brother. It happens because that's the way time and space are.

Imagine a universe that is completely empty. In such a universe it would be meaningless to talk about time and space. If there is only one object in the universe, it would be meaningless to ask whether that object is moving or at rest. It's just there. If there are two objects in the universe, then it makes sense to ask if they are getting closer together or moving farther apart. In other words, some questions about relative motion make sense. If the universe contains a lot of objects, we can talk about relative motion, and we might be able to devise some way of measuring how big an object is, or how far apart objects are from each other. But there is no reason whatsoever to choose one of those objects and say, "That one is at rest and all others are in motion, and so all others should try to come to rest with respect to the privileged one." There's no reason to do that, and so in the simplest of all universes – which is ours of course – a body in any state of motion will retain that state of motion. That's why the law of inertia makes sense. Moreover, it must continue to make sense in a universe in which light can propagate through the void. However, if there's only one frame of reference in which light has a certain speed, all of the foregoing argument collapses.

Other logically consistent universes may or may not exist. But in the one we live in, the speed of light is the same for all observers. Because of that, absolute rest has no meaning, the law of inertia works, and all the physics we understand works. Knowing that alone, it's possible to deduce that the traveling astronaut will come back younger than her twin brother who stayed behind. The nature of the metabolism of the traveler, why one ages, how clocks work, or what makes crystal oscillators tick – none of that is relevant. The argument we presented is all that's needed to conclude that the traveler will come back younger than the twin who stayed behind. That is an absolutely breathtaking feat of pure logic – one of the remarkable consequences of the special theory of relativity.

CHAPTER 48

MASS, MOMENTUM, ENERGY

> Once when I was with Einstein in order to read with him a work that contained many objections against his theory... he suddenly interrupted the discussion of the book, reached for a telegram that was lying on the windowsill, and handed it to me with the words, 'Here, this will perhaps interest you.' It was Eddington's cable with the results of measurements of the eclipse expedition. When I was giving expression to my joy that the results coincided with his calculations, he said quite unmoved, 'But I knew that the theory was correct'; and when I asked, what if there had been no confirmation of his prediction, he countered: 'Then I would have been sorry for the dear Lord – the theory *is* correct.
>
> A Student of Einstein (1919)

48.1 INERTIA AND RELATIVITY

The law of inertia, one of the building blocks of classical mechanics, states that a body in motion tends to remain in motion with constant velocity. Newton incorporated this principle in his first law of motion. The second law focused on changes in motion. Calling force **F** the agent of change, we usually write it in the form

$$\mathbf{F} = m\frac{d\mathbf{v}}{dt},$$

where the mass m is a measure of the difficulty in changing a body's velocity. The second law became the focal point of Newton's dynamical description of the world.

The law of inertia, in turn, is part of a deeper idea – the principle of relativity, which says that there is no such thing as absolute rest, so there is no reason for a body to come to rest. It keeps moving according to the law of inertia. The principle works well for mechanics, but if relativity is to be a general principle, it must work for all of physics. In particular, the principle of relativity must also describe the propagation of light. That leads to difficulties and contradictions unless we assume that every inertial observer will measure the same speed of light. Einstein started with the principle of relativity and the constancy of the speed of light as fundamental postulates and from them deduced the compelling consequences we have seen in Chapters 46 and 47. One of these is that nothing can move faster than the speed of light. Once we accept this fact, it's easy to see that Newton's second law $\mathbf{F} = m\mathbf{a}$ can't be valid at speeds approaching c.

To demonstrate this, apply a constant force that moves a body along a straight line and increases its speed. If the force is applied continuously, the speed of the body keeps increasing at a rate proportional to the force and therefore becomes arbitrarily large, eventually exceeding the speed of light, because there is no mechanism in Newton's second law to prevent it. But exceeding the speed of light contradicts one of the major deductions in Einstein's special theory of relativity.

There is a way to resolve this contradiction and at the same time preserve Newton's second law. First, we write Newton's second law more nearly as Newton actually stated it,

$$\mathbf{F} = \frac{d}{dt}(m\mathbf{v}),$$

and realize that the mass m is not necessarily constant but may vary with speed. To understand why mass might vary with speed we turn to the principle of conservation of momentum.

48.2 MOMENTUM AND MASS

In classical mechanics the principle of conservation of momentum follows from Newton's second law of motion and it incorporates the law of inertia. Every body in motion possesses something that it can't get rid of, and that's what gives it its inertia. That something is Newtonian momentum, the product of mass and velocity. In Chapter 45 we showed that the law of conservation of Newtonian momentum is preserved under the Galilean transformation. That is, if we have two inertial frames S and S' related by the Galilean transformation, then in either frame the total Newtonian momentum of a system before collision is equal to that after collision. This was deduced from two facts: In either frame the total mass of the system is conserved, and velocities combine in a direct manner.

Because velocities do not combine in the same way under the Lorentz transformation, Newtonian momentum is not conserved in relativistic mechanics. This can be demonstrated by a simple example, illustrated in Fig. 48.1. An observer A on the ground (in frame S) sees two identical balls crash into each other with equal and opposite constant velocities u and $-u$. Each ball has mass m, and when they collide

48.2 MOMENTUM AND MASS

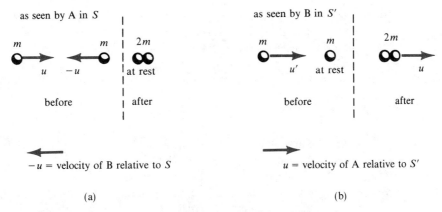

Figure 48.1 Newtonian momentum is conserved in (a) but not in (b).

they stick together and form a body of mass $2m$ at rest, as shown in Fig. 48.1a. Before collision, one ball has Newtonian momentum mu, the other $-mu$, so the total Newtonian momentum is zero. This is equal to the total Newtonian momentum after collision, $(2m)0 = 0$, so Newtonian momentum is conserved in frame S.

A second observer B in an airplane (frame S') flies over the same experiment with constant velocity $-u$ relative to the ground, as shown in Fig. 48.1a. Before the collision he sees one of the masses moving with velocity 0, as shown in Fig. 48.1b, and the other moving with velocity u' given by the relativistic composition law

$$u' = \frac{2u}{1 + u^2/c^2}. \tag{48.1}$$

According to B, the total Newtonian momentum before collision is mu' while the total Newtonian momentum after collision is $2mu$, and these are not equal because $u' \neq 2u$. In other words, Newtonian momentum is not conserved in relativistic mechanics. Therefore, if we want to have a law of conservation of momentum in relativistic mechanics that is consistent with the Lorentz transformation, we must adopt a different definition of momentum.

The solution of the problem turns out to be very simple, once we incorporate an idea proposed by Einstein. He pointed out that it's a matter of choice to redefine momentum or mass. We can regard the mass of a moving body in a given frame not as constant, but as a quantity that varies with its speed u. When $u = 0$ the body has Newtonian inertial mass m_0, which we refer to as the *rest mass*. But when the body has speed u it has relativistic mass $m(u)$, a function of u with $m(0) = m_0$. We will show presently that this function is given by the formula

$$m(u) = \gamma(u)m_0, \tag{48.2}$$

where

$$\gamma(u) = \frac{1}{\sqrt{1 - u^2/c^2}}. \tag{48.3}$$

Despite similarity in appearance, this is not the dilation factor γ that occurs in the Lorentz transformation, because u is not the speed of one reference frame relative to another but rather the speed of the particle in a given frame.

Having defined relativistic mass by Eq. (48.2), we now define the relativistic momentum of the particle in frame S to be the product of the relativistic mass and its velocity **u**,

$$\mathbf{p} = m(u)\mathbf{u}. \tag{48.4}$$

Note that relativistic momentum is a vector quantity completely analogous to Newtonian momentum. The only difference is that the mass is no longer constant but varies with speed.

In Newtonian mechanics, mass is a conserved quantity, in the same sense that electric charge is conserved: The mass of each body is always the same. But if we regard the mass of a body to be a function of its speed, we've obviously given up that simple aspect of mass conservation. We need to examine mass conservation in terms of variable mass. We'll assume that mass continues to be conserved in the same sense that momentum is conserved: The total relativistic mass of a collection of bodies doesn't change so long as there's no outside force acting on them. Later we will see the deeper significance of this assumption.

To see why the function $m(u)$ is chosen as indicated in Eq. (48.2), we refer once again to the inelastic collision of Fig. 48.1 and recalculate the total relativistic momentum before and after the collision in each of the two frames. In this discussion we assume that $m(u)$ is some unspecified function of u with $m(0) = m_0$, the rest mass. Then we show that if relativistic mass and momentum are conserved in both frames S and S' we must necessarily take for $m(u)$ the function defined by Eq. (48.2).

In frame S, shown in Fig. 48.2a, the two masses are traveling with the same speed because they have equal and opposite velocities, so they have the same relativistic mass $m(u)$. Before collision the sum of the relativistic momenta is $m(u)u - m(u)u = 0$, and after collision there is one body of mass M_0 at rest so its relativistic momentum is also zero. In other words, relativistic momentum is conserved in frame S. (We do not use vector notation in this discussion because the motion is along a line.)

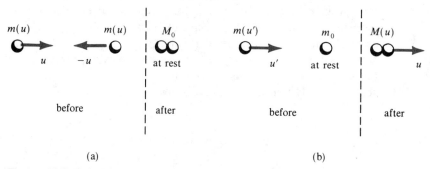

Figure 48.2 Relativistic momentum is conserved in both frames.

48.2 MOMENTUM AND MASS

In frame S', before collision we have one body of mass m_0 at rest and one with relativistic mass $m(u')$ moving with velocity u', as indicated in Fig. 48.2b. Therefore, the total relativistic momentum before collision is

$$m(u')u'. \tag{48.5}$$

After collision the composite body has relativistic mass $M(u)$, say, and travels at velocity u in frame S', so its relativistic momentum is

$$M(u)u. \tag{48.6}$$

To conserve relativistic momentum in frame S' we need

$$m(u')u' = M(u)u, \tag{48.7}$$

and to conserve the total relativistic mass of the system we have

$$M(u) = m(u') + m_0. \tag{48.8}$$

Substituting this into Eq. (48.7) we find

$$m(u')u' = (m(u') + m_0)u,$$

which, when solved for $m(u')$, gives us

$$m(u') = \frac{um_0}{u' - u} = \frac{m_0}{u'/u - 1}. \tag{48.9}$$

Now the velocities u' and u are related by Eq. (48.1).

$$u' = \frac{2u}{1 + u^2/c^2}. \tag{48.1}$$

Simple algebraic manipulation, which is described below in Example 1, shows that Eq. (48.1) implies the relation

$$u'/u - 1 = \sqrt{1 - (u'/c)^2}, \tag{48.10}$$

so Eq. (48.9) becomes

$$m(u') = \frac{m_0}{\sqrt{1 - (u'/c)^2}}.$$

Thus we see that conservation of mass and momentum in each frame implies that the relativistic mass function $m(u)$ must be given by Eq. (48.2).

For the example of inelastic collision of two bodies, relativistic mass and momentum are conserved both in frame S and in frame S'. More generally, it can also be shown that the definition of relativistic mass in Eq. (48.2) together with the definition of relativistic momentum in Eq. (48.4) implies that the laws of conservation of relativistic mass and momentum hold true under the Lorentz transformation for any system of bodies.

The fact that mass depends on speed should not be viewed with alarm, because from the outset mass was a somewhat mysterious quantity. Mass, a numerical quantity associated with a body, is a measure of how difficult it is to change the inertia of the body. Relativistic mass still serves this purpose, but now the quantity

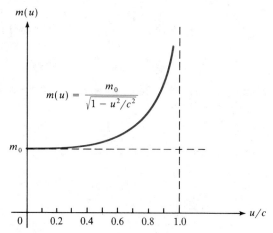

Figure 48.3 Relativistic mass $m(u)$ as a function of the ratio u/c.

changes with speed and becomes very large if the speed is greatly increased. In other words, as the speed of a body gets larger, it becomes increasingly difficult to further increase the body's speed. Figure 48.3 shows the graph of relativistic mass $m(u)$ as a function of the ratio u/c.

Example 1
Verify that Eq. (48.1) implies Eq. (48.10).
From Eq. (48.1) we find

$$\frac{u'}{u} = \frac{2}{1 + u^2/c^2} = \frac{2c^2}{c^2 + u^2},$$

and therefore

$$\frac{u'}{u} - 1 = \frac{2c^2}{c^2 + u^2} - 1 = \frac{c^2 - u^2}{c^2 + u^2}. \tag{48.11}$$

On the other hand, Eq. (48.1) also gives us

$$\frac{u'}{c} = \frac{2uc}{c^2 + u^2},$$

and hence

$$1 - \left(\frac{u'}{c}\right)^2 = 1 - \frac{4u^2c^2}{(c^2 + u^2)^2} = \left(\frac{c^2 - u^2}{c^2 + u^2}\right)^2.$$

48.2 MOMENTUM AND MASS

Taking the positive square root of both members we get

$$\sqrt{1 - \left(\frac{u'}{c}\right)^2} = \frac{c^2 - u^2}{c^2 + u^2}.$$

Comparing this with Eq. (48.11) we obtain Eq. (48.10).

Example 2

Refer to the collision in Fig. 48.2.

(a) Show that the rest mass M_0 of the composite body in frame S is not $2m_0$, as might be expected, but rather

$$M_0 = 2\gamma(u)m_0.$$

(b) Show that the relativistic mass of the composite body in frame S' is given by

$$M(u) = \gamma(u)M_0.$$

(a) In frame S, the composite body is formed from two bodies, each with relativistic mass $m(u)$, so $M_0 = 2m(u)$. But $m(u) = \gamma(u)m_0$, hence $M_0 = 2\gamma(u)m_0$.

(b) In frame S' we have

$$M(u) = m_0 + m(u') = m_0(1 + \gamma(u')).$$

But from the last equation of Example 1 we find that

$$\gamma(u') = \frac{1 + u^2/c^2}{1 - u^2/c^2},$$

hence

$$1 + \gamma(u') = 1 + \frac{1 + u^2/c^2}{1 - u^2/c^2} = \frac{2}{1 - u^2/c^2} = 2\gamma^2(u),$$

so

$$M(u) = 2m_0\gamma^2(u) = \gamma(u)(2m_0\gamma(u)) = \gamma(u)M_0.$$

Note that when $u = 0$ we get $M(0) = \gamma(0)M_0 = M_0$, so the rest mass $M(0)$ in frame S' is equal to the rest mass M_0 in frame S. This tells us that the rest mass of an object is an intrinsic and invariant property of the object, independent of the motion of the object. By contrast, the relativistic mass depends on the reference frame from which it is viewed. Increasing the speed in that frame increases its relativistic mass in that frame.

Figure 48.4 shows a graph of $m(v)v$, the magnitude of relativistic momentum, plotted as a function of the ratio v/c. In Newtonian mechanics, with mass assumed

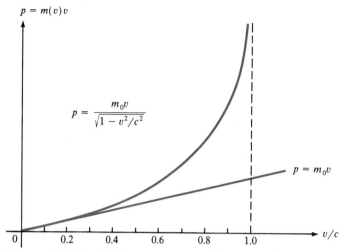

Figure 48.4 Relativistic momentum as a function of the ratio v/c.

constant, the graph of the momentum is just a straight line $p = m_0 v$ of constant slope. But the relativistic momentum behaves differently. At low speeds it very nearly obeys Newtonian mechanics and its graph is nearly a straight line of constant slope. At speeds exceeding about $0.3c$ the graph starts to bend upward, and as $v \to c$ it becomes asymptotic to the vertical line $v/c = 1$. So as you push a particle you keep increasing its momentum, but instead of only moving faster it also becomes more massive.

In Newtonian mechanics the force **F** with which you have to push the particle is related to its momentum **p** by the equation

$$\mathbf{F} = \frac{d\mathbf{p}}{dt} = \frac{d}{dt}(m\mathbf{v}). \tag{48.12}$$

In relativistic mechanics this equation is also adopted as the *definition* of force, with the understanding that **p** is relativistic momentum. This preserves Newton's second law of classical mechanics and it turns out to be a useful definition in relativistic mechanics as well.

When the mass m is not constant, the product rule for derivatives shows that the derivative of the product $m\mathbf{v}$ consists of two terms, so Eq. (48.12) becomes

$$\mathbf{F} = \frac{d}{dt}(m\mathbf{v}) = m\frac{d\mathbf{v}}{dt} + \mathbf{v}\frac{dm}{dt}. \tag{48.13}$$

In other words, the force required to change an object's momentum consists of two components, one parallel to the acceleration $d\mathbf{v}/dt$, due to the change in velocity, and another parallel to the velocity **v**, due to the change in mass. For an alternate form of Eq. (48.13), see Question 7 at the end of this section.

48.2 MOMENTUM AND MASS

Example 3
If $\gamma = (1 - v^2/c^2)^{-1/2}$, show that

$$v\frac{dv}{d\gamma} = \frac{c^2}{\gamma^3}. \tag{48.14}$$

The definition of γ implies

$$\gamma^2\left(1 - \frac{v^2}{c^2}\right) = 1. \tag{48.15}$$

Differentiate both members of this equation with respect to γ, using the product rule, to obtain

$$2\gamma\left(1 - \frac{v^2}{c^2}\right) + \gamma^2\left(-\frac{2v}{c^2}\right)\frac{dv}{d\gamma} = 0.$$

Multiply by $\gamma/2$ and use Eq. (48.15) to rewrite this as

$$1 = \frac{\gamma^3 v}{c^2}\frac{dv}{d\gamma}.$$

This is equivalent to Eq. (48.14).

Example 4
If $\mathbf{p} = m\mathbf{v}$, where $m = \gamma m_0$ and $\gamma = (1 - v^2/c^2)^{-1/2}$, show that

$$\mathbf{v} \cdot \frac{d\mathbf{p}}{d\gamma} = m_0 c^2. \tag{48.16}$$

This equation, which shows that the dot product $\mathbf{v} \cdot d\mathbf{p}/d\gamma$ is constant, will be used in the discussion of relativistic kinetic energy in the next section.
    Write $\mathbf{p} = \gamma m_0 \mathbf{v}$ and differentiate with respect to γ, using the product rule. This gives us

$$\frac{d\mathbf{p}}{d\gamma} = m_0 \mathbf{v} + \gamma m_0 \frac{d\mathbf{v}}{d\gamma}.$$

Dot multiplication by $\mathbf{v}$ transforms this to

$$\mathbf{v} \cdot \frac{d\mathbf{p}}{d\gamma} = m_0 v^2 + \gamma m_0 \mathbf{v} \cdot \frac{d\mathbf{v}}{d\gamma} \tag{48.17}$$

because

$$\mathbf{v} \cdot \mathbf{v} = v^2.$$

Differentiating this last equation with respect to γ we find

$$\mathbf{v} \cdot \frac{d\mathbf{v}}{d\gamma} + \frac{d\mathbf{v}}{d\gamma} \cdot \mathbf{v} = 2v \frac{dv}{d\gamma}$$

or

$$\mathbf{v} \cdot \frac{d\mathbf{v}}{d\gamma} = v \frac{dv}{d\gamma}.$$

Using this in the last term of Eq. (48.17) we obtain

$$\mathbf{v} \cdot \frac{d\mathbf{p}}{d\gamma} = m_0 v^2 + \gamma m_0 v \frac{dv}{d\gamma}.$$

Because of Eq. (48.14) in Example 3, this becomes

$$\mathbf{v} \cdot \frac{d\mathbf{p}}{d\gamma} = m_0 v^2 + m_0 \frac{c^2}{\gamma^2} = m_0 \left(v^2 + \frac{c^2}{\gamma^2} \right). \tag{48.18}$$

But Eq. (48.15) implies

$$v^2 + \frac{c^2}{\gamma^2} = c^2,$$

so Eq. (48.18) reduces to (48.16).

Questions

1. An electron's speed cannot be greater than c. Is there an upper limit to its momentum? Explain why or why not.

2. A proton of rest mass 1.67×10^{-27} kg has a relativistic mass which is three times its rest mass. What is its speed?

3. At what speed will an object's mass be 1% greater than its rest mass?

4. Derive a formula that gives the density of an object as a function of speed.

5. An electron used in an experiment at the Stanford Linear Accelerator (SLAC) has $\gamma(u) = 10^4$. What fraction of the speed of light is its speed u?

6. A proton ($m_0 = 1.67 \times 10^{-27}$ kg) travels at a speed of $0.8c$.

    (a) Determine its relativistic momentum.
    (b) Compare its relativistic momentum to its Newtonian momentum.

7. Let $\mathbf{a} = d\mathbf{v}/dt$, $m = \gamma m_0$, and $\gamma = (1 - v^2/c^2)^{-1/2}$. Show that:

    (a) $d\gamma/dv = v\gamma^3/c^2$.
    (b) $v(dv/dt) = \mathbf{v} \cdot \mathbf{a}$.
    (c) $dm/dt = m\gamma^2(\mathbf{a} \cdot \mathbf{v})/c^2$.

(d) Use part (c) to show that Eq. (48.13) can be written in the form

$$\mathbf{F} = m\mathbf{a} + \frac{m\gamma^2(\mathbf{a} \cdot \mathbf{v})}{c^2}\mathbf{v}.$$

This equation shows that the velocity component of $\mathbf{F}$ is zero if $\mathbf{v} \cdot \mathbf{a} = 0$, and Eq. (48.13) shows that this component is zero if the mass is constant.

48.3 RELATIVISTIC KINETIC ENERGY

In Chapter 45 we learned that the law of conservation of kinetic energy is preserved under the Galilean transformation. That is, if the Newtonian energy $mv^2/2$ of a system (with m constant) is conserved in an inertial frame S, then it is also conserved in any other frame S' related to S by the Galilean transformation. But conservation of Newtonian kinetic energy doesn't survive under the Lorentz transformation whether m is taken as the rest mass m_0 or as the relativistic mass $m(v) = \gamma(v)m_0$. This can be seen by considering examples involving elastic collision of two identical masses from the point of view of two different observers. Such examples show that a new definition of kinetic energy is needed in relativistic mechanics.

In Newtonian mechanics, energy is related to work. If a force $\mathbf{F}$ moves a particle from point A to point B then the work done by this force is equal to the change in its kinetic energy $K_B - K_A$, or, in other words,

$$K_B - K_A = \int_A^B \mathbf{F} \cdot d\mathbf{r}, \tag{48.19}$$

where $\mathbf{r}$ is the vector function describing the path of the particle from A to B. We will use this integral to define the work in relativistic mechanics as well, and this will suggest a natural definition for relativistic kinetic energy. For the force $\mathbf{F}$ we take the rate of change of relativistic momentum, given by Eq. (48.12),

$$\mathbf{F} = \frac{d\mathbf{p}}{dt}. \tag{48.12}$$

In a moment we will show that line integration of this force yields a surprisingly simple result. When the force $\mathbf{F}$ brings a particle from rest to a final speed v the work done along any path turns out to be

$$(m - m_0)c^2, \tag{48.20}$$

where m_0 is the rest mass and $m = \gamma m_0$ is the relativistic mass. We then use this quantity as the *definition* of relativistic kinetic energy. Although this seems to be markedly different from the Newtonian formula $m_0v^2/2$ for kinetic energy, in Example 5 we show that it reduces to the Newtonian formula when the speed v is small compared to c.

To derive (48.20) from Eq. (48.19) we write

$$\int_A^B \mathbf{F} \cdot d\mathbf{r} = \int_a^b \mathbf{F} \cdot \frac{d\mathbf{r}}{dt} dt,$$

where $r(a) = A$ and $r(b) = B$. Using $v = dr/dt$ and Eq. (48.12) for F we find

$$\int_A^B F \cdot dr = \int_a^b \frac{dp}{dt} \cdot v \, dt$$

$$= \int_a^b v \cdot \frac{dp}{d\gamma} \frac{d\gamma}{dt} \, dt.$$

(48.21)

But in Example 4 we showed that

$$v \cdot \frac{dp}{d\gamma} = m_0 c^2,$$

(48.16)

a constant, so Eq. (48.21) becomes

$$\int_A^B F \cdot dr = m_0 c^2 \int_a^b \frac{d\gamma}{dt} \, dt.$$

The last integral can be evaluated by the second fundamental theorem of calculus, giving us

$$\int_A^B F \cdot dr = m_0 c^2 (\gamma(b) - \gamma(a))$$

$$= (m_B - m_A) c^2,$$

where m_B is the relativistic mass at B and m_A the relativistic mass at A. In particular, if the particle is at rest at A and has speed v at B, the value of the integral reduces to (48.20).

The foregoing calculation suggests that we define the relativistic kinetic energy K by the formula

$$K = (m - m_0) c^2.$$

(48.22)

Note that this assigns zero kinetic energy to a body at rest. In Example 5 we will show that it assigns Newtonian kinetic energy $m_0 v^2/2$ to a body moving with a speed v that is small compared to c.

Example 5

Show that when the speed v is small compared to c, the relativistic kinetic energy K in Eq. (48.22) reduces to the Newtonian kinetic energy $m_0 v^2/2$.

Let $\delta = 1/\gamma$ so that $\delta^2 = 1/\gamma^2 = 1 - v^2/c^2$ and $1 - \delta^2 = v^2/c^2$. Then Eq. (48.22) can be written as

$$K = m_0 c^2 (\gamma - 1) = m_0 c^2 \left(\frac{1}{\delta} - 1 \right) = m_0 c^2 \left(\frac{1 - \delta}{\delta} \right).$$

(48.23)

But

$$\frac{1 - \delta}{\delta} = \frac{1 - \delta}{\delta} \frac{1 + \delta}{1 + \delta} = \frac{1 - \delta^2}{\delta(1 + \delta)} = \frac{v^2/c^2}{\delta(1 + \delta)},$$

48.3 RELATIVISTIC KINETIC ENERGY

so (48.23) becomes

$$K = m_0 \frac{v^2}{\delta(1 + \delta)}.$$

But if v/c is small, then γ is nearly 1, so δ is nearly 1, and $1 + \delta$ is nearly 2, hence K is nearly equal to $m_0 v^2/2$, the Newtonian kinetic energy.

For those familiar with infinite series there is an alternate derivation using the binomial expansion

$$\frac{1}{\sqrt{1-x}} = 1 + \frac{1}{2}x + \frac{3}{8}x^2 + \cdots,$$

valid for $|x| < 1$, where $+ \cdots$ refers to terms involving higher powers of x. Taking $x = v^2/c^2$ we find

$$\gamma = 1 + \frac{1}{2}\frac{v^2}{c^2} + \cdots.$$

Neglecting higher powers we obtain

$$\gamma - 1 \approx \tfrac{1}{2}v^2/c^2 \quad \text{and} \quad K = m_0 c^2(\gamma - 1) \approx \tfrac{1}{2}m_0 v^2.$$

The discussion that led to our definition of kinetic energy in Eq. (48.22) shows how an applied force increases both the mass and the speed of a body and thereby increases its kinetic energy. The difference $m - m_0$ is the increase in the mass over the rest mass m_0, and $(m - m_0)c^2$ is the corresponding increase in kinetic energy. Now the quantity mc^2 itself has units of energy, and when the mass increases from m_0 to m the quantity mc^2 changes by an amount equal to the kinetic energy acquired. Einstein called the quantity mc^2 the *total energy* E associated with a particle,

$$E = mc^2. \tag{48.24}$$

This is the famous Einstein equation, which says that mass and energy are equivalent. The quantity

$$E_0 = m_0 c^2 \tag{48.25}$$

is called the *rest-mass energy* or simply the *rest energy*. Equation (48.22) states that the kinetic energy K represents the change in the total energy of a particle,

$$K = E - E_0. \tag{48.26}$$

The equation $E = mc^2$ is one of the most important discoveries in twentieth century physics. When Einstein first proposed this equation in 1905, he wrote

> It is not impossible that with bodies whose energy content is variable to a high degree (e.g. with radium salts) the theory may be successfully put to the test.

The theory has indeed been abundantly put to the test. We'll discuss specific examples in the next section. Let us review briefly the line of reasoning verified by the successful tests.

The equation $E = mc^2$ arose out of our assumptions that relativistic momentum and relativistic mass are conserved. The final conclusion was that, in relativity, mass and energy are equivalent quantities, related by a constant of conversion, just as miles and kilometers are equivalent quantities, related by a constant of conversion. Thus, when we assumed that the total relativistic mass would be conserved, it turned out that this assumption was logically equivalent to assuming that energy would be conserved. So the underlying principle is the law of conservation of energy.

This point is not trivial, especially in the example we used – an *inelastic* collision, one in which kinetic energy is not conserved in any frame of reference. In classical mechanics that kind of collision is described by saying that all the kinetic energy is converted into heat or some other internal energy of the system. In relativity all energy, including heat, contributes to an increase in the rest mass of the impacted objects. All forms of energy, potential, kinetic, thermal, and every other kind of energy contribute to the relativistic mass of a body. The law of conservation of mass–energy states that the total energy in the universe is constant.

According to Newtonian mechanics, the nonrelativistic kinetic energy $m_0 v^2/2$ of a body increases with the square of the speed, as shown on the lower curve in Fig. 48.5, reaching the value $m_0 c^2/2$ when $v = c$. The total relativistic energy of a particle is plotted in the upper curve in Fig. 48.5. Its dependence on v/c is given by

$$E = mc^2 = \frac{m_0 c^2}{\sqrt{1 - v^2/c^2}}.$$

At zero speed the total energy is $E_0 = m_0 c^2$, and as the speed increases towards c

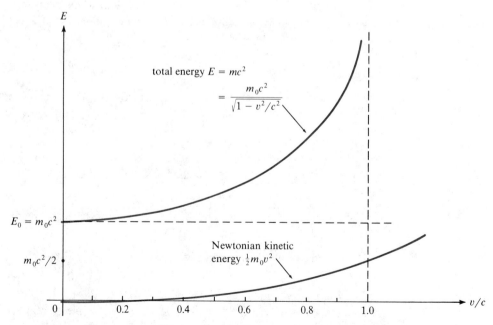

Figure 48.5 Comparison of Newtonian kinetic energy (lower curve) and relativistic total energy (upper curve) as functions of v/c.

48.3 RELATIVISTIC KINETIC ENERGY

the total energy becomes arbitrarily large. This indicates that to propel an object to the speed of light would require giving the object an infinite amount of energy. At small speeds the upper and lower curves have nearly the same shape because $E = E_0 + K$ and, as shown in Example 5, $K \approx m_0 v^2/2$ for small v, so $E - m_0 v^2/2 \approx E - K = E_0$, a constant.

We can gain greater insight by expressing the total energy E in terms of the magnitude of momentum, $p = mv$. Start with the equation $m = \gamma m_0$, square both members, solve for m_0^2, and use $1/\gamma^2 = 1 - v^2/c^2$ to get

$$m_0^2 = (1 - v^2/c^2) m^2.$$

Multiplication by c^4 gives us

$$m_0^2 c^4 = m^2 c^4 - m^2 v^2 c^2.$$

Substituting $p = mv$ and $E = mc^2$, we obtain

$$m_0^2 c^4 = E^2 - p^2 c^2$$

or

$$E^2 = (pc)^2 + (m_0 c^2)^2. \tag{48.27}$$

This result gives a relationship between energy and momentum that does not involve the speed. It shows that part of the energy is associated with the momentum and part with the rest mass. The graph of E as a function of p is shown in Fig. 48.6. This graph is compared with the corresponding relation from classical mechanics, $E - E_0 = m_0 v^2/2$, which can be written in terms of p as

$$E = E_0 + \frac{1}{2m_0} p^2.$$

The graph reveals what happens in the two extreme cases of very low speeds and very high speeds.

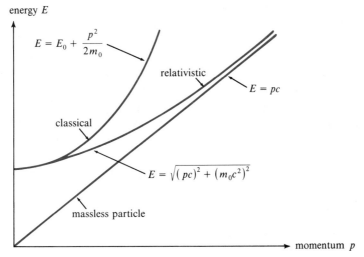

Figure 48.6 Total relativistic energy as a function of momentum.

For $v \ll c$ we have the classical region of a nonrelativistic particle. The relativistic equations for mass, momentum, and kinetic energy reduce to

$$m \approx m_0, \qquad p \approx m_0 v, \qquad K \approx \tfrac{1}{2} m_0 v^2.$$

In this region the kinetic energy is much less than the rest energy, as is seen by examining the ratio

$$\frac{K}{E_0} = \frac{\tfrac{1}{2} m_0 v^2}{m_0 c^2} = \frac{1}{2} \frac{v^2}{c^2} \ll 1.$$

At the opposite extreme, when v is close to c, $p^2 c^2$ is nearly $m^2 c^4$ and, because m is large compared to m_0, the term $m_0^2 c^4$ in Eq. (48.27) is small compared to $p^2 c^2$ and can be ignored. Then the relationship between energy and momentum is $E \approx pc$. At this extreme we also have $m \gg m_0$, $E \gg E_0$, and $K \approx E$.

There's a third interesting case – that of a particle with *zero* rest mass, $m_0 = 0$. Particles of this type include photons (particles of light) and neutrinos, subatomic particles produced by the weak nuclear force. For such a particle, energy and momentum make no sense in the classical viewpoint, but they are meaningful in relativity theory. From Eq. (48.27) we see that a particle with zero rest mass has total energy given by

$$E = pc, \qquad (48.28)$$

which is also equal to its kinetic energy, $K = E$. Moreover, the equation $E = mc^2$ becomes $E = (p/v)c^2$, so the speed of such a particle is given by

$$v = pc^2/E, \qquad (48.29)$$

which, when combined with Eq. (48.28) gives $v = c$. In other words, particles that have zero rest mass must travel at the speed of light.

One additional note regarding units. A convenient unit of energy for subatomic particles is the mega (million) electron volt, denoted by MeV, where

$$1 \text{ MeV} = 10^6 \text{ eV}.$$

We recall that $1 \text{ eV} = 1.6 \times 10^{-19}$ J, so $1 \text{ MeV} = 1.6 \times 10^{-13}$ J. Another unit is the giga electron volt, GeV, where

$$1 \text{ GeV} = 1000 \text{ MeV} = 10^9 \text{ eV} = 1.6 \times 10^{-10} \text{ J}.$$

The Einstein relation $E = mc^2$ implies $m = E/c^2$, which gives a convenient measure of mass in terms of energy because c is a universal constant. A convenient unit of mass is MeV/c^2. In these units an electron has a mass of 0.511 MeV/c^2, and a proton has mass 938 MeV/c^2. To convert to kg we use the fact that 1 eV/c^2 equals 1.78×10^{-36} kg. Similarly, Eq. (48.29) implies $p = Ev/c^2 = (v/c)E/c$, so a convenient unit of momentum is that of energy divided by the speed of light, such as MeV/c or GeV/c.

Example 6

A particle of rest mass m_0 and kinetic energy $5m_0 c^2$ strikes a stationary particle of rest mass m_0. The resulting composite particle has rest mass M_0 and moves with a speed V, as shown.

48.3 RELATIVISTIC KINETIC ENERGY

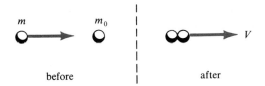

before | after

(a) What is the total momentum of the system before the collision?
(b) What is the total energy of the system before the collision?
(c) What is the speed V of the composite particle?
(d) What is the mass M_0 in terms of m_0?

(a) The total energy of the moving particle is related to its momentum by the equation

$$E^2 = (pc)^2 + (m_0 c^2)^2, \qquad (48.27)$$

so the momentum of the moving particle before the collision is

$$p = \frac{1}{c}\sqrt{E^2 - (m_0 c^2)^2},$$

where E, the total energy of the particle, is given by

$$E = K + m_0 c^2 = 6 m_0 c^2.$$

Using this in the expression for p we find that $p = \sqrt{35}\, m_0 c$. But the other particle is not moving before collision, so this is also the total momentum p_T of the system before the collision.

$$p_T = \sqrt{35}\, m_0 c.$$

(b) The total energy E_T of the system before the collision is the sum of the total energies of the two particles.

$$E_T = 6 m_0 c^2 + m_0 c^2 = 7 m_0 c^2.$$

(c) After collision, the composite particle has momentum p_T and energy E_T. According to Eq. (48.29), the speed of the particle is given by

$$V = p_T c^2 / E_T = \left(\sqrt{35}\, m_0 c\right) c^2 / \left(7 m_0 c^2\right) = 0.85 c.$$

(d) Applying Eq. (48.27) to the composite particle and solving for M_0 we have

$$M_0 c^2 = \sqrt{E_T^2 - (p_T c)^2}.$$

Substituting for E_T and p_T we find $M_0 = \sqrt{14}\, m_0$. Note that $M_0 \neq 2 m_0$, so rest mass is not conserved. Instead, kinetic energy of the incident particle has been converted into mass of the final particle.

Questions

8. If mass is a form of energy, does a spring have more mass when it is compressed than when it is relaxed? Explain.

9. Does the equivalence of mass and energy depend on the nature of the forces involved? That is, does $E = mc^2$ hold for nuclear, electric, and gravitational forces? Explain.

10. How many grams of rest mass are needed to produce 1 J of energy if all the rest mass is transformed?

11. An electron ($m_0 = 9.1 \times 10^{-31}$ kg) travels at a speed of $0.6c$. What is its kinetic energy? How does this compare with the Newtonian kinetic energy at the same speed?

12. An electron has a total energy that is five times its rest-mass energy. Determine (a) its speed and (b) its momentum.

13. A particle of rest mass m_0 initially moves at a speed of $0.4c$.

    (a) If its speed is doubled, by what factor does its kinetic energy increase?
    (b) If the total energy is increased by the factor 100, by what factor is its speed increased?

14. Show that the momentum p of a particle is related to its kinetic energy K and rest mass m_0 by the equation $pc = \sqrt{K^2 + 2Km_0c^2}$.

15. A particle has a total energy of 5 GeV and a momentum of 3 GeV/c according to measurements in a certain frame.

    (a) What is the energy of the particle in a frame in which its momentum is 5 GeV/c?
    (b) What is the rest mass of the particle?
    (c) What is the relative speed of the two frames?

16. (a) Show that in classical mechanics a particle having Newtonian kinetic energy K and momentum p has speed $v = dK/dp$.
    (b) Show that the speed of a relativistic particle is given by dE/dp, where E is the total energy.

48.4 APPLICATIONS OF CONSERVATION OF RELATIVISTIC ENERGY AND MOMENTUM

The interconversion of mass and energy was a stunning prediction of relativity theory. In the decades since this prediction was made by Einstein, the equivalence of mass and energy has been confirmed by a host of experiments. The interconversion is most easily detected in processes involving subatomic particles

For example, consider the decay of an elementary particle known as a neutral pion, or pi zero (π^0). This subatomic particle, which has a rest mass of 135 MeV/c^2 and an average proper decay time of 0.83×10^{-16} s, decays into two photons. If the decay occurs in a frame in which the pion is at rest, what are the energies of the photons? We analyze the decay by applying conservation of relativistic momentum and energy.

In the rest frame of π^0 the particle is at rest and has zero momentum. Conservation of momentum requires that the total momentum of the photons in

48.4 APPLICATIONS OF CONSERVATION OF RELATIVISTIC ENERGY AND MOMENTUM

that frame must be zero. Therefore, the photons have momenta that are equal in magnitude but have opposite directions. But the momentum pc and energy E_γ of a photon are related by Eq. (48.28), $E_\gamma = pc$, so the photons also have equal energies.

According to the conservation of mass–energy, the mass of the pion appears entirely as energy of the photons because they have zero rest mass. If m_π is the rest mass of the pion and E_γ is the energy of either photon, conservation of mass–energy requires that

$$m_\pi c^2 = 2E_\gamma,$$

so that each photon has an energy

$$E_\gamma = m_\pi c^2/2,$$

a result that has been confirmed by experiment.

The reverse process – the conversion of energy into mass – is also commonly observed by nuclear physicists. Under certain conditions, photons passing near nuclei disappear, with the simultaneous creation of particles with rest mass.

On a larger scale, the energy produced in nuclear reactors is the result of the conversion of the rest mass of uranium nuclei into energy in a process called fission. On an even grander scale, the sunlight received on Earth results from fusion reactions taking place in the core of the sun, where, through a series of reactions, hydrogen is converted into helium; the difference in rest mass between the initial reactants and the final products appears as energy. Consequently, the mass of the sun is actually decreasing as it radiates energy into space.

Example 7

An elementary particle called the neutral K meson, or kaon, decays into two neutral pions. The rest mass of a kaon is 498 MeV/c^2 and that of a neutral pion is 135 MeV/c^2.

(a) If a kaon decays at rest in the laboratory frame, what is the kinetic energy of the resulting pions?

(b) What is the speed of each pion?

(a) Conservation of energy requires that the energy of the kaon, which is entirely rest-mass energy, be equal to the total energy of the resulting pions. Thus, we have

$$m_K c^2 = K_T + 2m_\pi c^2,$$

where K_T is the total kinetic energy of the pions. Solving for K_T and substituting for the rest masses, we find $K_T = 228$ MeV.

(b) Because the initial kaon is at rest and has zero momentum, conservation of momentum requires that the pions have equal and opposite momenta, and because the particles have identical rest masses, it follows that they also are produced with equal kinetic energies. Therefore, the kinetic energy of either pion is $K = \tfrac{1}{2}K_T =$

114 MeV. Applying Eq. (48.27) to a single pion, we find

$$p = \sqrt{E^2 - (m_\pi c^2)^2}/c = 209 \text{ MeV}/c,$$

where we used $E = K + m_\pi c^2$. Then by Eq. (48.29), we find the speed to be $v = pc^2/E = 0.84c$.

Questions

17. Would it be possible for a pi zero at rest to decay into two identical particles of rest mass 70 MeV/c^2? Would the decay be possible if the pi zero is moving? Explain.

18. Radiation from the sun reaches the earth at a rate of 1400 W/m² at the equator at noon on the first day of spring.

 (a) What is the loss in mass of the sun each second?
 (b) What is the percentage decrease in the mass of the sun over 10 billion years? (The distance from the earth to the sun is 1.5×10^{11} m and the present mass of the sun is 2.0×10^{30} kg.)

19. An experimenter observes a proton and an antiproton (a particle having the same rest mass as a proton, 938 MeV/c^2, but opposite charge) approach each other from opposite directions. Each particle has kinetic energy equal to twice its rest-mass energy. The particles collide and annihilate each other, and two photons are created. In the frame of the experimenter, determine each of the following:

 (a) the momentum of the proton;
 (b) the speed of the proton;
 (c) the total energy of each photon;
 (d) the momentum of each photon.
 (e) Can the proton and antiproton annihilate each other and in the process create only a single photon?

20. A particle of rest mass m_0 and kinetic energy $3m_0c^2$ strikes and becomes bound to an initially stationary particle of rest mass $5m_0$. The composite particle moves off with a speed V.

 (a) Find the speed of the composite particle in terms of the speed of light.
 (b) Determine the rest mass of the composite particle in terms of m_0.
 (c) Find the speed of the incoming particle.

21. A pi meson of rest mass m_π decays at rest into a muon (rest mass m_μ) and a neutrino (of zero rest mass). Show that the kinetic energy of the muon is given by $K(m_\pi - m_\mu)^2c^2/(2m_\pi)$.

22. The Σ^+ particle is an unstable elementary particle with a rest mass of 1.19 GeV/c^2 and a mean proper lifetime of 0.8×10^{-10} s. What is the minimum kinetic energy such a particle must have to travel a distance of 10 cm before decaying?

23. Show that it is dynamically impossible for a single photon to strike a stationary electron and give up all its energy to the electron.

48.5 A FINAL WORD

When Einstein wrote his famous paper on the theory of relativity in 1905, he was only 26 years old and was employed as a patent clerk in Berne, Switzerland. The paper changed the meaning of space and time, and superseded Newtonian mechanics, which had been the basis of all physics, and indeed of all philosophy, for 200 years. What sort of man was this patent clerk who had the audacity to throw out everything that was known about the world and invent a complex theory that contained no internal contradictions? This may be an appropriate time to say a few words about his background.

Einstein was born in 1879 and died in 1955. His father, an unsuccessful Munich businessman, was an owner of an electrochemical plant that failed. The family eventually left Germany, spending one year in Milan, Italy, and then moving to Switzerland, where Einstein finished high school. As a child, Albert was a slow learner and didn't learn to speak until after he was 3 years old. Although he was an indifferent student, he finished high school in Switzerland and went to the famous Swiss Federal Polytechnic (ETH) in Zurich, from which he graduated in 1900.

From 1902 to 1909 he was employed as a patent clerk in Berne, and during that period laid some of the foundations for twentieth century physics. His contributions were not only to relativity. During the epic year 1905 he published three landmark papers in the same volume of the journal *Annalen der Physik*. For the first of these, on the photoelectric effect, he was awarded the Nobel prize in physics in 1921. The second paper, an outgrowth of his Ph.D. thesis for the University of Zurich, was an analysis of the phenomenon known as Brownian motion. If he had done nothing else, this work by itself would have made him famous. The third paper presented the special theory of relativity. During his years as a patent clerk he also invented the quantum theory of solid state physics.

The period from 1902 to 1909, and especially the year 1905, can be compared to the period Newton spent on his farm in Lincolnshire in the plague years 1665–6. It was an immensely rich and productive period in Einstein's life. By 1909 the academic world had heard of him, and he was appointed Associate Professor of Physics at the University of Zurich where he had done his graduate work. In 1911 he moved to Prague, then back to his earlier alma mater, ETH, in 1912, and in 1914 he was appointed director of the Kaiser Wilhelm Institute in Berlin.

From 1909 to 1916 he tried to extend his theory of relativity to encompass the phenomenon of gravity, and by 1916 he succeeded, and produced what is known today as the general theory of relativity.

In a letter to George Ellery Hale of Caltech dated 1913, Einstein states, "...some simple theoretical considerations have led me to believe that...light passing near the sun would be deflected by the sun." That is, the gravitational pull

Figure 48.7 Photograph of Einstein circa 1916. (By permission of the Hebrew University of Jerusalem, Israel.)

of the sun would deflect a light beam passing near it. Einstein predicted the angle of deflection as 0.84 s of arc. It turned out that 0.84 was the wrong numerical value. There are two possible results: one deflection derived from special relativity and another twice as large predicted from general relativity neither of which is 0.84 s of arc. Einstein knew there would be a deflection, but arithmetic was not one of his strong points and he made an arithmetical error in his calculation. He asked Hale to use the Mt. Wilson Observatory telescope (the world's largest telescope at the time) to look toward the sun and see if stars near the sun seem to be displaced a little bit.

Hale replied, in essence, that they wouldn't point the telescope at the sun because it might cost the observer his eyesight. He also suggested that the ideal time to conduct such an experiment is during a solar eclipse. Because of World War I, the experiment could not be carried out until 1919. Eclipses are visible only in certain regions, and in 1919 two expeditions were sent by the Royal Astronomical Society, one to northern Brazil, and one to West Africa. Both found the displacement of light near the sun to be close to what Einstein's general theory had predicted. Sir Arthur Eddington, leader of the African expedition, cabled the news to Einstein and to the Society. The announcement had an effect that is very difficult to understand today. It was not only the scientific community that was impressed,

48.5 A FINAL WORD

but the world at large. Although it was an obscure, arcane discovery, which very few people understood, the newspapers and all the media gave it wide publicity and, as a result, Einstein immediately became an international folk hero. In later years the experiment was refined and repeated many times. The average deflection observed at 11 different eclipses agrees with the predicted value to within one part in 500.

It's hard to explain the sudden surge of public adulation of Einstein. Perhaps the public, weary from World War I, craved good news and was desperate for a nonpolitical hero. Or perhaps it was because Einstein resembled everyone's favorite uncle. In any case, Einstein's name became a household word and he became a legend in his own time.

CHAPTER 49

ATOMS

But I must confess I am jealous of the term *atom*: for though it is very easy to talk of atoms, it is very difficult to form a clear idea of their nature, especially when compound bodies are under consideration.

Michael Faraday, *Experimental Researches in Electricity* (1833)

49.1 EARLY HISTORY OF ATOMIC THEORY

The famous ninth edition of the *Encyclopaedia Brittanica*, published in 1875, contains an article entitled "Atom" by James Clerk Maxwell. He writes, in part:

> Atom (ἄτομος) is a body which cannot be cut in two. The atomic theory is a theory of the constitution of bodies, which asserts they are made up of atoms. The opposite theory is that of the homogeneity and continuity of bodies, and asserts, at least in the case of bodies having no apparent organization, such, for instance, as water, that we can

divide a drop of water into two parts which are each of them drops of water, so we have reason to believe that these smaller drops can be divided again, and the theory goes on to assert that there is nothing in the nature of things to hinder this process of division from being repeated over and over again, times without end. This is the doctrine of the infinite divisibility of bodies and it is in direct contradiction with the theory of atoms.

The atomists assert that after a certain number of such divisions the parts would no longer be divisible, because each of them would be an atom. The advocates of the continuity of matter assert that the smallest conceivable body has parts, and that whatever has parts may be divided.

In ancient times Democritus was the founder of the atomic theory, while Anaxagoras propounded that of continuity.... The arguments of the atomists and their reply to the objections of Anaxagoras are to be found in Lucretius....

In modern times the study of nature has brought to light many properties of bodies which appear to depend on the magnitude and motions of their ultimate constituents, and the question of the existence of atoms has once more become conspicuous amongst scientific inquiries....

Apparently Maxwell was not aware that the concept of atomic or granular structure of matter was formulated by Indian philosophers as early as 1200 B.C. The idea was further expounded in the fifth century B.C. by the Greek philosopher Leucippus and his distinguished pupil Democritus, who proposed that matter consisted of eternal, impenetrably hard atoms that moved inertially in a vacuum. Their views were immortalized in verse by Lucretius, the famous Roman poet of the first century B.C., in what has been described as "the greatest philosophical poem of all times," *De Rerum Natura* (On the nature of things). Although much of the qualitative nature of the early atomic theory has survived to modern times, the theory was based on pure philosophical speculation rather than controlled quantitative measurements. Twenty-two centuries were to elapse before these speculations could be substantiated by experimental evidence.

The opposing view of Anaxagoras, that matter is continuous and can be subdivided indefinitely, was also supported by Aristotle, to whom the concepts of a vacuum and self-moving bodies were anathema. However, Aristotle did admit to a practical lower limit to material division – *minima naturalis*, somewhat analogous to the nineteenth century "molecule." As the early civilized world succumbed to the attacks of barbarians, the atomic theory all but vanished and the Aristotelian view prevailed through the Middle Ages.

It was not until the beginning of modern science that atomic theory was revived by Galileo, Descartes, Boyle, and Newton. Arguments in favor of the atomic view remained qualitative until the late eighteenth century when a number of experiments were performed that could best be explained on the basis of an atomic hypothesis. We turn now to a brief discussion of some of these experiments.

49.2 EXPERIMENTAL EVIDENCE SUPPORTING ATOMIC THEORY

Credit for the quantitative form of atomic theory is generally ascribed to the English chemist and physicist John Dalton, whose work in the period 1803–10 culminated in his treatise *A New System of Chemical Philosophy*. A number of experiments on

49.2 EXPERIMENTAL EVIDENCE SUPPORTING ATOMIC THEORY

chemical reactions had been performed prior to Dalton's work by the French chemists Antoine-Laurent Lavoisier in 1775 and Joseph Louis Proust in 1800. Dalton's atomic theory provided a simple explanation of the results of these experiments, and the theory was further verified by subsequent experiments in chemistry and physics.

Lavoisier found that if a substance A combines completely with a substance B to form a substance C, then the weight of C that is produced is equal to the sum of the weights of A and B. He referred to his experimental findings as evidence of a *law of conservation of matter*.

Proust's experiments led him to formulate the *law of definite proportions*, which states that the proportions by weight in which elements enter into a given compound are constant. For example, 1 g of hydrogen will combine with 35.2 g of chlorine to form hydrogen chloride. If more than 35.2 g of chlorine is used, some free chlorine will be left over. If less than 35.2 g of chlorine is used, some free hydrogen will be left over. Sometimes a certain amount of substance A will combine with an amount of substance B to form substance C, but the same amount of substance A will combine with a different amount of B to form a different substance D. Dalton found that when this happens one of the amounts of B required must be a rational multiple of the other (a rational number being the ratio of two integers). For example, 16 g of oxygen can combine with 14 g of nitrogen to form 30 g of nitric oxide, but 16 g of oxygen can also combine with twice as much nitrogen (28 g) to form 44 g of nitrous oxide. Also, 32 g of oxygen can combine with 14 g of nitrogen to form 46 g of nitrogen dioxide. This illustrates the *law of simple and multiple proportions* (Fig. 49.1).

Dalton explained these empirically determined laws by assigning to each element an experimental number that by itself, or when multiplied by some small integer, expresses the mass by which the element enters into combination with other elements. He called this experimental number the *atomic weight* of the element. Producing a consistent scheme of relative atomic weights was a slow process that took nearly half of the nineteenth century to complete. (A list of the 109 elements known in 1986 with their atomic weights is given in Table 49.1. Dalton's concept of

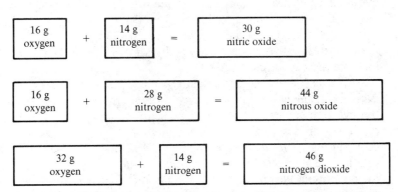

Figure 49.1 The law of multiple proportions applied to oxygen and nitrogen.

atomic weight has been replaced by the concept of atomic mass, described in Section 49.3.) In 1815 an English chemist, William Prout, observed that the known relative atomic weights were all very nearly integer multiples of that of hydrogen, and he suggested that all elements might be composed of hydrogen atoms. This idea became known as *Prout's hypothesis*. It was later discredited when more accurate measurements revealed that some atomic weights were not exact integer multiples of that of hydrogen, but today it is known that the observed atomic weights are actually averages over several isotopes, which are atoms constaining the same number of protons but different numbers of neutrons in their nucleus. Prout's hypothesis is not far from the modern idea that all atoms of a particular element consist of the constituents of hydrogen atoms, plus neutrons. It is interesting to note that the proton was named after William Prout.

Support for the atomic theory also came from experiments dealing with volumes rather than weights. Dalton performed experiments on various gases that led him to formulate a simple law of combining volumes. For example, when 1 L of nitrogen is combined with 1 L of oxygen, 2 L of nitric oxide are obtained.

In 1808, experiments performed on gases by another French chemist and physicist, Joseph Louis Gay-Lussac, led to results that did not conform to Dalton's idea that the elementary gases are composed of single atoms. For example, 2 L of hydrogen combine with 1 L of oxygen to form 2 L of water vapor, or 1 L of nitrogen gas combines with 3 L of hydrogen gas to form 2 L of ammonia gas. Gay-Lussac concluded that Dalton's law of combining volumes must be modified to state that, under conditions of equal pressure and temperature, reacting gases combine in simple volumetric proportions.

In an obscurely phrased essay published in 1811, the Italian physicist Amedeo Avogadro proposed that for some elements the smallest particles were not single atoms but groups of two or more atoms. He suggested that one must distinguish between two fundamental structures, one of which is the atom and the other, that which is now called the *molecule*. Moreover, Avogadro proposed that equal volumes of different gases at the same temperature and pressure must contain equal numbers of molecules. This hypothesis, now known as *Avogadro's law*, together with the idea that gases of some elements such as hydrogen, oxygen, and nitrogen consist of diatomic molecules, gave a clear explanation of Gay-Lussac's experiments and his empirical law of combining volumes. But nearly 50 years passed before the implications of Avogadro's ideas were realized. In 1858, the Italian chemist Stanislao Cannizzaro drew attention to Avogadro's ideas and showed that they provided a reasonable basis for chemistry.

Michael Faraday's experiments on electrolysis provided another source of evidence in favor of the atomicity of matter. When an electric current passes through certain chemical solutions, it seems to carry with it matter that is deposited at the electrodes (which become plated) or that is liberated in gaseous form. Experiments performed by Faraday in 1832 and 1833 revealed that for a given solution the amount of matter deposited on the electrodes is directly proportional to the quantity of electricity passed through the solution. Today we know that this is because electricity is composed of individual particles (electrons), and quantity of electricity represents numbers of electrons. The nature of electrons was not well known in 1833, but the laws of electrolysis suggested to both Dalton and Faraday that matter was atomic in structure.

Further evidence in support of the atomic theory was found in the interpretation of thermal phenomena in terms of random molecular motions. Research in thermodynamics suggested that heat energy resided in the kinetic and potential energy of atomic and molecular motions. The second law of thermodynamics and the concept of entropy were explained as a natural tendency for molecular motions to proceed from ordered to disordered forms. By the second half of the nineteenth century, it had become possible to estimate the size of an atom from measurements of the rates at which one gas diffuses through another, and from measurements of the viscosities of gases. James Clerk Maxwell was prominent in this field of research, and the estimate he cited in his 1875 *Encyclopaedia Brittanica* article was within a factor of 2 of the modern value of the typical size of an atom.

Perhaps the most striking evidence supporting atomic theory was the discovery in 1869 by Dmitri Mendeleev of the periodic classification of the elements. He made a thorough study of the relation between atomic weights of elements and their physical and chemical properties, exploiting the concept of valence that had been introduced in 1852 by Edward Frankland. He arranged the known elements in order of increasing atomic weight and discovered a periodic recurrence of elements having similar chemical properties. This similarity of properties implied a similarity of atomic structure, and was the first convincing evidence that atoms of different elements have a common pattern in their structure.

The modern views of atomic structure were built largely from a series of discoveries made at the very end of the nineteenth century. They included Roentgen's discovery of X rays in 1895, Becquerel's discovery of radioactivity in 1896, and J. J. Thomson's discovery in 1897 that the electron was a common constituent of many kinds of matter. Incidentally, the name *electron* for the unit of electrical charge was suggested by G. Johnstone Stoney in 1874. Thomson wanted to use the name *corpuscle*, but his wish did not prevail.

49.3 THE ATOMIC STRUCTURE OF MATTER

Today it is common knowledge that all ordinary matter consists of atoms of some 100 elements, an *element* being matter represented by one kind of atom. Substances that consist of atoms of two or more different kinds are called *compounds*. Atoms themselves are made of even smaller particles: electrons, protons, and neutrons. However, the atom is the unit that retains its identity when chemical reactions take place, and is the structural member of all solids, liquids, and gases. A list of the known elements is given in Table 49.1.

No human eye has ever seen an atom, so there is only indirect evidence of how an atom is put together. The commonly accepted model, that of a *nuclear* atom, assumes that every atom consists of a central nucleus containing electrically positive protons and neutral neutrons, surrounded by enough negatively charged electrons to make the atom electrically neutral. The simplest atom is that of hydrogen. Its nucleus (just one proton) has positive charge $e = 1.60219 \times 10^{-19}$ C and a mass about 1836 times that of an electron.

Table 49.1 A List of the Known Elements.

Atomic number	Symbol	Name	Atomic mass	Atomic number	Symbol	Name	Atomic mass
1	H	hydrogen	1.01	55	Cs	cesium	132.91
2	He	helium	4.00	56	Ba	barium	137.33
3	Li	lithium	6.94	57	La	lanthanum	138.91
4	Be	beryllium	9.01	58	Ce	cerium	140.12
5	B	boron	10.81	59	Pr	praseodymium	140.91
6	C	carbon	12.01	60	Nd	neodymium	144.24
7	N	nitrogen	14.01	61	Pm	promethium	(145)
8	O	oxygen	16.00	62	Sm	samarium	150.36
9	F	fluorine	19.00	63	Eu	europium	151.96
10	Ne	neon	20.18	64	Gd	gadolinium	157.25
11	Na	sodium	22.99	65	Tb	terbium	158.93
12	Mg	magnesium	24.31	66	Dy	dysprosium	162.50
13	Al	aluminum	26.98	67	Ho	holmium	164.93
14	Si	silicon	28.09	68	Er	erbium	167.26
15	P	phosphorus	30.97	69	Tm	thulium	168.93
16	S	sulphur	32.06	70	Yb	ytterbium	173.04
17	Cl	chlorine	35.45	71	Lu	lutetium	174.97
18	Ar	argon	39.95	72	Hf	hafnium	178.49
19	K	potassium	39.10	73	Ta	tantalum	180.95
20	Ca	calcium	40.08	74	W	tungsten	183.85
21	Sc	scandium	44.96	75	Re	rhenium	186.2
22	Ti	titanium	47.88	76	Os	osmium	190.2
23	V	vanadium	50.94	77	Ir	iridium	192.2
24	Cr	chromium	52.00	78	Pt	platinum	195.08
25	Mn	manganese	54.94	79	Au	gold	196.97
26	Fe	iron	55.85	80	Hg	mercury	200.59
27	Co	cobalt	58.93	81	Tl	thallium	204.38
28	Ni	nickel	58.69	82	Pb	lead	207.19
29	Cu	copper	63.55	83	Bi	bismuth	208.98
30	Zn	zinc	65.39	84	Po	polonium	(209)
31	Ga	gallium	69.72	85	At	astatine	(210)
32	Ge	germanium	72.59	86	Rn	radon	(222)
33	As	arsenic	74.92	87	Fr	francium	(223)
34	Se	selenium	78.96	88	Ra	radium	226.03
35	Br	bromine	79.90	89	Ac	actinium	227
36	Kr	krypton	83.80	90	Th	thorium	232.04
37	Rb	rubidium	85.47	91	Pa	protactinium	231
38	Sr	strontium	87.62	92	U	uranium	238.03
39	Y	yttrium	88.91	93	Np	neptunium	(237)
40	Zr	zirconium	91.22	94	Pu	plutonium	(244)
41	Nb	niobium	92.91	95	Am	americium	(243)
42	Mo	molybdenum	95.94	96	Cm	curium	(247)
43	Tc	technetium	(98)	97	Bk	berkelium	(247)
44	Ru	ruthenium	101.07	98	Cf	californium	(251)
45	Rh	rhodium	102.91	99	Es	einsteinium	(252)
46	Pd	palladium	106.4	100	Fm	fermium	(257)
47	Ag	silver	107.87	101	Md	mendelevium	(258)
48	Cd	cadmium	112.41	102	No	nobelium	(259)
49	In	indium	114.82	103	Lr	lawrencium	(260)
50	Sn	tin	118.71	104	Rf	rutherfordium	(261)
51	Sb	antimony	121.75	105	Ha	hahnium	257–262
52	Te	tellurium	127.60	106	not named; 1974		259–263
53	I	iodine	126.91	107	not named; 1981		262
54	Xe	xenon	131.29	108	not named; 1984		265
				109	not named; 1982		266

Numbers in parentheses are mass numbers of most stable isotope of that element.

49.3 THE ATOMIC STRUCTURE OF MATTER

The number of protons in the nucleus (or, equivalently, the number of electrons surrounding the nucleus) is called the *atomic number* of the atom, and is usually denoted by Z. This is also called the atomic number of the element, all of whose atoms have atomic number Z. Hydrogen, the lightest element, has atomic number $Z = 1$. Helium, the next lightest element, has $Z = 2$, and so on. The known elements represent all the atomic numbers from 1 to 109.

In 1932, James Chadwick discovered that nuclei contain particles in addition to protons. These particles, which are electrically neutral and have slightly greater mass than protons, were later named *neutrons*. Collectively, protons and neutrons are referred to as nucleons because they are the building blocks of nuclei. The total number of nucleons in a nucleus is called the *atomic mass number A*, and is equal to $N + Z$, where N is the number of neutrons in a nucleus. For example, the mass number of helium is $A = 4$, because helium contains two protons and two neutrons. However, the number of protons Z and the number of neutrons N in the nucleus of a given element are not always equal. Nuclei of an element that have the same atomic number Z but different numbers of neutrons N are known as *isotopes*. For example, hydrogen has an isotope called tritium that has $Z = 1$ and $N = 2$. This isotope is unstable and decomposes spontaneously. In Chapter 51 we'll see how studies of isotopes have led to a better understanding of the physics of the nucleus.

Masses of atoms are conveniently expressed in *atomic mass units*. By definition, 1 atomic mass unit (1 u), also known as the unified mass unit, is 1/12 the mass of the most abundant isotope of carbon, which has $A = 12$. A glance at Table 49.1 reveals that relative to this standard a remarkable number of elements have atomic masses that are nearly integers, these masses being nearly equal to the mass number.

The term *molecule* refers to the smallest unit into which a substance (element or compound) can be divided and still retain the chemical properties of that substance. Most molecules contain at least two atoms, but a few are monatomic. For example, an atom of helium is also a molecule of helium, but each molecule of hydrogen or of oxygen contains two atoms. The molecules of some chemical substances contain huge numbers of atoms bonded tightly together.

The *molecular weight* of a compound is the sum of the atomic masses of the atoms in a molecule of the compound. A *mole* of a substance is M grams of the substance, where M is numerically equal to its molecular weight.

Example 1
Prove that any two substances contain the same number of molecules in 1 mole.

For a given substance A, let $N(A)$ denote the number of molecules in 1 mole, let $M(A)$ denote the molecular weight of A, and let $w(A)$ denote the weight of a single molecule of A. Then

the weight of 1 mole of $A = w(A)N(A)$.

Similarly,

the weight of 1 mole of $B = w(B)N(B)$,

and therefore the ratio of these weights is

$$\frac{\text{the weight of 1 mole of } A}{\text{the weight of 1 mole of } B} = \frac{w(A)N(A)}{w(B)N(B)}.$$

But by the definition of mole we also have

$$\frac{\text{the weight of 1 mole of } A}{\text{the weight of 1 mole of } B} = \frac{M(A)}{M(B)},$$

and hence

$$\frac{M(A)}{M(B)} = \frac{w(A)N(A)}{w(B)N(B)}.$$

On the other hand, the molecular weight of each substance is proportional to the weight of one molecule of the substance, so we also have

$$\frac{M(A)}{M(B)} = \frac{w(A)}{w(B)}.$$

Comparing this with the foregoing equation we see that $N(A) = N(B)$. In other words, any two substances A and B contain the same number of molecules in 1 mole.

It should be noted that the calculation in Example 1 does not depend on whether the substances are solids, liquids, or gases. This means that there is a universal integer that represents the number of molecules in 1 mole of any substance in any state. This number is called *Avogadro's number* N_A and it can be determined experimentally. The currently accepted value is

$$N_A = 6.0220 \times 10^{23} \text{ molecules/mole.}$$

Example 2

Measurement shows that 1 mole of ordinary table salt (NaCl) has a volume of 27.0 cm^3. Patterns of X rays scattered by salt crystals suggest that the atoms of sodium (Na) and chlorine (Cl) are arranged in a cubical lattice. Use this information together with Avogadro's number to estimate the radius of an atom of sodium or chlorine.

According to Example 1, 1 mole of NaCl should contain $2N_A = 12.04 \times 10^{23}$ atoms. Because these are distributed in 27.0 cm^3 of space, each atom occupies (on the average) a volume of

$$27.0/(12.04 \times 10^{23}) = 22.4 \times 10^{-24} \text{ cm}^3.$$

A cube with this volume would have an edge equal to 2.82×10^{-8} cm, and a sphere inscribed in this cube would have a radius of

$$r = 1.46 \times 10^{-8} \text{ cm}.$$

This figure is remarkably close to the radius of an atom such as sodium or chlorine as determined by experimental methods.

Questions

1. One mole of liquid water occupies a volume of 18 cm^3. Assume that each molecule is at the center of a small cube.

   (a) Find the length of each edge of this small cube.
   (b) Compare your result with the diameter of an atom as given in Example 2.

2. One mole of an ideal gas occupies 22.4 L at standard temperature and pressure (273 K and 1 atm).

   (a) Assuming each molecule to be at the center of a small cube, find the length of each edge of the cube.
   (b) Explain why your answer does not agree with the estimated size of an atom as given in Example 2.

3. When a drop of oil forms a film on a water surface, the thickness of the film is the size of one molecule. Suppose you know the amount of oil in a drop and can measure the radius of the circular film that forms after the drop spreads out over a water surface. Show how you can use these measurements to determine Avogadro's number.

49.4 RUTHERFORD'S MODEL OF THE ATOM

The nuclear model, proposed in 1911 by Ernest Rutherford of the University of Manchester, superseded an earlier model of J. J. Thomson in which each atom was thought to be a spherical globule of homogeneous positive charge with electrons imbedded in it. Thomson's model was not consistent with results of experiments performed by Rutherford and his colleagues Hans Geiger and Ernest Marsden, who studied the scattering of energetic alpha particles from radioactive elements passing through thin foils of heavy metals, such as gold.

If Thomson's model of the atom were correct, firing alpha particles into gold foil would be somewhat like firing bullets into a bag of marshmallows. You wouldn't expect the bullets to deviate very much from their path. To describe what actually happened in these experiments, we use Rutherford's own words:

> I would like to use this example to show you how often facts are stumbled upon by accident. In the early days, I had observed the scattering of alpha particles, and Dr. Geiger in my laboratory had examined it in detail. He found in thin pieces of heavy metal that the scattering was usually small, of the order of one degree. One day Geiger came to me and said, "Don't you think that young Marsden, whom I am training in

Figure 49.2 Photograph of Ernest Rutherford. (Courtesy of the Archives, California Institute of Technology.)

radioactive methods, ought to begin a small research?" Now I had thought that, too, so I said, "Why not let him see if any alpha particles can be scattered through a large angle?" I may tell you in confidence that I did not believe that they would be, since we knew that the alpha particle was a very fast, massive particle, with a great deal of energy, and you could show that if the scattering was due to the accumulated effect of a number of small scatterings, the chance of an alpha particle's being scattered backwards is very small. Then I remember two or three days later Geiger coming to me in great excitement and saying, "We have been able to get some of the alpha particles coming backwards. ..." It was quite the most incredible event that has happened in my life. It was almost as incredible as if you fired a 15-inch shell at a piece of tissue paper and it came back and hit you. On consideration, I realized the scattering backward must be the result of a single collision, and when I made calculations, I saw that it was impossible to get anything of that order of magnitude unless you took a system in which the greater part of the mass of the atom was concentrated in a minute nucleus. It was then that I had the idea of an atom with a minute massive center carrying a charge. I worked out mathematically what laws the scattering should obey and I found that the number of particles scattered through a given angle should be proportional to the thickness of the scattering foil, the square of the nuclear charge, and inversely proportional to the fourth power of the velocity. These deductions were later verified by Geiger and Marsden in a series of beautiful experiments. [*Background to Modern Science*, Macmillan (1938)]

49.4 RUTHERFORD'S MODEL OF THE ATOM

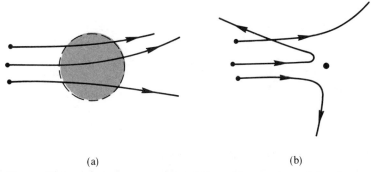

Figure 49.3 Scattering (a) as expected from Thomson's model and (b) as observed by Geiger and Marsden.

These experiments led Rutherford to formulate the nuclear model of the atom, with electrons moving around a massive nucleus. This is sometimes called the *planetary model* because it suggests that the electrons move around the nucleus like planets orbiting the sun. This model has great appeal but it is not entirely consistent with classical physics. To understand some of the difficulties this model implies, we consider the simplest type of atom, the hydrogen atom, and assume that its single electron moves with constant speed v in a circular orbit of radius r, as shown in Fig. 49.4.

We know from our study of uniform circular motion in Chapter 9 that the electron must undergo centripetal acceleration v^2/r and hence must be acted on by a centripetal force of magnitude

$$F = \frac{m_e v^2}{r},$$

where m_e is the mass of the electron. The source of this force is the electrical attraction between the electron and the nucleus. By Coulomb's law, Eq. (32.1), this

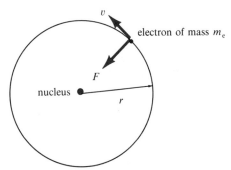

Figure 49.4 Planetary model of a hydrogen atom with one electron moving at constant speed in a circular orbit around the nucleus.

force is given by

$$F = K_e \frac{e^2}{r^2},$$

where e denotes the magnitude of the charge of the electron. Equating the two expressions for F we find

$$m_e v^2 = K_e \frac{e^2}{r}. \tag{49.1}$$

This equation shows that any orbital radius can provide a stable orbit, with the velocity varying inversely as the square root of the radius. Now Maxwell's electromagnetic theory tells us that an accelerated charge always emits electromagnetic energy. To see what this implies about the orbiting electron we calculate the total (nonrelativistic) energy of the electron.

Example 3

Calculate the total (nonrelativistic) energy possessed by a hydrogen atom in which an electron orbits the proton at constant speed v along a circle of radius r.

The total energy consists of two parts,

$$E = K + U,$$

where K is the kinetic energy of the electron and U is the potential energy of the system. By Eq. (49.1), the kinetic energy is given by

$$K = \frac{1}{2} m_e v^2 = K_e \frac{e^2}{2r},$$

whereas the potential energy is the amount of work done on the system against the attractive force F in moving the electron from infinity to the distance r from the proton,

$$U = -\int_\infty^r F(x)\, dx.$$

At a distance x from the nucleus the force is

$$F(x) = -K_e \frac{e^2}{x^2}.$$

Integrating this force to obtain the potential energy, we find

$$U = K_e e^2 \int_\infty^r x^{-2}\, dx,$$

hence

$$U = -K_e \frac{e^2}{r}. \tag{49.2}$$

49.4 RUTHERFORD'S MODEL OF THE ATOM

Therefore, the total energy is

$$E = K + U = -K_e \frac{e^2}{2r}. \qquad (49.3)$$

For an orbiting electron in a hydrogen atom to lose energy as it accelerates, the term on the right of Eq. (49.3) must become more negative, which means the radius r of the orbit must decrease. In other words, a constantly radiating electron will spiral down into the nucleus, thereby destroying the planetary nature of the model. Moreover, as the electron spirals in, its frequency of rotation will vary continuously, and this implies that the atom should emit electromagnetic waves with continuously varying frequencies. Much that is known about atomic structure is deduced from a study of the light that is emitted and absorbed by atoms. Observations show that the hydrogen atom emits energy only at certain fixed *discrete* frequencies, so the predictions of continuously varying frequencies from the planetary model do not agree with experimental observations.

Questions

4. On the basis of the kinetic theory of gases it is known that the diameter of a molecule of hydrogen is about 2×10^{-10} m. Use the planetary model of the hydrogen atom to show that the total energy of the electron is -23×10^{-19} J if the diameter of the electron orbit is 10^{-10} m.

5. Calculate the orbital speed of the electron in Question 4.

6. Show that if a particle of charge q is placed in the field of a fixed charge Q at a point a distance r from it, the electric potential energy of q is given by

$$U(q) = K_e \frac{qQ}{r}.$$

7. An alpha particle has rest mass 6.6×10^{-27} kg and positive charge $2e$, where e is the charge of a proton.

   (a) Calculate the kinetic energy of an alpha particle moving at two thirds the speed of light. [*Hint:* Use the relativistic equations (48.3) and (48.22).]

   (b) Calculate the distance r_0 of closest approach of the alpha particle in part (a) to the nucleus of a gold atom, given that the nucleus of the gold atom has positive charge $79e$. [*Hint:* Use the result of Question 6.]

8. Alpha particles emitted in the decay of radium having an average kinetic energy of 4.7 MeV are scattered from a target consisting of gold foil. Calculate the distance of closest approach that an alpha particle makes with the nucleus of the gold atom.

9. The gravitational potential energy of a hydrogen atom due to the gravitational attraction of the electron and the proton is given by $U(r) = -Gm_e m_p/r$, where

m_e and m_p are the respective masses. Compare this to the electrical potential energy and explain whether or not we can safely ignore the contribution of gravitational potential energy to the total energy of a hydrogen atom.

49.5 SPECTRA OF ELECTROMAGNETIC RADIATION

An atomic model that replaced Rutherford's model was introduced by Niels Bohr in 1913. To understand what led Bohr to propose his model, it is necessary to describe some parallel developments that were taking place in both physics and chemistry at the turn of the century. These have to do with electromagnetic radiation.

We recall that every electromagnetic wave has associated with it a wavelength λ and a frequency f related by the equation

$$f\lambda = c,$$

where c is the speed of light. The fundamental properties of electromagnetic waves are basically the same at all frequencies. But methods for detecting these waves

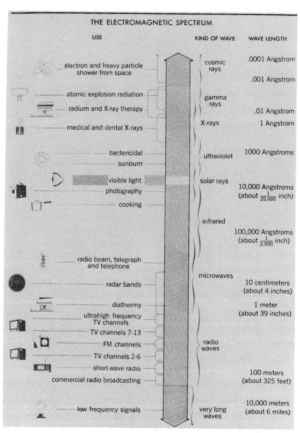

Figure 49.5 The electromagnetic spectrum of frequencies. (Reprinted from Britannica Junior Encyclopaedia by permission of Encyclopaedia Brittanica, Inc.)

49.5 SPECTRA OF ELECTROMAGNETIC RADIATION

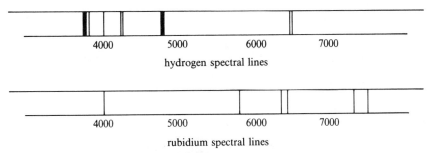

Figure 49.6 Discrete line spectra of (a) hydrogen and (b) rubidium.

depend on the frequencies. For example, semiconductor diodes detect radio waves at low frequencies, human eyes detect light waves at intermediate frequencies, and photographic emulsions detect X rays at higher frequencies. As already mentioned in Chapter 44, the collection of all frequencies is called the *electromagnetic spectrum*, and it is usually divided into overlapping regions with special names, as indicated in Fig. 49.5. The boundaries in Fig. 49.5 are only approximate. For example, the limits of visible light will vary from individual to individual.

When a substance in gaseous or vaporized form is heated or excited by the passage of an electric spark the atoms in the substance emit light of definite wavelengths, called the *emission spectrum* of the substance. The light produced by many sources (such as the sun, a flame, or a glowing red hot body) has a continuous spectrum, indicating that it is produced by thermal emission. But some sources produce what is called a discrete or line spectrum, representing relatively few wavelengths. Figure 49.6a shows the line spectrum of atomic hydrogen, while Fig. 49.6b shows that of rubidium, an element that was discovered spectroscopically by Robert Bunsen in 1860. The emission spectra of elements are like fingerprints, unique to each element, and can be used for identifying them.

The spectroscope is an instrument, invented in 1859 by Kirchhoff, which uses a prism or a ruled grating to analyze light into its constituent wavelengths. (See Fig. 49.7.)

Spectroscopic measurements in the latter half of the nineteenth century helped identify many of the known elements and also led to the discovery of new ones. The

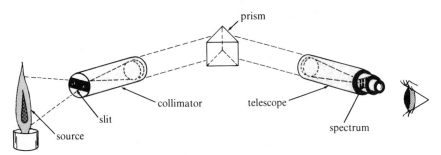

Figure 49.7 Schematic drawing of a spectroscope.

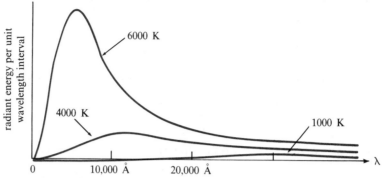

Figure 49.8 Spectral distributions of black body radiation at different temperatures.

most accurate measurements were made by a Swedish spectroscopist, A. J. Ångström, in honor of whom the angstrom unit is named, where $1 \text{ Å} = 10^{-10}$ m. Visible light has a wavelength in the vicinity of 5000 Å.

In practice, the easiest part of the spectrum to observe is the visible spectrum and a portion of the infrared spectrum. A body heated until it glows red hot will emit heat radiation in the infrared band as well as a dull red light in the visible band. Conversely, these wavelengths are also easily absorbed by many bodies. For example, light impinging on a dark body will seem to disappear and the body will become heated. In general, electromagnetic waves can be radiated and absorbed by bodies with a corresponding loss or gain of energy.

One way to study the energy conveyed by electromagnetic waves radiating from a body is to project its spectral distribution on a screen. By placing thermometers at various parts of the screen and observing the rise in temperature, it is revealed that not all wavelengths convey the same amount of energy.

For example, if a black body is heated to 1000 K it would reveal a spectral distribution indicated by the lowest curve in Fig. 49.8. But when heated to 4000 or 6000 K the distribution curves change dramatically as shown in the figure. One of the open questions at the beginning of the twentieth century was to find an explanation for the dependence on temperature of the spectral distribution of a heated black body.

Some physicists tried unsuccessfully to account for these curves on the basis of emission and absorption of light by vibrating molecules in the heated body. Wilhelm Wien proposed one explanation, which agreed with experimental data up to about 30,000 Å but failed progressively at longer wavelengths. Lord Rayleigh and James Jeans used another approach, which agreed with experimental data above 70,000 Å but failed progressively at shorter wavelengths.

In an attempt to find a formula that agreed with Wien's results at short wavelengths and with Rayleigh–Jeans at long wavelengths, Max Planck discovered that a satisfactory explanation could be formulated by assuming that a hot body does not emit energy of a given wavelength in arbitrary amounts, but rather in fixed amounts or *quanta* at that wavelength. Once emitted, the quantum of energy

49.5 SPECTRA OF ELECTROMAGNETIC RADIATION

Table 49.2 Wavelengths of Some Hydrogen Spectrum Lines (in angstroms).

n	λ calculated from Balmer's formula	Observed value of λ
3	6562.08	6562.10
4	4860.8	4860.74
5	4340	4340.1
6	4101.3	4101.2

assumes the usual wave form with a frequency $f = c/\lambda$. He also related the amount of energy in his quantum to the frequency by the remarkably simple formula

$$E = hf,$$

where h is a constant of proportionality independent of the body and its temperature. This constant is one of the fundamental constants of nature and lies at the heart of quantum theory. It is now called Planck's constant and has the value

$$h = 6.6262 \times 10^{-34} \text{ J s}.$$

Planck's theory did not require that light itself consist of bundles of energy, but it was soon pointed out by Einstein that other evidence supported the concept of *light quanta* or *photons*.

Another question that puzzled nineteenth century scientists was to explain the spacing of the lines in discrete spectra, such as that of the hydrogen atom in Fig. 49.6. In 1885, a Swiss high school teacher named Johann Balmer concocted an empirical formula that seemed to fit the measured lines for hydrogen very well. He wrote

$$\lambda = 3645.6 \frac{n^2}{n^2 - 4}, \tag{49.4}$$

where λ is the wavelength in angstroms and n is a positive integer ≥ 3. The success of this formula is revealed in Table 49.2, which compares calculated values with observed values as given in Balmer's paper.

When instruments sensitive to ultraviolet light are used to examine the spectrum of atomic hydrogen, more lines are found that do not correspond to those given by Balmer's formula. In fact, Balmer himself speculated that there might be such lines and suggested that their wavelengths could be found by replacing the 4 in the denominator of (49.4) by the square of other integers. This stimulated others to search for additional spectral lines and it led the Swedish spectroscopist Johannes R. Rydberg to propose a generalization of Balmer's formula.

If Balmer's formula is written in terms of the reciprocal of the wavelength, it becomes

$$\frac{1}{\lambda} = \frac{4}{3645.6} \left(\frac{1}{2^2} - \frac{1}{n^2} \right).$$

In this form it is a special case of a more general formula proposed by Rydberg:

$$\frac{1}{\lambda} = R\left(\frac{1}{m^2} - \frac{1}{n^2}\right), \qquad (49.5)$$

where R is a constant, known as the Rydberg constant, which has the value

$$R = 109{,}677.58 \text{ cm}^{-1}$$

for all lines of the hydrogen spectrum. In 1908, F. Paschen found two new lines in the infrared region whose wavelengths correspond to $m = 3$ and $n = 4$ and 5 in Rydberg's formula, and other investigators found many more lines as experimental techniques improved.

Neither Balmer nor Rydberg proposed a physical mechanism to explain why these formulas predicted wavelengths so accurately. That came later when Niels Bohr found a connection between these formulas and Planck's concept of the quantization of energy.

Questions

10. Determine the minimum wavelength in angstroms of the spectral lines arising from the Balmer formula (49.4).

11. Calculate the wavelengths of the two infrared spectral lines that were found by Paschen.

49.6 THE BOHR MODEL OF THE ATOM

Niels Henrik David Bohr was born in Copenhagen in 1885, the year Balmer published his formula on the hydrogen spectrum. His father was a professor of physiology at the University of Copenhagen, his brother Harrald became a distinguished mathematician, and his son Aghe Bohr won a Nobel prize in physics.

After receiving his doctorate at the University of Copenhagen with a thesis on the electron theory of metals, Niels Bohr spent a few months in 1911 with J. J. Thomson at Cambridge, then went to Manchester to join Rutherford's group. This was about the time that Rutherford had proposed his planetary model of the atom, and Bohr was aware of the difficulties inherent in that model.

After returning to Copenhagen in 1913, Bohr learned about the Rydberg formula during a casual conversation with a friend. The existence of integers in Rydberg's formula was consistent with the fact that the emission of hydrogen lines was somehow quantized. Bohr was also aware of the new quantum concepts that had been introduced by Planck in 1900 and extended by Einstein in 1905. From these clues Bohr modified Rutherford's planetary model of atomic structure.

First, Bohr assumed that the classical electromagnetic theory did not necessarily apply to atomic phenomena. Specifically, he asserted that the orbiting electron in the planetary model of the hydrogen atom does not *normally* radiate energy while being accelerated. He needed this to explain why the planetary model could survive. Radiation does occur, but only in special circumstances. Bohr postulated that not all

49.6 THE BOHR MODEL OF THE ATOM

Figure 49.9 Photograph of Niels Bohr. Reprinted from *Niels Bohr–The Man, His Science, and The World They Changed* by Ruth Moore. Published by Alfred A. Knopf.

orbital radii are possible but only certain discrete values of radii, $r_1, r_2, r_3, \ldots$, with corresponding energies in those orbits $E_1, E_2, E_3, \ldots$. He referred to these stable orbits as *stationary states* because electrons in those orbits do not radiate. He also suggested that an electron could change from one stable orbit of radius r_n to another of larger radius r_m by acquiring sufficient energy $E_m - E_n$, or, it could at a later time fall back from the larger orbit to the smaller, and in so doing would emit the same amount of energy $E_m - E_n$. He also asserted that this energy would be radiated as a *photon* or light quantum whose frequency f would satisfy the relation

$$E_m - E_n = hf, \tag{49.6}$$

where h is Planck's constant. This assumption is one form of energy conservation, but it deviates from the classical theory that requires the frequency of radiation to be the same as that of the motion of the charged particle.

Bohr's model is indicated schematically in Fig. 49.10. The arrow between the two allowable orbits suggests the electron dropping from a higher energy level to a lower energy level. The wavy line represents the photon that is emitted as a consequence of this change.

Bohr's model provided a mechanism for spectral emission. Because the allowable energy levels were discrete, the corresponding spectral lines would also be discrete.

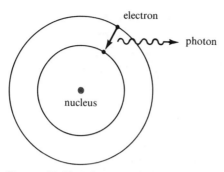

Figure 49.10 Schematic representation of the Bohr model of the atom.

The acceptability of any model depends on its ability to make predictions that can be verified by experiment. We now carry out two calculations based on Bohr's model. One of them predicts the allowable radii of the Bohr orbits and the other derives Rydberg's empirical formula for the hydrogen spectrum.

Bohr was motivated to find a quantum condition that would imply Rydberg's formula for the spectral lines of hydrogen. After much trial and error he found such a condition in terms of the angular momentum L of the electron. He postulated that in any stable orbit the angular momentum must be an integer multiple of Planck's constant $\hbar$,

$$L = n\hbar, \tag{49.7}$$

where $\hbar = h/(2\pi)$ and $n = 1, 2, 3, \ldots$.

This was a key assumption in Bohr's theory so it is useful to see how he might have arrived at it. Suppose that an electron in a stable orbit of relatively large radius r suddenly moves down to the next stable orbit of radius r', where the difference $r - r'$ is small compared to r. According to Eq. (49.6) the change in energy ΔE caused by the electron moving from one stable orbit to the next lower orbit satisfies

$$|\Delta E| = hf, \tag{49.8}$$

where f is the frequency of the photon radiated. In Example 3 we calculated the total energy E of an electron in a circular orbit of radius r and found that

$$E = -K_e \frac{e^2}{2r}. \tag{49.3}$$

This energy can be expressed in terms of the angular momentum L. We know from Chapter 23 that the angular momentum of the orbiting electron is given by

$$L = m_e rv,$$

where m_e is the mass of the electron and v its speed. From Eq. (49.1) we find

$$K_e e^2 = m_e v^2 r = vL.$$

Using this in Eq. (49.3) we obtain

$$E = -\frac{vL}{2r}. \tag{49.9}$$

The next step is to relate this to the frequency of orbital rotation of the electron. An electron in orbit along a circle of radius r with constant speed v and frequency f_r travels a distance $2\pi r$ in time $T = 1/f_r$, so $v = 2\pi r/T = 2\pi r f_r$. Using this in (49.9) gives us

$$E = -\pi f_r L. \tag{49.10}$$

Now we come to the crucial step, finding a relation between the orbital frequency f_r and the frequency f of the emitted photon. Classical physics would say that the frequency of the radiation emitted will be equal to the orbital frequency of the electron going around the orbit. But that can't be right in this case because Bohr's basic assumption is that when the electron is in orbit it doesn't give up any energy. As the electron jumps from one orbit of radius r to another of radius r' its orbital frequency changes from f_r to $f_{r'}$. Perhaps, then, the frequency f of the radiation emitted is the average of f_r and $f_{r'}$. Bohr applied this speculation to the capture of a free electron by a hydrogen ion (a hydrogen atom with its electron missing). In other words, assume that the electron is initially free and then falls into one of the stable orbits of radius r. In that case, it starts with initial frequency zero and ends up with orbital frequency f_r. Therefore, the frequency f of the photon given off is $\frac{1}{2}f_r$, the average of 0 and f_r. Replacing f_r by $2f$ in Eq. (49.10) we obtain

$$E = -2\pi f L. \tag{49.11}$$

Consequently, the change in energy ΔE is related to the corresponding change in angular momentum ΔL by the equation

$$\Delta E = -2\pi f \Delta L.$$

Bohr's assumption in Eq. (49.8) that energy is emitted in quanta with $|\Delta E| = hf$ now implies that

$$\Delta L = \frac{h}{2\pi} = \hbar,$$

where $\hbar = h/(2\pi)$. In other words, the angular momentum of the electron in its Bohr orbit must have one of the values

$$L_n = n\hbar, \quad \text{where } n = 1, 2, 3, \ldots.$$

This is Bohr's assumption of Eq. (49.7). The corresponding integer n is called the *quantum number* of the Bohr orbit.

To determine the corresponding orbital radius r_n with quantum number n we write the last equation in the form

$$n\hbar = m_e r_n v \tag{49.12}$$

and use Eq. (49.1) to express r_n in terms of v. This gives us

$$r_n = \frac{K_e e^2}{m_e v^2}. \tag{49.13}$$

But Eq. (49.12) can be rewritten in the form

$$\frac{1}{v^2} = \left(\frac{m_e r_n}{n\hbar}\right)^2.$$

Using this in Eq. (49.13) we find

$$r_n = \frac{K_e e^2}{m_e}\left(\frac{m_e r_n}{n\hbar}\right)^2,$$

which, when solved for r_n, gives

$$r_n = \frac{n^2 \hbar^2}{e^2 m_e K_e}. \qquad (49.14)$$

The constants $\hbar$, e, m_e, and K_e are known, so this formula expresses the allowable radii of the Bohr orbits explicitly in terms of the quantum number n. The smallest permissible orbit (called the ground state) corresponds to $n = 1$ and has the value

$$r_1 = 0.529 \times 10^{-10} \text{ m} = 0.529 \text{ Å}.$$

This is called the *Bohr radius* and is usually denoted by a_0. The allowed radii can be expressed as $r_n = n^2 a_0$.

It is now an easy matter to derive Rydberg's formula. First we use Eq. (49.3) to express the total energy of the electron in terms of the quantum number n and we find

$$E_n = -K_e \frac{e^2}{2r_n} = -\frac{K_e^2 e^4 m_e}{2n^2 \hbar^2}. \qquad (49.15)$$

Therefore, the change in energy from orbit n to orbit m is

$$E_n - E_m = \frac{K_e^2 e^4 m_e}{2\hbar^2}\left(\frac{1}{m^2} - \frac{1}{n^2}\right).$$

Using the quantum relation $E_n - E_m = hf$, and replacing the frequency f by c/λ, we obtain

$$\frac{1}{\lambda} = \frac{K_e^2 e^4 m_e}{2hc\hbar^2}\left(\frac{1}{m^2} - \frac{1}{n^2}\right).$$

When the known values of the constants are inserted on the right we find

$$\frac{1}{\lambda} = R\left(\frac{1}{m^2} - \frac{1}{n^2}\right),$$

where $R = 1.09 \times 10^7 \text{ m}^{-1}$.

Thus, Rydberg's formula, which was originally concocted to fit experimental results, was shown by Bohr to be a logical consequence of placing electrons in their appropriate orbits. This remarkable triumph for the Bohr model caused great excitement in the scientific community. Many investigators immediately began to extend the theory and to check it with further experiments. The importance of this contribution was recognized when Niels Bohr was awarded the Nobel prize in physics in 1922.

Questions

12. The series of spectral lines for hydrogen corresponding to $m = 1$ in Rydberg's formula is called the Lyman series, after Theodore Lyman who found these lines in the period 1906 to 1916. Determine the longest wavelengths in the Lyman series.

13. Show that the entire Lyman series lies in the ultraviolet region.

14. Determine the wavelength of the lowest energy photon that will ionize a hydrogen atom, that is, that will cause an electron in the ground state ($n = 1$) to be excited to the state $n = \infty$.

15. A hydrogen atom absorbs a photon, which causes an electron in a state corresponding to $n = 2$ to jump to the state corresponding to $n = 5$. Determine (a) the energy and (b) the wavelength of the absorbed photon.

16. A hydrogen atom with its electron in the ground state absorbs a photon of wavelength 1025 Å. A short time later it emits two photons. Find the wavelength of these photons.

17. For an electron in a hydrogen atom in its lowest energy state, show that the ratio of its orbital speed v to that of light c, is approximately $v/c = 1/137$.

18. A helium ion He$^+$ consists of a helium nucleus plus a single electron. Determine the energy of the electron in its lowest orbit.

19. Calculate the wavelength of a photon that will excite an electron in a helium ion from the ground state $n = 1$ to the state with quantum number $n = 2$.

20. A muon is an unstable particle that has the same charge as an electron but a mass 207 times that of an electron. In some experiments a muon can be substituted for the electron in a hydrogen atom to form a muonic atom. Determine the energy needed to ionize a muonic atom if the muon is in its ground state.

21. If the muon in a muonic atom makes a transition from the state with $n = 2$ to that with $n = 1$, what is the wavelength of the light emitted?

22. Show that the orbital frequency f_n of an orbiting electron in the nth state of a hydrogen atom is given by

$$f_n = \frac{K_e^2 e^4 m_e}{2\pi \hbar^3 n^3}.$$

49.7 A FINAL WORD

The ultimate test of any theory is whether or not it agrees with experiment. The Bohr model of the atom certainly had striking success in predicting and explaining the physical basis for Rydberg's formula giving the spectral lines of atomic hydrogen. It also provided a great stimulus to the field of spectroscopy and encouraged the development of high-resolution instruments. These instruments soon revealed that the hydrogen spectrum was more complex than originally thought. Some of the spectral lines were composed of a number of fine lines so close together that they appeared to coincide when viewed with less accurate instruments. Bohr's circular orbits could not account for all these lines. Working independently and almost simultaneously, Niels Bohr, Arnold Sommerfeld, and others were able to explain some of the fine structure by modifying the Bohr model to allow for elliptical orbits as well as circular orbits.

Another striking success of the Bohr model was the value it produced for the Bohr radius a_0, about $\frac{1}{2}$ Å. This agreed quite well with earlier results derived from the kinetic theory of gases that suggested the diameter of a hydrogen atom should be about 1 Å.

The Bohr model received further support when the theory was extended to more complicated structures containing only one electron in orbit, in particular ionized helium and ionized lithium. Again the observed spectral lines for these substances agreed fairly well with those predicted by the Bohr model.

Despite these and many other successful applications of the Bohr theory, the swift and relentless progress in the physical sciences that took place in the early decades of this century have demonstrated that Bohr's theory is incorrect. There is a wealth of physical phenomena that cannot be explained by the Bohr model and, moreover, some of its predictions are wrong. For example, one of the key assumptions in the Bohr model is that the lowest orbit in the hydrogen atom has angular momentum $L = \hbar$. But modern quantum mechanics tells us that the correct answer for the lowest orbit is $L = 0$, something no one could have known at the time Bohr developed his model. This and other incorrect predictions tell us that Bohr's model is not merely the first step in a sequence of ever-improving models, but is actually wrong. But even though the Bohr theory is wrong, it was an essential step leading to the development of modern quantum mechanics. In fact the Bohr theory is often referred to as the *old quantum theory*.

The modern view of the atomic model still has some features of the Bohr model, such as the concepts of energy levels and quantization. But the intuitively appealing picture of electrons in definite orbits jumping from one level to another turns out to be misleading for some purposes. In fact, the current model no longer provides a simple physical picture to guide the imagination. An important step toward the development of the current model was a new description of the atom made in 1924 by the French physicist Louis de Broglie. This story is told in the next chapter.

CHAPTER 50

PARTICLES AND WAVES

The energy of a ponderable body cannot be divided into indefinitely many indefinitely small parts, whereas the energy emitted by a point light source is regarded on the Maxwell theory (or more generally according to every wave theory) as continuously spread over a continuously increasing volume.

Such wave theories of light have given good representation of purely optical phenomena and will surely not be replaced by any other theory. It is to be remembered that the optical observations refer to time mean values, not to instantaneous values, and it is quite conceivable that, in spite of complete success in dealing with diffraction, reflection, dispersion, etc. such a theory of continuous fields, could lead to contradictions with experience when applied to phenomena of light emission and absorption.

    Albert Einstein, "A Heuristic Viewpoint Concerning the Emission and Transformation of Light" (1905)

50.1 BLACK BODY RADIATION

Early in the twentieth century, a small number of scientists began to realize that bold new ideas would be needed to solve the perplexing problem of the internal structure of the atom. But even before that problem came to a head, there were clues that something was wrong in the house of physics. Those clues came from entirely different observed phenomena. One of them, called *black body radiation*, had to do with the color of a hot glowing body. The other, called *the photoelectric effect*, is widely used today to open automatic elevator doors. Let's begin with the problem of black body radiation.

Every physical object gives off electromagnetic radiation of all wavelengths – including visible light – all the time. The hotter a body is, the more it radiates, and if it gets hot enough, it starts to radiate enough visible light to be seen or even to illuminate a room. That's precisely how an incandescent lamp works. However, if the body is in thermodynamic equilibrium, it absorbs just as much as it radiates, at each wavelength, so it doesn't get hotter, or cooler, or change in any other way. That's the meaning of equilibrium.

A black body is one that absorbs all the radiation that falls on it, in contrast to a perfect mirror, which reflects everything. The radiation given off by a heated black body is exactly the same as that it would absorb from the electromagnetic field in the space around it, if the body and the field were in equilibrium at the same temperature. This suggests that the electromagnetic field itself must be, in some sense, like an object that can change its temperature in order to be in equilibrium with a radiating body. Otherwise, thermodynamic equilibrium would be impossible. In fact, at any given temperature, the electromagnetic field must have just enough radiant energy at each wavelength to balance the radiation of a black body. Near the end of the nineteenth century, physicists were striving to determine how much energy was just enough for equilibrium, and how that energy changed with temperature. That quest was called the black body problem.

Example 1

From common visual observation of the color of glowing, heated bodies (such as embers in a fireplace or red hot iron) determine whether the intensity of radiation increases or decreases with increasing wavelength in the visible part of the spectrum.

When the glow of a heated body first becomes visible, it appears red, turning more toward white as the body gets hotter. That means there's more energy in the red spectrum (long wavelengths) and less in the blue part of the spectrum (low wavelengths) as sketched in Fig. 50.1.

Some discussion of the black body problem was given in Chapter 49. In Germany, Wilhelm Wien made measurements of the intensity of radiation from a heated black body at various wavelengths, and concluded that the radiation decreases sharply at short wavelengths. But in England, Lord Rayleigh and Sir James Jeans, arguing on purely theoretical grounds, concluded that the intensity of radiation should increase with decreasing wavelength (or increasing frequency). Even without Wien's evidence to the contrary, that would have been a disconcerting conclusion: at very high frequencies, each body in equilibrium would be exchanging arbitrarily large amounts of energy with the electromagnetic field, a phenomenon referred to as the *ultraviolet catastrophe*.

We know from the discussion in Chapter 49 that Max Planck reconciled the results of Wien with those of Rayleigh–Jeans by invoking a new idea. Planck

50.1 BLACK BODY RADIATION

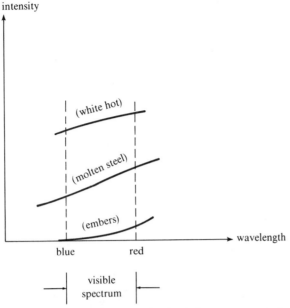

Figure 50.1 Qualitative graph of relative energy as a function of wavelength in the visible spectrum at various temperatures.

proposed that a hot body emits energy E in quanta proportional to the frequency f,

$$E = hf, \tag{50.1}$$

where h is Planck's constant,

$$h = 6.6262 \times 10^{-34} \text{ J s}.$$

This was a revolutionary idea, of which Planck himself was suspect. Maxwell's theory, at that time the most perfect theory in all of physics, implied that energy in the electromagnetic field depends on the field strength, and not the frequency. Planck's strategy was to repeat the Rayleigh–Jeans argument, using Eq. (50.1) as a temporary convenience, then at the end of the calculation let h harmlessly shrink to zero, in the hope that a sensible result would emerge.

Instead, Planck arrived at a result that fitted Wien's data splendidly, with a definite nonzero value of h. He obtained a formula for the energy density u_λ (radiant energy per unit wavelength interval) of the form

$$u_\lambda = \frac{8\pi hc}{\lambda^5} \frac{1}{e^{hc/(\lambda kT)} - 1}, \tag{50.2}$$

where c is the speed of light, k is Boltzmann's constant from the ideal gas law (see Chapter 15), and T is the absolute temperature of the body. (Because $c = \lambda f$, the exponential can also be written as $e^{hf/(kT)}$.) The general shape of the graph of u_λ as a function of the wavelength λ is shown in Fig. 50.2.

For large values of λ the exponential term in Eq. (50.2) can be estimated by using the linear approximation to the exponential

$$e^x \approx 1 + x,$$

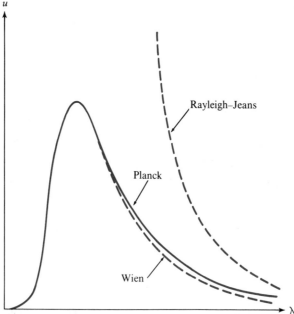

Figure 50.2 Planck's radiation law at a fixed temperature $T = 1646$ K, shown in comparison with radiation laws of Wien and Rayleigh–Jeans.

valid for small x. Taking $x = hc/(\lambda kT)$, which is small for large λ, we find that Eq. (50.2) gives the approximate formula

$$u_\lambda = \frac{8\pi kT}{\lambda^4}, \quad \lambda \text{ large}, \tag{50.3}$$

a result predicted by Rayleigh–Jeans. But for small values of λ the term -1 is negligible compared to the exponential and Eq. (50.2) gives the approximation

$$u_\lambda = \frac{8\pi hc}{\lambda^5} e^{-hc/(\lambda kT)}, \quad \lambda \text{ small}, \tag{50.4}$$

which is in accordance with Wien's observations.

Example 2

The kinetic theory of gases (Chapter 15) relates the average energy of a gas atom to the temperature of the gas. Think of the electromagnetic field as being a gas of light quanta (photons) in equilibrium. Does the theory of black body radiation suggest any relation between the energy and temperature of photons in the electromagnetic field?

50.2 THE PHOTOELECTRIC EFFECT

In Chapter 15 we found that the mean kinetic energy of an atom of gas at absolute temperature T is

$$\bar{K} = \tfrac{3}{2}kT,$$

where k is Boltzmann's constant. Using Planck's relation (50.1) and ignoring factors like $\tfrac{3}{2}$, we might guess that the average energy of light quanta in a gas of photons is given by

$$\bar{E} = h\bar{f} \approx kT,$$

where $\bar{f}$ is the average frequency. In fact, by setting the derivative of the right-hand member of Eq. (50.4) equal to zero it can be shown that the maximum of the function u_λ in Wien's law occurs at

$$hf_{\max} = 5kT.$$

Assuming the maximum energy is roughly proportional to the average energy of a light quantum, we see that the behavior of light quanta does bear some resemblance to the behavior of atoms of a gas.

Questions

1. Estimate the most efficient temperature for an incandescent lamp filament (the temperature at which the greatest possible radiation is visible). Do incandescent lamps operate at that temperature?

2. A prediction of the "big bang" theory of cosmology is that the initially hot universe would have cooled by adiabatic expansion, reaching a temperature by now of 3 K. This prediction has recently been confirmed by observing the "background" radiation in the universe. In what part of the electromagnetic spectrum is this radiation to be found? Would you expect the discovery to have been made by optical astronomers? Radio astronomers? Someone else entirely?

3. Show that the derivative of the function u_λ defined by Eq. (50.4) has the value 0 when $hf = 5kT$.

4. Show that the derivative of the function u_λ defined by Eq. (50.2) has the value 0 when

$$hf = 5kT(1 - e^{-hf/(kT)}).$$

5. Astrophysicists predict that in 5 billion years the sun will swell out to the orbit of Mars and become a red giant. If the present temperature of the surface of the sun is 6000 K and in the red giant phase the temperature will be 4500 K, determine the ratio of the energy densities u_λ in the mid range of the visible spectrum for these two temperatures.

50.2 THE PHOTOELECTRIC EFFECT

According to Planck's theory, matter can emit or absorb radiant energy only in units of hf. After Planck announced his theory in 1900 there were several attempts to explain the physical meaning of Planck's equation

$$E = hf. \tag{50.1}$$

Some thought it reflected a property of the atoms of matter, about which relatively little was known at the time. Others suggested that for some reason energy existed in the electromagnetic field in units of hf. At the time, the first of these two ideas seemed more likely than the second, because it would leave Maxwell's theory unaffected.

The possibility that the mystery of black body radiation might be explained by some undiscovered property of the atom was shattered with the explanation of the phenomenon mentioned in the opening paragraph of this chapter, the photoelectric effect, which we now discuss in more detail.

It had been known for some years that ultraviolet light made it easier for a spark to jump between two metallic poles, and that a negatively charged piece of zinc metal could be discharged by shining ultraviolet light on it. In 1905, young Albert Einstein proposed an explanation based on Planck's equation $E = hf$. According to this equation, ultraviolet radiation, which has higher frequency than visible light, should have more energy, in fact, enough energy to expel electrons from zinc. When radiation of a given frequency f falls on a metal, it gives up energy in units of hf. If hf is greater than the energy ϕ binding the electrons to the metal, then electrons can be emitted, with the surplus energy in the form of kinetic energy K, so that

$$K = hf - \phi. \tag{50.5}$$

This relation, called Einstein's photoelectric effect equation, was later confirmed by Robert A. Millikan, who measured the voltage required to stop all the electrons emitted from various metals as a function of the frequency, or color, of the light incident on them.

Figure 50.3 gives a schematic drawing of the apparatus Millikan used to study photoelectric emission. Ultraviolet light of frequency f passes through window W and falls on cathode C, ejecting electrons with residual kinetic energy K. The stopping voltage of the electrons, V, is related to their maximum kinetic energy by the equation

$$eV = K.$$

The stopping voltage V is what is needed to make the current gathered at anode A equal to zero.

When Millikan plotted V versus the incident frequency, he got a series of straight lines, one for each metal used as the cathode. They had the same slope, as shown in Fig. 50.4, but different intercepts, and they obeyed Einstein's Eq. (50.5) with K replaced by eV:

$$V = \frac{hf}{e} - \frac{\phi}{e}.$$

Thus all the lines had slope h/e, a universal constant, the same for all metals, but the intercept, $-\phi/e$, was different for different metals. The quantity ϕ is called the work function (see Chapter 36).

50.2 THE PHOTOELECTRIC EFFECT

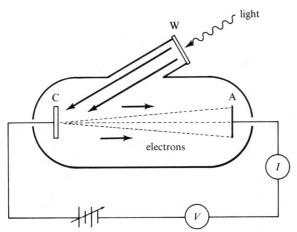

Figure 50.3 Schematic of apparatus for studying the photoelectric effect.

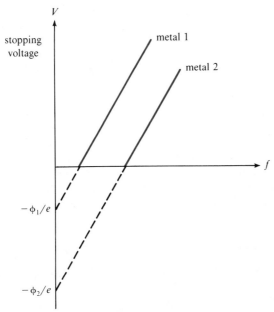

Figure 50.4 Linear relations established by Millikan's data.

Question

6. Estimate the work function of a metal that's found to emit photoelectrons when irradiated by visible light.

Einstein's theory of the photoelectric effect (for which he was later awarded the Nobel prize) was much more than a mere application of Planck's earlier idea. It meant that Planck's explanation of black body radiation could no longer be attributed to some mysterious property of matter. Electrons would absorb energy hf from the incident radiation only if the radiation had that amount of energy. In other words, energy in the electromagnetic field must already exist in units of hf before entering matter. These units, or quanta of energy, were later called *photons*.

More than two centuries earlier, Isaac Newton asserted that light was a kind of particle or corpuscle, while his rival Christian Huygens thought it was a wave. In the nineteenth century, the wave theory triumphed, culminating in Maxwell's famous equations. Now, at the dawn of the twentieth century, particles were making a comeback. But there was no question that, in other experiments, those showing diffraction for example, light did behave as a wave. (See Chapter 44.) How could the same thing be both particle and wave? That question captured the attention of the world's physicists for the next quarter of a century.

Questions

7. An electromagnetic plane wave passes through a certain region of free space. Its electric component is given by

   $$\mathbf{E} = E_0 \sin(kx - \omega t)\hat{\mathbf{k}},$$

   where $E_0 = 1$ V/m and the angular frequency $\omega = 1 \times 10^{15}$ s^{-1}.

   (a) Calculate the average energy per cubic meter in the electromagnetic field due to the wave.
   (b) If the energy really exists as photons, how many are there in a cubic meter? Does the answer have to be an integer?

8. Photoelectrons from a certain metal are stopped by a stopping voltage of 0.92 V when the metal is irradiated by light of wavelength 250 nm. Determine the work function of the metal.

9. Light of wavelength 410 nm is incident on lithium, which has a work function of 2.13 eV. Find the maximum kinetic energy of the photoelectrons emitted.

10. What wavelength of incident light will cause photoelectrons to be emitted from lithium at a speed of $0.6c$?

11. The threshold wavelength for the emission of photoelectrons from calcium is 384 nm. Determine the work function of calcium.

12. A uniform beam of yellow light ($\lambda = 600$ nm) has an intensity of 8.0×10^{-8} W/m^2.

    (a) Determine the energy of a single photon of this wavelength.
    (b) Calculate how many photons each second are transported across a surface of area 1 cm^2 perpendicular to the beam.

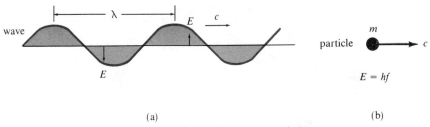

Figure 50.5 The dual nature of light as (a) a wave and (b) a particle.

50.3 THE DUAL NATURE OF LIGHT

In 1922 a debate was raging about the dual nature of light. Experiments on diffraction showed that light acted like a wave, while others on the photoelectric effect showed that light behaved like particles. Figure 50.5 illustrates the two points of view. In the wave theory, light has a definite wavelength and frequency, as in Fig. 50.5a. Actually, light is a composite of electric and magnetic waves. The sinusoidal curve in Fig. 50.5a shows how the electric field strength varies with time or with distance. As the light advances, it follows a straight-line path represented by the x axis in the figure. At a given instant, light exists everywhere along this line, with the sinusoidal curve representing its electric field intensity. In this interpretation, there is no precise location for the light. We have an intuitive feeling for what light is, but we cannot specify where it is.

Figure 50.5b shows the particle interpretation. Viewed as a particle, light has mass, momentum, and energy. Because it travels with the speed of light c, it has no rest mass. As a particle or bundle of energy, it has a precise location so we know where it is, but we don't know how to describe in detail the bundle of energy.

The story of how the two points of view were reconciled is described in the following excerpt from *fables of physics*:

> Once upon a time, there was a handsome young prince named Louis de Broglie. One day in 1924 the prince submitted his doctoral thesis in physics at the University of Paris.
>
> His thesis committee was beside itself with anxiety. For one thing, a prince of the realm is rarely a physics doctoral student. But to make matters worse, Prince Louis' thesis not only suggested that light waves might be particles, but also that real particles, like electrons, might be waves.
>
> What to do? The committee hit upon a brilliant solution. They would ask the wisest of the wise, Albert Einstein himself, to evaluate the thesis. If Einstein said it was nonsense, they could reject the thesis, even though it came from a prince.
>
> But when Einstein read the thesis, he didn't think it was nonsense at all. In fact, he thought it had merit. The thesis was accepted.
>
> Prince Louis de Broglie won the Nobel prize for his thesis in 1929, became a professor of physics at the University of Paris, and lived happily ever after.

Like all fables, this one has some truth in it. Prince de Broglie noted that the particle nature of light was more apparent for light of very short wavelength, and

Figure 50.6 Photograph of Prince Louis de Broglie. (From AIP Niels Bohr Library.)

that the wave nature was more obvious for light of very long wavelength. This led him to propose a sort of compromise that suggested that light consists of *wave trains*, as illustrated in Fig. 50.7. A wave train is like a wave, except it occupies a finite portion of space, with a beginning and an end.

For light of large wavelengths, the beginning and end of the wave train are far apart, and the train resembles our usual picture of a wave, as in Fig. 50.7a. But if the wavelength is very short (or, equivalently, the frequency is very high), the wave train occupies a relatively small portion of space, as in Fig. 50.7b, and the whole train resembles a particle, even though there is a wave structure inside it. At intermediate wavelengths, shown in Fig. 50.7c, the wave train is extended but still has some characteristics of a particle.

To reconcile the two points of view we examine some of the physical quantities associated with waves and with particles. First, each light wave has a wavelength λ and a frequency f related by the equation

$$\lambda = c/f. \tag{50.6}$$

A quantum of energy E associated with a wave of frequency f is given by Planck's equation

$$E = hf. \tag{50.1}$$

50.3 THE DUAL NATURE OF LIGHT

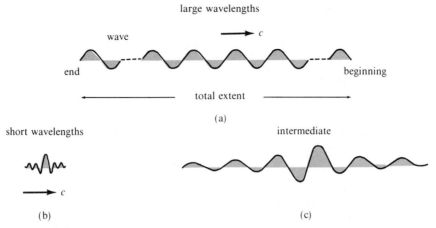

Figure 50.7 Light represented as wave trains.

On the other hand, Einstein's theory of relativity implies that

$$E = mc^2,$$

which, when combined with Planck's relation, gives $hf = mc^2$, or

$$\frac{c}{f} = \frac{h}{mc}.$$

Comparing this with Eq. (50.6) we see that if a particle is to act like a light wave, its "wavelength" must satisfy the equation

$$\lambda = \frac{h}{mc}. \tag{50.7}$$

At this point, de Broglie made a bold leap of imagination. He argued that if a particle of mass m moving with speed v has a wave structure, then by analogy with Eq. (50.7) its "wavelength" λ would satisfy the equation

$$\lambda = \frac{h}{mv}. \tag{50.8}$$

Although he gave no description of the wave structure of a moving particle, de Broglie referred to the wave as a "matter wave."

When this equation is applied to the particle of smallest known mass, the electron, it says that the matter wave associated with the electron will have the largest wavelength of all matter waves at the same speed.

Example 3
Using Eq. (50.8), calculate the wavelength of an electron moving in the Bohr orbit of the hydrogen atom with quantum number n.

To apply Eq. (50.8) we need to know the product of the mass and the speed of the electron, or in other words, its momentum mv. By Eq. (49.12) the momentum in the Bohr orbit with quantum number n is

$$m_e v = \frac{n\hbar}{r_n} = \frac{nh}{2\pi r_n},$$

where r_n is the corresponding radius of the Bohr orbit. Substituting this in Eq. (50.8) we find

$$\lambda = \frac{2\pi r_n}{n}. \tag{50.9}$$

For example, when $n = 1$ we know that $r_1 = 0.529$ Å, so the corresponding wavelength is $\lambda = 3.32$ Å

50.4 THE DE BROGLIE MODEL OF THE HYDROGEN ATOM

Equation (50.9) in Example 3 shows that the circumference of the Bohr orbit with quantum number n is exactly equal to $n\lambda$. In other words, each Bohr orbit contains exactly n wavelengths, where n is an integer, the quantum number. This is the key element in de Broglie's model of the atom. He conceived the idea that the electron appears as a standing wave when in orbit, with the wave wrapped around in a circle rather than being along a straight line. Figure 50.8a shows how this idea becomes translated into a picture of the various permissible orbits of the electron. In Fig. 50.8b the orbits are shown unwrapped along a line.

The fact that the orbit must close on itself in a smooth way is a physical condition that forces the wavelengths and frequencies to be quantized by Eq. (50.9). It is often described by saying that the electron *resonates* in such an orbit.

To help understand the de Broglie model it is useful to think of the unwrapped orbits in Fig. 50.8b as being analogous to the standing waves of a vibrating string stretched between two fixed endpoints. Such a string can only vibrate in various discrete modes, because in each mode the length of the string must be an integer multiple of the wavelength. If the length of the string is not an integer multiple of the wavelength, interference would cause the vibrations to be quickly damped. Similarly, an electron in an orbit whose circumference is not an integer multiple of the wavelength would also be quickly damped by destructive interference and would not resonate.

Returning to Eq. (50.8), we note that the denominator is $mv = p$, the momentum of the particle, so Eq. (50.8) can also be written in the form

$$p = \frac{h}{\lambda}. \tag{50.10}$$

This equation relates the momentum of a particle to the wavelength of the associated matter wave. For an electron in a Bohr orbit of radius r_n, the product pr_n

50.4 THE DE BROGLIE MODEL OF THE HYDROGEN ATOM

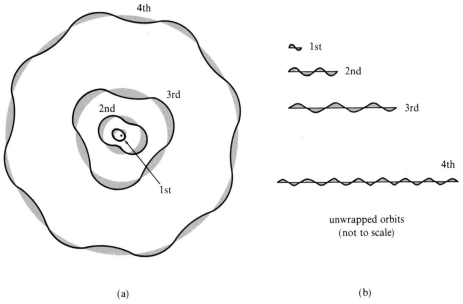

Figure 50.8 The de Broglie model of the hydrogen atom showing (a) some of the possible orbits and (b) the unrolled orbits as standing waves.

is the angular momentum L_n in that orbit. Thus, Eq. (50.10) implies

$$L_n = pr_n = \frac{hr_n}{\lambda}.$$

Using (50.9) for λ we find

$$L_n = \frac{nh}{2\pi} = n\hbar,$$

which is Bohr's fundamental assumption concerning angular momentum. In other words, Bohr's quantization of the energy levels in terms of angular momentum, which seemed somewhat arbitrary at the time he introduced it, is now seen to be a logical consequence of a physical requirement that the orbit should contain an integer number of wavelengths.

The de Broglie model also explains something that the Bohr model did not account for. In the Bohr model, the electron does not radiate while in orbit but only when it changes from one allowable orbit to another. The Bohr theory predicts the correct energy change, as observed by measuring spectral emission, but it does not explain how the electromagnetic waves are produced. The de Broglie model offers an explanation by again using the analogy with vibrating strings.

Consider two strings vibrating at slightly different frequencies f_1 and f_2, each with the same amplitude, represented by two standing waves as shown in Fig. 50.9a.

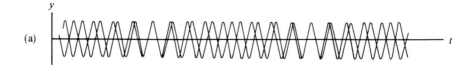

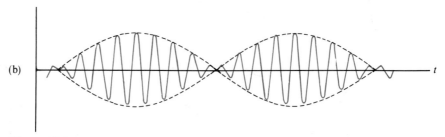

Figure 50.9 (a) Two waves of slightly different frequencies with the same amplitude. (b) Beats produced by a superposition of the two waves.

At time $t = 0$ the waves are in phase, at a later time t_1 they are out of phase by 180°, at time t_2 they are back in phase, and at t_3 they are out of phase again.

Figure 50.9b shows a superposition of the two waves. In a moment we will show that the resultant frequency f is the average of f_1 and f_2, but the amplitude is modulated as indicated by the dotted curve, with its maximum value at times $t = 0$ and $t = t_2$, and its minimum (zero) at times t_1 and t_3. In a vibrating string of a musical instrument, such as a violin or guitar, these variations of amplitude give rise to variations of loudness that are called *beats*. The ear hears a tone whose intensity varies alternately between loud and soft, and, as we will presently show, the beat frequency is the difference of the two original frequencies.

Example 4

Given two waves of the same amplitude represented by the two cosine functions

$$y_1 = A \cos(2\pi f_1 t), \qquad y_2 = A \cos(2\pi f_2 t),$$

show that the sum $y = y_1 + y_2$ is given by the equation

$$y = A(t) \cos(2\pi f t), \tag{50.11}$$

where $f = (f_1 + f_2)/2$ and

$$A(t) = 2A \cos\left(2\pi \frac{f_1 - f_2}{2} t\right). \tag{50.12}$$

50.4 THE DE BROGLIE MODEL OF THE HYDROGEN ATOM

To deduce Eq. (50.11) we simply add the two cosine functions representing y_1 and y_2 and use the trigonometric identity

$$\cos a + \cos b = 2 \cos \frac{a-b}{2} \cos \frac{a+b}{2},$$

with $a = 2\pi f_1$ and $b = 2\pi f_2$.

Note that the resultant amplitude $A(t)$ in Eq. (50.12) varies with time t at a frequency $(f_1 - f_2)/2$. If f_1 and f_2 are nearly equal, and if t is not too large, the argument in the cosine term in Eq. (50.12) is small so the cosine itself is nearly 1 and the amplitude changes very slowly. A beat, or maximum absolute amplitude $|A(t)|$ occurs when the cosine term in Eq. (50.12) has the value $+1$ or -1. Each of these values occurs once in each cycle, so the number of beats per second is equal to twice the frequency $(f_1 - f_2)/2$. In other words, the beat frequency (the number of beats per second) is the difference of the two given frequencies.

Now we apply the phenomenon of beats to the de Broglie model. Suppose an electron in a hydrogen atom drops from an orbit of energy E_m to a lower orbit of energy E_n. These energies are related to the corresponding frequencies of the electron waves by the equations

$$E_m = hf_m, \qquad E_n = hf_n.$$

Therefore, the energy change is

$$E_m - E_n = h(f_m - f_n). \tag{50.13}$$

However, the change from one orbit to the other does not take place instantaneously. It requires a short interval of time, during which the two modes of vibration of the electron coexist, one at frequency f_m and the other at frequency f_n. During that short time interval the beat frequency f is also present, and because $f = f_m - f_n$, Eq. (50.13) shows that

$$E_m - E_n = hf.$$

This is Eq. (49.6), the result that Bohr assumed in formulating his model of the hydrogen atom. But without de Broglie's hypothesis, nothing in the Bohr model actually oscillates at frequency f. In the de Broglie model the frequency f is that of a beat that exists during the transition process. Before or after the transition the beat does not exist and there is no electromagnetic emission.

Questions

13. Determine the de Broglie wavelength of each of the following:

    (a) An electron moving at a speed of $0.8c$.
    (b) A baseball of mass 0.2 kg moving at 28 m/s.

14. Determine the de Broglie wavelength of a hydrogen atom moving at 1.8 km/s.

15. At what kinetic energy will the de Broglie wavelength of a hydrogen atom be equal to its diameter (1.0 Å)?

16. Determine the de Broglie wavelength of a charged particle of charge q and mass m moving in a circular path of radius R in a uniform magnetic field of intensity B perpendicular to the plane of the motion.

17. Use relativistic momentum to determine the wavelength of a 2-GeV electron.

50.5 THE BIRTH OF QUANTUM MECHANICS

The de Broglie model of 1924 occupied a position analogous to that of the Bohr model 10 years earlier. The new model was better than the old one and it gave impetus to a host of investigations, both experimental and theoretical.

de Broglie's idea that the electron could be considered as a matter wave immediately raised the question of whether one could actually detect such a wave. In fact, when de Broglie was asked this very question by a member of his doctoral thesis committee, he replied that they should be detectable by diffraction experiments with electrons scattered by crystals. Shortly thereafter two Americans, C. J. Davisson and L. H. Germer, provided the necessary experimental verification by reflecting electrons from the face of a crystal, although when they made their initial observations in 1925 they did not know of de Broglie's theory. Further verification was also given in England by G. P. Thomson (the son of J. J. Thomson) with a different type of experiment that studied the transmission of electrons through a thin crystalline film.

One of the theoretical questions that arose from the de Broglie electron-wave idea was to determine the actual nature of the wave, and to find how the mass and charge of the electron were distributed along the wave. The search for solutions to these problems gave birth to one of the most productive periods in theoretical physics and it led to the discovery of new aspects of nature now referred to as quantum phenomena. The mathematical laws governing these phenomena are known as *quantum mechanics*, and the new atomic model that came out of this theory is called the quantum mechanical model.

The birth of quantum mechanics came from two independent streams of thought. One was de Broglie's concept of matter waves, which, however, only had a major impact after Erwin Schrödinger published a series of papers on wave mechanics early in 1926. Schrödinger's principal contribution was to formulate a differential equation that was satisfied by the amplitude of the de Broglie wave. The Schrödinger equation, which we will say more about later in this section, expressed particle mechanics in wavelike equations.

The other stream of thought was a treatment of the radiative properties of an atom in late 1925 by Werner Heisenberg, Max Born, and P. Jordan, who used the mathematics of matrices together with ideas from mathematical probability to develop what came to be known as *matrix mechanics*. It was soon realized that the two streams were simply different mathematical approaches to the same physical ideas. In a famous paper published in 1926, Born established the fundamental basis

50.5 THE BIRTH OF QUANTUM MECHANICS

Figure 50.10 Photograph of Erwin Schrödinger. (Courtesy of the Archives, California Institute of Technology.)

for applying wave mechanics to collision problems and introduced a probability interpretation that, to this date, has not been superseded.

It is not possible to describe the mathematical formulation of quantum mechanics in any detail without introducing mathematical concepts and techniques that are far beyond the scope of an introductory text such as this. For example, Schrödinger's equation is a complicated differential equation involving second-order derivatives with respect to three space directions and with respect to time. But by restricting the discussion to one spatial dimension and ignoring variations with time we can give an intuitive argument that leads to the correct form of the Schrödinger equation in this special case.

Figure 50.11a shows an enlarged version of the wave train of Fig. 50.7b, which was used to suggest that light of very short wavelength resembles a particle. But in this diagram we interpret the wave train to be the matter wave associated with a particle of mass m and velocity v moving along the x axis, as shown in Fig. 50.11b. Because the wave represents the particle, it is reasonable to assert that the amplitude of the wave, which we denote by Ψ, is large near those points where the particle is located and very small elsewhere. Now if the wave were a pure sine or cosine curve with wavelength λ, say $\Psi = \sin(2\pi x/\lambda)$ or $\Psi = \cos(2\pi x/\lambda)$, or a linear combina-

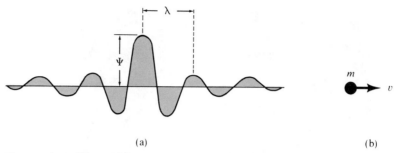

(a) (b)

Figure 50.11 The matter wave (a) associated with a particle (b).

tion of two such curves.

$$\Psi = A \sin \frac{2\pi x}{\lambda} + B \cos \frac{2\pi x}{\lambda},$$

the function Ψ would satisfy the equation of simple harmonic motion,

$$\frac{d^2\Psi}{dx^2} + \frac{4\pi^2}{\lambda^2}\Psi = 0. \qquad (50.14)$$

Because the wave in Fig. 50.11a dies out away from the location of the particle, the function Ψ shown there is not a combination of sine and cosine curves with the same wavelength λ, so Ψ does not satisfy Eq. (50.14). Nevertheless, the curve shown in Fig. 50.11a retains some geometrical features of a cosine curve, such as peaks and valleys (maxima and minima) which we will now analyze and arrive at a slight variation of Eq. (50.14), which might be satisfied by Ψ. The idea is to see how Ψ and its second derivative $d^2\Psi/dx^2$ must be related to the wavelength λ in order to contain the wave in a definite region of space.

The second derivative $d^2\Psi/dx^2$ of any function Ψ is the rate of change of the first derivative $d\Psi/dx$, and therefore it tells us how the slope of the curve changes. If the slope is increasing, as in Fig. 50.12a, the second derivative is positive, but if the slope is decreasing, as in Fig. 50.12b, the second derivative is negative. Near a sharp peak of the curve, such as that at the center of the curve in Fig. 50.11a, the slope decreases rapidly from positive to negative, so the second derivative has a large negative value. The sharper the peak, the more rapid the decrease of the slope and the more negative the second derivative becomes. The sharpness of the peak is determined by the size of Ψ. If Ψ is large and positive the second derivative will be large and negative, so it is reasonable to expect that the second derivative is proportional to $-\Psi$. Near a valley, where Ψ is negative, the slope increases from negative to positive so the second derivative is positive, which again is consistent with the second derivative being proportional to $-\Psi$.

Now let's see how the second derivative varies with the wavelength. Figure 50.13a shows one wave with wavelength λ, while Fig. 50.13b shows another with wavelength $\lambda/2$. Shortening the wavelength from λ to $\lambda/2$ has two separate effects on the changing slope. First, the peak in Fig. 50.13b is sharper, which about doubles the rate of change of the slope, and secondly a region occupied by one peak in

50.5 THE BIRTH OF QUANTUM MECHANICS

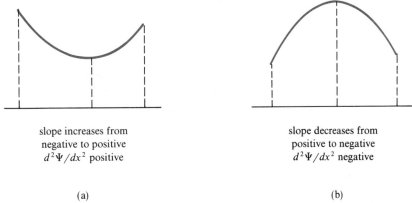

Figure 50.12 Geometric meaning of the sign of the second derivative.

Fig. 50.13a is now occupied by two peaks in Fig. 50.13b, which means the slope has to change twice as often in the same region. The effect of cutting the wavelength in half is to multiply the second derivative by 4. In general we can expect each of these effects to be proportional to $1/\lambda$, so when taken together the rate of change of the slope would be proportional to $1/\lambda^2$.

In summary, this discussion suggests that the second derivative of Ψ is proportional to $-\Psi$ and also to $1/\lambda^2$, so we can write

$$\frac{d^2\Psi}{dx^2} = -C\frac{\Psi}{\lambda^2}, \tag{50.15}$$

where C is a constant of proportionality.

Up to this point, this discussion has used only the wave properties of the particle and not any of its mechanical properties, such as mass or speed. These are introduced into the equation by using de Broglie's relation

$$\lambda = \frac{h}{mv}. \tag{50.8}$$

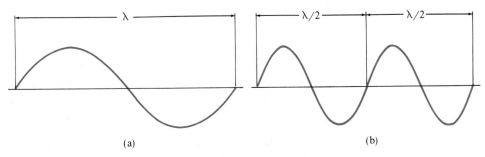

Figure 50.13 Change in second derivative due to decrease in wavelength.

Equation (50.15) now becomes

$$\frac{d^2\Psi}{dx^2} = -C\frac{(mv)^2}{h^2}\Psi. \qquad (50.16)$$

The quantity in the numerator, $(mv)^2$, is related to the kinetic energy of the particle, $K = mv^2/2$. If we express the kinetic energy as the difference of the total energy E of the particle minus its potential energy U we have

$$K = E - U$$

or

$$mv^2 = 2(E - U),$$

and hence

$$(mv)^2 = 2m(E - U).$$

Using this in Eq. (50.16) we obtain

$$\frac{d^2\Psi}{dx^2} = -C\frac{2m(E-U)}{h^2}\Psi.$$

This is the form taken by Schrödinger's equation for a particle in one dimension (the x axis) with time variation ignored. In the actual Schrödinger equation the constant of proportionality $C = 4\pi^2$, the same constant that occurs in Eq. (50.14), the equation of simple harmonic motion. Thus, the Schrödinger equation is written as

$$\frac{d^2\Psi}{dx^2} = -\frac{2m(E-U)}{\hbar^2}\Psi, \qquad (50.17)$$

where $\hbar = h/(2\pi)$.

Example 5

Show that for the one-dimensional classical harmonic oscillator the Schrödinger equation assumes the form

$$\frac{d^2\Psi}{dx^2} = (\alpha^2 x^2 - \beta)\Psi, \qquad (50.18)$$

where α and β are constants.

The harmonic oscillator corresponds to the situation in which a particle is attracted toward the origin by a force $F = -kx$ proportional to its displacement x from the origin, where k is a positive constant. The corresponding potential energy is $U(x) = kx^2/2$ (see Chapter 14), so Schrödinger's equation becomes

$$\frac{d^2\Psi}{dx^2} = -\frac{2m(E - kx^2/2)}{\hbar^2}\Psi,$$

50.5 THE BIRTH OF QUANTUM MECHANICS

which is of the form of Eq. (50.18), with

$$\alpha^2 = \frac{mk}{\hbar^2} \quad \text{and} \quad \beta = \frac{2m}{\hbar^2} E. \tag{50.19}$$

To gain insight into the nature of the solutions of the Schrödinger equation we shall discuss the special equation (50.18) in further detail. A function that nearly satisfies Eq. (50.18) is the rapidly decaying exponential function

$$G(x) = e^{-\alpha x^2/2}, \tag{50.20}$$

because its derivatives are given by

$$G'(x) = -\alpha x \, e^{-\alpha x^2/2} = -\alpha x G(x) \tag{50.21}$$

and

$$G''(x) = -\alpha x G'(x) - \alpha G(x).$$

Using Eq. (50.21) for $G'(x)$ we can rewrite this as

$$G''(x) = (\alpha^2 x^2 - \alpha) G(x). \tag{50.22}$$

Note that the right member of Eq. (50.22) is nearly the same as that in Eq. (50.18), except that α appears instead of β. So in the special case that $\beta = \alpha$ the function $G(x)$ is a solution of (50.18). The graph of $G(x)$ is shown in Fig. 50.14.

Of course, if $\beta \neq \alpha$, the function $G(x)$ will not satisfy Eq. (50.18). In this case we modify $G(x)$ and try a solution of the form

$$\Psi(x) = u(x) G(x),$$

for some choice of $u(x)$. Substituting this for $\Psi(x)$ in Eq. (50.18) and making use of Eqs. (50.21) and (50.22) we find that Eq. (50.18) is satisfied by $\Psi(x)$ if and only if $u(x)$ satisfies the differential equation

$$u''(x) - 2\alpha x u'(x) + (\beta - \alpha) u(x) = 0. \tag{50.23}$$

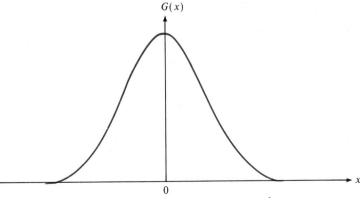

Figure 50.14 Graph of the exponential $G(x) = e^{-\alpha x^2/2}$.

There is a standard method for solving this equation by writing $u(x)$ as a power series in x,

$$u(x) = a_0 + a_1 x + a_2 x^2 + \cdots + a_m x^m + \cdots,$$

and determining the constant coefficients $a_0, a_1, a_2, \ldots$ so that $u(x)$ will satisfy (50.23). We won't discuss this method here but we summarize what happens in this particular case. For most choices of the constants α and β the function $u(x)$ determined in this manner grows so rapidly for large x that, even when it is multiplied by the decaying exponential $G(x)$, the product grows too fast to give physically meaningful solutions for Ψ. But if the constants α and β are related by an equation of the form

$$\beta = (2n + 1)\alpha, \quad n = 0, 1, 2, \ldots, \tag{50.24}$$

then for appropriate choices of a_0 and a_1, the infinite series for $u(x)$ terminates after n terms and becomes a polynomial of degree n, $u_n(x)$. When this happens the resulting product $\Psi_n(x) = u_n(x)G(x)$ decays exponentially for large x and gives a physically meaningful wave function.

Using the definitions of β and α given in Eq. (50.19), we find that the condition in Eq. (50.24) requires that the corresponding energies E_n satisfy

$$E_n = \left(n + \tfrac{1}{2}\right)\hbar\sqrt{k/m}. \tag{50.25}$$

In Chapter 20 we learned that for simple harmonic motion the quantity

$$f = \frac{1}{2\pi}\sqrt{\frac{k}{m}}$$

is the vibrational frequency of the motion. For this reason, Eq. (50.25) is usually written in the form

$$E_n = \left(n + \tfrac{1}{2}\right)hf. \tag{50.26}$$

This equation displays the quantization of the permissible state energies. In this example the corresponding polynomials $u_n(x)$ are given by

$$u_0(x) = a_0, \quad u_1(x) = \sqrt{2\alpha}\, a_0 x, \quad u_2(x) = \tfrac{1}{2}\sqrt{2}\, a_0 (2\alpha x^2 - 1),$$

where $a_0 = (\alpha/\pi)^{1/4}$. The energies and the wave functions for the ground state and the first two excited states are therefore given by

$$E_0 = \tfrac{1}{2}hf, \quad \Psi_0(x) = (\alpha/\pi)^{1/4} \exp(-\alpha x^2/2), \tag{50.27}$$

$$E_1 = \tfrac{3}{2}hf, \quad \Psi_1(x) = (\alpha/\pi)^{1/4} \sqrt{2\alpha}\, x \exp(-\alpha x^2/2), \tag{50.28}$$

$$E_2 = \tfrac{5}{2}hf, \quad \Psi_2(x) = (\alpha/\pi)^{1/4} \sqrt{2}\left(\alpha x^2 - \tfrac{1}{2}\right)\exp(-\alpha x^2/2). \tag{50.29}$$

Graphs of these wave functions are shown in Fig. 50.15. The constant $a_0 = (\alpha/\pi)^{1/4}$ has been chosen so that each wave function $\Psi_n(x)$ will satisfy the condition

$$\int_{-\infty}^{+\infty} \Psi_n^2(x)\, dx = 1.$$

The reason for this restriction will be described in the next section.

50.5 THE BIRTH OF QUANTUM MECHANICS

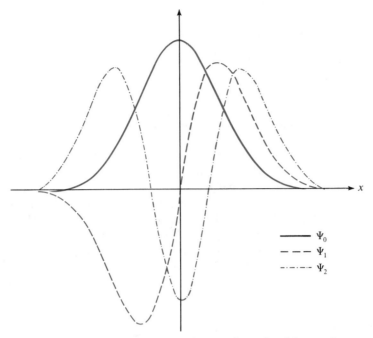

Figure 50.15 Wave functions for the one-dimensional harmonic oscillator with quantum numbers $n = 0$, 1, and 2.

Example 6

We have already shown that if $\beta = \alpha$ the function $G(x)$ satisfies Eq. (50.18). Show that (a) the function $\Psi_1(x) = xG(x)$ satisfies Eq. (50.18) if $\beta = 3\alpha$ and (b) that $\Psi_2(x) = (2\alpha x^2 - 1)G(x)$ satisfies (50.18) if $\beta = 5\alpha$.

(a) Differentiating the product $\Psi_1 = xG$ and using (50.21) we find

$$\Psi_1' = xG' + G = (1 - \alpha x^2)G.$$

Differentiating once more and using (50.21) again we obtain

$$\Psi_1'' = (1 - \alpha x^2)G' - 2\alpha xG$$
$$= (1 - \alpha x^2)(-\alpha xG) - 2\alpha xG$$
$$= (\alpha^2 x^2 - 3\alpha)xG$$
$$= (\alpha^2 x^2 - 3\alpha)\Psi_1.$$

This shows that Ψ_1 satisfies (50.18) if $\beta = 3\alpha$.

(b) If $\Psi_2(x) = (2\alpha^2 - 1)G(x)$, an argument similar to that in (a) shows that

$$\Psi_2' = (5\alpha x - 2\alpha^2 x^3)G$$

and
$$\Psi_2'' = (2\alpha^3 x^4 - 11\alpha^2 x^2 + 5\alpha)G$$
$$= (\alpha^2 x^2 - 5\alpha)(2\alpha x^2 - 1)G$$
$$= (\alpha^2 x^2 - 5\alpha)\Psi_2,$$
so Ψ_2 satisfies (50.18) if $\beta = 5\alpha$.

The general Schrödinger equation, of which Eq. (50.17) is a very special case, applies to matter waves associated with a particle moving in three-dimensional space subject to an external force field, such as that which binds electrons in an atom. Schrödinger was led to his equation by using formal analogies between optics and classical particle mechanics. In quantum mechanics, the Schrödinger equation is taken as a fundamental law, just as Newton's law $F = ma$ is taken as a fundamental law in classical mechanics. Its importance rests on the success of its predictions, rather than on the validity of its derivation. With the introduction of Schrödinger's equation, wave properties were used to explain particle mechanics in a new way and often led to conclusions not predicted by classical mechanics.

One of the questions that arose in connection with the Schrödinger equation was the physical meaning of the amplitude Ψ. For an electromagnetic wave the amplitude is related to the intensity of the electric and magnetic fields. But for a matter wave associated with a moving particle there is no obvious physical quantity that can be associated with Ψ. Nevertheless, as we shall see in the next section, Max Born found a new and fruitful interpretation for the amplitude.

Questions

18. Calculate the ground state energy of a pendulum with period 2 s.

19. Many diatomic molecules (consisting of two atoms bound to each other) can exhibit vibrational spectra, which result from transitions between vibrational energy levels of the system. Using the simple harmonic oscillator as a model and drawing on your knowledge that light is emitted when a system jumps from one energy level to another, describe how you would expect the spectral lines of a diatomic molecule to be spaced.

20. Calculate the vibrational energy levels for the carbon monoxide diatomic molecule (mass = 28 u) assuming a force constant $k = 1.87$ J/m. Will the transition from state $n = 1$ to state $n = 0$ result in the emission of visible light?

21. If $G(x) = \exp(-\alpha x^2/2)$ and $\Psi(x) = (2\alpha x^3 - 3x)G(x)$, show that $\Psi(x)$ satisfies the differential equation (50.18) if $\beta = 7\alpha$.

50.6 THE QUANTUM MECHANICAL MODEL OF THE ATOM

In trying to reconcile the Born–Heisenberg–Jordan approach with that used by Schrödinger, Max Born introduced a new idea that turned out to be of fundamental importance in the history of physics. It is best described in Born's own words translated from his landmark paper entitled "Quantenmechanik der Stossvorgange" [*Z. Physik* **38** (1926)]:

> I wish to try to give here a third interpretation and to investigate its usefulness in treating collision processes. For this I build on a remark of Einstein concerning the relation of the wave field and light quanta. He said in effect that the waves are only there to point the way to the corpuscular light quanta, and in this sense spoke of a "ghost-field." This ghost-field determines the probability that a light quantum, as the carrier of energy and momentum, shall take a definite path. No energy or momentum belongs to the field itself.
>
> ... But in view of the complete analogy between light quanta and electrons, one may undertake to formulate the laws of electron motion in a similar way. In this way it is natural to regard the de Broglie–Schrödinger waves as the "ghost-field" or better the "guiding field."
>
> I wish therefore to follow up the view: that the guiding field, represented by a scalar function Ψ of the coordinates of all particles in the system, is propagated according to the Schrödinger differential equation. Momentum and energy however are transported as if corpuscles (electrons) are actually flying about. The paths of the particles are only determined to the extent that they are limited by the conservation of momentum and energy theorems. After that the choice of a particular path is determined by probability in accordance with the field of values of Ψ. One can express this somewhat paradoxically: the motion of the particles follows probability laws, but the probability itself is propagated in accordance with a causal law.

In this paper, Born proposed that the wave amplitude Ψ in the Schrödinger equation is related to the probability of finding the particle at the coordinates where Ψ is evaluated, and, in fact, Born stated that the exact value of this probability is determined by Ψ^2.

In mathematics, probability is a number lying in the interval [0, 1], which measures the likelihood of a given event. Unlikely events have probability close to 0, while highly likely events have probability close to 1. In applications of the type considered here, such as a particle executing simple harmonic motion, we do not assign a positive probability to the particle being at any particular point on the x axis. Instead, the probability that it lies in a given interval, say the interval $a \leq x \leq b$, is equal to the integral of Ψ^2 over this interval,

$$\int_a^b \Psi^2(x)\, dx = \text{the probability that the particle lies in } [a, b]. \quad (50.30)$$

When probabilities are calculated in this way, the function under the integral sign is called the *probability density*. In other words, the square of the wave function is the probability density. The fact that the particle must lie somewhere in the interval $(-\infty, +\infty)$ is expressed probabilistically by writing

$$\int_{-\infty}^{+\infty} \Psi^2(x)\, dx = 1. \quad (50.31)$$

This condition puts a restriction on the wave function Ψ. If Ψ is any solution of the Schrödinger equation, then any constant times Ψ is also a solution. Equation (50.31) determines the constant factor that must be used to express Ψ^2 as a probability density.

Figure 50.16 Photograph of Max Born. (From AIP Niels Bohr Library.)

Figure 50.17 displays the graph of Ψ^2 for the first three wave functions of the one-dimensional harmonic oscillator shown earlier in Fig. 50.15. The probability that the particle in a given energy state E_n lies in a given interval, is equal to the area of the region between this interval and the graph of the corresponding probability density Ψ_n^2. These graphs reveal a significant difference between classical mechanics and quantum mechanics. For example, Fig. 50.18 shows the probability density Ψ_0^2 together with the probability density function computed from classical mechanics for a harmonic oscillator with the same energy. The two graphs are quite different. In classical mechanics, a particle executing simple harmonic motion will always be found somewhere between the extremities of its motion, and for a given energy these boundaries are quite sharply defined. In Fig. 50.18 the motion has been scaled to lie in the interval $[-1, +1]$ for the sake of the discussion. Classical mechanics predicts that the particle is most likely to be near the extremities of the motion, which, in Fig. 50.18, are the points -1 and $+1$. But quantum mechanics permits some possibility of penetration into the forbidden region outside the interval because there is a positive amount of area under the graph of Ψ_0^2 outside the interval $[-1, +1]$.

50.6 THE QUANTUM MECHANICAL MODEL OF THE ATOM

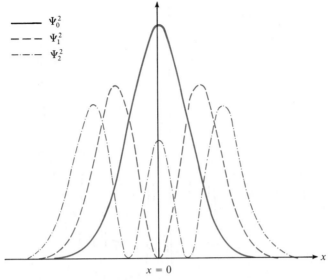

Figure 50.17 Graphs of the first three probability density functions Ψ_n^2 for the one-dimensional harmonic oscillator.

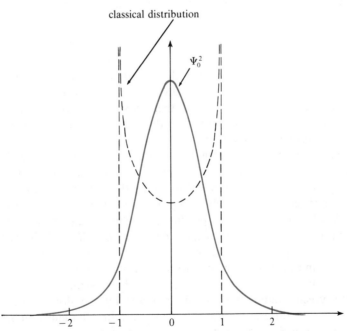

Figure 50.18 Comparison of the probability density Ψ_0^2 with that of the classical distribution for an oscillator with the same total energy.

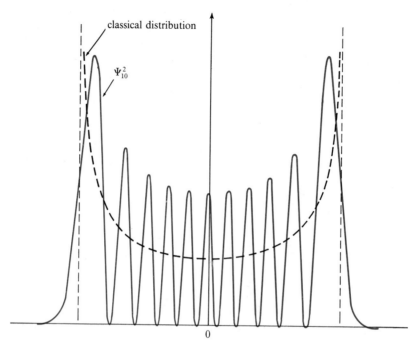

Figure 50.19 The density function Ψ_{10}^2 compared with the density function from classical mechanics for an oscillator with the same total energy.

Figure 50.19 illustrates that the quantum mechanical picture approaches the classical model as the quantum number n increases. It shows the density function Ψ_n^2 with $n = 10$, together with the density function from classical mechanics (dashed curve) for an oscillator with the same total energy. The average value of Ψ_n^2 closely approximates the dashed curve.

Max Born's fruitful idea led to the replacement of the de Broglie "matter waves" with Born's "waves of probability." For the hydrogen atom, it provided a new quantum mechanical model in which the orbiting electron occupies a position that is not specified precisely – only its probable position is known.

Whereas Bohr's theory specifies the orbits of an electron in a hydrogen atom quite precisely, quantum mechanics only provides the probability that an electron in a particular quantum state will be found in an interval of distances from the nucleus, and provides similar information with regard to the speed with which the electron moves. The average distance and the root-mean-square speed are the same as the precise radius and speed predicted by the Bohr theory. But the angular momentum is different in the two theories. Bohr's fundamental assumption was that an electron in the ground state has angular momentum $\hbar$, but quantum mechanics predicts that in the ground state the electron has zero angular momentum.

In the quantum mechanical model of the hydrogen atom, the motion of the electron is described by mathematical functions, called orbital wave functions. Each electron is said to occupy an *orbital* rather than an orbit, a different name being used to emphasize that the motion is not as precise as that specified in the Bohr model. The orbitals play the role of the "ghost fields" suggested by Einstein. For example, for the hydrogen atom there is exactly one orbital with quantum number $n = 1$. It is called the $1s$ orbital and is said to constitute the K shell. For the quantum number $n = 2$ there are four orbitals. One of these, called the $2s$ orbital, has zero angular momentum. The other three are called $2p$ orbitals, and each has orbital angular momentum equal to $\hbar$. The four orbitals with quantum number $n = 2$ constitute the L shell. Quantum mechanics also implies a corresponding theory of orbitals for all atoms, and the distribution of the electrons among these orbitals governs the electronic structure of all atoms. The orbital structure will be described in further detail in Chapter 51.

Questions

22. Suppose you make repeated measurements of the position of a simple harmonic oscillator. Use Fig. 50.17 to determine in what region the oscillator is most likely to be found in the $n = 0$ state. How does your finding compare with what you would expect from classical mechanics?

23. The graphs in Fig. 50.17 extend beyond the limits of motion of a classical harmonic oscillator. Describe the probabilistic interpretation of this fact.

24. Refer to the graphs in Fig. 50.17. For what states is the quantum mechanical simple harmonic oscillator most likely to be found near the classical limits of motion?

50.7 THE HEISENBERG UNCERTAINTY PRINCIPLE

With the new quantum mechanics, the wave nature of a particle could be described with great precision, but knowledge concerning location of the particle was less precise because the new theory only told where the particle was *likely* to be. At first it was thought that this lack of precision was an imperfection in the theory, but in 1927 Heisenberg formulated a principle, now known as the *Heisenberg uncertainty principle*, which asserted that there is an uncertainty inherent in nature and that it can be quantified.

In its most elementary form, Heisenberg's uncertainty principle refers to the probable errors that can occur when measuring the position x of a particle and its corresponding momentum component p. If Δx denotes the probable error in measuring x and Δp the probable error in measuring p, the Heisenberg uncertainty principle states that

$$|\Delta x| \, |\Delta p| \geq \hbar/2. \tag{50.32}$$

In other words, increasing precision in specifying x implies increasing uncertainty in knowledge of p, and conversely. We will presently describe the type of argument used by Heisenberg to derive (50.32), but first we make some informal remarks about how this inequality fits into the theoretical framework of quantum mechanics.

Figure 50.20 Photograph of Werner Heisenberg. (From AIP Niels Bohr Library, W. F. Meggers Collection.)

Viewed from a purely mathematical standpoint, quantum mechanics is like any deductive system. It consists of a small number of postulates that involve mathematical symbols subject to various laws of combination. The symbols are intended to represent physical concepts, such as wave functions, energy, and momentum, while the postulates are chosen to reflect basic physical principles, such as the Schrödinger equation and conservation laws. Theorems are then deduced as logical consequences of these postulates. One of these theorems is the Heisenberg uncertainty principle, an inequality that reduces to the form in (50.32) for a particle moving along the x axis.

Heisenberg arrived at this inequality from the following type of argument. Consider a wavelike disturbance whose amplitude is appreciably different from zero only in a bounded region, such as that shown in Fig. 50.21. This disturbance can be thought of, for example, as the matter wave associated with an electron in motion along the x axis. As the electron moves about, the matter wave changes its size and shape. The position x of the electron is known to within a certain accuracy Δx. This means that the actual position is somewhere in the interval from $x - |\Delta x|$ to $x + |\Delta x|$, where $|\Delta x|$ is positive. The velocity of the electron corresponds to that of the matter wave, but cannot be specified exactly. This uncertainty in the velocity

50.7 THE HEISENBERG UNCERTAINTY PRINCIPLE

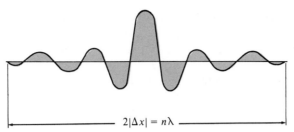

Figure 50.21 Example used to motivate the uncertainty principle.

causes an uncertainty in the momentum $p = mv$ in the amount Δp, which means the actual momentum lies between $p - |\Delta p|$ and $p + |\Delta p|$, where $|\Delta p| > 0$. Now recall the de Broglie equation, which relates the momentum of a particle to the wavelength of the associated matter wave:

$$p = \frac{h}{\lambda}. \tag{50.10}$$

The derivative of p with respect to λ is

$$\frac{dp}{d\lambda} = -\frac{h}{\lambda^2},$$

so a small change Δp is related to a small change $\Delta \lambda$ by the approximate equation

$$\Delta p = -\frac{h}{\lambda^2} \Delta \lambda. \tag{50.33}$$

Next, suppose that the matter wave is a superposition of sinusoidal waves, with wavelengths differing by $\Delta \lambda$, but all nearly equal to λ. In other words, the superposition of such waves confines the matter wave to a region of space of length $2|\Delta x|$, but at the same time introduces an uncertainty Δp into the momentum because of Eq. (50.33). A certain integral number of peaks or valleys of these waves will lie within the region of uncertainty of length $2|\Delta x|$, so for some positive integer n we have the approximate equation $2|\Delta x| = n\lambda$, or

$$n = \frac{2|\Delta x|}{\lambda}.$$

This equation is approximate because λ is not well determined. If at any instant the interval Δx were shifted slightly to the right or to the left, the number of peaks or valleys in the interval might change. Because λ is not the same everywhere, more might move into one end of the interval than leave out of the other. Thus, there's an uncertainty $\Delta \lambda$ in λ that introduces a corresponding uncertainty Δn in n. To relate Δn to $\Delta \lambda$ we differentiate the approximate equation for n with respect to λ and we find

$$\frac{dn}{d\lambda} = -\frac{2|\Delta x|}{\lambda^2},$$

so a small change $\Delta\lambda$ in λ produces a corresponding change Δn given by

$$\Delta n = -\frac{2|\Delta x|}{\lambda^2}\Delta\lambda.$$

Comparing this with Eq. (50.33) we see that

$$\Delta n = \frac{|\Delta x|\Delta p}{h/2}. \tag{50.34}$$

But a change in λ means that n increases or decreases by 1 because the length $|\Delta x|$ is fixed. Therefore $|\Delta n| \geq 1$ and Eq. (50.34) implies the inequality in (50.32).

The foregoing discussion is not to be regarded as a proof or derivation of (50.32), but only as an intuitive argument showing that such an inequality seems reasonable. A proof would require precise definitions of the "uncertainties" Δx and Δp, which we have left vague. As mentioned earlier, the Heisenberg uncertainty principle is an inequality that can be deduced from the axioms of quantum mechanics, but a precise mathematical statement of the inequality involves mathematical concepts beyond the scope of this text.

The physical interpretation of the inequality usually refers to statistical dispersion of probable errors Δx and Δp arising from a sequence of measurements. The situation is somewhat analogous to recording the location and speed of a large number of automobiles traveling along a long straight road. Statistical theory tells us something about the distribution or scatter of the automobiles around a certain average position and speed. In particular it tells us that we cannot reduce the scatter below a certain limit. This is the kind of information the Heisenberg uncertainty principle conveys.

A commonly accepted physical interpretation of the inequality is that it imposes some kind of upper limit to precision of measurements, or lower limits to their imprecision. In particular, it states that, regardless of improvements in experimental techniques, it is not possible to obtain exact knowledge of the position and momentum of a particle simultaneously. An inherent characteristic of nature is that only probable values can be obtained.

Despite some initial philosophical objections to the Heisenberg uncertainty principle, it provided great support for quantum mechanics by showing that there was an uncertainty inherent in nature, just as predicted by the theory. Quantum mechanics itself has become one of the most successful theories in all of physics, not yet contradicted by any experimental evidence.

Example 7

Show that the Heisenberg uncertainty principle is satisfied by a harmonic oscillator in its lowest energy state.

In Section 50.5 we showed that the lowest energy state of the harmonic oscillator has wave amplitude

$$\Psi_0(x) = a_0 e^{-\alpha x^2/2}$$

50.7 THE HEISENBERG UNCERTAINTY PRINCIPLE

and energy

$$E_0 = \frac{1}{2}hf = \frac{h}{4\pi}\sqrt{\frac{k}{m}} = \frac{1}{2}\hbar\sqrt{\frac{k}{m}}.$$

The corresponding probability density is given by

$$\Psi_0^2 = a_0^2 e^{-\alpha x^2}.$$

The integral of this function from $-\infty$ to $+\infty$ is equal to 1, but the integral from $-\pi/\sqrt{2\alpha}$ to $+\pi/\sqrt{2\alpha}$ is about 0.999, which can be interpreted as saying that the particle spends most of its time in this interval. So we will arbitrarily define the uncertainty Δx in the position x to be $\pi/\sqrt{2\alpha}$,

$$\Delta x = \frac{\pi}{\sqrt{2\alpha}}.$$

The energy of a classical harmonic oscillator is a constant, the sum of its potential and kinetic energies. In each cycle, the kinetic energy $p^2/(2m)$ varies from 0 to E_0, where E_0 is the total energy, so the maximum momentum is

$$p_{max} = \sqrt{2mE_0}. \tag{50.35}$$

At any instant the oscillator can have any value of momentum between 0 and p_{max}. Assuming that p_{max} is about the same as that predicted by classical mechanics, Eq. (50.35), with E_0 the lowest state energy, we can say that the uncertainty Δp is some constant times $\sqrt{2mE_0}$,

$$\Delta p = C\sqrt{2mE_0},$$

where $C \approx 1$ and $E_0 = \frac{1}{2}\hbar\sqrt{k/m}$. Thus the product $\Delta x \Delta p$ is given by

$$\Delta x \Delta p = \pi C\sqrt{\frac{mE_0}{\alpha}}.$$

Because $\alpha = \sqrt{mk/\hbar}$ this becomes

$$\Delta x \Delta p = \pi C \frac{\hbar}{\sqrt{2}} = \frac{C}{\sqrt{2}} \frac{h}{2}.$$

This is consistent with the Heisenberg inequality (50.32) if $C \geq \sqrt{2}$, a number of order 1, as expected.

Questions

25. Discuss how the concept of Bohr orbits violates the uncertainty principle.

26. A particle moves parallel to the x axis. The uncertainty in its x component of momentum is 6.6×10^{-30} J s/m.

 (a) What is the minimum uncertainty in its x coordinate?
 (b) What is the uncertainty in its y component of momentum?
 (c) What is the uncertainty in its y coordinate?

27. A particle is confined to an interval of length 1 Å, the diameter of an atom. Determine the uncertainty in its momentum.

28. In its ground state ($n = 0$) a simple harmonic oscillator has an energy $hf/2$, called the zero point energy. Use the uncertainty principle to explain why the oscillator must always have some positive energy.

29. Wavelengths can be measured with an accuracy of one part in 10^6. What is the uncertainty in position of a 0.1-nm X-ray photon when its wavelength is simultaneously measured?

30. Suppose a particle moving along the x axis with speed v has uncertainty Δx in its position. The uncertainty in time at which the particle will arrive at some point is $\Delta t = \Delta x/v$.

  (a) Find the uncertainty in the momentum of the particle.
  (b) Determine the uncertainty ΔE in the energy of the particle.
  (c) Show that $\Delta E \Delta t \geq h/2$. This is another version of the uncertainty principle proposed by Einstein.

31. The shutter of a camera remains open for 1 μs. Use the result of Question 30 to determine the uncertainty in energy of any one photon passing through the camera lens.

32. Use the result of Question 30 to calculate the uncertainty in energy of the $n = 2$ state of a hydrogen atom that has an energy of -3.4 eV and a lifetime of 10^{-9} s.

50.8 A FINAL WORD

Louis Cesar Victor Maurice de Broglie – known as Maurice – was born in Paris in 1875, scion of a distinguished family that had supplied France with several generations of soldiers, diplomats, and politicians. Maurice chose the navy – the most technically oriented of the services – and had a brilliant career. One of his noteworthy accomplishments was installing the first wireless radio on a French warship.

In 1904, Maurice approached his grandfather, the Duke de Broglie, for permission to retire from the navy and take up science as a career. "Science," the old man said, "is an old lady, content with the attentions of old men." Science wasn't a dashing enough career for a de Broglie.

Nevertheless, a compromise was worked out, no doubt facilitated by the Duke's demise, and in 1908 Maurice was permitted to retire to the family estate where a personal laboratory was built for him. He presented a thesis to the University of Paris on Brownian motion (also the subject of Einstein's thesis in 1905), under the distinguished theorist Paul Langevin, and then turned his attention to the new field of X rays.

His principal competitor was the young English physicist H. G. J. Moseley. Both worked on determining the X-ray spectra of the elements, but while Maurice's style was an exhaustively careful measurement of the spectrum of each element, Moseley dashed rapidly through many elements. Moseley's technique turned out to

50.8 A FINAL WORD

be the more fruitful because he found systematic tendencies that lent support to Niels Bohr's new and controversial model of the atom.

In 1914, this intense but peaceful competition was interrupted by the outbreak of World War I. Both Moseley and de Broglie saw action. Moseley died in the battle of Gallipoli. Maurice de Broglie survived, but by the time the war ended, his work had been superseded by scientists in Sweden, a country that had remained neutral in the war. Nevertheless, after the war Maurice de Broglie took his place as one of France's most respected scientists.

All of this would be no more than a minor historical footnote, except that Maurice had a younger brother named Louis. Born in 1892, seventeen years after Maurice, Louis too grew up wanting to be a scientist. With Maurice setting the family precedent, Louis found no opposition to his peculiar aspiration. The rest, as we have seen, is history.

CHAPTER 51

ATOMS TO QUARKS

It seems probable to me, that God in the Beginning formed matter in solid, massy, hard, impenetrable, movable particles, of such sizes and figures, and with such other properties, and in such proportions, as most conduced for the end which He formed them; and that these primitive particles being solids, are incomparably harder than any other porous bodies compounded of them; even so very hard as never to wear or break into pieces, no ordinary power being able to divide what God Himself made one in the first Creation.

Isaac Newton

When I was your age, I always did it for half an hour a day. Why sometimes I've believed as many as six impossible things before breakfast.

The Red Queen in *Through the Looking-Glass* by Lewis Carroll

51.1 THE NATURE OF MATTER

Physics derives its name from φύσις, the Greek word for nature, and the goal of physics is to seek the essential nature of things. As we've seen in earlier chapters, philosophers for centuries have debated the fundamental constituents of matter. During that time many conjectures were offered concerning the makeup of all types of matter, ranging from water to atoms. One school of ancient Greek philosophers, the hylozoists, believed that life is a property of all matter, including stones.

The most widely accepted view emerging through the centuries was that atoms were lifeless particles moving in a void. The cause of their motion, the subject of *The Mechanical Universe*, was eventually explained by Isaac Newton. But the subtle mysteries of the nature of matter began to be revealed only in the twentieth century.

In the last chapter we learned that matter exhibits both particle and wave characteristics. An electron, for example, can behave like a hard, impenetrable particle, as it does in a television tube, but when it is diffracted by crystals it behaves like a wave. This chapter probes more deeply into the wave and particle nature of matter and extends our knowledge from the atomic to the subatomic level.

51.2 QUANTUM STATES OF THE HYDROGEN ATOM

Thomson, Rutherford, Bohr, de Broglie, and others devised ingenious models to explain the structure of the simplest atom – hydrogen. The most successful model, however, comes from applying the Schrödinger equation to describe electron waves about a proton. Although a comprehensive quantum mechanical treatment of the hydrogen atom is beyond the level of this text, nonetheless we can explore features of the hydrogen atom that provide insight into other atoms as well.

Like Newton's second law, the Schrödinger equation is a second-order differential equation with widespread applicability. It holds for *all* atomic systems, as long as the speed of electrons is not comparable with the speed of light. Specific applications usually involve consideration of forces influencing a system, together with initial conditions. One quantity that enters into the Schrödinger equation is the potential energy. A specific potential energy function, $V(\mathbf{r}, t)$ corresponds to a definite physical situation. For example, in Chapter 50 we described the solution of the equation for a one-dimensional simple harmonic oscillator. There the potential energy function was given by $V(x) = kx^2/2$, and the boundary conditions required that the wave functions be finite for large x. This led to wave functions describing the state of the oscillator. Knowledge of the wave functions, in turn, permits predictions concerning the position, momentum, and energy of the oscillator – all physically measurable quantities.

Because atoms are not confined to a line but exist in a three-dimensional world, we need to consider the three-dimensional form of the Schrödinger equation. For the wave function $\Psi(x, y, z)$ of an electron of mass m bound to a nucleus, the time-independent Schrödinger equation takes the form

$$-\frac{\hbar^2}{2m}\left(\frac{\partial^2 \Psi}{\partial x^2} + \frac{\partial^2 \Psi}{\partial y^2} + \frac{\partial^2 \Psi}{\partial z^2}\right) + V(x, y, z)\Psi = E\Psi.$$

If the electron is in the attractive field of a nucleus of charge $+Ze$, the potential energy is given by a familiar expression, the classical potential energy of two charges:

$$V(x, y, z) = -\frac{ZK_e e^2}{r},$$

where $r = (x^2 + y^2 + z^2)^{1/2}$ is the distance from the nucleus to the electron.

51.2 QUANTUM STATES OF THE HYDROGEN ATOM

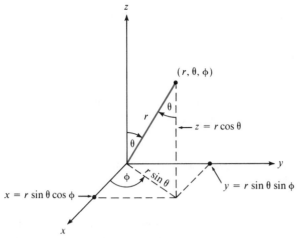

Figure 51.1 A spherical coordinate system and its connection to rectangular coordinates.

Because of the spherical symmetry of the potential, it is natural to use spherical coordinates (r, θ, ϕ) to locate the electron, as illustrated in Fig. 51.1.

We won't attempt to solve the Schrödinger equation, but we will describe the nature of the solutions. First, we recall some facts concerning the one-dimensional simple harmonic oscillator. The Schrödinger equation for that case involves two parameters α and β. A physically meaningful solution $\Psi(x)$ occurs only if the parameters are related by an equation of the form

$$\beta = (2n + 1)\alpha, \quad n = 0, 1, 2, \ldots . \tag{50.24}$$

This assures that the wave function $\Psi(x)$ approaches 0 for large x. One result of this condition is the quantization of the energy:

$$E_n = \left(n + \tfrac{1}{2}\right)hf. \tag{50.26}$$

In other words, the quantum number n specifies the permissible energies that the oscillator can have in one-dimensional motion. In the language of quantum theory, the quantum number n is said to specify the *state* of the oscillator. When Ψ_n is known, everything else we wish to know about the oscillator can be calculated.

In the three-dimensional case, we again require the wave function $\Psi(x, y, z)$ to approach 0 for large r because we are interested in the bound states of the electron. This requirement makes sense physically because the Coulomb force becomes weaker at large distances, decreasing as $1/r^2$, so the electron bound in a hydrogen atom is less likely to be found at large distances from the nucleus. In fact, the probability of finding the electron in some three-dimensional region is equal to the integral of Ψ^2 over that region, and this probability approaches zero as the region recedes from the nucleus. Because of the three-dimensional nature of the problem, three orbital quantum conditions are imposed, and these are expressible in terms of three quantum numbers, denoted by n, l, and m. We turn now to the physical significance of the various quantum numbers and their interrelations.

(1) The *principal quantum number n*, with possible values $n = 1, 2, 3, 4, \ldots$, determines the energy of the state according to the equation

$$E_n = -\frac{mK_e^2 Z^2 e^4}{4\hbar^2 n^2} = -13.6 \frac{Z^2}{n^2}, \tag{51.1}$$

where E_n is measured in electron volts (eV). This is the same result that Bohr found for the energy levels of a hydrogen atom ($Z = 1$). Of course, the Schrödinger equation had to reproduce those features of the Bohr theory that conformed to experimental evidence.

(2) For a given principal quantum number n, the *orbital quantum number l*, takes the possible values

$$l = 0, 1, 2, \ldots, n - 1.$$

and it determines the angular momentum of the state. A state with a given value of l has an orbital angular momentum L equal to

$$L = \left(l(l+1)\right)^{1/2} \hbar. \tag{51.2}$$

Note that this is *not* the same result obtained by the Bohr theory, which predicts an angular momentum $L = n\hbar$.

(3) For a given orbital quantum number l, the *magnetic quantum number m* (sometimes written m_l) specifies the component of the angular momentum along some fixed direction, usually chosen as the z direction (although it should be realized that atoms don't have coordinate axes attached to them). For each l there are $2l + 1$ allowed values for m:

$$m = 0, \pm 1, \pm 2, \pm 3, \ldots, \pm l.$$

The component of the angular momentum along the z direction is given by

$$L_z = m\hbar.$$

The prefix "magnetic" is used because this quantum number serves to describe the splitting of spectral lines into discrete components from atoms placed in an intense magnetic field. This splitting is known as the *Zeeman effect*, after Pieter Zeeman, who observed it in 1896. The orbital angular momentum vector cannot have an arbitrary direction in space in a magnetic field. Instead, it is restricted to particular orientations for which the component in the direction of the magnetic field is an integral multiple of $\hbar$.

The state of a hydrogen atom is labeled by giving the three quantum numbers (n, l, m) that specify the energy, angular momentum, and z component of angular momentum, respectively. We label the corresponding wave functions by Ψ_{nlm}. Distinct triples of quantum numbers give distinct wave functions.

Example 1

List all the orbital states of a hydrogen atom corresponding to the quantum numbers $n = 1, 2,$ and 3.

According to the rules outlined above, we see that for $n = 1$ there is only one state

$$(n, l, m) = (1, 0, 0),$$

and hence only one wave function Ψ_{100}. From Eq. (51.1) we see that the energy of this state is $E_1 = -13.6$ eV and is the lowest possible energy – the ground state energy of the hydrogen atom.

For $n = 2$ there are four states,

$$(n, l, m) = (2, 0, 0),$$
$$(2, 1, 1), (2, 1, 0), (2, 1, -1).$$

For $n = 3$ there are nine states,

$$(n, l, m) = (3, 0, 0),$$
$$(3, 1, 1), (3, 1, 0), (3, 1, -1),$$
$$(3, 2, 2), (3, 2, 1), (3, 2, 0), (3, 2, -1), (3, 2, -2).$$

In the foregoing display, each row contains those states with the same orbital quantum number l.

Example 2

How many states of a hydrogen atom can have energy equal to -1.51 eV?

From the allowed energy values given by Eq. (51.1), we find that the energy $E = -1.51$ eV corresponds to a value $n = 3$. From Example 1 we know that there are nine possible states. In the absence of a magnetic field, all these states have the same energy. However, if a strong magnetic field is present, the atoms in different states will interact with the field according to the magnetic quantum number and their energies will be slightly different. That's how Zeeman was able to detect them.

In the early years of quantum theory, physicists labeled spectral lines as sharp, principal, diffuse, fine, and so on. That terminology has survived, and the first letters of these words are used to identify states of different values of orbital angular momentum. For example, all states with $l = 0$ are called s states (from *s*harp). States with $l = 1$ are called p states (from *p*rincipal), and the state with $n = 1$ and $l = 0$ is called the $1s$ state. Similarly, the state with $n = 2$ and $l = 1$ is called the $2p$ state, and so on. Table 51.1 summarizes this labeling scheme.

51.3 WAVE FUNCTIONS FOR THE HYDROGEN ATOM

In our treatment of the one-dimensional harmonic oscillator in Chapter 50 we learned that the wave function corresponding to the quantum number n has the

Table 51.1 Labeling Scheme for States with Orbital Quantum Number $l = 0, 1, 2,$ and 3.

l	Label
0	s
1	p
2	d
3	f

form

$$\Psi_n(x) = u_n(x) e^{-\alpha x^2/2}$$

where $u_n(x)$ is a polynomial in x of degree n. The probability of finding the oscillator in an interval between $x = a$ and $x = b$ is

$$\text{probability} = \int_a^b \Psi(x)^2 \, dx. \tag{50.27}$$

Corresponding results for the hydrogen atom can be obtained by solving the three-dimensional Schrödinger equation when the potential energy V is a function of r alone, where r is the distance from the nucleus to the electron. We will not attempt to derive the solution here, but will simply describe the nature of the resulting wave functions.

When the Schrödinger equation with V a function of r is transformed to spherical coordinates (r, θ, ϕ), solutions can be obtained in which the wave function Ψ is a product of three functions, each of which depends on only one of the spherical coordinate variables:

$$\Psi = R(r)\Theta(\theta)\Phi(\phi).$$

This separation of the variables results in three different second-order differential equations, one for each of the factors R, Θ, Φ. When these are solved separately, the wave function Ψ_{nlm} that corresponds to the quantum state (n, l, m) takes the form

$$\Psi_{nlm} = R_{nl}(r)\Theta_{lm}(\theta)\Phi_m(\phi).$$

The simplest of the factors is $\Phi_m(\phi)$, which, aside from a constant factor $1/\sqrt{2\pi}$ or $1/\sqrt{\pi}$, is a linear combination of $\cos m\phi$ and $\sin m\phi$, where m is the magnetic quantum number. Each factor $\Theta_{lm}(\theta)$ is a polynomial in $\sin\theta$ and $\cos\theta$, the polynomial depending on the quantum numbers l and m. Each radial factor $R_{nl}(r)$ is equal to a polynomial in r multiplied by an exponential factor of the form

$$\exp\left\{-\frac{r}{na_0}\right\},$$

51.3 WAVE FUNCTIONS FOR THE HYDROGEN ATOM

where a_0 is the Bohr radius and where $\exp(t) = e^t$. The polynomial factor depends on the two quantum numbers n and l, whereas the exponential factor depends only on the principal quantum number n.

In particular, the wave function of the hydrogen atom in the quantum state $(1, 0, 0)$ is equal to

$$\Psi_{100} = \frac{1}{\sqrt{\pi a_0^3}} \exp\left\{\frac{-r}{a_0}\right\}.$$

The probability density is the square of the wave function,

$$\Psi_{100}^2 = \frac{1}{\pi a_0^3} \exp\left\{\frac{-2r}{a_0}\right\}.$$

Because this function is independent of θ and ϕ, the normal or ground state of the hydrogen atom has spherical symmetry. This property was not possessed by the ground state of the Bohr atom because the Bohr orbit was assumed to lie in a plane.

To calculate the probability that the electron lies inside a spherical shell about the nucleus with inner radius a and outer radius b, the probability density must be integrated over the region between the two spheres. Calculation of the integral is best carried out in spherical coordinates. The details of the calculation will not be discussed here, but the final result is that the probability is

$$\text{probability of electron in shell} = \frac{4}{a_0^3} \int_a^b r^2 \exp\left\{\frac{-2r}{a_0}\right\} dr. \tag{51.3}$$

This can be written as

$$\text{probability} = F(b) - F(a),$$

where

$$F(r) = \frac{4}{a_0^3} \int_0^r \rho^2 \exp\left\{\frac{-2\rho}{a_0}\right\} d\rho.$$

The function $F(r)$ cannot be expressed in terms of the elementary functions of calculus, but it can be evaluated approximately by numerical methods, and it has been tabulated. The probability that the electron is in the spherical shell with radii a and b is numerically equal to the area of the shaded region in Fig. 51.2, the region above the interval $[a, b]$ and under the graph of the derivative $F'(r)$,

$$F'(r) = \frac{4}{a_0^3} r^2 \exp\left\{\frac{-2r}{a_0}\right\}. \tag{51.4}$$

The electron is most likely to be found near values of the radius where $F'(r)$ is large, and unlikely to be found where $F'(r)$ is small. The maximum value of $F'(r)$ occurs where the second derivative $F''(r) = 0$. In Example 3 we show that this is at the point $r = a_0$. In other words, an electron in the ground state of a hydrogen atom spends most of its time at a distance from the nucleus near the Bohr radius.

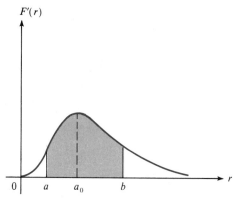

Figure 51.2 Graph of the derivative $F'(r)$. The area of the shaded region is the probability that the electron is in the spherical shell with radii a and b.

Example 3

Show that the derivative of the function in Eq. (51.4) is zero when $r = a_0$.

Using the product rule to differentiate $F'(r)$ we find

$$F''(r) = \frac{4}{a_0^3} r^2 \exp\left\{\frac{-2r}{a_0}\right\}\left(-\frac{2}{a_0}\right) + \frac{4}{a_0^3} 2r \exp\left\{\frac{-2r}{a_0}\right\}.$$

Equating this to zero and solving for r we obtain $r = a_0$.

The foregoing analysis referred to the ground state of the hydrogen atom. The same type of argument can also be carried out for an arbitrary state (n, l, m). The principal difference is that the integrand in the probability integral corresponding to (51.3) is the product $4\pi r^2 R_{nl}(r)^2$. Figure 51.3 shows the graph of this product as a function of r for several different states. The peaks on the graphs reveal how the distances from the nucleus where the electron is most likely to occur can vary with the quantum numbers n and l. The product $4\pi r^2 R_{nl}(r)^2$ is referred to as the *radial probability density*. It tells us the probability of finding the electron in a thin shell whose inner and outer radii are nearly equal to r.

In the quantum mechanical picture of the atom, electrons are not revolving about the nucleus along specific orbits. Because of their wave nature, electrons are not localized. We cannot specify the exact position of an electron but only the probability of finding it in a specified region.

If we could make repeated measurements of the position of the electron and plot a point for each position, we would eventually build up a collection of points in space. Regions where the points cluster together correspond to regions in which the electron is more likely to be found. Figure 51.4 illustrates this idea for several states.

51.3 WAVE FUNCTIONS FOR THE HYDROGEN ATOM

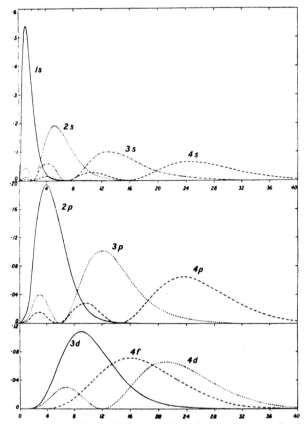

Figure 51.3 The radial probability density for several values of the quantum numbers n and l. (From *The Theory of Atomic Spectra* by E. U. Condon and G. G. Shortley, Cambridge University Press, 1953.)

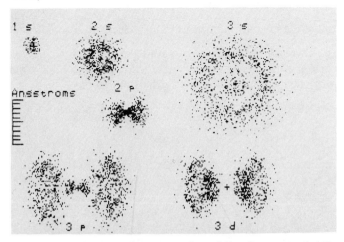

Figure 51.4 Computer representation of the electron probability density for various states. (Courtesy of Cross Educational Software.)

Alternatively we can think of the electron as having its charge spread out, or distributed, as a kind of *electron cloud*. Although we usually consider the charge of an electron to be indivisible, it is sometimes useful to imagine the charge as being spread out to produce a charge distribution. In terms of the electron cloud, the charge is greater where a large number of points tend to accumulate, as suggested by Fig. 51.4. As we've seen above, that corresponds to regions where $r^2 R_{nl}(r)^2$ is large. The idea of an electron cloud is a fruitful one. From it we can analyze the chemical, electrical, and magnetic properties of atoms in terms of their charge distributions.

Questions

1. Determine which of the following statements are true and which are false for hydrogen in (i) the Bohr theory and (ii) in the Schrödinger theory. Correct any statements that are false.

    (a) In a given state the electron can exist at an arbitrary distance from the nucleus.
    (b) The total energy can assume any negative value.
    (c) The orbital angular momentum of the electron can be any integral multiple of $\hbar$.
    (d) The orbital angular momentum of the ground state is $1\hbar$.
    (e) The ground state energy (orbital number $n = 1$) is -13.6 eV.

2. With the aid of Fig. 51.3, determine whether an electron is more likely to be found nearer the nucleus or farther away from it as the principal quantum number increases.

3. Determine how many states of a hydrogen atom have an energy equal to -0.85 eV.

4. Use Fig. 51.3 to rank the following states from greatest to least probability of finding the electron close to the nucleus: $3p, 2s, 3d, 1s, 2p$.

5. Determine the angular momentum of an electron in the $4d$ state of hydrogen.

6. How much more probable is it to find the electron in the ground state of hydrogen in a thin shell of radius nearly equal to the Bohr radius than in a thin shell of radius nearly equal to twice the Bohr radius?

7. The $2s$ state of a hydrogen atom has the radial factor

    $$R(r) = (1/a_0)^{3/2}(2 - r/a_0) \exp\{-r/2a_0\}.$$

    Determine the value of r that makes the radial probability density $4\pi r^2 R(r)^2$ largest.

51.4 ATOMS WITH MANY ELECTRONS

In its normal state, a hydrogen atom has its electron in the lowest energy quantum state. We might ask, what are the normal configurations of more complex atoms?

51.4 ATOMS WITH MANY ELECTRONS

For example, are all 92 electrons of a uranium atom in the same quantum state, to be envisioned as perhaps all in the same electron cloud? Many lines of reasoning suggest that this is unlikely. First, this would mean that atoms differ greatly in size, becoming smaller as the atomic number Z increases. Experimental evidence, on the other hand, indicates that all atoms are roughly the same size, on the order of 10^{-10} m. Moreover, if all electrons were in the same state, chemical properties would not be periodic and would not undergo abrupt changes with the addition of one electron to the atom. Analysis of spectra indicates that not all electrons in complex atoms are in the same state, that is, they do not have the same values of n, l, and m. To better understand the structure of atoms, we investigate an additional property of electrons.

With the development of spectroscopic instruments of high resolution, it was soon revealed that the hydrogen spectrum was more complex than originally thought. Some of the spectral lines were composed of a number of fine lines so close together that they appeared to coincide when viewed with less accurate instruments. In an attempt to explain this structure, George Uhlenbeck and Samuel Goudsmit in 1925 proposed that the electron in a hydrogen atom has an intrinsic angular momentum in addition to its orbital angular momentum. Thus the idea of *spin* was introduced.

Returning for a moment to the planetary model of the hydrogen atom, we can visualize the electron as a small charged sphere spinning about an axis as it revolves in orbit about the nucleus. This is suggested by planetary motion, where each planet not only revolves about the sun but also spins on an axis that is tilted at some angle with respect to the plane of its orbit. But the quantum mechanical electron is not envisioned as a rigid sphere spinning on an axis. Instead spin represents another way in which an electron can interact with other particles and with electric and magnetic fields. As it turns out, spin helps explain the fine structure of spectra and also the structure of atoms with many electrons.

The spin of an electron, or proton, or any particle can, in principle, be measured by observing the deflection of the particle in an inhomogeneous magnetic field. Any charged particle with spin behaves like a tiny magnet in that it possesses a magnetic moment that is directly proportional to its spin. In an inhomogeneous magnetic field, the magnetic moment experiences an unbalanced force and accelerates in response to it. Experimentally this is evidenced by the splitting of a beam of particles as the beam passes through the field, as illustrated in Fig. 51.5.

As we have seen, the orbital angular momentum of an electron is given by

$$L = (l(l+1))^{1/2}\hbar, \qquad (51.2)$$

where l is an integer. Uhlenbeck and Goudsmit attributed to the electron an intrinsic angular momentum vector **S** and a corresponding magnetic moment vector **μ**, such as would be associated with the spinning motion of an electrically charged body about an axis through it. By analogy with Eq. (51.2) they ascribed to the angular momentum vector **S** a magnitude S given by

$$S = (s(s+1))^{1/2}\hbar. \qquad (51.5)$$

where the number s, called the spin number, or simply the spin, is required by the

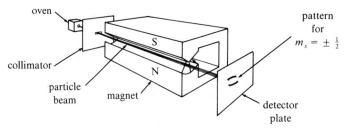

Figure 51.5 The intrinsic spin of a particle is responsible for the particle possessing a magnetic moment that interacts with an inhomogeneous magnetic field and splits a fine beam of particles.

experimental data to have the value $\frac{1}{2}$. The component of the angular momentum vector along any prescribed axis is either $\frac{1}{2}\hbar$ or $-\frac{1}{2}\hbar$ and therefore can be written as $m_s \hbar$, where m_s is $\pm \frac{1}{2}$. The number m_s can be regarded as a new quantum number and is often referred to as *spin up* or *spin down*, according as m_s is $+\frac{1}{2}$ or $-\frac{1}{2}$. The up–down terminology is suggested by the diagram in Fig. 51.6.

The concept of spin was later extended to apply to other particles by allowing s to be any positive integer multiple of $\frac{1}{2}$ in Eq. (51.5). Electrons, protons, and neutrons each have intrinsic spin $\frac{1}{2}$. Particles that have half-integral units of spin are classified as *fermions*, after the physicist Enrico Fermi. Photons have spin 1 and fall into a classification known as *bosons*, named after the Indian physicist, S. Bose.

We now return to the hydrogen atom to see how the spin of the electron modifies our picture. Within the framework of Schrödinger theory, electron spin is a phenomenon that must be added to the theory because it does not arise from solving the Schrödinger equation. In constructing a quantum mechanics compatible with the theory of relativity, P. A. M. Dirac obtained a set of equations different from those obtained from the nonrelativistic Schrödinger equation. When these were solved for the one-electron system, Dirac found that the spin of the electron was obtained automatically, without the need of a separate hypothesis. His equations led to a complete description of the energy levels for the hydrogen atom including the fine structure.

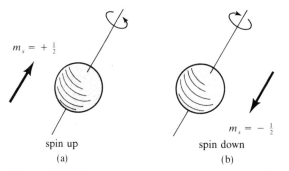

Figure 51.6 Schematic representation of an electron's spin orientation. (a) Spin up; (b) spin down.

51.4 ATOMS WITH MANY ELECTRONS

It is now known that the complete state of an electron in the hydrogen atom requires *four* quantum numbers, (n, l, m, m_s). There are, therefore, twice as many states of an electron in a hydrogen atom as we would have expected from the Schrödinger equation alone. For example, for $n = 1$ there are two states, corresponding to quantum numbers $(1, 0, 0, +\frac{1}{2})$ and $(1, 0, 0, -\frac{1}{2})$. The concept of electron spin explains far more than the fine structure of spectral lines. It is mainly responsible for magnetism in materials such as iron and is needed to understand elementary properties of molecules and solids.

Once we incorporate electron spin we find that quantum theory provides a wealth of information about chemical and physical properties of elements, molecules, and compounds. One of the greatest achievements of quantum theory is that it explains the basis for the observed ordering of the chemical elements as they appear in the periodic table. Originally the periodic table was cleverly constructed by Dmitri Mendeleev in 1869 by listing the elements in order of increasing atomic weights. By doing so, he found remarkable periodicities in their properties. The key principle that led to the quantum theoretical understanding of the periodic table was proposed by Wolfgang Pauli in 1925 and is known as the *Pauli exclusion principle*.

Pauli exclusion principle: *A given quantum state can be either unoccupied or occupied by at most one electron.*

In other words, no two electrons in an atom can have the same set of quantum numbers n, l, m, and m_s. The principle is supported by spectroscopic evidence indicating that atoms simply never occur with two electrons occupying the same state. It is analogous, but not equivalent, to the classical assertion that two hard spheres cannot be at the same place at the same time. Later it was learned that the same principle applies to protons, neutrons, and all fermions.

Although the Schrödinger equation could, in principle, be solved exactly for multielectron atoms, it presents a formidable mathematical task. Nonetheless, we can understand the behavior of atoms by knowing how the possible electron states are occupied. Two basic rules are needed:

1. Only one electron is in any particular quantum state of an atom.
2. An atom is stable if its total energy is a minimum. Consequently, electrons occupy states with the lowest possible energy consistent with the Pauli exclusion principle.

To see how these rules determine the atomic structure of the elements, we first consider the states of a hydrogen atom. Figure 51.7 shows a representation of the possible energy states available to an electron in a hydrogen atom. Corresponding to the two spin orientations, there are two states for each horizontal line in the diagram. In its ground state, the electron can have $n = 0$, $l = 0$, $m = 0$, and $m_s = \frac{1}{2}$ or $-\frac{1}{2}$.

Now imagine adding another electron to the atom and also increasing the charge of the nucleus to $+2|e|$ so as to construct a neutral atom of helium. The

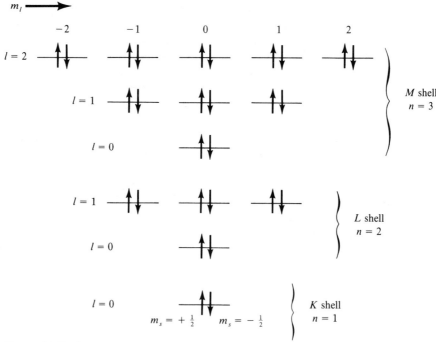

Figure 51.7 Diagram representing possible energy states for a hydrogen atom.

separation between the levels will not be the same as indicated in Fig. 51.7 because the charge of the nucleus is greater and the two electrons experience mutual repulsion from each other besides attraction to the nucleus. Nevertheless, a similar diagram will still represent the possible states of electrons. According to the Pauli exclusion principle, the two electrons in the ground state of helium are in the states $(1, 0, 0, +\frac{1}{2})$ and $(1, 0, 0, -\frac{1}{2})$. No more electrons can be added to the atom with quantum numbers $n = 1$, $l = 0$, and $m = 0$, and we say that the $n = 1$ shell, also known as the K shell, is full. Because the electron spins are in opposite directions, a helium atom has no net magnetic moment, and because the two electrons are tightly bound to the nucleus, a considerable amount of energy is required to excite them to a higher state. For these reasons helium is chemically inactive.

By specifying the quantum numbers of all electrons in a given atom, we specify the electron configuration of the atom. A convention used to specify electron configurations is to specify the principal quantum number n, followed by the letter denoting the orbital angular momentum quantum number of individual electrons, and then use a superscript to give the number of electrons having the particular values of n and l. For example, the electron configuration for hydrogen is $1s^1$, while that for helium is $1s^2$.

The element with the next lowest atomic number, $Z = 3$, is lithium. It has three electrons, but only two can occupy the states $n = 1$, $l = 0$, and $m = 0$. The Pauli exclusion principle requires that, in the ground state, the third electron be in the

51.4 ATOMS WITH MANY ELECTRONS

Table 51.2 Electron Configurations for Elements with $Z = 1$ through $Z = 30$.

Element	K 1s	L 2s	L 2p	M 3s	M 3p	M 3d	N 4s	N 4p	N 4d	N 4f
^{1}H	1									
^{2}He	2									
^{3}Li	2	1								
^{4}Be	2	2								
^{5}B	2	2	1							
^{6}C	2	2	2							
^{7}N	2	2	3							
^{8}O	2	2	4							
^{9}F	2	2	5							
^{10}Ne	2	2	6							
^{11}Na	2	2	6	1						
^{12}Mg	2	2	6	2						
^{13}Al	2	2	6	2	1					
^{14}Si	2	2	6	2	2					
^{15}P	2	2	6	2	3					
^{16}S	2	2	6	2	4					
^{17}Cl	2	2	6	2	5					
^{18}Ar	2	2	6	2	6					
^{19}K	2	2	6	2	6		1			
^{20}Ca	2	2	6	2	6		2			
^{21}Sc	2	2	6	2	6	1	2			
^{22}Ti	2	2	6	2	6	2	2			
^{23}V	2	2	6	2	6	3	2			
^{24}Cr	2	2	6	2	6	5	1			
^{25}Mn	2	2	6	2	6	5	2			
^{26}Fe	2	2	6	2	6	6	2			
^{27}Co	2	2	6	2	6	7	2			
^{28}Ni	2	2	6	2	6	8	2			
^{29}Cu	2	2	6	2	6	10	1			
^{30}Zn	2	2	6	2	6	10	2			

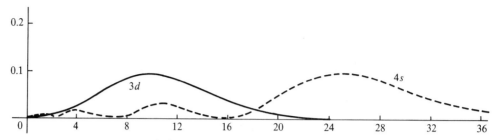

Figure 51.8 Graphs of r^2R^2 for the $3d$ and $4s$ hydrogen states reveal that in the $4s$ state the electron ventures nearer the nucleus.

state $n = 2$, $l = 0$, known as the L shell. Consequently, the ground state configuration of lithium is $1s^2 2s^1$. The outer electron in the $2s$ subshell is shielded from the full charge of the nucleus by the two inner electrons. Due to this screening the electron is held relatively weakly to the nucleus. Therefore, we expect that the electron in the L shell will be easy to excite and that the atom interacts chemically much like hydrogen, which it does.

By adding one electron to the electron cloud and increasing the nuclear charge by one unit, and applying the Pauli exclusion principle, we can construct the ground state electron configurations of the other atoms one by one. Table 51.2 lists the ground state configurations for elements up to $Z = 30$ (zinc). The shells from $n = 1$ onward are labeled alphabetically by the letters K, L, M, etc. The values of l in each shell form subshells. From the table, we see that the K shell contains 2 electrons, the L shell 8, the M shell 18, and so on. This grouping is reflected in the structure of the periodic table, which is given in Appendix C.

As the atomic number increases, subshells are filled continuously in the order of increasing energy states: $1s$, $2s$, $2p$, $3s$, and then $3p$. Spectroscopic evidence indicates that the $4s$ subshell is filled before the 10 states in the $3d$ subshell. We can understand this by examining the graphs of r^2R^2 for the $3d$ and $4s$ states of hydrogen in Fig. 51.8. In the $4s$ states the electron spends more time nearer the nucleus than in the $3d$ states. In multielectron atoms an electron in the $4s$ state is less shielded from the charge of the nucleus by the electrons in the filled subshells, and consequently its energy is slightly lower than in the $3d$ state. Therefore, electron subshells are filled in the following order, that has been verified by spectroscopic analysis of emission lines from various elements: $1s, 2s, 2p, 3s, 3p, 4s, 3d, 4p, 5s, 4d, 5p, 6s, 4f$.

From the electron configurations alone we can understand some characteristic properties of atoms. Atoms with closed electron subshells have both their orbital angular momentum and spin orbital angular momentum paired off so that these quantities are zero. These closed subshells correspond, for example to $1s^2, 2s^2$ and $2p^6, 3p^6$, and so on. Because of this structure we expect atoms with closed subshells to be chemically inert because there is neither excess nor deficiency of electrons in other subshells. From Table 51.2 we see that these closed subshells correspond to the elements helium ($Z = 2$), neon ($Z = 10$), and argon ($Z = 18$) – the noble gases that ordinarily fail to form chemical compounds.

51.4 ATOMS WITH MANY ELECTRONS

At the other end of the periodic table are the alkali metals that easily form positive ions by losing an electron. These elements include lithium, sodium, and potassium. From Table 51.2 we see that these elements have one electron in a subshell outside a closed subshell core that is like one of the noble gases. In an atom of atomic number Z, this unpaired electron is shielded from the charge of the nucleus by the $Z - 1$ electrons in the closed subshell. Effectively this outer electron experiences a nuclear charge of $Z - (Z - 1) = 1$ and is weakly bound to the atom. We would expect the spectra of the alkali metals to resemble that of hydrogen, which they indeed do.

The structure of the periodic table can be understood from the idea of electron shells and wave functions. A search for a more detailed understanding takes us into the realm of the chemical universe, and we leave that topic for a separate course.

Questions

8. List the quantum numbers of each electron in the ground state of lithium.

9. Refer to the periodic table in Appendix C and describe the electronic configuration of the ground state of krypton ($Z = 36$). Explain why it is a noble gas.

10. The accompanying figure shows a graph of $r^2 R^2$ as a function of r for sodium ($Z = 11$). The occupied $n = 1$ and $n = 2$ shells are the shaded regions.

    (a) For $n = 3$, what respective values of the orbital angular momentum quantum number l do the curves 1, 2, and 3 represent?
    (b) Of the three states represented by the curves 1, 2, and 3, which would require the greatest amount of energy to ionize an electron in it? Which would require the least?
    (c) Describe the electronic configuration of sodium in the ground state.

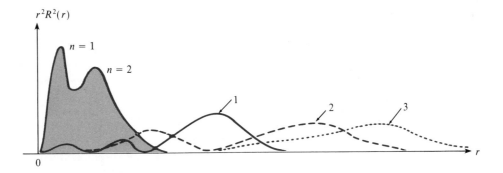

11. Carbon and silicon have similar chemical behavior. Use electron configurations for the ground states of these atoms to explain their similarity.

12. The periodic table has a periodicity of 8. That is, elements differing in atomic number by 8 have similar chemical properties. Draw on your knowledge of electronic subshells to explain this periodicity.

51.5 NUCLEI AND RADIOACTIVITY

More than a decade before Rutherford's pioneering work on the structure of the atom, Antoine Henri Becquerel, working across the Channel in Paris, discovered emissions from deep within the very hearts of atoms. In 1896, while exploring the possibility that sunlight might cause crystals of a uranium compound to emit X rays, which had been discovered by Wilhelm Roentgen a year earlier, Becquerel found that these crystals themselves caused the fogging of photographic plates on which they were resting. These mysterious rays, which became known as beta rays, were emitted by the compound in an unending stream, radiating in all directions. Becquerel correctly attributed the source of these rays to uranium.

In 1898, Madame Marie Sklodowska Curie found that similar rays emanated from the element thorium. Later she and her husband, Pierre Curie, discovered the element radium, which was much more active than uranium. Madame Curie coined the term *radioactivity* to describe the phenomenon. Researchers were confounded by the fact that there were three types of radiation. Rutherford's research in 1895-8 indicated two types, which he labeled alpha (α) and beta (β) rays. He found beta rays to be as penetrating as X rays, and alpha rays to be easily stopped by the thinnest foil. The third type of radiation was later discovered by Rutherford and labeled gamma (γ) rays.

Subsequent investigations by Becquerel revealed that beta rays could be deflected by a magnetic field – an observation that indicated the radiation might consist of charged particles. He measured the charge-to-mass ratio of the particles and concluded, in 1900, that the radiation consisted of negatively charged particles, identical to the cathode rays identified by J. J. Thomson. Beta rays were electrons with speeds much greater than those in cathode rays.

In 1901, Becquerel discovered that the source of the radiation was uranium atoms, and energy requirements indicated that the electrons came from deep within the nucleus. With that startling discovery, Becquerel ushered in a new branch of physics – nuclear physics. For the first time there was conclusive evidence that the nucleus itself was not a fundamental, indivisible particle, but might contain other particles, for example, electrons.

Example 4

Electrons in the decay of a radioactive nucleus typically have energies in the range of 1.0 MeV; that is, the kinetic energy of the electrons streaming from a sample is on the order of 1 MeV. The diameter of a typical radioactive nucleus is about 18.0 fm = 18.0×10^{-15} m. Use the uncertainty principle to calculate the minimum kinetic energy that an electron would need to have if it is confined inside the nucleus.

From the uncertainty principle, as expressed in (50.29), we know that

$$\Delta p \, \Delta x \geq h/2,$$

if Δx and Δp are positive. The momentum of an electron must be greater than the uncertainty in its momentum,

$$p \geq \Delta p \geq \frac{h}{2\,\Delta x}.$$

To find this minimum momentum and the corresponding energy, we take the maximum uncertainty in position to be the radius of the nucleus, $\Delta x = 9.0 \times 10^{-15}$ m. The relativistic kinetic energy is given by

$$K = \sqrt{p^2c^2 + m^2c^4} - mc^2,$$

so we have

$$K_{min} \geq \sqrt{\frac{h^2c^2}{4(\Delta x)^2} + m^2c^4} - mc^2.$$

Using $\hbar c = 197$ MeV fm and $mc^2 = 0.511$ MeV we get

$$K_{min} \geq \sqrt{\frac{(197 \text{ MeV fm})^2 4\pi^2}{4(9.0 \text{ fm})^2} + (0.511 \text{ MeV})^2} - 0.511 \text{ MeV}$$

$$\geq 68 \text{ MeV}.$$

This value is more than 60 times the observed value of the energy of the electron. We conclude that electrons cannot exist within the nucleus. Additional experimental evidence also supports this conclusion.

Alpha rays were more difficult to deflect with electric or magnetic fields, but Rutherford succeeded in measuring the deflection and obtaining a value for the charge-to-mass ratio. In 1903, he identified alpha rays as helium nuclei, which we now know consist of two protons and two neutrons. Later he was even able to collect a sample of alpha particles from the decay of radium and analyze their spectra, which he found to be identical to the spectra of helium in the sun. It was in the sun that the element helium was first observed in 1895, where it makes up approximately 25% of the mass. Rutherford didn't realize that helium should be abundant in the universe because it is the most tightly bound of the light nuclei.

In 1903, Rutherford published two classic papers entitled "The Cause and Nature of Radioactivity" in which he argued that radioactivity is a change of one chemical element into another that is accompanied by the emission of a charged alpha or beta particle. This hypothesis attacked the sacred premise of chemists on the immutability of atoms. Consequently, Rutherford's explanation was at first largely ignored, but was later accepted as the experimental evidence mounted.

The nature of the third type of radiation was eventually resolved in 1914 when Rutherford measured the wavelength of gamma rays scattered from crystals. He found that gamma rays were electromagnetic radiation of extremely short wavelength.

Once the different kinds of radiation are identified, we can understand the changes that accompany the decay of a radioactive element. The particles emitted in the decay do not exist "inside" the nucleus but rather are created in the decay. The radioactive element is commonly called the *parent* nucleus and its product the *daughter*. (Daughter rather than son because the daughter can spawn other nuclei by decaying.) We shall denote a parent nucleus by $^A_Z P$, where Z is the atomic number and A is the atomic mass number, and a daughter nucleus similarly by $^A_Z D$. The atomic number Z represents the number of protons in the nucleus and specifies the element. The atomic mass number is the total number of protons and neutrons in the nucleus and gives an indication of the mass of the element isotope. A given element (Z) can have more than one isotope (A). Because electrons are not nucleons but have charge, they have no atomic mass number and are assigned an atomic number of -1; in this scheme they are denoted by $_{-1}^{0}e$. Alpha particles, being helium nuclei, are denoted by $^4_2 He$.

In the decay of a nucleus all the conservation laws of physics remain valid. Energy, momentum, angular momentum, and charge are all conserved. A general description of each of the three basic modes of nuclear decays can be given as follows:

Beta Decay:

$$^A_Z P \rightarrow \,^A_{Z+1} D + \,_{-1}^{0}e.$$

An example is the decay of a radioactive isotope of potassium into calcium:

$$^{40}_{19}K \rightarrow \,^{40}_{20}Ca + \,_{-1}^{0}e.$$

Alpha Decay:

$$^A_Z P \rightarrow \,^{A-4}_{Z-2} D + \,^4_2 He.$$

An example is a decay of uranium:

$$^{238}_{92}U \rightarrow \,^{234}_{90}Th + \,^4_2 He.$$

Gamma Emission: Because gamma rays have no rest mass or charge (but they do carry off energy, momentum, and spin), a nucleus is unchanged in atomic and mass numbers. Gamma rays accompany the spontaneous decay of a nucleus in an excited state, much like visible, ultraviolet, and infrared photons accompany the transition of an atom from an excited to lower energy state. Denoting an excited nucleus by a superscripted asterisk (*), we have

$$^A_Z P^* \rightarrow \,^A_Z D + \gamma.$$

An example is the decay of tellurium:

$$^{208}_{81}Tl^* \rightarrow \,^{208}_{81}Tl + \gamma.$$

Table 51.3 provides the decay modes of some radioactive isotopes.

51.5 NUCLEI AND RADIOACTIVITY

Table 51.3 Decay Modes and Half-lives of Several Naturally Occurring Radioactive Nuclei.

Nuclide	Decay mode	Half-life
Beryllium-8 ($^{8}_{4}$Be)	α	1×10^{-16} s
Polonium-213 ($^{213}_{84}$Po)	α	4×10^{-6} s
Carbon-16 ($^{16}_{6}$C)	β	0.75 s
Aluminum-28 ($^{28}_{13}$Al)	β	2.24 min
Magnesium-28 ($^{28}_{12}$Mg)	β	21 h
Iodine-131 ($^{131}_{53}$I)	β	8 d
Cobalt-60 ($^{60}_{27}$Co)	β	5.3 yr
Strontium-90 ($^{90}_{38}$Sr)	β	28 yr
Radium-226 ($^{226}_{88}$Ra)	α	1600 yr
Carbon-14 ($^{14}_{6}$C)	β	5730 yr
Uranium-238 ($^{238}_{92}$U)	α	4.5×10^{9} yr
Rubidium-87 ($^{87}_{37}$Rb)	β	4.7×10^{10} yr

Example 5

In the alpha decay of $^{238}_{92}$U the alpha particle has an energy of 4.2 MeV. Determine the recoil velocity of the daughter nucleus assuming that the uranium nucleus is initially at rest.

According to the law of conservation of momentum, the daughter nucleus must recoil because the alpha particle has energy and momentum. The momentum of the daughter is equal in magnitude and opposite in direction to that of the alpha particle: $P = p$, where P is the momentum of the daughter, $^{234}_{90}$Th, and p the momentum of the alpha particle. Assuming nonrelativistic particles, we have $K = p^2/(2m)$ for the energy of the alpha particle, which indicates that the momentum of the particle is $p = (2mK)^{1/2}$. The momentum of the thorium nucleus is just MV, its mass times its velocity, where the mass is approximately 234 times the mass of a proton. Expressing masses in MeV/c^2 and energy in MeV, we have

$$V = (2mK)^{1/2}/M$$
$$= [2(4 \times 940 \text{ MeV}/c^2)(4.2 \text{ MeV})]^{1/2}/(234 \times 940 \text{ MeV}/c^2),$$
$$= 8.1 \times 10^{-4} c = 240 \text{ km/s}.$$

The study of radioactive thorium provided one crucial insight into radioactivity. Rutherford noticed that the amount of radioactivity coming from a sample of thorium decreased rapidly. After approximately 1 min (actually 54.5 s), a given sample had only half of its initial activity; after 2 min it had one-fourth; after 3 min, one-eighth; and so on. Rutherford interpreted this observation to indicate that each nucleus of thorium has a 50% probability of emitting an α particle in each minute, regardless of how long the nucleus has already survived or how many other nuclei are present.

Other radioactive nuclei were soon found to follow the same decay law. Each radioactive nucleus has a characteristic *half-life* – the time in which a nucleus has a 50% chance of undergoing radioactive decay. Equivalently the half-life represents the time it will take a sample to lose half its activity. We can formulate this as a mathematical law as follows: Let T denote the half-life of a certain radioactive nucleus, let N_0 be the initial number of such nuclei present, and let $N(t)$ be the number present at time t. Then the number surviving after any time t is simply given by

$$N(t) = N_0 \left(\tfrac{1}{2}\right)^{t/T}. \tag{51.6}$$

The power of $\tfrac{1}{2}$ can be written as an exponential, using $x^{-n} = \exp\{-n \ln x\}$, and we find that Eq. (51.6) becomes

$$N(t) = N_0 \exp\{-t(\ln 2)/T\}.$$

The constant $\lambda = (\ln 2)/T$ is called the *decay constant*, and the decay law can be written as

$$N(t) = N_0 e^{-\lambda t}. \tag{51.7}$$

In other words, the number of radioactive nuclei decreases exponentially. Figure 51.9 illustrates the decay curve.

The activity of a radioactive sample is measured by the number of disintegrations per second. The common unit employed is named in honor of the Curies: The definition of the curie (Ci) is 1 Ci = 3.7×10^{10} decays/s. This definition stems from the activity of 1 g of radium, $^{226}_{88}$Ra. Since it represents a rather large unit, millicuries are often used to measure radiation levels. The SI unit of radioactivity is the becquerel (Bq), defined as the number of decays per second: 1 Bq = 1 decay/s. The *activity A* of a sample is the rate of disintegration and is defined by

$$A = -\frac{dN}{dt} = \lambda N(t). \tag{51.8}$$

51.5 NUCLEI AND RADIOACTIVITY

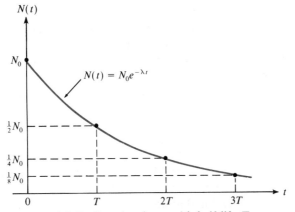

Figure 51.9 Radioactive decay with half-life T.

As time marches on, the radioactivity decreases because the number of nuclei decreases. If a sample decays very slowly, that is, if it has a long half-life, the number of nuclei present and hence the activity remain constant.

Example 6
Uranium has two long-lived isotopes, $^{238}_{92}U$ and $^{235}_{92}U$, with half-lives of 4.51×10^9 and 7.1×10^8 yr, respectively. It is believed that these isotopes were produced in roughly equal amounts in the supernova explosions of an earlier generation of stars. Today on Earth we observe that $^{235}_{92}U$ is only about 0.0072 times as abundant as $^{238}_{92}U$ because most of the shorter-lived isotope has decayed. How long ago was the uranium formed?

We'll assume that most of the original $^{238}_{92}U$ is still present because it has such a long lifetime. Then the original number of $^{235}_{92}U$ nuclei was the same, so the ratio today is given by $N/N_0 = 0.0072$. Using Eq. (51.7) we can solve for the time that has passed:

$$t = -\ln(N/N_0)/\lambda = -\ln(0.0072)(7.1 \times 10^8 \text{ yr})/(\ln 2) = 5.0 \times 10^9 \text{ yr}.$$

This agrees with other estimates that assert that the sun was formed in interstellar space 5 billion years ago from gas and dust that was enriched by the explosions of an earlier generation of stars.

Because of their fixed characteristic decay rates, radioactive nuclei can be used to determine the age of objects. One method uses the radioactive carbon isotope $^{14}_{6}C$, which has a half-life of 5740 yr. Carbon-14 is formed in the upper atmosphere by the bombardment of cosmic rays. The radioactive atom interacts chemically just like nonradioactive carbon-12, and it is taken up by living organisms, such as trees and animals, through carbon dioxide in the atmosphere. While an organism is alive it continues to intake carbon-14 and hence the activity of that isotope remains constant at 15.3 Bq/g of carbon. However, when the organism dies, the carbon-14 in it is no longer replaced and the activity drops. By measuring the activity level of carbon-14, the age of the object can be determined. Because of the short lifetime of carbon-14, dating by this method can only be used for objects up to about 30,000 years old.

For two decades after the discovery of natural radioactivity, most physicists held the view that the nuclei of all elements consisted of positive particles (later called protons) and negative particles (electrons), which were observed in beta rays and thought to have originated from inside the nucleus. A helium nucleus, for example, has atomic number 2 and atomic mass number 4, and so it was thought to contain 4 protons and 2 electrons, giving it the observed charge and mass. Likewise, heavier elements were believed to be composed of appropriate numbers of protons and electrons. To discover what actually makes up a nucleus required the splitting of the nucleus.

The model of a nucleus with electrons contained within it raised a major question: Why are some electrons confined inside the nucleus while others orbit it at great distances compared to the size of the nucleus? No one offered a description of the force responsible for binding the electrons inside at the extremely small dimensions of a nucleus.

In 1932, James Chadwick, a former student of Rutherford's, discovered a neutral nuclear particle. In a study of rays of great penetrating power that are emitted by the bombardment of light elements with energetic alpha particles, Chadwick concluded that the rays consisted of neutral particles, appropriately named *neutrons*. Because the mass difference between a neutron and a proton was approximately equal to that of an electron, Chadwick considered the neutron to be composed of a proton and an electron.

Chadwick offered no explanation of the role of neutrons in nuclei, but a proposal was made by Werner Heisenberg. In a series of papers published in 1932, Heisenberg proposed that nuclei consisted of protons and neutrons held together by the exchange of electrons. A proton could absorb an electron and become a neutron, and a neutron could emit an electron and change into a proton. His idea was that the nucleus is held together through a continual exchange of momentum, energy, and charge in the form of an electron. Despite the ingenuity of this idea, energy considerations such as those given in Example 4 show that electrons cannot exist inside neutrons. Nonetheless, his idea of forces being represented by the exchange of particles helped shape the current view of forces. Later work on molecular spectra suggested that the neutron was a fundamental particle like the proton and the electron, and not a composite particle.

Questions

13. Drawing upon your knowledge of the differences among alpha, beta, and gamma rays, why would you expect alpha particles to be the least penetrating?

14. Radioactive particles from different elements are passed through a region where there is an intense magnetic field, as illustrated in the accompanying sketch. Identify the types of radioactive particles in the three beams emerging at the top.

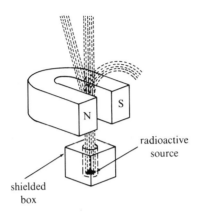

15. Radioactive radon, $^{222}_{86}$Rn undergoes alpha decay.

    (a) Find the daughter nucleus.
    (b) The daughter nucleus undergoes both beta and alpha decay. Determine the "granddaughter" in both cases.

16. Gamma rays emitted from radioactive thorium have an energy of 1.11 MeV.

    (a) What is their wavelength?
    (b) How does your answer in (a) compare to the size of a thorium nucleus, which is about 17 fm?
    (c) What would be the wavelength of an electron with the same energy?

17. Alpha particles emitted from radioactive radium have an energy of 6.9 MeV. The diameter of a radium nucleus is 17 fm. Use the uncertainty principle to show that the observed energy of the alpha particles is in accordance with the idea of alpha particles existing inside the radium nucleus.

18. Radioactive plutonium $^{239}_{94}$Pu decays by the emission of an alpha particle of energy 5.15 MeV.

    (a) Determine the daughter nucleus.
    (b) Estimate the recoil velocity of the daughter.

19. A $^{8}_{4}$Be* nucleus decays to its ground state with the emission of a 17.6-MeV gamma ray. Determine the kinetic energy of the recoiling nucleus.

20. The noble gas radon is commonly used in cancer therapy and has a half-life of 3.8 days. How can a radioactive isotope of such short half-life occur naturally?

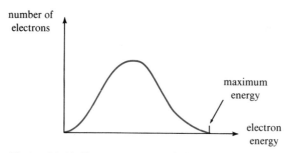

Figure 51.10 Energy spectrum of electrons emitted in beta decay of a nucleus.

21. The half-life of $^{235}_{92}\text{U}$ is 2.25×10^{16} s. Calculate the fraction remaining since its production by the first generation of stars 10 billion years ago.

22. A radioactive isotope has a half-life of 13 h. How long will it take until 5% of the original amount is left?

23. A sample of petrified wood has an activity of 2 Bq/g of carbon. Calculate the age of the petrified wood.

24. If an elephant died today, how long would it take before there would be an activity of 4 Bq/g from the carbon in its bones?

51.6 PARTICLES AND MORE PARTICLES

Early researchers of beta decay noticed that the electron emitted in this decay mode does not emerge with a definite kinetic energy. Instead the electron has a range of energies, as shown in Fig. 51.10. In stark contrast, alpha particles and gamma rays always have definite kinetic energies in the decay of a particular type of nucleus. An attractive explanation of this observation was that the energy of the decay was shared between the electron and an undetected particle. In 1930, Wolfgang Pauli boldly hypothesized that another particle besides the electron is emitted in beta decay. Although the particle is electrically neutral, it is not a gamma ray. Pauli's particle became known as the *neutrino*, the little neutral one. Besides its neutrality, the neutrino was believed to be so highly penetrating that its energy could not be detected by its absorption.

Rather than introduce a new particle, physicists at first attempted to explain the energy spectrum of emitted electrons by an undetected gamma ray. Conservation of angular momentum applied to the spins of the particles involved in the beta decay required that the undetected particle have a spin of $\frac{1}{2}$. However, the photon has spin 1, so energetic photons were ruled out. Conservation of angular momentum required the neutrino to carry spin $\frac{1}{2}$.

In addition to its neutrality, the neutrino was believed to be massless and therefore, like the photon, to travel at the speed of light. So neutrinos are distinguished from photons by their spin and by their extremely weak interactions.

Enrico Fermi incorporated the neutrino into his theory of beta decay in 1933. The fundamental process in Fermi's theory was one in which a neutron spontaneously changed into a proton, electron, and a neutrino:

$$^1_0 n \rightarrow {}^1_1 p + {}^0_{-1} e + \bar{\nu}_e.$$

Free neutrons are observed to decay with a mean half-life of 15 min. The neutrino emitted in the decay of a neutron is what later became known as an electron antineutrino. We shall return to this later. Fermi's theory made it possible to calculate the probability of neutrinos interacting with matter. Because neutrinos interact so weakly, those produced in beta decay could travel through billions of kilometers of lead before being absorbed. Although neutrinos are exceedingly difficult to detect, they are produced in enormous numbers in the sun and in nuclear reactions. In 1955, Clyde L. Cowan, Jr. and Frederick Reines succeeded in detecting neutrinos streaming from a nuclear reactor. Modern theories hypothesize that neutrinos may be as common as photons.

Another surprising advance in understanding the ultimate constituents of matter was made by Dirac in 1930. In searching for a version of quantum theory that would be consistent with relativity and would apply to relativistic electrons, Dirac found an equation that naturally incorporated the half-integer spin of the electron but also predicted the existence of a particle with the same mass but opposite charge as the electron. His theory predicted particles that were *antiparticles* of the electrons. If an electron and antielectron collide, they annihilate each other and create a gamma ray, and conversely, a gamma ray striking an electron can produce an electron–antielectron pair, as shown in Fig. 51.11. This prediction was experimentally confirmed in 1932 when Carl Anderson observed tracks of cosmic ray particles much like electron tracks but that curve in the opposite direction in a magnetic field. The antiparticle of the electron became known as the positron. Because positrons quickly annihilate electrons, they are not found in ordinary matter.

Soon after the discovery of the positron, physicists realized that to each kind of particle there is a corresponding antiparticle. In general, antiparticles have the same mass as the particle but opposite charge. Despite the fact that the laws of nature do not distinguish between particles and antiparticles, we have not observed any appreciable amounts of antimatter, composed entirely of antiparticles, anywhere in the universe. Possible theoretical explanations on the forefront of research invoke an asymmetry in the early universe to explain the predominance of matter over antimatter.

In 1933, the number of fundamental particles had grown from two particles (protons and electrons) to four particles and their corresponding antiparticles. However, a nagging question remained unanswered by the discovery of new particles: What holds protons and neutrons together to form nuclei? In other words, what is the nature of the nuclear force?

A significant step toward unraveling the mystery of the nuclear force was taken by the Japanese physicist Hideki Yukawa in 1935. Yukawa realized that a simple relation exists between the range of a force and the mass of a particle whose exchange produces the force. For example, the electric force has infinite range;

Figure 51.11 A gamma ray collides with an electron of a hydrogen atom in a bubble chamber and creates a positron–electron pair. The paths of the positron and electron proceed in opposite directions because the chamber is immersed in a strong magnetic field. The spiral paths are due to the subsequent loss of energy by the electron and positron. (Courtesy of Lawrence Berkeley Laboratory, University of California.)

because it diminishes according to $1/r^2$, no matter how far apart two charges are, they always experience a force. This infinite-ranged force can be thought of as due to the exchange of massless photons, the particles of electromagnetism. The nuclear force rapidly diminishes to zero outside a nucleus, which is known to be about 10^{-15} m in diameter, so the nuclear force must be due to the exchange of a new particle intermediate in mass between electrons and protons. Such a particle was called a *meson*, from the Greek word, μέσο, for middle.

The relationship between the range R of a force and the mass m of the particle transmitting the force is simply

$$R = \frac{\hbar}{mc}. \tag{51.9}$$

51.6 PARTICLES AND MORE PARTICLES

Taking $R = 10^{-15}$ m as the range of the nuclear force, we find that mesons should have a mass of about 200 MeV/c^2, or about 20% the mass of a proton. Equation (51.9) can be thought of as an application of the Heisenberg uncertainty principle.

Two years after Yukawa's proposal, a new particle was identified in cosmic rays. Although it was widely assumed to be Yukawa's meson, the observed particle interacted too weakly with protons and neutrons to be responsible for the force binding them inside nuclei. Later, another type of meson was observed that was slightly heavier and that interacted more strongly with nuclei. The lighter particle was called the muon and subsequent studies have revealed it to interact similarly to the electron but with 207 times the mass: muons don't participate in the strong nuclear force. The heavier particle was called the pi meson, or pion, and identified as Yukawa's particle. Later, three types of pions were identified: the positive (π^+), the negative (π^-), and the neutral pion (π^0).

Muons and pions are both unstable particles. They both decay spontaneously in characteristic ways. Associated with the muon is a type of neutrino that is different from the neutrino associated with electrons in beta decay and it is known as a muon neutrino. The muon decays into an electron, a muon neutrino, and an electron antineutrino:

$$\mu^- \rightarrow e^- + \nu_\mu + \bar{\nu}_e.$$

Pions decay into muons or antimuons that themselves decay into electrons or antielectrons.

From the analysis of cosmic rays, physicists turned to creating particles out of energy in laboratories. The advent of particle accelerators in the 1950s and 1960s ushered in a rapid increase in the number of so-called elementary particles. Figure 51.12 illustrates the growth in the number of particles over time and includes the names of a few milestone particles. Intimately connected with the quest for ultimate constituents of matter was the very nature of the fundamental forces of nature, which we introduced in Chapter 10: the weak nuclear force, the strong nuclear force, the electromagnetic force, and the gravitational force. Despite the profusion of particles, three major classifications of particles emerged from the debris of accelerator collisions: (1) Baryons – particles that are as massive or more massive than the proton and neutron and include these particles. (2) Mesons – particles intermediate in mass between protons and electrons that experience the strong nuclear force. (3) Leptons – light particles, such as electrons, muons, and their neutrinos, that are immune to the strong nuclear force.

Experiments indicate that baryons and mesons are subject to all the fundamental forces of nature. Since they interact through the strong nuclear force, they are generically referred to as *hadrons*. Leptons, on the other hand, interact only through the weak force, the electromagnetic force (if they have electric charge), and the gravitational force. Experiments determining the intrinsic spins revealed something even more: Baryons have half-integer spin, $\frac{1}{2}, \frac{3}{2}, \frac{5}{2}$, and so on. Mesons have integral spin, such as 0 for the pion, 1 for the K meson, and so on, while leptons have half-integer spin. Table 51.4 lists the breakdown of particles into subclassifications.

Several attempts were made to bring some order to and reduce the proliferation of the ever-expanding family of "elementary" particles. In the early 1960s Murray

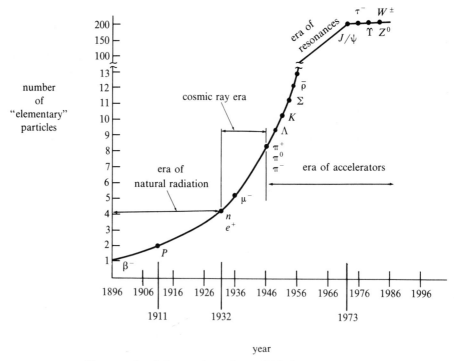

Figure 51.12 The growth of the number of so-called elementary particles as a function of time.

Gell-Mann and George Zweig independently proposed that the hadrons were composed of truly elementary building blocks called *quarks*. The name was adapted by Gell-Mann from the line "Three quarks for Muster Mark!" in *Finnegan's Wake* by James Joyce. The motivation for inventing quarks came from the symmetries in masses, charges, and decays of particles. Originally, three quarks were needed to explain the observed strongly interacting particles. Quarks are distinguished from other particles by their fractional charge: they have either $\frac{2}{3}$ or $-\frac{1}{3}$ the magnitude of the charge of an electron. In addition, quarks have spin $\frac{1}{2}$ and, therefore, must obey the Pauli exclusion principle. The types of quarks are called *flavors*, and the original three were up (u), down (d), and strange (s). Table 51.5 lists the quark flavors, their charges, and masses.

According to quark theory, all baryons are composed of three quarks, denoted by qqq, and all antibaryons of three antiquarks, denoted by $\bar{q}\bar{q}\bar{q}$ with bars over the qs. Much like the way electrons fill energy levels in multielectron atoms, quarks fill nuclear levels and make up the baryons and their excited states (known as resonances, of which there are hundreds). The proton is composed of two up quarks and one down quark, denoted as uud. A neutron has a quark structure of udd, that is, one up quark and two down quarks. One of the astounding successes of the theory was the prediction of the mass and decay mode of a particle composed of three strange quarks, sss. In 1964, a particle that fit the description, known as the omega minus, was observed by means of tracks in a hydrogen bubble chamber.

51.6 PARTICLES AND MORE PARTICLES

Table 51.4 Subclassifications of Elementary Particles with Examples.

Family	Particle	Symbol	Charge	Spin	Mass (MeV/c^2)	Lifetime (s)
Photon	photon	γ	0	1	0	stable
Leptons	electron	e	-1	$\frac{1}{2}$	0.511	stable
	muon	μ	-1	$\frac{1}{2}$	105.7	2.2×10^{-6}
	electron neutrino	ν_e	0	$\frac{1}{2}$	0?	stable
	muon neutrino	ν_μ	0	$\frac{1}{2}$	0?	stable
Mesons	charged pion	π^+	1	0	139.6	2.6×10^{-8}
	neutral pion	π^0	0	0	135.0	0.8×10^{-16}
	charged kaon	K^+	1	0	493.8	1.2×10^{-8}
	neutral kaon	K^0	0	0	497.7	0.9×10^{-10} or 5.2×10^{-8}
	eta	η	0	0	548.7	$\sim 2 \times 10^{-19}$
Baryons	proton	p	1	$\frac{1}{2}$	938.3	stable
	neutron	n	0	$\frac{1}{2}$	939.6	917
	lambda	Λ	0	$\frac{1}{2}$	1115.4	2.6×10^{-10}
	sigma plus	Σ^+	1	$\frac{1}{2}$	1189.4	7.9×10^{-11}
	neutral sigma	Σ^0	0	$\frac{1}{2}$	1192.3	5.8×10^{-20}
	sigma minus	Σ^-	-1	$\frac{1}{2}$	1197.2	1.5×10^{-10}
	neutral xi	Ξ^0	0	$\frac{1}{2}$	1314.3	2.9×10^{-10}
	xi minus	Ξ^-	-1	$\frac{1}{2}$	1320.8	1.6×10^{-10}

The omega-minus particle, according to quark theory, consists of three strange quarks all in the same state, and that violates the Pauli exclusion principle. In the 1970s, a theory of the strong force emerged that advanced the idea that quarks are held together by a force attributed to a property that quarks possess called *color*. Color is just as whimsical name of the physical quantity responsible for the

Table 51.5 Quark Flavors, Charges, and Masses.

Quarks	Charge	Mass (MeV/c^2)
u (up)	$\frac{2}{3}$	~ 100
d (down)	$-\frac{1}{3}$	~ 100
c (charm)	$\frac{2}{3}$	~ 4000
s (strange)	$-\frac{1}{3}$	~ 500
t (top)	$\frac{2}{3}$	?
b (bottom)	$-\frac{1}{3}$	~ 5000

attraction of quarks and does not refer to the usual meaning of the word. Just as electric charge is responsible for the force binding electrons to nuclei in atoms, the color force, stemming from the colors of quarks, binds quarks into hadrons. The theory, known as quantum chromodynamics, proposes three colors, which are often referred to as red, blue, and green. The strong force, as mentioned earlier, does not extend appreciably outside the domain of the nucleus. Within the framework of quantum chromodynamics, the lack of long-ranged nuclear forces is partially explained by the observation that ordinary matter is colorless and consists of a superposition of all three color combinations of quarks making up the particles. So the omega-minus particle consists of a red, blue, and green strange quark. The combination of all three colors produces a colorless particle, and the fact that the three strange quarks are different resolves the dilemma with the Pauli exclusion principle so that they can all be in the same state.

Although color seems contrived, it explains many other features of the strong force. The theory proposes that colored quarks exchange gluons that are responsible for the strong force. Thus, the strong force is represented by the exchange of gluons, the electromagnetic force by the exchange of photons, the gravitational force by the exchange of gravitons, and the weak force by the exchange of W and Z bosons. Collectively, the particles that are exchanged as manifestations of a fundamental force form a separate class of truly elementary particles known as *gauge particles*.

Returning to hadrons, according to quark theory, mesons are bound states of a quark and an antiquark of the same color, denoted by $q\bar{q}$, where q stands for a flavor of quark. For example, Yukawa's π^+ meson is composed of an up and an antidown quark, $u\bar{d}$. Its antiparticle, the π^-, has the "antistructure" $\bar{u}d$. Because mesons are composed of quarks and antiquarks of the same color, they are colorless.

The discovery of a new, relatively long-lived particle in 1974, known as the J/Ψ required the introduction of a new quark flavor, called the charm quark. In 1977, the discovery of the upsilon particle (Υ) required a fifth quark flavor, called the bottom quark, and evidence is mounting for a sixth quark flavor, the top quark.

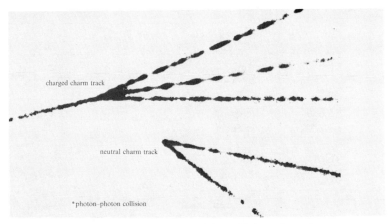

Figure 51.13 Computer re-creation of the collision of an electron and positron producing jets of particles. (Courtesy of The Stanford Linear Accelerator Center, Stanford University.)

Thus quarks come in 6 flavors and 3 colors comprising a total of 18 fundamental particles (and 18 fundamental antiparticles).

Despite the undeniable success of quark theory, some physicists are reluctant to embrace the quark model because free quarks have never been observed. An explanation is offered by quantum chromodynamics. The strong force is much like the spring force: rather than decreasing with distance between two quarks, it increases. Consequently, as energy is pumped into a particle by collisions from other elementary particles in an effort to liberate the quarks, the force between the quarks increases dramatically. Instead of freeing the enslaved quarks, the energy goes into the creation of new quark–antiquark pairs, in accordance with $E = mc^2$. Thus, other particles – baryons and mesons – are created in the process. Figure 51.13 shows a computer re-creation of a electron–positron collision with the creation of matter from the energies of the incident particles.

Besides quarks and gauge particles, the other truly fundamental constituents of matter are the leptons. In perfect symmetry with the six quark flavors, there are six flavors of leptons. As far as experiments can determine, leptons are point particles – they have no spatial extent. The most recent lepton was discovered in 1975 and is known as the tau lepton. Each charged lepton (the electrons, muon, and tau) has its own type of neutrino. The three pairs of quarks and leptons, known as generations, are listed in Table 51.6. The symmetry that is reflected in the table, where the particles are grouped in pairs of charge 0 and charge -1, is believed by physicists to be fundamental to the structure of matter.

Theories that seek to unite the four fundamental forces, known as grand unified theories (GUTS) and supersymmetric theories, seek to exploit the symmetries between quarks and leptons. In these theories, the fundamental forces are unified at fantastically high energies, or equivalently, temperatures on the order of 10^{28} K. Such energies were present in the first 10^{-35} s of the universe after the big bang.

Table 51.6 Fundamental Particles (circa 1986).

	Quarks	Charge	Mass (MeV/c^2)	Discovered
First generation	u (up)	$\frac{2}{3}$	~ 100	1963
	d (down)	$-\frac{1}{3}$	~ 100	1963
Second generation	c (charm)	$\frac{2}{3}$	~ 4000	1974
	s (strange)	$-\frac{1}{3}$	~ 500	1964
Third generation	t (top)	$\frac{2}{3}$	?	...
	b (bottom)	$-\frac{1}{3}$	~ 5000	1977

	Leptons	Charge	Mass (MeV)	Discovered
First generation	ν_e (electron neutrino)	0	0?	1955
	e (electron)	-1	0.511	1898
Second generation	ν_μ (muon neutrino)	0	0?	1962
	μ (muon)	-1	106	1936
Third generation	ν_τ (tau neutrino)	0	0?	?
	τ (tau lepton)	-1	1784	1975

Field Quanta

Force	Field quantum	Mass (MeV/c^2)	Discovered
Electromagnetic	photon (γ)	0	1905
Weak	intermediate vector boson ($W^\pm$, Z^0)	80,800 for $W^\pm$ 92,600 for Z^0	1983 1984
Strong	gluon	?	...
Gravity	graviton	0	...

Thus, the quest for the fundamental constituents of matter leads us full circle to the search for understanding our origin in the initial stages of the universe.

Questions

25. Find the maximum possible kinetic energy of the electron emitted in the decay of $^{14}_{6}$C. (Refer to Table 49.1 for masses.)

26. The pion has a mass of 139 MeV/c^2. Determine the range of the force it represents.

27. The W boson that transmits the weak force has a mass of 80,800 MeV/c^2. Determine the range of the weak force.

28. Grand unified theories predict that the fundamental forces are unified near a distance scale of 10^{-35} m, known as the Planck length. Calculate the mass of the hypothetical particle associated with this unified force in MeV and in kg.

29. Determine all the possible combinations that can be constructed from the three original quarks. Refer to baryons in Table 51.4.

30. Electrons and positrons annihilate each other. Why do you suppose that the quark and antiquark inside a meson don't annihilate each other?

51.7 A FINAL WORD

Throughout this text we have seen that science progresses by constructing models and inventing theories. Some philosophers assert that the only way science can progress is by proving that previously held theories are inadequate descriptions of nature. In other words, we can never prove a theory is right, but we can prove that it is wrong. No matter how many experimental facts agree with it, the theory may still be wrong, while if only one experimental fact disagrees with it, we know the theory can't be correct. As Aldous Huxley stated: "Many a beautiful theory is spoiled by an ugly fact."

However, this conjecture on theories is itself a theory that may be wrong. In reality, some theories become so well verified by observation that they are promoted to the status of fact. For example, the theory of relativity is still called a theory for historical reasons, but today it is a widely accepted fact, used systematically in the design and construction of accelerators.

Another example is the theory of evolution in biology. These are facts in the sense that we can be more confident of them than of just about anything else in all of human knowledge. In this chapter we have explored the idea of the fundamental constituents of nature. Whether or not a future generation will find quarks to be the ultimate constituents – the hard, massy spheres of Newton or the six impossible things of the Red Queen in the lines that opened this chapter – or whether they will be replaced remains to be seen. Quark theory may end up like relativity theory or the theory of evolution – a widely accepted fact about how the world works. The quest for quarks symbolizes the deepest goal of science: to understand nature in the simplest possible terms.

CHAPTER 52

THE QUANTUM MECHANICAL UNIVERSE

> Where is the good, old-fashioned solid matter that obeys precise, compelling mathematical laws? The stone that Dr. Johnson once kicked to demonstrate the reality of matter has become dissipated in a diffuse distribution of mathematical probabilities.
>
> Morris Kline, *Mathematics in Western Culture* (1953)

52.1 INTRODUCTION

The time has come for one, last, brief glance backward at where we've been.

It all started with Nicolaus Copernicus, who thought the universe might look simpler from the sun's point of view. Galileo and Kepler filled in some of the essential details, and in the end Isaac Newton brought it all together, uniting heaven and earth by inventing the science of mechanics.

In his old age, Isaac Newton presided over meetings of the Royal Society at which the mysterious phenomenon of electricity was demonstrated. Gilbert,

Franklin, Faraday, and others unearthed gleaming bits of scientific truth, putting each in place for yet another grand synthesis, by yet another Cambridge professor, James Clerk Maxwell, who brought forth light out of electricity and magnetism combined.

There was, however, an inner contradiction between the two great sciences of mechanics and electromagnetism. The principle of inertia, the very starting point of mechanics, made perfect sense provided there was no absolute state of rest in the universe. It followed that there could be no absolute speed either, because one inertial frame of reference is as good as any other. The speed of any object depended on the speed of the observer. But Maxwell's theory had an absolute speed in it – the speed of light – and that meant there would be a preferred frame in which light would have that speed, and that frame would serve as a state of absolute rest. If that were true, then the principle of inertia was in jeopardy.

Once again a synthesis was needed, this time between Newton's mechanics and Maxwell's electromagnetism. It was provided by Albert Einstein with his special theory of relativity.

Meanwhile, another story was unfolding, the study of the structure of matter. It began in the murky depths of alchemy, out of which arose the modern science of chemistry. Thermodynamics, the study of matter and energy, was part of that story, which had its own heroes, and in which Newton, Maxwell, and Einstein all played important roles. In the end it brought forth the atom, but once again there was a problem: Newton's mechanics would not suffice to explain the inner working of the atom. This time, what was needed was not another synthesis, but something entirely new, and so we arrived at the Quantum Mechanical Universe.

As we've seen in the previous chapter, the quest for the ultimate nature of matter has by no means come to an end. Neutrons and protons, the inner constituents of atoms, themselves have inner constituents, quarks, which also combine to make other particles, and so, the story of the structure of matter is far from complete.

Today, as at any time in history, we can see pretty clearly where we've been, but much less clearly where we're going. Do we understand where we are now? What is the real nature of the "Quantum Mechanical Universe"? In this final chapter of our book, we pause to look around us and ask: What does it all mean?

52.2 DIFFRACTION OF MATTER WAVES

We begin by recalling the essential ideas introduced in Chapter 50. First of all, particles such as electrons have associated with them waves whose behavior is governed by the Schrödinger equation. These waves represent the probability of an electron being found in a given place at a given time. What exactly does that mean?

Very often in science, progress depends on asking the right question. For example, when Galileo discovered that an object in motion tends to stay in motion, he might have decided to investigate what keeps the object moving. That would have led nowhere. Instead, he accepted the principle of inertia and worked out its consequences. That led to results that we have discussed at some length.

52.2 DIFFRACTION OF MATTER WAVES

Now we are faced with a similar situation. Given the fact that particles, such as electrons, act like waves, the obvious question is, What's waving? Exactly what is the stuff whose amplitude is the function Ψ of Chapter 50? It is not a new question. When Thomas Young demonstrated in 1803 that light is a wave phenomenon, he took the first step in a century-long quest for the luminiferous aether – the name assigned to the fictitious medium that was supposed to be waving. Without knowing what's waving, nineteenth century physicists developed a sophisticated science of optics and spectroscopy because they knew the general properties of waves. We can do some of the same for electrons.

The sine qua non of wave motion is interference – the ability of waves to combine constructively, making more intense waves, or destructively, canceling each other out. The most important role of the electron is in the formation of the atoms and molecules that comprise all matter. We've already seen, in Section 50.4, that the electron-wave idea provides a crucial ingredient in understanding the structure of the atom.

Of course, it would be nice to have some more direct evidence of electron interference – something like the two-slit experiment of 1803 by which Thomas Young showed that light is a wave. Such experiments have indeed been done, and they do result in diffraction patterns strikingly similar to those obtained with light. They are very difficult to do, however, because electron wavelengths are typically orders of magnitude smaller than light wavelengths. To begin the discussion, it's helpful to have an idea of the order of magnitude of electron wavelengths.

Example 1

Find the wavelength associated with an electron accelerated from rest to a potential of 1 V.

The relativistic formula relating energy and momentum is Eq. (48.30),

$$E^2 = E_0^2 + c^2p^2.$$

At a kinetic energy of 1 eV, it's safe to use the nonrelativistic formula

$$E = \frac{p^2}{2m_0} + m_0c^2.$$

The kinetic energy K is

$$K = \frac{p^2}{2m_0},$$

and by Eq. (50.10) we have $p = h/\lambda$, where λ is the de Broglie wavelength of the electron, so the energy is given by

$$K = \frac{h^2}{2m_0\lambda^2}.$$

Solving for λ we find

$$\lambda = \sqrt{\frac{h^2}{2m_0 K}}.$$

Putting in numerical values, we obtain

λ = 12.2 Å.

Because λ is inversely proportional to $K^{1/2}$, we can give a few more values for the de Broglie wavelengths at other energies:

K	λ
1 eV	12.2 Å
100 eV	1.2 Å
10^4 eV	0.12 Å

As the next example shows, these values explain the difficulty of designing a two-slit interference experiment.

Example 2

Design a two-slit electron diffraction experiment, making rough estimates of the sizes of important design parameters.

A simple design might look like that shown in the accompanying diagram.

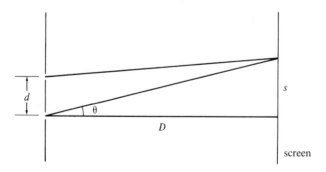

Electrons are emitted by a heated filament cathode, accelerated through a voltage V to an anode (positive electrode) with two slits a distance d apart. Those electrons passing through the slits fall on a phosphorescent screen (a television or oscilloscope screen or any cathode ray tube screen) a distance D away. The whole thing is inside a vacuum tube. We want to find the values of the quantities d, D, and V.

52.2 DIFFRACTION OF MATTER WAVES

Analysis of the problem of two-slit diffraction was given in Chapter 44, leading to Eq. (44.11). The first diffraction fringe comes out at an angle θ given by

$$\frac{\lambda}{d} = \sin\theta = \frac{s}{D},$$

where s is the spacing between fringes as shown in the figure. The problems inherent in this design are revealed by the table of values in Example 1. The electron wavelengths λ are of the order of a few angstroms, and because $s = \lambda D/d$, it is difficult to make the factor D/d large enough to produce a value of s that can be resolved on a phosphorescent cathode ray tube screen.

For example, suppose we require $s = 1$ mm and $D = 1$ m (it might be hard to get a good vacuum if the system is much bigger). Optical gratings of the size $d = 0.02$ mm are commonly made. This would require electron wavelengths equal to

$$\lambda = \frac{sd}{D} = 10^{-6} \text{ cm} = 10^2 \text{ Å}.$$

Referring again to the table of values in Example 1, that would mean a kinetic energy of 10^{-2} eV.

But the experiment cannot be done with these parameters because electrons come out of the heated cathode with a range of energies much larger than 10^{-2} eV. In fact, like the ideal gas of Chapter 15, they have mean kinetic energy

$$\bar{K} \approx kT,$$

where $T \approx 10^3$ K for a cathode, so the mean kinetic energy is about 0.1 eV. That is, roughly speaking, not only the average energy but also the range of energies, and it's 10 times the energy needed for the experiment. By contrast, to have a sharp diffraction pattern, all electrons should have the same energy, so that they'll have the same momentum, and therefore the same wavelength.

Questions

1. What is the de Broglie wavelength of an electron accelerated through 10^8 V? At what approximate energy does the nonrelativistic approximation become invalid?

2. Construct a table similar to that in Example 1 for a proton wave.

3. An actual two-slit electron diffraction experiment, performed by Claus Jonsson in 1961, used the following parameters:

$$V = 50 \text{ kV}, \quad d = 2 \times 10^{-3} \text{ mm}, \quad D = 35 \text{ cm}.$$

Estimate λ and s. Also find $\Delta s/s$, the smearing of the diffracted line due to the spread of 0.1 eV in the energies of the electrons from the cathode. [*Note:* Because s is very small in Jonsson's experiment, powerful electrostatic lenses, of the type used in electron microscopes, were used to magnify the image. The diffraction pattern obtained by Jonsson is shown in the accompanying figure.]

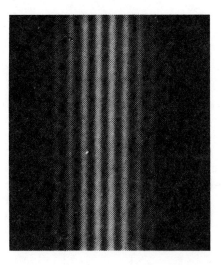

Two-slit experiment. Courtesy of *The American Journal of Physics*, Volume 42, January, 1974. (Photograph by Claus Jonsson.)

The two-slit electron diffraction experiment is difficult to carry out because the size of the slits anyone is capable of manufacturing is very large compared to the usable wavelengths of electrons. Ideally, diffraction should be done with much smaller spacing, on the order of a few angstroms. Therefore, electron diffraction, like X-ray diffraction discussed in Chapter 44, is best done by the atoms of a crystal lattice.

As mentioned in Chapter 50, the first electron diffraction experiment was in fact done that way, by C. J. Davisson and L. H. Germer, in 1925, only a year after de Broglie advanced his particle–wave hypothesis. When they started their experiment, Davisson and Germer hadn't heard of de Broglie, but they learned about him, and triumphantly confirmed his strange idea.

Today, electron diffraction is a routine technique used in laboratories everywhere. Electron diffraction, like X-ray diffraction, can be used to find how the atoms of the target material are arranged. Because electrons are charged particles (oops! we mean waves), they don't penetrate matter as easily as X rays do, so they are principally used to study the region near the surface of a material. Electron diffraction patterns look very much like the X-ray diffraction patterns in Chapter 44. An example is shown in Fig. 52.1.

Because they're charged, electrons, unlike X rays, can also be focused by electromagnetic fields to form sharp images rather than diffraction patterns. That's the principle of the electron microscope. Electron microscopes can produce images of much smaller objects than can light microscopes, because electron waves have much smaller wavelengths than light waves.

Electrons and X-ray photons are by no means the only particles whose wavelike properties are routinely used in modern diffraction experiments. Neutrons from nuclear reactors, and even atoms such as hydrogen and helium are commonly used, each because of its own special properties. (See Fig. 52.2.)

52.3 DUAL NATURE OF WAVES AND PARTICLES

Figure 52.1 An electron diffraction pattern. (Courtesy of B. Fultz and R. Whitt, California Institute of Technology.)

Question

4. Neutron diffraction experiments use neutrons that are produced in nuclear reactors. Before emerging, the neutrons are "thermalized," that is, cooled to "room temperature" inside the reactor. To get neutrons of a single energy from the thermal spread of energies, they are first diffracted from a crystal, each wavelength (and the corresponding radiant energy) coming out in a different direction. Estimate the wavelengths of thermal neutrons.

52.3 ELECTRON DIFFRACTION AND THE DUAL NATURE OF WAVES AND PARTICLES

Although electron diffraction by atoms is of more practical importance, the two-slit experiment can be used to illustrate the dual nature of waves and particles. Without worrying about annoying experimental detail, let's consider the simple two-slit experiment shown in Fig. 52.3.

The source emits electrons, one at a time, all of very nearly the same energy and momentum (this much does happen in a modern electron microscope). An electron that passes through the slits arrives at the screen, where its arrival is detected, say, by a flash of phosphorescence.

Now, there is no doubt at all that what arrives at the screen is a particle. For one thing, it makes a dot of light at one point on the screen. For another, it carries with it a definite amount of electric charge – the quantum unit of electric charge measured so carefully by Robert Millikan (Chapter 12). The arrival of one electron at the screen gives no hint at all of the two-slit diffraction pattern, already shown in Fig. 52.1. The next electron that arrives also makes a dot, generally somewhere else on the screen, and so on.

Suppose, however, we take time-lapse photographs of the screen. Then a record gradually forms of where dots have been and where they go. As time goes on, the

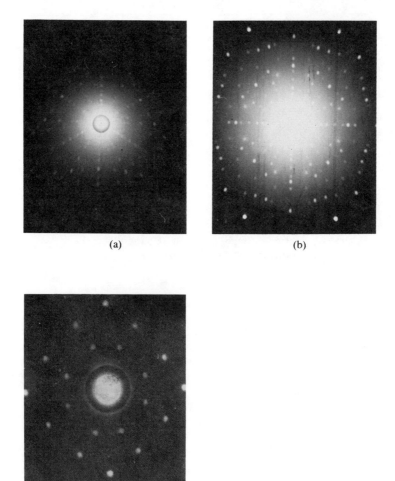

Figure 52.2 Diffraction patterns produced by (a) X rays and (b) electrons on an aluminum foil target. (Courtesy of B. Fultz and R. Whitt, California Institute of Technology.) (c) Laue pattern of diffraction of neutrons from a nuclear reactor by a single sodium chloride crystal. *Quantum Physics of Atoms, Molecules, Solids, Nuclei, and Particles* by Robert Eisberg and Robert Resnick, copyright© by John Wiley & Sons. Reprinted by permission of John Wiley & Sons, Inc.

two-slit diffraction pattern gradually begins to emerge, as shown in Fig. 52.4. The electrons arrive at the screen, each at a seemingly random position, uncorrelated with those that came before, but when enough of them have arrived, an orderly pattern emerges from the confusion. That pattern is a wave diffraction pattern, even though it's created by the arrival of many particles.

Strange though this behavior may seem, much the same sort of thing happens whenever any photograph is made. An optical system manipulates light waves

52.3 DUAL NATURE OF WAVES AND PARTICLES

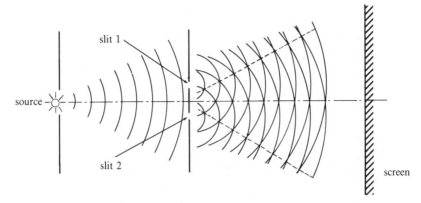

Figure 52.3 The two-slit experiment.

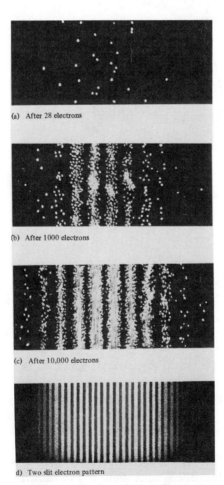

(a) After 28 electrons

(b) After 1000 electrons

(c) After 10,000 electrons

(d) Two slit electron pattern

Figure 52.4 The two-slit interference pattern produced by successively increasing numbers of electrons. (Courtesy of E. R. Huggins, *Physics I* © 1968, Benjamin/Cummings Publishing Company.)

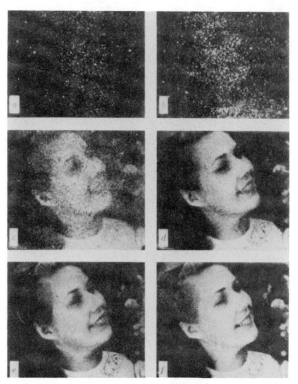

Figure 52.5 A series of photographs showing how the quality of an image improves as an increasing number of photons impinge on the photographic emulsion. (Figure reproduced courtesy of Albert Rose, *Journal of the Optical Society of America*, Vol. 43, 715, 1953 with permission from the Optical Society of America.)

through lenses, apertures and so on, but the image is formed because photon particles arrive at the emulsion, each with enough energy to cause a chemical reaction precipitating metallic silver out of a silver bromide molecule. If the image is made with a very weak light source – one photon at a time – the picture builds up much the way the electron diffraction image does in the two-slit experiment. This is shown in Fig. 52.5.

Returning to the two-slit electron experiment, we can now imagine analyzing and manipulating the situation a bit to try to understand just what's going on.

The two-slit interference pattern is a consequence of the wave fronts starting, in phase, *from both slits* simultaneously. Then bright fringes occur on the screen wherever one wave has traveled an integer number of wavelengths more than the other, so that they arrive in phase. Different electrons make the trip at different times, each creating a dot of light when it arrives at the screen, so they can't interfere with each other. The interference pattern can only come about if *each* electron-wave passes through *both* slits.

This idea is easily tested. We can prevent each electron from passing through both slits simply by covering up one of the slits, and repeating the experiment. If that's done, however, the result is an entirely different pattern on the screen. (If the slit is very narrow, and $D \gg d$, the dots will be uniformly spread out on the screen.) To get a two-slit diffraction pattern, the electron must be allowed to pass through both slits.

But how can one electron pass simultaneously through both slits? After all, it most certainly is a particle, arriving, for example, at a single point on the phosphorescent screen, and carrying a definite electric charge. Isn't it true that each electron *must* have gone through one slit or the other?

Speculating on the answer to that question – trying to arrive at it by pure reason – gets us nowhere at all. On the other hand, if we make any attempt experimentally to determine which of the two slits an electron passed through, the very act of performing the experiment will destroy the two-slit pattern just as effectively as we destroyed it by covering up one slit. For example, we might imagine trying to illuminate the slits with a light source between the slits and the screen, so that we can see which slit the electron emerges from. But in order to "see" the electron, light must be reflected by the electron. That means, at the very least, scattering a photon from the electron. But the scattering event must conserve energy and momentum, and so the electron will be deflected, and not go to its proper point in the two-slit pattern on the screen.

The general principle is this: The two-slit pattern will emerge only if we make no attempt to find out which slit each electron went through.

Much time and thought has been spent trying to find a sensible and satisfying way to explain the inner meaning of this and many analogous experiments. The interpretation that emerged, slowly and painfully, like the pattern on the screen, has been confirmed by many dramatic experiments. It is this: The electron (or photon, or atom, or any other particle) has associated with it a wave describing its *probability* of being near a given place at a given time. In particular, the quantity $\Psi(x)$ introduced in Eq. (50.3) is called the *probability amplitude*. It can be positive or negative, and its square, $\Psi^2(x)$, which is never negative, is the probability density, the function whose integral over an interval gives the probability that the electron will be detected in that interval.

Thus, the two-slit experiment would work this way: The probability amplitude for an electron to pass through slit 1 and wind up at height x on the screen is, say $\Psi_1(x)$. The corresponding quantity for passing through slit 2 and winding up at x is $\Psi_2(x)$. The probability density that the electron makes a dot at x is

$$P(x) = |\Psi_1(x) + \Psi_2(x)|^2.$$

This function, $P(x)$, is a quantification of the two-slit diffraction pattern: $P(x)$ is large near $x = 0$ and near integer multiples of $x = \pm s = \pm \lambda D/d$, meaning bright fringes occur near those positions, and dark fringes in between. That is a measure of the likelihood that the electron arrives at x after going through either slit 1 or slit 2. But if we block slit 2, the result is

$$P_1(x) = |\Psi_1(x)|^2.$$

If, instead, we merely determine which slit the electron went through, those going through slit 1 give $|\Psi_1(x)|^2$ as above, and those going through slit 2 give $|\Psi_2(x)|^2$.

The probability density of going through slit 1 *or* slit 2 (but not both) is

$$P_1(x) + P_2(x) = |\Psi_1(x)|^2 + |\Psi_1(x)|^2.$$

Neither P_1 alone, nor P_2, nor $P_1 + P_2$ gives the probability density for the two-slit interference that requires going through both slits.

Example 3

Explain how $|\Psi_1(x) + \Psi_2(x)|^2$ leads to a two-slit diffraction pattern. What is the pattern if one slit is blocked?

Start with one slit blocked. The situation is shown in the figure. A cylindrical wave, whose crests and troughs are sketched in the figure, comes out of the slit. At the instant shown, bright fringes appear at the locations of the arrows, but these fringes sweep across the screen (upward above the middle of the pattern, and downward below the middle) so that after a few crests and troughs have arrived, the average illumination (or probability) is essentially the same everywhere.

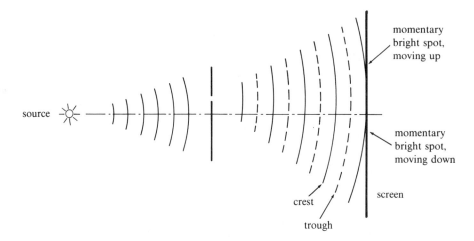

If both slits are open, each sends out cylindrical waves, which start in phase at the slits. They always arrive in phase at the point $x = 0$, which is equidistant from both slits, so that location oscillates from bright to dark and back with twice the intensity of $P_1(x)$ alone. Thus, on the average, this location winds up with some illumination. The same thing happens at $x = \pm \lambda D/d$, places where the wave from one slit has traveled a full wavelength more than the other, and at integer multiples of those values of x.

However, halfway between these locations, the arrival of a crest from one slit always corresponds to the arrival of a trough from the other and vice versa, so this value of x is never illuminated. These locations are always dark (the probability density of an electron arriving there is zero).

52.3 DUAL NATURE OF WAVES AND PARTICLES

Questions

5. Is it possible to make use of the fact that electrons are charged to devise a means of determining which slit they pass through? If so, can it be done without affecting the electron's trajectory, so that the diffraction pattern will remain unchanged?

6. Suppose the Michelson–Morley experiment (Chapter 45) were done with a very low level of illumination, so that the photons arrive at the detecting telescope one at a time.

    (a) Would each photon give an interference pattern?
    (b) Would the aggregate of many photons give an interference pattern?
    (c) If your answer to (b) is yes, does a single photon split in two at the beam splitter? Recall that the two paths go quite far apart (meters) before rejoining. Do you still expect an interference pattern?

7. In what way would the arguments of this section change if the two-slit experiment were done with a high intensity beam of electrons, instead of one at a time? (Neglect the direct electric and magnetic forces between electrons.)

In the two-slit experiment we have been discussing, we cannot say which slit the electron passes through, nor, for that matter, can we predict exactly where on the screen any particular electron will arrive. We can only assign probabilities, which lead to an increasingly accurate prediction as the number of events becomes larger and larger. This is characteristic of all of quantum mechanics. For example, we can never say where an electron is in its orbit around the nucleus. The most we can say is with what probability one will be found in a given interval of distances from the nucleus.

Many from the generation of physicists that discovered quantum mechanics did not find this situation satisfactory. Strict causality was the essence of Newtonian physics. Every experiment, in principle at least, had a precisely determined result. Now, suddenly, causality was gone and we were left only with probable outcomes. The power of quantum mechanics to explain and predict correctly was impressive, but nevertheless many illustrious scientists, including Albert Einstein himself, held out the hope that it would eventually be superseded by a more profound theory in which, for example, the electron definitely went through one slit or the other after all. However, time has proven to be on the side of the believers, not the doubters. All attempts to find some experimentally verifiable flaw in quantum mechanics have failed. Today the theory has no serious opposition among physicists.

How, then, do we answer the question, "Through which slit did the electron really go?" The answer is that the question is meaningless from the point of view of physics. The job of physics is to predict the results of well-defined experiments. If we do the experiment without attempting to answer the question, then quantum mechanics predicts a two-slit diffraction pattern, the prediction is confirmed by experiment, and physics is doing its job. If we do attempt an experiment to answer the question, that changes the result, again in a predictable way, and again physics is doing its job. But to ask what happens when we aren't looking, that is to say, when no experiment is being done to find out, is a question that simply doesn't have an operational meaning.

Some people find that view philosophically unsatisfying, but none can deny that it's been hugely successful.

52.4 IS NEWTONIAN MECHANICS OBSOLETE?

In Newtonian mechanics, every particle has a precise energy, momentum, and position at every instant of time. Any physical object is made up of particles, the aggregate of whose instantaneous values of those quantities at any given time determines completely the future history of the object. If particles A and B are headed toward one another, they will collide at a predictable time, and the results of that collision can be predicted with equal certainty. There may be 10^{23} atoms in a macroscopic bit of matter, far too many for us to track individually in this way, but in principle, at least, the future is completely determined by the present. That was one of the central ideas of *The Mechanical Universe*: The universe is a machine that, once set in motion, would require no further intervention by its creator as it evolved toward its inexorable destiny.

That idea stands in stark contrast to the discussion of the last section. If a beam of electrons passes through one or two narrow slits, it's impossible to predict where on the screen the next electron will arrive. The best we can do is calculate the probability density that the electron will emerge in any given direction from the slits.

This astonishing unpredictability in the behavior of an electron (or any other particle) is closely associated with the wavelike properties Prince Louis de Broglie supposed they might have. If electrons behaved like Newtonian particles, each would either pass through a slit or be blocked. The result would be a sharp image of the slits on the screen. However, instead of localizing the electrons into a sharp image, as we would expect of particles, the effect of the slits has been to splay them out, into a spreading beam, complete with interference if there are two (or more) slits. The attempt to localize the electrons in space, by making them pass through the slits, has brought out their wavelike properties and introduced unpredictability, or uncertainty in their behavior.

Our success in describing the experiment that way raises a profound dilemma. Newtonian mechanics worked very well in its time, and it still does. Regardless of this or any other discovery we might make in a laboratory, the moon will continue to fall a perfectly predictable 1/20 of an inch toward the earth every second, and other, more mundane phenomena, such as collisions of billiard balls, will continue to obey precisely the laws we have worked out for them. Yet, each of these objects, from the billiard ball to the moon, is comprised entirely of atoms, made up of protons, neutrons, and electrons, all of which display wavelike imprecision in their behavior. Apparently, the predictable Newtonian behavior of large scale, macroscopic objects somehow grows out of the combined behavior of its many less predictable, quantum mechanical constituents. How is that possible?

We've already seen a good model for how this might occur in the behavior of ordinary visible light (Chapter 44). Visible light has a wavelength of the order of $\lambda \approx 5000$ Å $\approx 5 \times 10^{-5}$ cm. As long as it's used to view an object very large

52.4 IS NEWTONIAN MECHANICS OBSOLETE?

compared to that size, it makes sharp, clear images just as if it were made up of perfectly Newtonian particles being reflected or absorbed. Indeed, as we have seen, Newton believed that light was made up of perfectly Newtonian particles. However, if light is used in an attempt to view an object whose dimensions are more nearly of the same order as the wavelength, its wavelike characteristics clearly emerge. The classic example is the two-slit diffraction of light, the same experiment we have used to illustrate the wavelike characteristics of electrons. Thus, what seems particlelike for large objects seems wavelike for small ones.

This idea is actually embodied in the Heisenberg uncertainty principle,

$$\Delta x \, \Delta p_x \geq h/2. \tag{50.29}$$

(The subscript x reminds us that the uncertainty is in that component of momentum.) As we saw from the argument in Section 50.7, the inequality in (50.29) is an expression of the behavior of waves. Thus, when equality nearly holds, that is, when

$$\Delta x \, \Delta p_x \approx h/2, \tag{52.1}$$

then the wavelike nature of the object is manifesting itself.

On the other hand, when the restriction imposed by the inequality in (50.29) is unimportant, the object is behaving in a classical, Newtonian manner.

These points are best understood through examples. To begin with, it's easy to convince yourself that the uncertainty principle doesn't apply in an important way to objects big enough to see or hold.

Example 4

A good machinist can measure the size of an object being fabricated with a tolerance of 0.1 mm. Estimate the uncertainty in the momentum and velocity of an object introduced by the measurement.

The uncertainty in momentum is

$$\Delta p_x \approx \frac{h/2}{\Delta x} = \frac{3 \times 10^{-34}}{1 \times 10^{-4}} \text{ kg m/s}$$
$$= 3 \times 10^{-30} \text{ kg m/s}.$$

For an object whose mass is about 0.01 kg, the uncertainty in velocity is about 10^{-29} m/s, much too small to observe in any conceivable way.

The restrictions imposed by the uncertainty principle are unimportant, not only for large objects, but very often also for quite small ones. In fact, Newtonian mechanics can often be applied as a reasonable approximation, even to atoms, as shown by the following examples.

Example 5

In Chapter 15, we deduced the properties of an ideal gas by applying Newtonian mechanics to the atoms of the gas, assuming, for example, that they had definite positions and momenta at each instant. In view of the Heisenberg inequality (50.29), was that procedure justified?

To answer this question, we'll assume that the uncertainty in the position of an atom is of the same order as its size, and ask whether the resulting uncertainty in momentum, Δp_x, is large or small compared to the corresponding component of momentum we expect it to have, p_x. If $\Delta p_x/p_x \ll 1$, the calculation in Chapter 50 cannot be seriously affected by the small error introduced by Eq. (52.1). Assuming

$$\Delta x \approx 2 \text{ Å} = 2 \times 10^{-10} \text{ m}$$

(all atoms are roughly this size), we have

$$\Delta p_x \approx \frac{h/2}{\Delta x} = \frac{3 \times 10^{-34} \text{ J/s}}{2 \times 10^{-10} \text{ m}} = 1.5 \times 10^{-24} \text{ kg m/s}.$$

The momentum is computed using

$$\frac{\bar{p}^2}{2m} = K = \frac{3}{2}kT.$$

To be precise, $p^2 = p_x^2 + p_y^2 + p_z^2 = 3p_x^2$ on the average, so

$$\frac{\bar{p}_x^2}{2m} = \frac{1}{2}kT,$$

or

$$\bar{p}_x = \sqrt{mkT}.$$

For example, consider argon at room temperature ($T \approx 300$ K, and $m \approx 40 m_p$, where m_p is the mass of a proton). We find

$$\bar{p}_x = \sqrt{(40 \times 1.7 \times 10^{-27} \text{ kg})(1.4 \times 10^{-23} \text{ J/K})(300 \text{ K})}$$
$$= 1.7 \times 10^{-23} \text{ kg m/s}.$$

This is 10 times larger than Δp_x above, so we conclude that argon gas at room temperature should obey classical mechanics reasonably well, as indeed it does.

Question

8. Determine whether Newtonian mechanics may be safely applied to each of the following systems:

(a) Solid argon at 10 K.
(b) Liquid helium at 1 K.
(c) The electrons in a metal at room temperature.
(d) Hydrogen and deuterium nuclei in the sun.

[*Hint:* In each case assume that $K = (3/2)kT$. An alternative test is to assume $\Delta p_x \approx p_x = \sqrt{mkT}$ and determine whether Δx is less than the distance between neighbors; when it is not, quantum mechanics becomes very important.]

52.5 QUANTUM MECHANICAL ESTIMATES

The conclusion to be reached from these examples is clear: Although quantum mechanics is the underlying truth, we have not wasted our time studying classical mechanics. If you want to understand the world, $F = ma$ is still a very good place to start.

52.5 QUANTUM MECHANICAL ESTIMATES

In those instances where quantum mechanics is important, the solution of the problem generally begins by writing Schrödinger's equation (of which Eq. (50.17) is a simple, special case) and finding its solutions. However, one of the nicest tricks of the physicist's trade is the fact that the approximate equality

$$\Delta p_x \Delta x \approx \hbar \tag{52.2}$$

often describes (approximately) the solutions of Schrödinger's equation. (Because this is only an approximation, we have written $\hbar$ instead of $h/2$.)

To illustrate the point, we'll use it to estimate one of the most important quantities in physics, the size of a hydrogen atom. We already know the correct answer, which comes out of Eq. (49.14) in Niels Bohr's theory,

$$a_0 = \frac{\hbar^2}{e^2 m K_e} = 0.529 \text{ Å}, \tag{52.3}$$

where m is the mass of the electron and e its charge.

Now we shall approach the same problem in a different way. We imagine that the electric force of a proton (a hydrogen nucleus) tries to form an atom by confining an electron to its immediate vicinity. According to Eq. (52.2), the better it succeeds, the more momentum the electron may have, and therefore, the more kinetic energy. In other words, the closer the electron allows itself to be attracted to the proton, the lower its electric potential energy will be, but the higher the kinetic energy it has, merely as a consequence of being confined in space. The most stable position of the electron will be that which has the least total energy.

The potential energy of an electron a distance r from a proton (assuming both are pointlike particles occupying no space themselves) is given by

$$U(r) = -K_e \frac{e^2}{r}.$$

If the electron is to be confined within a distance r of the nucleus, it can have any momentum up to about

$$p \approx \hbar/r,$$

so that we estimate its kinetic energy to be

$$K(r) = \frac{p^2}{2m} = \frac{\hbar^2}{2mr^2}.$$

The total energy is the sum of these,

$$E(r) = \frac{\hbar^2}{2mr^2} - K_e \frac{e^2}{r}. \tag{52.4}$$

The functions $U(r)$, $K(r)$, and $E(r)$ are sketched in Fig. 52.6.

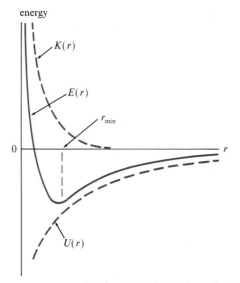

Figure 52.6 Graphs of $U(r)$, $K(r)$, and $E(r)$.

To find the most stable spacing, we set $dE/dr = 0$ to locate the minimum in $E(r)$. The result is

$$r_{min} = \frac{\hbar^2}{e^2 m K_e},$$

which is identical to the Bohr radius in Eq. (52.3).

The exact agreement with the Bohr theory is fortuitous, because the argument is a crude one, full of approximations. However, the idea behind it is profoundly correct, much more so, in fact, than the Bohr theory itself.

The Bohr theory, either in its original form, or as modified by the de Broglie hypothesis, was a crucially important stepping stone between Newtonian physics and quantum mechanics, but it is not quite right. For example, as long as it describes electrons in accelerated orbital motion, without radiating energy, the theory is not consistent with Maxwells equations, and, of course, as discussed in Chapter 49, there are other problems as well. On the other hand, the argument we have given here, although less precise than the Bohr theory, is based on exactly the right idea. Because that idea is so important, we shall reiterate it one more time.

The essential problem of the nuclear atom is that the electric force tends to make the electron fall into the nucleus, and there is no countervailing force to prevent that from happening. Starting from Chapter 10, we have stressed that electricity is the force that holds matter together. The trouble is, it does so altogether too well. Without quantum mechanics, the crucial question is, what keeps every atom from shrinking to a point, as the electric force would have it do?

The answer to the question is that, owing to its wavelike nature, the smaller the region an electron is confined to, the more kinetic energy it must have. It is that cost in kinetic energy that keeps the atom from shrinking below about 1 Å in diameter.

52.5 QUANTUM MECHANICAL ESTIMATES

Example 6
Apply the approximation

$$\Delta p_x \Delta x \approx h \qquad (52.2)$$

to estimate the angle θ of the first bright fringe in a two-slit electron diffraction experiment:

We take the direction of propagation along the y axis and assume the slits are a distance d apart along the x axis. The uncertainty is that we cannot know through which slit the electron passes. Thus, we write

$$\Delta x = d.$$

It follows that

$$\Delta p_x \approx \frac{h}{d}.$$

Before passing through the slits, the electron was moving in the y direction with momentum $p = p_y$ and $p_x = 0$. After diffraction, the electron may have been deflected through the angle θ, as shown in the following figure, so we have

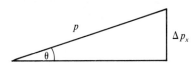

$$\sin\theta = \frac{\Delta p_x}{p}.$$

But $p = h/\lambda$, so

$$\sin\theta = \frac{\lambda \Delta p_x}{h} = \frac{\lambda}{h}\frac{h}{d} = \frac{\lambda}{d}.$$

This is the same result found in Example 2.

Questions

9. Analyze carefully the arguments of the last section for the size of an atom. Can you find places where dimensionless factors like 2π or $\sqrt{3}$ might reasonably have crept in to change the final result?

10. Compare the argument given for the size of the hydrogen atom with the situation described in Section 27.5 for the size of a planet's orbit, pointing out both the similarities and the differences between the two cases.

11. What is the angular momentum of the electron in the model developed in this section? [*Hint:* Question 10 should help you find the answer.] What is the corresponding angular momentum of the lowest energy state in the Bohr atom?

52.6 A FINAL FINAL WORD

The story of *The Mechanical Universe* is an epic tale that began with one scientific revolution and ended with another. The idea was that the universe is a machine, wondrously complicated but preprogrammed by its initial conditions to evolve through its inevitable history according to precise, mechanical laws. Knowingly or not, that idea is the silent foundation of all Western thought, from the time of Newton up to and including the present. But it is no longer believed by those who invented it, the physicists.

To be sure, much is left unchanged by the second revolution, and the same was true of the first. After Galileo, heavy bodies continued to fall faster than lighter ones, just as they had always done, even long before the time of Aristotle. But it was no longer a fundamental law. Instead, it was the combined effects of two perfectly respectable Newtonian forces, gravity and viscosity, that produced that Aristotelian effect. In a similar way, the combined quantum mechanical behaviors of 10^{24} atoms can conspire today to form a billiard ball that will invariably award victory to the player who is the better master of Newtonian, not quantum, mechanics.

There are other interesting and important examples of ideas that survived from classical physics into modern physics, not as mere approximations, but as deep truths. Three of these, as promised many chapters ago, are the great conservation laws. Energy, momentum, and angular momentum are rigorously conserved quantities in quantum physics. In fact, they have become even more important than they were. For example, the idea of energy is a relative newcomer in classical physics, and its conservation serves mainly as a kind of bookkeeping device to help sort out complicated situations. Certainly it is a derived quantity, a consequence of the fundamental law, $F = ma$. By contrast, in modern physics, both in quantum mechanics and in relativity, energy conservation is a central and fundamental concept. Much the same could be said of momentum and angular momentum.

Another interesting carryover from the old physics into the new is the nature of the mathematics it uses. Isaac Newton taught us that physics is done by solving differential equations. Erwin Schrödinger showed that the day to day work of using quantum mechanics could largely be reduced to the same familiar task. However, it

52.6 A FINAL FINAL WORD

may be worth remembering that Newton didn't present his new mechanics in the language of his new calculus. He expressed himself in terms of geometric figures and ratios of quantities, the classical mathematical language familiar to seventeenth century scientists. Schrödinger's equation itself may turn out, in the long run, to be a similar sort of thing, a way of presenting new ideas in familiar language.

As we have seen in this chapter, quantum mechanics is indispensable for a profound and complete understanding of the world, but for most instances of common experience, classical physics continues to do very well as a means of description. Moreover, classical physics is not contradicted by modern physics, but rather arises out of modern physics. That is, when an object is large and massive, the predictions of quantum mechanics become indistinguishable from those of classical mechanics, and when speeds are much less than that of light, Einstein's relativity predicts the same results as Newton's mechanics.

Nevertheless, there are profound and irreconcilable differences between the new physics and the old. These tend to be not so much practical issues as philosophical ones; not what things are, but what they mean. For example, Einstein's theory is not a theory of the behavior of clocks; it is a theory about the nature of time. In a similar way, quantum mechanics is not merely a theory of measurement (although we often express it that way); it is really a theory of what is, and it says that one of our oldest and most cherished beliefs, the logical belief in strict causality, is no longer tenable.

In Aristotelian physics, every motion had to have an immediate cause. In Newtonian physics, motion no longer needed a cause, but every change in motion did. In modern physics, however, a given cause doesn't always produce the same effect. This is the real departure from the past: The results of a given set of circumstances are not completely determined. There are alternative possible outcomes. Only the probabilities of the alternatives are precisely determined.

Physicists tend to express that thought in slightly different language. They say the results of a given experiment cannot be predicted with better than a given uncertainty. Saying it that way helps to obscure somewhat the revolutionary nature of quantum mechanics because, in a sense, the statement has always been true. Real experiments in the real world have always been subject to quantifiable error, and if we can't predict exactly the result of a single experiment, we certainly can't know the future of the world. But that's not quite the point. The more interesting question is not whether we can predict the future history of the universe – there was never any hope of doing that – but rather whether it is predetermined by the laws of physics and the initial conditions. In classical physics it was. In quantum mechanics it isn't.

Depending on your taste in these matters, that may or may not seem to be a very important distinction. But then, the changes that came about in the first scientific revolution may not have seemed so very important at the time. After all, no matter what Copernicus said, the sun continued to rise in the east and set in the west every day. Didn't it?

In the year 1600, Copernicus's disciple, Giordano Bruno, was burned to death at the stake in the Piazza Campo di Fiori (Field of Flowers) in Rome. That was 57 years, or two generations, after the publication of Copernicus's book. A generation later, in 1634, Galileo Galilei was put on trial in the same city, to be humiliated and imprisoned for the rest of his life. The sun continued to rise and set, but knowledge of the meaning of things had become positively dangerous.

About the same amount of time, two generations, has passed since the discovery of quantum mechanics in modern times. Today we live in a world in which physics and its discoveries are no longer restricted to mere academic or intellectual interest. But now as then, the real consequences of the revolution may have just begun.

APPENDIX A

THE INTERNATIONAL SYSTEM OF UNITS

Basic Units	Definitions
Length	The meter (m) is currently defined as the distance that light travels in 1/299,792,458th of a second.
Time	The second (s) is the duration of 9,192,631,770 periods of the radiation emitted in a transition between two specified energy levels of the cesium-133 atom.
Mass	The kilogram (kg) is the mass of a particular cylinder of platinum–iridium alloy preserved in a vault at Sèvres, France.

Current
: The ampere (A) is that current in two very long parallel wires 1 m apart that gives rise to a magnetic force per unit length of 2×10^{-7} N/m.

Temperature
: The kelvin (K) is 1/273.16 of the thermodynamic temperature of the triple point of water.

Names and Symbols for the SI Units

Quantity	Name of unit	Symbol	Definition
length	meter	m	
time	second	s	
mass	kilogram	kg	
current	ampere	A	
temperature	kelvin	K	
force	newton	N	$1 \text{ N} = 1 \text{ kg m/s}^2$
work, energy	joule	J	$1 \text{ J} = 1 \text{ N m}$
power	watt	W	$1 \text{ W} = 1 \text{ J/s}$
frequency	hertz	Hz	$1 \text{ Hz} = 1 \text{ s}^{-1}$
electric charge	coulomb	C	$1 \text{ C} = 1 \text{ A s}$
electric potential	volt	V	$1 \text{ V} = 1 \text{ J/C}$
electric field	volt per meter	V/m	$1 \text{ V/m} = 1 \text{ V m}^{-1}$
electric resistance	ohm	Ω	$1 \text{ Ω} = 1 \text{ V/A}$
capacitance	farad	F	$1 \text{ F} = 1 \text{ C/V}$
inductance	henry	H	$1 \text{ H} = 1 \text{ J/A}^2$
magnetic field strength	tesla	T	$1 \text{T} = 1 \text{Ns/Cm}$
magnetic flux	weber	Wb	$1 \text{ Wb} = 1 \text{ T m}^2$
entropy	joule per kelvin	J/K	$1 \text{ J/K} = 1 \text{ J K}^{-1}$
specific heat	joule per kg kelvin	J/kg K	
pressure	pascal	Pa	$1 \text{ Pa} = 1 \text{ N/m}^2$

APPENDIX B

CONVERSION FACTORS

Length

1 in. = 2.54 cm

1 ft   = 12 in. = 30.48 cm

1 yd   = 3 ft = 91.44 cm

1 km   = 0.6215 mi

1 mi = 1.609 km
1 Å = 0.1 nm

Time

1 min = 60 s
1 h = 60 min
1 d = 24h = 1440 min
1 yr = 365.24 d ≈ $\pi \times 10^7$ s

Mass

1 kg = 1000 g
1 slug = 14.59 kg
1 kg = 6.852×10^{-2} slugs

Area

1 m² = 10^4 cm²
1 in.² = 6.4516 cm²
1 m² = 10.76 ft²
1 acre = 43,560 ft²

Volume

1 m³ = 10^6 cm³
1 L = 1000 cm³ = 10^{-3} m³
1 gal = 3.786 L
1 gal = 4 qt = 8 pt = 128 oz = 231 in.³
1 ft³ = 1728 in.³ = 28.32 L

Force

1 N = 0.2248 lb = 10^5 dyn
1 lb = 4.4482 N

Energy

1 ft lb = 1.356 J
1 cal = 4.1840 J
1 Cal = 1000 cal
1 Btu = 778 ft lb = 252 cal
1 eV = 1.602×10^{-19} J

CONVERSION FACTORS

$$1 \text{ erg} = 10^{-7} \text{ J}$$
$$1 \text{ J} = 1 \text{ W s}$$

Power

$$1 \text{ hp} = 550 \text{ ft lb/s} = 745.7 \text{ W}$$
$$1 \text{ W} = 1.341 \times 10^{-3} \text{ hp}$$

Pressure

$$1 \text{ atm} = 101.325 \text{ kPa}$$
$$1 \text{ atm} = 14.7 \text{ lb/in.}^2 = 1.01 \times 10^6 \text{ dyn/cm}^2 = 1.01 \times 10^6 \text{ erg/cm}^3$$
$$1 \text{ atm} = 760 \text{ mm Hg} = 38.8 \text{ ft H}_2\text{O}$$
$$1 \text{ lb/in.}^2 = 6.895 \text{ kPa}$$
$$1 \text{ torr} = 1 \text{ mm Hg} = 133.32 \text{ Pa}$$
$$1 \text{ bar} = 100 \text{ kPa}$$

Angles

$$\pi \text{ rad} = 180°$$
$$1 \text{ rad} = 57.30°$$
$$1° = 1.745 \times 10^{-2} \text{ rad}$$

APPENDIX C

THE PERIODIC TABLE OF THE ELEMENTS

THE PERIODIC TABLE OF THE ELEMENTS

n	Group I	Group II							
1	1 H 1.00								
2	3 Li 6.94	4 Be 9.01							
3	11 Na 22.99	12 Mg 24.31							
4	19 K 39.10	20 Ca 40.08	21 Sc 44.96	22 Ti 47.88	23 V 50.94	24 Cr 52.00	25 Mn 54.94	26 Fe 55.85	27 Co 58.93
5	37 Rb 85.47	38 Sr 87.62	39 Y 88.91	40 Zr 91.22	41 Nb 92.91	42 Mo 95.94	43 Tc (98)	44 Ru 101.07	45 Rh 102.91
6	55 Cs 132.91	56 Ba 137.33	57–71 *	72 Hf 178.49	73 Ta 180.95	74 W 183.85	75 Re 186.2	76 Os 190.2	77 Ir 192.2
7	87 Fr (223)	88 Ra 226.03	89–103 **						

*Lanthanides	57 La 138.91	58 Ce 140.12	59 Pr 140.91	60 Nd 144.24	61 Pm (145)	62 Sm 150.36	63 Eu 151.96
**Actinides	89 Ac 227	90 Th 232.04	91 Pa 231	92 U 238.03	93 Np (237)	94 Pu (244)	95 Am (243)

THE PERIODIC TABLE OF THE ELEMENTS

			Group III	Group IV	Group V	Group VI	Group VII	Group VIII
								2 He 4.00
			5 B 10.81	6 C 12.01	7 N 14.01	8 O 16.00	9 F 19.00	10 Ne 20.18
			13 Al 26.98	14 Si 28.09	15 P 30.97	16 S 32.06	17 Cl 35.45	18 Ar 39.95
28 Ni 58.69	29 Cu 63.55	30 Zn 65.39	31 Ga 69.72	32 Ge 72.59	33 As 74.92	34 Se 78.96	35 Br 79.90	36 Kr 83.8
46 Pd 106.4	47 Ag 107.87	48 Cd 112.41	49 In 114.82	50 Sn 118.71	51 Sb 121.75	52 Te 127.60	53 I 126.91	54 Xe 131.29
78 Pt 195.08	79 Au 196.97	80 Hg 200.59	81 Tl 204.38	82 Pb 207.19	83 Bi 208.98	84 Po (209)	85 At (210)	86 Rn (222)

64 Gd 157.25	65 Tb 158.93	66 Dy 162.50	67 Ho 164.93	68 Er 167.26	69 Tm 168.93	70 Yb 173.04	71 Lu 174.97
96 Cm (247)	97 Bk (247)	98 Cf (251)	99 Es (252)	100 Fm (257)	101 Md (258)	102 No (259)	103 Lw (260)

Numbers in parentheses are mass numbers of most stable isotope of that element

APPENDIX D

ASTRONOMICAL DATA

Earth

    mass $M_E = 5.975 \times 10^{24}$ kg
    radius   6371 km
    acceleration due to gravity   9.80665 m/s^2

Sun

mass 1.987×10^{30} kg
radius   696,500 km
Earth–Sun mean distance 1.496×10^{11} m = 1 AU

Moon

mass 7.343×10^{22} kg
radius   1738 km
Earth–Moon mean distance   384,400 km

Planet	Semimajor axis (10^6 km)	Orbital period (d)	Mass/M_E	Eccentricity
Mercury	57.9	87.96	0.055	0.2056
Venus	108.2	224.68	0.81	0.0068
Earth	149.6	365.24	1.00	0.0167
Mars	227.9	686.95	0.11	0.0934
Jupiter	778.3	4,337	318	0.0483
Saturn	1427.0	10,760	95	0.0560
Uranus	2871.0	30,700	15	0.0461
Neptune	4497.1	60,200	17	0.0100
Pluto	5983.5	90,780	0.0023	0.2484

APPENDIX E

PHYSICAL CONSTANTS

Gravitational constant	G	6.672×10^{-11} N m^2/kg^2
Speed of light	c	2.997925×10^8 m/s
Electron's charge	e	1.60219×10^{-19} C
Coulomb constant	K_e	8.98755×10^9 N m^2/C^2
Permittivity of free space	ε_0	8.85419×10^{-12} C^2/N m^2
Magnetic constant	K_m	10^{-7} N/A^2

Permeability of free space	μ_0	$4\pi \times 10^{-7}$ N/A^2
Boltzmann's constant	k	1.3807×10^{-23} J/K
		8.617×10^{-5} eV/K
Avogadro's number	N_A	6.0220×10^{23} particles/mol
Gas constant	$R = N_A k$	8.314 J/mol K
		1.9872 cal/mol K
		8.206×10^{-2} L atm/mol K
Planck's constant	h	6.6262×10^{-34} J s
		4.1357×10^{-15} eV s
	$\hbar = h/2\pi$	1.05459×10^{-34} J s
		6.5822×10^{-16} eV s
Mass of the electron	m_e	9.1095×10^{-31} kg
		0.511 MeV/c^2
Mass of the proton	m_p	1.67265×10^{-27} kg
		938.28 MeV/c^2

SELECTED BIBLIOGRAPHY

Apostol, Tom M., *Calculus*, Vol. 1, Second Edition (John Wiley and Sons, New York, 1967).

Apostol, Tom M., *Calculus*, Vol. 2, Second Edition (John Wiley and Sons, New York, 1969).

Cheney, Margaret, *Tesla: Man Out of Time* (Prentice-Hall, Englewood Cliffs, NJ, 1981).

Clark, Ronald W., *Einstein: The Life and Times* (The World Publishing Co., New York, 1971).

Cohen, I. Bernard, *Revolution in Science* (Harvard University Press, Cambridge, MA, 1985)

Condon, E. U., and Odabasi, Halis, *Atomic Structure* (Cambridge University Press, Cambridge, 1980).

Davies, Paul, *Other Worlds* (Simon and Schuster, New York, 1980).

Einstein, Albert, and Infeld, Leopold, *The Evolution of Physics* (Simon and Schuster, New York, 1966).

Epstein, L. C., and Hewitt, P., *Thinking Physics Part 2* (Insight Press, San Francisco, 1979).

Everitt, C. W. F., *James Clerk Maxwell, Physicist and Natural Philosopher* (Charles Scribner's Sons, New York, 1975).

Faraday, Michael, *Experimental Researches in Electricity, 1839–1855*, 3 Vols. (Everyman Edition, London, 1951).

French, A. P., *Special Relativity* (W. W. Norton and Co., New York, 1968).

Goldberg, Stanley, *Understanding Relativity* (Birkhauser, Boston, 1984).

Harré, Rom, *Great Scientific Experiments* (Phaedon Press, Oxford, 1981).

Heisenberg, Werner, *The Physical Principles of the Quantum Theory* (Dover Publications, New York, 1949).

Holton, Gerald, *The Scientific Imagination* (Cambridge University Press, New York, 1978).

Hoffmann, Banesh, *Relativity and Its Roots* (W. H. Freeman and Co., San Francisco, 1983).

Josephson, Matthew, *Edison* (McGraw-Hill, New York, 1959).

Kerwin, Larkin, *Atomic Physics* (Holt, Rinehart and Winston, Inc., New York, 1963).

Magie, William Francis, *A Source Book in Physics* (McGraw-Hill, New York, 1935).

Maxwell, James Clerk, *The Scientific Papers of James Clerk Maxwell*, ed. by W. D. Niven (Dover Publications, New York, 1966).

Michelson, A. A., and Morley, E. W., "The Motion of the Earth Relative to the Luminiferous Ether," *American Journal of Science*, Vol. 34, p. 333 (1887).

Norwood, Joseph, Jr., *Twentieth Century Physics* (Prentice-Hall, Inc., Englewood Cliffs, NJ, 1976).

Pauling, Linus, and Wilson, E. Bright, *Introduction to Quantum Mechanics* (McGraw-Hill, New York, 1935).

Purcell, Edward M., *Electricity and Magnetism*, Berkeley Physics Course, Vol. 2, Second Edition (McGraw-Hill, New York, 1985).

Shankland, R. S., "Michelson–Morley Experiment," *American Journal of Physics*, Vol. 32, p. 52 (1964).

Speilberg, Nathan, and Anderson, Bryon D., *Seven Ideas That Shook the Universe* (John Wiley and Sons, New York, 1985).

Turner, D. M., *Makers of Science: Electricity and Magnetism* (Oxford University Press, London, 1927).

Walker, Jearl, *The Flying Circus of Physics* (John Wiley and Sons, New York, 1975).

Weinberg, Steven, *The Discovery of Subatomic Particles* (Scientific American Library, New York, 1983).

Whittaker, E. T., *History of the Theories of Aether and Electricity* (University of Edinburgh Press, Edinburgh, 1951).

Wheaton, Bruce R., *The Tiger and the Shark: Empirical Roots of Wave-Particle Dualism* (Cambridge University Press, Cambridge, 1983).

INDEX

Absolute temperature, 457
Accelerated charges, 312
Accelerators, 519
Activity of radioactive substance, 512
Addition of velocities, relativistic, 395
Aether, 323, 341, 356
Alkali metals, 507
Alpha decay, 510

Alpha particle, 182, 439, 508, 509
Alternating current-circuit, 276
  power and energy in, 288
Alternating electromotive force, 276
Amber, 13, 14
  conductivity of, 136
  resistivity of, 136
American Philosophical Society, 63

Ampère, André-Marie, 132, 191, 192, 296
Ampere (SI unit of current), 132, 206
Ampère's law, 209, 297
Anaxagoras, 432
Anderson, Carl D., 517
Ångström, A. J., 446
Angstrom (unit of distance), 9, 446
Angular momentum of electron
   in Bohr model, 450
   in quantum mechanic model, 454
Annihilation, 517
Anode, 460, 530
Antielectron, 517
Antineutrino, 519
Antiparticle, 517
Aristotle, 356, 432
Aston, F. W., 180
Atomic mass, 437
Atomic mass number, 437
Atomic model
   Bohr, 449
     de Broglie, 466
     Rutherford, 441
     quantum mechanical, 479
     Thomson, 439
Atomic nucleus, 437
Atomic number, 437
Atomic theory, 234, 432
Atomic weight(s), 433
Aurora Borealis, 179
Average of a function, 288
Average power, 290
Avogadro, Amedeo, 8, 193, 434
Avogadro's law, 193, 434
Avogadro's number, 9, 10

B, magnetic field, 164
Balmer, Johann, 447
Banks, Sir Joseph, 113
Baryon, 519
Battery, 123
   lemon, 125
   storage, lead–sulfuric acid, 127
Beats, 468
   electron, 469
Becquerel, Henri, 435, 508
Becquerel (SI unit of radioactivity), 512
Beta decay, 510, 516
Beta particles (rays), 508
Biot, Jean Baptiste, 193
Biot–Savart formula, 194, 195
Black-body radiation, 455
Bohr, Niels, 9, 448, 454
Bohr model, 449
Bohr orbits, 449

Bohr radius a_0, 452
Boltzmann's constant k, 457
Born, Max, 470, 478
Boscovich, Roger, 233
Bose, G. M., 27
Bose, S. N., 502
Boson, 502
Bradley, James, 398
Brahe, Tycho, 160
Brownian motion, 370, 488
Bruno, Giordano, 160, 188, 547
Bubble chamber, 176
Bunsen, Robert, 445

Cannizzaro, Stanislao, 434
Capacitance, 78
   SI unit of, 79
Capacitive reactance, 278, 284
Capacitor(s), 77
   energy stored in, 88
   in parallel, 84
   in series, 83
   parallel-plate, 64
   spherical, 80
Carroll, Lewis, 491
Cathode, 460, 530
Cathode rays, 508
Cavendish, Henry, 5, 21, 314
Chadwick, Sir James, 437, 514
Chappe, Claude, 129
Charge, 15
   by induction, 19
   conservation of, 15
   density, 37, 42
   distribution on conductors, 107–109
   negative, 15
   polarization, 18
   positive, 15
   separation, 17
Christie, Samuel, 148
Circulation, 223
Collinson, Peter, 63, 91
Compass needle, 160, 169
Compounds, 435
Conductivity, 136
   of various materials, 136
Conductors, 14, 15
   field at surface of, 59
   field inside of, 57
Conservation
   of electric charge, 15
   of kinetic energy, 350
   of momentum, 349
   of relativistic energy, 420
   of relativistic mass, 410, 411
   of relativistic momentum, 410

INDEX

Conservative field, 66, 222
Constructive interference, 333, 354
Cooke, William Fothergille, 130
Copernicus, Nicolaus, 5, 343, 527, 547
Copper, resistivity of, 136
Corona discharge, 105
Cosmic rays, 320, 400
Coulomb, Charles Augustin, 20, 162, 191
Coulomb (SI unit of charge), 23, 132, 175, 206
Coulomb force, 32
Coulomb's law, 21
Cowan, Clyde L., Jr., 517
Critical angle, 331
Cunaeus, Andreas, 27
Curie (unit of radioactive decay), 512
Curie, Marie, 508
Curie, Pierre, 508
Current, 132
  alternating, 276
  electric, 133
Current density, 133
Current loop
  magnetic dipole moment of, 186, 200
  magnetic field of, 199
  torque on, 186
Cuthbertson, John, 29
Cyclotron, 176
Cyclotron frequency, 176

Dalton, John, 110, 233, 271, 432
Damping of resonant circuit, 283
Davisson, C. J., 470, 532
Davy, Humphrey, 110, 193, 270, 271
de Broglie, Louis, 454, 463
de Broglie, Maurice, 488
de Broglie's model of the atom, 466
de Broglie's wavelength, 465
Del symbol ∇, 94
Democritus, 432
Dempster, A. J., 180
Destructive interference, 333, 354
Diffraction
  electron, 532
  neutron, 533
  X ray, 339, 532
Dipole field, 167
Dipole moment, 165
  magnetic, 167, 169
  of current loop, 186, 200
Dirac, Paul A. M., 502, 517
Directional derivative, 93
Displacement current, 298
Domains, magnetic, 165
Double-slit experiment, 333, 355
Dryden, John, 189
Dufay, Charles, 13, 14

Earth's magnetic field, 179
Eddington, Sir Arthur, 428
Eddy currents, 253
Edison, Thomas Alva, 273, 274, 294
Einstein, Albert, 11, 365, 369, 389, 407, 409, 419, 427, 528
Einstein's photoelectric equation, 460
Electric battery, 123
Electric charge, 15
  conservation of, 15
  distribution on conductors, 107–109
  sign of, 15
Electric constant K_e, 21, 23
Electric currents, 133
  alternating, 276
  and charge conservation, 146
  and energy conservation, 145
  energy dissipation in flow of, 138
Electric dipole, 36
Electric field, 33, 34
  energy stored in, 74
  flux of, 46
  Gauss's law for, 48, 50
  inside a conductor, 57
  inside a solid sphere, 56
  inside a spherical shell, 55
  integral for, 38
  intensity or strength, 33, 34
  of a circular loop, 41
  of a dipole, 36
  of a flat sheet of charge, 43
  of uniform linear charge, 40
  outside a solid sphere, 56
  outside a spherical shell, 55
  SI unit for field strength, 33
Electric field lines, 32
Electric potential, 65
  between parallel conducting plates, 70
  of a point charge, 66
  of a system of charges, 68
  of a uniformly charged spherical shell, 71
  of an ion, 102
Electrical potential energy, 74
  of a system of point charges, 75
Electrode, 123, 434
Electrolyte, 123
Electrolysis, 434
Electromagnet, 201
Electromagnetic force, 519
Electromagnetic induction, 236
Electromagnetic radiation, 318
Electromagnetic spectrum, 318
Electromagnetic wave, 320
Electromotive force, 125
  alternating, 276
  induced, 245

Electrostatic force, 32
Electron
  charge of, 8, 10, 66
  diffraction, 532
  e/m measurement of, 8
  mass of, 8
  spin, 501
  stable orbits of, 449
  velocity of in orbit, 451
  wavelength of, 466, 530
  wave nature of, 466
Electron shells, 504, 505
Electronvolt (energy unit), 66
Electroscope, 16
Electrostatic field, 33, 34
Electrostatic unit of charge, 23, 206
Elementary particles, 517, 520
Elements, 435, 436
  electron configuration of, 505
Energy
  in alternating-current circuit, 288
  dissipation of, in resistor, 137
  stored in a capacitor, 88
  stored in an inductor, 268
  stored in electric field, 74
  stored in magnetic field, 232
Energy density, 232
  in a capacitor, 88
  in an inductor, 269
  in a magnetic field, 232, 269
  of water, 232
Equipotential surfaces, 99
Equivalence of mass and energy, 419
Exclusion principle, 503

Farad (SI unit of capacitance), 79
Faraday, Michael, 11, 31, 61, 163, 218, 235, 236, 269, 296, 434, 528
Faraday's law, 247, 297
Fermat, Pierre de, 326
Fermat's principle of least time, 326
Fermi, Enrico, 502, 517
Fermion, 502
Field, 219
  electric, 33
  magnetic, 164
  scalar, 219
  vector, 219
Fine structure, 454
Fission, 425
Fitzgerald, George, 365, 367
Fitzgerald–Lorentz contraction, 365
Fizeau, Armand, 398
Fizeau experiment, 398

Flux, 44, 220
  electric, 46
  magnetic, 172, 220
Force
  between magnetic poles, 163
  between parallel currents, 205
  electric, 32, 34
  in relativistic mechanics, 414
  magnetic, 163, 193, 194
  on current-carrying wire, 184
  on moving charged particle, 174, 179
Foucault, Leon, 398
Frankland, Edward, 435
Franklin, Benjamin, 4, 11, 15, 20, 63, 77, 89, 91, 528
Frequency
  of a wave, 306
  of harmonic oscillator, 281
  of radiated photon, 451
  of resonance, 281
Fresnel, Augustin Jean, 216, 324, 333, 334
Fundamental particles, 517, 520
Fuel cell, 127
Fusion, 425

Galilean relativity, 345, 347
Galilean transformation, 346
Galileo, 11, 161, 343, 527, 548
Galvani, Luigi, 113
Gamma emission, 510
Gamma rays, 320, 509
Gauge particles, 522
Gauss, Carl Friedrich, 51, 130
Gauss (unit of magnetic field strength), 175
Gauss's law, 48, 173
  applications of, 52–56
  for electric flux, 48, 50, 297
  for magnetism, 173, 297
Gay-Lussac, Joseph Louis, 434
Gay-Lussac's law, 434
Geiger, Hans, 439
Gell-Mann, Murray, 520
Germer, L. H. 470, 532
Gilbert, William, 14, 160, 188, 296, 527
Gluon, 522
Goudsmit, Samuel, 501
Gradient vector, 94
Gravitational constant G, 10
Gray, Stephen, 14
Greek atomic theory, 432
Grounding, 19

Hadron, 19
Hale, George Ellery, 427

INDEX

Half-life, 512
Hall, D. B., 401
Heisenberg, Werner, 470, 514
Heisenberg uncertainty principle, 483
Helium atom, 437, 505
Helmholtz, Heinrich von, 233, 358
Henry, Joseph, 130, 257
Henry (SI unit of inductance), 257
Hertz, Heinrich, 314, 317
Hertz (SI unit of frequency), 306
Hooke, Robert, 354
Huygens, Christian, 217, 323, 354, 462
Huygens's principle, 324
Huxley, Aldous, 525
Hydrodynamics, 218
Hydrogen atom, 435
    energy levels of, 449
    fine structure of, 454
    ground state of, 452
    radius of, 9, 452
    spectrum, 445
Hylozoists, 491

Impedance, 283
Index of refraction, 329
Induced electromotive force, 245
Inductance
    mutual, 260
    self, 256
Induction, electrostatic charging by, 19
Inductive reactance, 279, 284
Inductor, 256
    energy stored in, 268
Inertial frame, 344
Infrared light, 319
Insulators, 14, 15
Intensity (strength)
    of electric field, 33, 34
    of magnetic field, 164
Interference, 333, 354
Interferometer, 358
Invariance of Newton's laws, 347
Ion(s), 451
Ionization potential, 103
Isotopes, 180, 437, 511

Jeans, Sir James, 446, 456
Jonsson, Claus, 531, 532
Jordan, P., 470
Joule heating, 137, 288
Joyce, James, 520

Kaon, 521
Kelvin, Lord (William Thomson), 7, 233

Kepler, Johannes, 160, 189, 527
Kinetic energy
    Newtonian, 417
    relativistic, 417
Kirchhoff, Robert Gustav, 144, 324, 445
Kirchhoff's laws, 145, 146
Kleist, E. J. von, 26
Kline, Morris, 527

Langevin, Paul, 488
Larmor radius, 176
Lavoisier, Antoine Laurent, 128, 271, 433
Law(s)
    of combining volumes, 434
    of conservation of matter, 433
    of definite proportions, 433
LCR circuit, 283
Leibniz, Gottfried Wilhelm, 217
Length contraction, 372, 378
Lenz, Emil, 251
Lenz's law, 251
Lepton(s), 519, 523
Leucippus, 432
Leyden jar, 27, 77
Light
    dual nature of, 332, 354, 463
    speed of, 7, 10, 301
Line integral, 64
    for a magnetic field, 195
    for circulation, 223
    of a magnetic field, 209
    of an electric field, 65
Linear charge density, 37
Lines of force, 32
Lithium atom, 504
Lodestone (magnetite), 159
Lorentz, Hendrik A., 179, 365, 368, 386
Lorentz contraction, 369
Lorentz force, 179
Lorentz invariant, 378
Lorentz transformation, 369, 376
    applications of, 378
LR circuits, 263
    time constant of, 264
Lucretius, 234, 432
Luminiferous aether, 323, 341, 356
Lyman, Theodore, 452
Lyman series, 452

Magnetic bottle, 178
Magnetic constant K_m, 163
Magnetic dipole, 165
Magnetic dipole moment, 165
Magnetic domains, 165

Magnetic field, 164
   around current-carrying wire, 197
   energy stored in, 232
   inside a solenoid, 213
   inside a toroid, 214
   line integral of, 208
   of current loop, 199
Magnetic flux, 172
Magnetic force, 163
   on currents, 183
   on moving charges, 174
Magnetic moment, 186
Magnetic monopole, 172
Magnetic quantum number, 494
Magnetite (lodestone), 159
Many-electron atoms, 501
Marconi, Guglielmo, 318
Marsden, Ernest, 439
Mass, relativistic, 409
Mass–energy equivalence, 419
Mass spectrograph, 180, 181
Matter wave, 465
Maxwell, James Clerk, 2, 11, 61, 217, 218, 234, 295, 314, 357, 431, 435, 528
Maxwell–Ampère law, 298
Maxwell's equations, 299
   in free space, 301
Mendeleev, Dmitri, 193, 271, 435, 503
Meson(s), 15, 518, 519
Michell, John, 21, 162, 164
Michelson, Albert A., 342, 357
Michelson–Morley experiment, 357, 367
Microfarad μF (unit of capacitance), 80
Microwaves, 319
Millikan, Robert Andrew, 8, 460, 533
Minkowski, Hermann, 351, 382
Mole, 8, 437
Molecular weight, 437
Molecule(s), 437
Monatomic molecule, 437
Moon radius, 6
Morley, Edward W., 361
Morse, Samuel F. B., 130
Moseley, H. G. J., 488
Muon (mu meson), 400, 519
Muon experiment, 401
Musschenbroek, Pieter van, 27
Mutual inductance, 260

Names of elements, 436
Napoleon, 128
Negative charge, 15
Neutrino, 516
Neutron, 435, 514

Newton, Sir Isaac, 1, 11, 217, 332, 354, 462, 491, 527
Nobel prize, 11, 294, 365, 370, 427, 448, 452, 461, 463
Noble gases, 506
Nucleon, 8
Nucleon mass, 10
Nucleus, atomic, 435

Oersted, Hans Christian, 130, 191, 236
Oersted's experiment, 130
Ohm, George Simon, 134
Ohm (SI unit of resistance), 135
Ohm's law, 134, 276
Orbital, 483
Orbital angular momentum, 494
Orbital quantum numbers, 493, 494
Orbital radius, 452
Oscillator, harmonic, 474

Parallel currents, force between, 206
Parallel-plate capacitor, 64
   capacitance of, 78
Partial derivative, 94
Paschen, F., 448
Pauli, Wolfgang, 503, 516
Pauli exclusion principle, 503
Peregrinus, Peter, 160
Periodic table, 503, 556
Permanent magnet, field of, 164
Permeability of free space, 200
   value in SI units, 201
Permittivity of free space, 23
Phase angle in alternating-current circuit, 283
Photoelectric effect, 460
Photoelectric work function, 460
Photons, 447, 462
Picofarad pF (unit of capacitance), 80
Pion, 519
Planck, Max, 9, 446, 456
Planck's constant, 9, 10, 447, 457
Planck's radiation law, 457
Plane waves, 302
Planté, Gaston, 127
Poincaré, Henri, 369, 386, 391
Point charge, 21
Polarization, 18
Positive charge, 15
Positron, 517
Potential, electric, 65
   between two conducting plates, 70
   derivation of field from, 94
   of charged spherical shell, 71
   of charged wire, 70

INDEX

Potential energy, electrical, 65
  of a conduction electron in a metal, 119
  of a system of charges, 68
Power in alternating-current circuit, 288
Poynting, John Henry, 320
Poynting vector, 320
Priestly, Joseph, 20, 271
Primary coil, 292
Principal quantum number, 494
Probability density, 479
Probability wave, 482
Proper length, 390
Proper time, 390
Proton, 8, 369, 434
Proust, Joseph Louis, 433
Prout, William, 434

Quantization of energy, 446, 462
Quantum chromodynamics, 522
Quantum mechanics, 470, 484
Quantum numbers, 451, 493
Quantum state, 493
Quantum theory, 484
Quark, 61, 520
  color, 521
  flavor, 522

Radial probability density, 498
  for hydrogen atom, 499
Radiation by accelerated charge, 313
Radio waves, 318
Radioactive carbon dating, 514
Radioactive decay, 513
Radioactive elements, 508
Radioactivity, 435, 508
Radium, 508
Rayleigh, Lord, 446, 456
Rayleigh–Jeans law, 446, 458
RC circuit, 151
  time constant of, 153
Reactance
  capacitive, 278
  inductive, 279
Reflection, law of, 326
Refractive index, 329
Relativistic kinetic energy, 417
Relativistic mass, 409
Relativistic momentum, 410
Relativity
  Einstein's postulates for, 370
  Galilean, 345, 347
  general theory of, 427
  special theory of, 370
Resinous electricity, 15

Resistance, 135
Resistivity, 136
  for various materials, 136
Resistor(s), 137
  in parallel, 142
  in series, 140
Resonance, 281
Resonant circuit, 281
Resonant frequency, 281
Rest mass, 409
Rest-mass energy, 419
Richmann, G. W., 89
Roemer, Ole, 397
Roentgen, Wilhelm Konrad, 435, 508
Root mean square, 289
Rossi, B., 401
Rubidium, 445
Rutherford, Ernest, 439, 448, 509
Rydberg, Johannes R., 447
Rydberg constant, 448
Rydberg formula, 448

Savart, Felix, 193
Scalar field, 219
Schilling, Paul, 130
Schrödinger, Erwin, 470
Schrödinger equation, 474, 492
Seawater, resistivity of, 139
Secondary coil, 192
Self-inductance, 256
Sharp spectral states, 495
Shells, electron, 504, 505
Sinks, 221
Snell, Willebrord, 326
Snell's law of refraction, 326, 329
Solenoid, 213, 263
Solid body rotation, 223
Sommerfeld, Arnold, 454
Sources, 221
Space–time diagrams, 351, 382
Special theory of relativity, 370
Spectra, atomic, 445
Spectral emission, 445
Spectrograph, 180, 181
Spectroscope, 445
Speed of light, 7, 10, 301
Spherical coordinates, 493
Spin of electron, 501
Standing wave, 306
Static electricity, 13
Stationary state, 449
Stoney, G. Johnstone, 435
Sun, radius of, 6
Superposition principle, 22

Surface charge density, 42
Surface integral, 46
Surfaces, equipotential, 99

Telegraph, 129
Tesla, Nikola, 273, 274, 294
Tesla (SI unit of magnetic field strength), 164, 175
Teylerian machine, 29
Thales, 13, 296
Thomson, Sir George Paget, 470
Thomson, Sir Joseph John, 8, 435, 439, 448, 470, 508
Thomson, Sir William (Lord Kelvin), 7, 233, 234
Thorium, 512
Time dilation, 371
Torque
  on current loop, 186
  on magnetic dipole, 169
Transformation
  Galilean, 346
  Lorentz, 369, 376
Transformer, 292
Twin paradox, 403

Uhlenbeck, George, 501
Ultraviolet light, 319
Uncertainty principle, 483
Unit of atomic mass (u), 437
Uranium, 508

Valence, 435
Van Allen radiation belts, 179
Van de Graaff generator, 26, 29, 67, 72, 82
Vector dipole moment, 167
Vector field, 219
Versorium, 14
Vitreous electricity, 15
Volt (SI unit of electric potential), 65
Volta, Alessandro, 111, 113, 127
Voltage, 65
Voltaic pile, 115

Von Kleinst, E. J., 26
Von Marum, Martinus, 29
Vortex field, 227
Vortex ring, 229, 233

Water
  conductivity of, 136
  resistivity of, 136
  index of refraction, 329
Wave amplitude, 306
Wave, electromagnetic, 314
Wave equation, 304, 310
Wave frequency, 306
Wave function, 305, 476
Wave nature
  of electron, 466
  of matter, 465
Wave theory of matter, 465
Wave speed, 305, 306
Wave train, 464
Wavelength, 306
Weber, Wilhelm Eduard, 130
Weber (SI unit of magnetic flux), 172
Westinghouse, George, 275
Wheatstone, Charles, 130, 135, 148, 158
Wheatstone bridge, 148
Wimshurst machine, 28
Whittaker, Sir Edmund, 386
Wien, Wilhelm, 446, 456
Wien's law, 458
Work function, 120
World line, 351, 382

X rays, 300, 435
  diffraction of, 339

Young, Thomas, 317, 333, 354, 529
Yukawa, Hideki, 517

Zeeman, Pieter, 494
Zeeman effect, 494
Zweig, George, 520